Y0-AAA-523

Location Theory

A Unified Approach

Stefan Nickel
Justo Puerto

Location Theory

A Unified Approach

With 116 Figures
and 36 Tables

Professor Dr. Stefan Nickel
Lehrstuhl für Operations Research und Logistik
Rechts- und Wirtschaftswissenschaftliche Fakultät
Universität des Saarlandes
66041 Saarbrücken
Germany
E-mail: s.nickel@orl.uni-saarland.de

Professor Dr. Justo Puerto Albandoz
Dpto. Estadística e Investigación Operativa
Faculdad de Matemáticas
Universidad de Sevilla
41012 Seville
Spain
E-mail: puerto@us.es

Cataloging-in-Publication Data
Library of Congress Control Number: 2005925027

ISBN 3-540-24321-6 Springer Berlin Heidelberg New York

Springer is a part of Springer Science+Business Media
springeronline.com

Printed in Germany

Cover design: Erich Kirchner
Production: Helmut Petri
Printing: Strauss Offsetdruck

SPIN 11376804 Printed on acid-free paper – 42/3153 – 5 4 3 2 1 0

This book is dedicated to our families.

Preface

Science is facts; just as houses are made of stones,
so is science made of facts; but
a pile of stones is not a house and a
collection of facts is not necessarily science.
Henri Poincaré (1854-1912)

Developing new topics is always divided into (at least) two phases. In the first phase relatively isolated research results are derived and at a certain stage papers are published with the findings. Only a small group of people typically constitutes the core of such a first phase. After a while, more people become attracted (if the topic is worth it) and the number of publications is starting to grow. Given that the first phase is successful, at a certain point the second phase of organizing and standardizing the material should start. This allows to use a unique notation and to make the theory accessible to a broader group of people. The two authors of this monograph work together since 1996 on what is now called ordered median problems and felt that after 7 years of publishing single results in research papers (and attracting other researchers to the topic), a book could be the right instrument to start phase two for ordered median problems. In this monograph we review already published results about ordered median problems, put them into perspective and also add new results. Moreover, this book can be used as an overview of the three main directions of location theory (continuous, network and discrete) for readers with a certain quantitative (mathematical) background. It is not explicitly a textbook which can directly be used in courses. However, parts of this monograph have been used with success in classroom by the authors.

To make things short, we had three main reasons to start this book project:

1. We wanted to present a unified theory for location problems, while keeping typical properties of classical location approaches.
2. We wanted to organize the published material on ordered median problems and put it into perspective to make the results easier accessible to the interested reader.
3. For the fun of working on ordered median problems.

From the above it is clear, that mainly researchers and advanced students will benefit from this book. In order to use this book also as a reference book we kept the structure as simple as possible. This means that we have separated the book into four main parts. Part 1 introduces the ordered median problem in general and gives some principal results. Parts 2-4 are devoted to continuous, network and discrete ordered median problems, respectively.

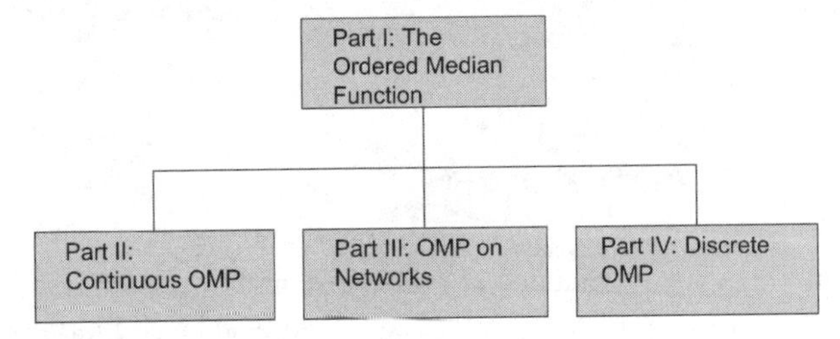

The above diagram shows that there are at least four itineraries approaching the material of the book. The standard one goes from the beginning to the end, but also a combination of Part I with any of the remaining parts is possible.

We would like to thank the Mathematisches Forschungsinstitut Oberwolfach for granting us two stays in the "Research in Pairs" program, which gave us the opportunity to work really intensive on some critical parts of the book. In addition, we would also like to thank the Spanish Ministry of Science and Technology and the German DAAD which gave us some financial support to have the necessary meetings. We also thank Werner A. Müller from Springer for sharing our opinion that this book would be interesting for the scientific community.

We would like to thank all people who have helped to develop some of the material included in this book: Natashia Boland, Patricia Domínguez-Marín, Francisco R. Fernández, Pierre Hansen, Yolanda Hinojosa, Jörg Kalcsics, Paula Lagares, Alfredo Marín, Nenad Mladenovic, Pier L. Papini, Dionisio Pérez, Antonio Rodríguez-Chía, Arie Tamir, Sebastian Velten, Ansgar Weißler.

For editorial support our thanks goes to Nadine Cwikla, Karin Hunsicker, Barbara Karg, Eva Nickel-Klimm, M. Paz Rivera, Maite Sierra and Roland Zimmer.

Finally, we thank everybody we forgot to mention in the preceding list. We hereby allow them to exclude us in their next acknowledgement.

On our book web page (*http://www.orl.uni-saarland.de/omp/*) you can find interesting links, software, news and the possibility to send us some feedback.

Saarbrücken, February 2005 *Stefan Nickel*
Sevilla, February 2005 *Justo Puerto*

Contents

Part IV The Discrete Ordered Median Location Problem

List of Figures

List of Tables

List of Algorithms

List of Algorithms

Part I

Location Theory and the Ordered Median Function

1

Mathematical Properties of the Ordered Median Function

1.1 Introduction

When talking about solving a location problem in practice at least the following phases have to be considered:

- Definition of the problem
- Identification of side constraints
- Choice of the right objective function(s)
- Data collection
- Data analysis (data mining)
- Optimization (actual resolution)
- Visualization of the results
- Discussion if the problem is solved or if the phases have to be started again.

Typically researchers in location theory have been concentrating very much on the resolution (optimization phase) of a given problem. The type of the problem and to some extent also the side constraints were motivated by classical papers in location theory (see [206, 209]). The idea of having a facility placed at a location which is in average good for each client, led to the median objective function (also called Weber or Fermat-Weber objective), see [209]. Finding a location which is even for the most remote client as good as possible, brought up the idea of the center objective (see [163]). The insight, that both points of view might be too extreme, led to the cent-dian approach (see [87]). In addition, researchers always distinguished between continuous, network and discrete location problems. Therefore, the main scope of researchers can be seen as picking a problem from the table in Figure 1.1 by selecting a row and a column, maybe adding some additional constraints to it, and then finding good solution procedures.

objective function	decision space		
	continuous	network	discrete
median			
center			
cent-dian			
...			

Fig. 1.1. A simplified classification of most of the classical location problems

Although modern location theory[1] is now more than 90 years old the focus of the researchers has been problem oriented. In this direction enormous advances have been achieved since the first analytical approaches. Today sophisticated tools from computer science, computational geometry and combinatorial optimization are applied. However, several years ago a group of researchers realized that although there is a vast literature of papers and books in location theory, a common theory is still missing (There are some exceptions, as for instance [184].) We found the question interesting and even more challenging. Thus, we started in 1996 to work on a unified framework. This monograph summarizes results we published in the last years in several papers. We want to present to the reader a neat picture of our ideas. Our aim is to have a common way for expressing most of the relevant location objectives with a single methodology for all the three major branches of locational analysis: continuous, network and discrete location problems. Or, looking at the table in Figure 1.1, we do not want to unify location problems with respect to the decision space, but with respect to the way the objective function is stated.

We will start in the next section with an illustrative example which will make our ideas transparent without going deep into the mathematics. After that, we give a rigorous mathematical foundation of the concept.

1.2 Motivating Example

Consider the following situation: Three decision makers (Mr. Optimistic, Mr. Pessimistic and Mrs. Realistic) are in a meeting to decide about a new service

[1] By modern location theory, we mean location theory with an economical background.

facility of a governmental institution. The purpose of the service facility is totally unclear but it is agreed on that its major clients live in five places ($P1, \dots, P5$) in the city. The three decision makers have offers for four different locations from the local real estate Mr. Greedy. The task of the decision makers is now to decide which of the four locations should be used to build the new service facility. The price of the four locations is pretty much the same and the area of each of the four locations is sufficient for the planned building. In a first step a consulting company (with the name WeKnow) was hired to estimate the cost for serving the customers at the five major places from each of the four locations under consideration. The outcome is shown in the following table:

	New1	New2	New3	New4
P1	5	2	5	13
P2	6	20	4	2
P3	12	10	9	2
P4	2	2	13	1
P5	5	9	2	3

Now the discussion is focussing on how to evaluate the different alternatives. Mr. Optimistic came well prepared to the meeting. He found a book by a person called Weber ([206]) in which it is explained that a good way is to take the location where the sum of costs is minimized.

The outcome is $New4$ with an objective value of 21.

Mr. Pessimistic argues that people might not accept the new service in case some of them have to spend too much time in reaching the new institution. He also had a look at some books about locational decisions and in his opinion the maximal cost for a customer to reach the new institution has to be minimized.

The outcome is $New1$ with an objective value of 12.

Mr. Optimistic, however, does not gives up so easily. He says: "Both evaluations are in some sense too extreme. Recently, a friend of mine told me about the Cent-Dian approach which in my understanding is a compromise between our two criteria". We simply have to take the sum (or median) objective (f), the center objective (g) and estimate an α between 0 and 1. The objective function is then calculated as $\alpha f + (1 - \alpha)g$. Mr. Pessimistic agrees that this is in principle a good idea. But having still doubts he asks

- How do we estimate this α?
- How much can we change the α before the solution becomes different?

Mr. Optimistic is a little confused about these questions and also does not remember easy answers in the books he read.

Suddenly, Mrs. Realistic steps in and argues: "The ideas I heard up to now sound quite good. But in my opinion we have to be a little bit more involved with the customers' needs. Isn't it the case that we could neglect the two closest customer groups?. They will be fine of anyway. Moreover, when looking at the table, I recognized that whatever solution we will take one group will be always quite far away. We can do nothing about it, but the center objective is determined by only this far away group. Couldn't we just have an objective function which allows to leave out the k_1 closest customers as well as the k_2 furthest customers." Both, Mr. Optimistic and Mr. Pessimistic, nod in full agreement. Then Mr. Pessimistic asks the key question: "How would such an objective function look like?" Even Mr. Optimistic has to confess that he did not come across this objective. However, he claims that it might be the case, that the objective function Mrs. Realistic was talking about is nothing else than a Cent-Dian function.

Mrs. Realistic responds that this might very well be the case but that she is not a mathematician and she even wouldn't know the correct α to be chosen. She adds that this should be the job of some mathematicians and that the book of Weber is now more than 90 years old and therefore the existence of some unified framework to answer these questions should be useful.

Mr. Pessimistic adds that he anyhow would prefer to keep his original objective and asks if it is also possible that each decision maker keeps his favorite objective under such a common framework.

The three of them decided in the end to stop the planning process for the new institution, since nobody knew what it was good for. Instead they took the money allocated originally to this project and gave it to two mathematicians (associated with the company NP) to work "hard" and to develop answers to their questions.

The core tasks coming out of this story:

- Find a relatively easy way of choosing an objective function.
- Are there alternatives which can be excluded independently of the specific type of objective function?

Of course, the story could have ended in another way leaving similar questions open: The discussion continues and Mrs. Realistic asks if they really should restrict themselves to the alternatives which Mr. Greedy provided. She suggests to use the streets (as a network) and compute there the optimal location in terms of distance. Then the neighborhood of this location should be checked for appropriate ground. Mr. Pessimistic adds that he knows that some streets are about to be changed and some new ones are built. Therefore he would be in favor of using a planar model which approximates the distances typical for the city and then proceed with finding an optimal location. After that he would like to follow the same plan as Mrs. Realistic.

1.3 The Ordered Median Function

In this section we formally introduce a family of functions which fulfils all the requirements discussed in Section 1.2. The structure of these functions involves a nonlinearity in the form of an ordering operation. It is clear that this step introduces a degree of complication into the function. Nevertheless, it is a fair price to be paid in order to handle the requirements shown in the previous section. We will review some of the interesting properties of these functions in a first step to understand their behavior. Then, we give an axiomatic characterization of this objective function.

We start defining the ordered median function (OMf). This function is a weighted average of ordered elements. For any $x \in \mathbb{R}^M$ denote $x_\leq = (x_{(1)}, \ldots, x_{(M)})$ where $x_{(1)} \leq x_{(2)} \leq \ldots x_{(M)}$. We consider the function:

$$\begin{aligned} sort_M : \mathbb{R}^M &\longrightarrow \mathbb{R}^M \\ x &\longrightarrow x_\leq . \end{aligned} \tag{1.1}$$

Let $\langle \cdot, \cdot \rangle$ denote the usual scalar product in $\mathbb{R}^M$.

Definition 1.1 *The function $f_\lambda : \mathbb{R}^M \longrightarrow \mathbb{R}$ is an ordered median function, for short $f_\lambda \in$ OMf(M), if $f_\lambda(x) = \langle \lambda, sort_M(x) \rangle$ for some $\lambda = (\lambda_1, \ldots, \lambda_M) \in \mathbb{R}^M$.*

It is clear that ordered median functions are nonlinear functions. Whereas the nonlinearity is induced by the sorting. One of the consequences of this sorting is that the pseudo-linear representation given in Definition 1.1 is pointwise defined. Nevertheless, one can easily identify its linearity domains (see chapters 2 or 3). The identification of these regions provides us a subdivision of the framework space where in each of its cells the function is linear. Obviously, the topology of these regions depends on the framework space and on the lambda vector. A detailed discussion is deferred to the corresponding chapters. Different choices of lambda lead also to different functions within the same family. We start showing that the most used objective functions in location theory, namely median, center, centdian or k-centrum are among the functions covered by the ordered median functions.

Some operators related to the ordered median function have been developed by other authors independently. This is the case of the ordered weighted operators (OWA) studied by Yager [214] to aggregate semantic preferences in the context of artificial intelligence; as well as SAND functions (isotone and sublinear functions) introduced by Francis, Lowe and Tamir [78] to study aggregation errors in multifacility location models.

We start with some simple properties and remarks concerning ordered median functions. Most of them are natural questions that appear when a family of functions is considered. Partial answers are summarized in the following proposition.

Proposition 1.1 *Let $f_\lambda(x), f_\mu(x) \in$ OMf(M).*

1. *$f_\lambda(x)$ is a continuous function.*
2. *$f_\lambda(x)$ is a symmetric function, i.e. for any $x \in \mathbb{R}^M$ $f_\lambda(x) = f_\lambda(sort_M(x))$.*

Table 1.1. Different instances of ordered median functions.

λ	$f_\lambda(x)$	Meaning
$(1/M,\ldots,1/M)$	$\frac{1}{M}\sum_{i=1}^{M} x_i$	mean average of x
$(0,\ldots,0,1)$	$\max_{1\le i\le M} x_i$	maximum component of x
$(1,0,\ldots,0,0)$	$\min_{1\le i\le M} x_i$	minimum component of x
$(\alpha,\ldots,\alpha,\alpha,1)$	$\alpha\sum_{i=1}^{M} x_i + (1-\alpha)\max_{1\le i\le M} x_i$	α-centdian, $\alpha\in[0,1]$
$(0,\ldots,0,1,\overset{k}{\ldots},1)$	$\sum_{i=M-k+1}^{M} x_{(i)}$	k-centrum of x
$(1,\overset{k}{\ldots},1,0,\ldots,0)$	$\sum_{i=1}^{k} x_{(i)}$	anti-k-centrum of x
$(-1,0,\ldots,0,1)$	$\max_{1\le i\le M} x_i - \min_{1\le i\le M} x_i$	range of x
$(\alpha,0,\ldots,0,1-\alpha)$	$\alpha \min_{1\le i\le M} x_i + (1-\alpha)\max_{1\le i\le M} x_i$	Hurwicz criterion $\alpha\in[0,1]$
$(0,\ldots,0,1,\ldots,1,0,\ldots,0)$	$\sum_{i=k_1+1}^{M-k_2} x_{(i)}$	(k_1,k_2)-trimmed mean
$\lambda_1 \gg \lambda_2 \gg \ldots \gg \lambda_M$	$\sum_{i=1}^{M} \lambda_i x_{(i)}$	lex-min of x in any bounded region
$\vdots$	$\vdots$	$\vdots$

3. *$f_\lambda(x)$ is a convex function iff $0 \le \lambda_1 \le \ldots \le \lambda_M$.*
4. *Let $\alpha \in \mathbb{R}$. $f_\alpha(x) \in$ OMf(1) iff $f_\alpha(x) = \alpha x$.*
5. *If c_1 and c_2 are constants, then the function $c_1 f_\lambda(x) + c_2 f_\mu(x) \in$ OMf(M).*
6. *If $f_\lambda(x) \in$ OMf(M) and $f_\alpha(u) \in$ OMf(1), then the composite function is an ordered median function of x on $\mathbb{R}^M$.*
7. *If $\{f_{\lambda^n}(x)\}$ is a set of ordered median functions that pointwise converges to a function f, then $f \in$ OMf(M).*
8. *If $\{f_{\lambda^n}(x)\}$ is a set of ordered median functions, all bounded above in each point x of $\mathbb{R}^M$, then the pointwise maximum (or sup) function defined at each point x is not in general an OMf (see Example 1.1).*
9. *Let $p < M-1$ and $x^p = (x_1,\ldots,x_p)$, $x^{\backslash p} = (x_{p+1},\ldots,x_M)$. If $f_\lambda(x) \in$ OMf(M) then $f_{\lambda^p}(x^p) + f_{\lambda^{\backslash p}}(x^{\backslash p}) \lesseqqgtr f_\lambda(x)$ (see Example 1.2).*

Proof.
The proof of (1) and (3) can be found in [179]. The proofs of items (2) and (4) to (6) are straightforward and therefore are omitted.
To prove (7) we proceed in the following way. Let $f(x)$ be the limit function. Since, $f_{\lambda^n}(x) = \langle \lambda^n, sort_M(x) \rangle$ then at x it must exist λ such that $\lim_{n \to \infty} \lambda^n = \lambda$. Hence, at the point x we have: $f(x) = \langle \lambda, sort_M(x) \rangle$. Now, since the above limit in λ^n does not depend on x then we get that $f(y) = \langle \lambda, sort_M(y) \rangle$, for all y.

Finally, counterexamples for the assertions (8) and (9) are given in the examples 1.1 and 1.2. □

Example 1.1
Let us consider the following three lambda vectors: $\lambda^1 = (-1, 0, 0, 1)$, $\lambda^2 = (0, 0, 1, 0)$, $\lambda^3 = (1, 1, 0, 0)$; and for any $x \in \mathbb{R}^4$ let

$$f_{max}(x) = \max_{i=1,2,3} \{f_{\lambda^i}(x)\}.$$

The following table shows the evaluation of the functions f_{λ^i} and f_{max} at different points.

x	$f_{\lambda^1}(x)$	$f_{\lambda^2}(x)$	$f_{\lambda^3}(x)$	$f_{max}(x)$
(1,2,3,4)	3	3	3	3
(0,1,2,4)	4	2	1	4
(0,0,1,2)	2	1	0	2
(1,3,3,3.5)	2.5	3	4	4

For the first three points the unique representation of the function f_{max} as an OMf(4) is obtained with $\lambda_{max} = (3, 0, -4, 3)$. However, for the fourth point $\hat{x} = (1, 3, 3, 3.5)$ we have

$$\langle \lambda_{max}, sort_M(\hat{x}) \rangle = 1.5 < f_{max}(\hat{x}) = 4.$$

Example 1.2
Let us take $x = (4, 1, 3)$ and $p = 1$. With these choices $x^1 = 4$ and $x^{\backslash 1} = (1, 3)$. The following table shows the possible relationships between $f_\lambda(x)$ and $f_{\lambda^1}(x^1) + f_{\lambda^{\backslash 1}}(x^{\backslash 1})$.

λ	$f_\lambda(x)$	λ^1	$f_{\lambda^1}(x^1)$	$\lambda^{\backslash p}$	$f_{\lambda^{\backslash 1}}(x^{\backslash 1})$	symbol
(1,1,1)	8	1	4	(1,1)	4	$=$
(1,2,3)	19	1	4	(2,3)	11	$>$
(4,1,2)	15	4	16	(1,2)	8	$<$

In order to continue the analysis of the ordered median function we need to introduce some notation that will be used in the following. We will consider in $\mathbb{R}^M_{0+}$ a particular simplex denoted by $S_M^{\leq}$ which is defined as

$$S_M^{\leq} = \Big\{(\lambda_1, \ldots, \lambda_M) : 0 \leq \lambda_1 \leq \ldots \leq \lambda_M, \sum_{i=1}^{M} \lambda_i = 1\Big\}. \tag{1.2}$$

Together with this simplex, we introduce two cones $\Lambda_M^{\leq}$ and $\Lambda_{r_1,\ldots,r_k}$

$$\Lambda_M^{\leq} = \{(\lambda_1, \ldots, \lambda_M) : 0 \leq \lambda_1 \leq \ldots \leq \lambda_M\} \tag{1.3}$$

$$\Lambda_{r_1,\ldots,r_k} = \Big\{\lambda \in \mathbb{R}^M_+ : \lambda = (\lambda^1, \ldots, \lambda^k) \in \mathbb{R}^M_+,\ \lambda^i = (\lambda^i_1, \ldots, \lambda^i_{r_i}) \in \mathbb{R}^{r_i}_+,$$
$$\max_j \lambda^i_j \leq \min_j \lambda^{i+1}_j \text{ for any } i = 1, \ldots, k-1\Big\} \quad 1 \leq k \leq M, \tag{1.4}$$
$$\Lambda_M = \mathbb{R}^M_+.$$

Notice that with this notation the extreme cases are Λ_M (which corresponds to $k = 1$) and $\Lambda_{1,\ldots,1}$ which is the set $\Lambda_M^{\leq}$.

In addition, let $\mathcal{P}(1 \ldots M)$ be the set of all the permutations of the first M natural numbers, i.e.

$$\mathcal{P}(1 \ldots M) = \{\sigma : \sigma \text{ is a permutation of } 1, \ldots, M\}. \tag{1.5}$$

We write $\sigma = (\sigma(1), \ldots, \sigma(M))$ or sometimes and if this is possible without causing confusion simply $\sigma = \sigma(1), \ldots, \sigma(M)$ or $\sigma = (1), \ldots, (M)$.

The next result, that we include for the sake of completeness, is well-known and its proof can be found in the book by Hardy et al. [98].

Lemma 1.1 *Let $x = (x_1, \ldots, x_M)$ and $y = (y_1, \ldots, y_M)$ be two vectors in $\mathbb{R}^M$. Suppose that $x \leq y$, then*

$$x_{\leq} = (x_{(1)}, \ldots, x_{(M)}) \leq y_{\leq} = (y_{(1)}, \ldots, y_{(M)}) \, .$$

To get more insight into the structure of ordered median functions we state some other properties concerning sorted vectors.

Theorem 1.1 *Let $x = (x_1, \ldots, x_r) \in \mathbb{R}^r$ and $y = (y_1, \ldots, y_s) \in \mathbb{R}^s$ be two vectors of real numbers with $r \leq s$. Let $\sigma^x \in \mathcal{P}(1 \ldots r)$ and $\sigma^y \in \mathcal{P}(1 \ldots s)$ such that*

$$x_{\sigma^x(1)} \leq \cdots \leq x_{\sigma^x(r)}, \tag{1.6}$$

and

$$y_{\sigma^y(1)} \leq \cdots \leq y_{\sigma^y(s)}. \tag{1.7}$$

Then, if

$$x_k \geq y_k, \quad \forall\, k = 1, \ldots, r, \tag{1.8}$$

we have that

$$x_{\sigma^x(k)} \geq y_{\sigma^y(k)}, \quad \forall\, k = 1, \ldots, r.$$

Proof.
Consider $k \in \{1, \ldots, r\}$. We use the fact that because σ^x and σ^y are permutations, the set $\{\sigma^x(1), \ldots, \sigma^x(k)\}$ consists of k distinct elements and the set $\{\sigma^y(1), \ldots, \sigma^y(k-1)\}$ consists of $k-1$ distinct elements, where for $k = 1$ the latter set is simply taken to be the empty set. Thus, there must exist $m \in \{1, \ldots, k\}$ such that $\sigma^x(m) \notin \{\sigma^y(1), \ldots, \sigma^y(k-1)\}$. (If $k = 0$ simply take $m = 1$.) Now by (1.7), it must be that $y_{\sigma^y(k)} \leq y_{\sigma^x(m)}$. Furthermore, by (1.8), we have $y_{\sigma^x(m)} \leq x_{\sigma^x(m)}$ and by (1.6) we have $x_{\sigma^x(m)} \leq x_{\sigma^x(k)}$. Hence $y_{\sigma^y(k)} \leq x_{\sigma^x(k)}$, as required. □

The following lemma shows that if two sorted vectors are given, and a permutation can be found to ensure one vector is (componentwise) not greater than the other, then the former vector in its original sorted state must also be not greater (componentwise) than the latter.

Lemma 1.2 *Suppose* $w, \hat{w} \in \mathbb{R}^M$ *satisfy*

$$w_1 \leq w_2 \leq \cdots \leq w_M, \tag{1.9}$$

$$\hat{w}_1 \leq \hat{w}_2 \leq \cdots \leq \hat{w}_M, \tag{1.10}$$

and

$$\hat{w}_{\sigma(i)} \leq w_i, \quad \forall\, i = 1, \ldots, M \tag{1.11}$$

for some $\sigma \in \mathcal{P}(1 \ldots M)$. *Then*

$$\hat{w}_i \leq w_i, \quad \forall\, i = 1, \ldots, M. \tag{1.12}$$

Proof.
We are using Theorem 1.1. Set $w'_i = \hat{w}_{\sigma(i)}$ for all $i = 1, \ldots, M$, and take $r = s = M$, $p = (w_1, \ldots, w_M)$, $q = (w'_1, \ldots, w'_M)$, σ^x to be the identity permutation and $\sigma^y = \sigma^{-1}$, the inverse permutation of σ. Note that in this case $w_{\sigma^x(i)} = w_i$ for all $i = 1, \ldots, M$, and furthermore

$$w'_{\sigma^y(i)} = w'_{\sigma^{-1}(i)} = \hat{w}_{\sigma(\sigma^{-1}(i))} = \hat{w}_i, \qquad \forall\, i = 1, \ldots, M. \tag{1.13}$$

It is obvious from (1.9) that the elements of p form an increasing sequence under the permutation σ^x. It is also clear from (1.10) and (1.13) that the elements of q form an increasing sequence under the permutation σ^y. Thus, the first two conditions of Theorem 1.1 are met. From (1.11) and the definition of w', we have that $w_i \geq w'_i$ for all $i = 1, \ldots, M$ and the final condition of Theorem 1.1 is met. From Theorem 1.1 we thus deduce that $w_{\sigma^x(i)} \geq w'_{\sigma^y(i)}$ for all $i = 1, \ldots, M$. Now for all $i = 1, \ldots, M$ we have that $w_{\sigma^x(i)} = w_i$ and $w'_{\sigma^y(i)} = \hat{w}_i$, so $\hat{w}_i \leq w_i$ as required. □

The next lemma shows the following: Take a sorted vector w of r real numbers that is componentwise not greater than r elements chosen from another sorted vector w' of $s \geq r$ real numbers. Then the first r entries of w' are componentwise greater or equal than w.

Lemma 1.3 *Let $w = (w_1, \ldots, w_s)$ be a vector of $s \geq 1$ real numbers with*

$$w_1 \leq \cdots \leq w_s.$$

Let $r \in \{1, \ldots, s\}$ and let $S \in \mathbb{R}^r$ be a vector with elements no less, componentwise, than r of the elements of w, say $w' = (w'_1, \ldots, w'_r)$ where for all $j = 1, \ldots, r$ there exists a unique $i(j) \in \{1, \ldots, s\}$ such that $w'_j \geq w_{i(j)}$, with

$$w'_1 \leq \cdots \leq w'_r.$$

Then

$$w'_j \geq w_j, \quad \forall\, j = 1, \ldots, r.$$

Proof.
The claim follows from Theorem 1.1, as follows. Take r and s as given. Take $y_j = w_{i(j)}$ for all $j = 1, \ldots, r$ and define $y_{r+1}, \ldots, y_s$ to be the components from the vector w which do not have index in $\{i(j) \ : j = 1, \ldots, r\}$. Note there is a one-to-one correspondence between the elements of w and y, i.e. y is a permutation of w. Take $x_j = w'_j$ for all $j = 1, \ldots, r$, so $x_j = w'_j \geq w_{i(j)} = y_j$ for all $j = 1, \ldots, r$. Also take σ^x to be the identity permutation, and note that σ^y can be taken so that that $y_{\sigma^y(i)} = w_i$ for all $i = 1, \ldots, s$, by the definition of w and since q is a permutation of w. Now by Theorem 1.1 it must be that for all $j = 1, \ldots, r$, $x_{\sigma^x(j)} \geq y_{\sigma^y(j)} = w_j$, and so $x_j = w'_j \geq w_j$, as required. □

To understand the nature of the OMf we need a precise characterization. Exactly this will be done in the following two results using the concepts of symmetry and sublinearity.

Theorem 1.2 *A function f defined over $\mathbb{R}^M_+$ is continuous, symmetric and linear over $\Lambda^{\leq}_M$ if and only if $f \in$ OMf(M).*

Proof.
Since f is linear over $\Lambda^{\leq}_M$, there exists $\lambda = (\lambda_1, \ldots, \lambda_M)$ such that for any $x \in \Lambda^{\leq}_M$ $f(x) = \langle \lambda, x \rangle$. Now, let us consider any $y \notin \Lambda^{\leq}_M$. There exists a permutation $\sigma \in \mathcal{P}(1 \ldots M)$ such that $y_\sigma \in \Lambda^{\leq}_M$. By the symmetry property it holds $f(y) = f(y_\sigma)$. Moreover, for y_σ we have $f(y_\sigma) = \langle \lambda, y_\sigma \rangle$. Hence, we get that for any $x \in \mathbb{R}^M$

$$f(x) = \langle \lambda, x_\leq \rangle.$$

Finally, the converse is trivially true. □

There are particular instances of the λ vector that make their analysis interesting. One of them is the convex case. The convex ordered median function corresponds to the case where $0 \leq \lambda_1 \leq \lambda_2 \leq \ldots \leq \lambda_M$. For this case, we can obtain a characterization without the explicit knowledge of a linearity region. Let $\lambda = (\lambda_1, \ldots, \lambda_M)$ with $0 \leq \lambda_1 \leq \lambda_2 \leq \ldots \leq \lambda_M$; and $\lambda_\pi = (\lambda_{\pi(1)}, \ldots, \lambda_{\pi(M)})$ being $\pi \in \mathcal{P}(1 \ldots M)$.

Theorem 1.3 *A function f defined over $\mathbb{R}^M$ is the support function of the set $S_\lambda = conv\{\lambda_\pi : \pi \in \mathcal{P}(1...M)\}$ if and only if f is the convex ordered median function*

$$f_\lambda(x) = \sum_{i=1}^{M} \lambda_i x_{(i)}. \tag{1.14}$$

Proof.
Let us assume that f is the support function of S_λ. Then, we have that

$$f(x) = \sup_{s \in S_\lambda} \langle s, x\rangle = \max_{\pi \in \mathcal{P}(1...M)} \langle \lambda_\pi, x\rangle = \max_{\pi \in \mathcal{P}(1...M)} \langle \lambda, x_\pi\rangle = \sum_{i=1}^{M} \lambda_i x_{(i)}.$$

Conversely, it suffices to apply Theorem 368 in Hardy, Littlewood and Polya [98] to the expression in (1.14). □

This result allows us to give global bounds on the ordered median function which do not depend on the weight λ.

Proposition 1.2

1. $f_{(1,0,\ldots,0)}(x) \leq f_\lambda(x) \leq f_{(0,\ldots,0,1)}(x) \quad \forall\, x \in X, \quad \forall\, \lambda \in S_M^{\leq}$.
2. $f_{(1/M,\ldots,1/M)}(x) \leq f_\lambda(x) \leq f_{(0,\ldots,0,1)}(x) \quad \forall\, x \in X, \quad \forall\, \lambda \in S_M^{\leq}$.

Proof.
We only prove (2.) since the proof of (1.) is similar.
It is well-known that $S_M^{\leq}$ is the convex hull of

$$\{e_M, 1/2(e_M + e_{M-1}), \ldots, 1/M \sum_{i=1}^{M} e_i\}$$

where e_i is the vector with 1 in the i-th component and 0 everywhere else (see e.g. Claessens et al. [41]). For each $x \in X$ the function $\lambda \to f_\lambda(x)$ is linear in λ. Thus, we have that the minimum has to be achieved on the extreme points of $S_M^{\leq}$, i.e.

$$\min_{\lambda \in S_M^{\leq}} f_\lambda(x) = \min\Big\{x_{(M)}, 1/2\Big(x_{(M)} + x_{(M-1)}\Big), \ldots, 1/M \sum_{i=1}^{M} x_{(i)}\Big\}$$

thus

$$\min_{\lambda \in S_M^{\leq}} f_\lambda(x) = \frac{1}{M} \sum_{i=1}^{M} x_{(i)}.$$

Analogously for the maximum we obtain that

$$\max_{\lambda \in S_M^{\leq}} f_\lambda(x) = x_{(M)}.$$

□

It is important to remark that these bounds are tight. For instance, for (2.) the lower bound is realized with $\bar{\lambda} = (1/M, \ldots, 1/M)$ and the upper bound with $\hat{\lambda} = (0, \ldots, 0, 1)$, since both $\bar{\lambda}$ and $\hat{\lambda}$ are in $S_M^{\leq}$.

Before continuing, it is worth introducing another concept that will help us in the following. For any vector $x \in \mathbb{R}^M$ and $k = 1, ..., M$, define $r_k(x)$ to be the sum of the k largest components of x, i.e.,

$$r_k(x) = \sum_{t=M-k+1}^{M} x_{(t)}. \tag{1.15}$$

This function is usually called in the literature k-centrum and plays a very important role in the analysis of the ordered median function. The reason is easily understood because of the representation in (1.16). For any $\lambda = (\lambda_1, \ldots, \lambda_M) \in \Lambda_M^{\leq}$ we observe that

$$f_\lambda(x) = \sum_{k=1}^{M} (\lambda_k - \lambda_{k-1}) r_{M-k+1}(x). \tag{1.16}$$

(For convenience we set $\lambda_0 = 0$.)

Convexity is an important property within the scope of continuous optimization. Thus, it is crucial to know the conditions that ensure this property. In the context of discrete optimization convexity cannot even be defined. Nevertheless, in this case submodularity plays a similar role. (The interested reader is referred to the Chapter by McCormick in the Handbook Discrete Optimization [130].) In the following, we also prove a submodularity property of the convex ordered median function.

Let $x = (x_i)$, $y = (y_i)$, be vectors in $\mathbb{R}^M$. Define the *meet* of x, y to be the vector $x \bigwedge y = (\min\{x_i, y_i\})$, and the *join* of x, y by $x \bigvee y = (\max\{x_i, y_i\})$. The meet and join operations define a lattice on $\mathbb{R}^M$.

Theorem 1.4 (**Submodularity Theorem [173]**) *Given $\lambda = (\lambda_1, ..., \lambda_M)$, satisfying $0 \leq \lambda_1 \leq \lambda_2 \leq ... \leq \lambda_M$. For any $x \in \mathbb{R}^M$ define the function $f_\lambda(x) = \sum_{i=1}^{M} \lambda_i x_{(i)}$. Then, $f_\lambda(x)$ is submodular over the lattice defined by the above meet and join operations, i.e., for any pair of vectors x, y in $\mathbb{R}^M$,*

$$f_\lambda(x \bigvee y) + f_\lambda(x \bigwedge y) \leq f_\lambda(x) + f_\lambda(y).$$

Proof.

To prove the above theorem it is sufficient to prove that for any $k = 1, ..., M$,

$$r_k(x \bigvee y) + r_k(x \bigwedge y) \leq r_k(x) + r_k(y). \tag{1.17}$$

Given a pair of vectors x and y in $\mathbb{R}^M$, let $c = x \bigwedge y$ and $d = x \bigvee y$. To prove the submodularity inequality for $r_k(x)$, it will suffice to prove that for any subset C' of k components of c, and any subset D' of k components of d, there exist a subset X' of k components of x and a subset Y' of k components of y, such that the sum of the $2k$ elements in $C' \cup D'$ is smaller than or equal to the sum of the $2k$ elements in $X' \cup Y'$. Formally, we prove the following claim:

Claim: Let I and J be two subsets of $\{1, 2, .., M\}$, with $|I| = |J| = k$. There exist two subsets I' and J' of $\{1, 2, ..., M\}$, with $|I'| = |J'| = k$, such that

$$\sum_{i \in I} c_i + \sum_{j \in J} d_j \leq \sum_{s \in I'} x_s + \sum_{t \in J'} y_t.$$

Proof of Claim: Without loss of generality suppose that $x_i \neq y_i$ for all $i = 1, ..., M$. Let $I_x = \{i \in I : c_i = x_i\}$, $I_y = \{i \in I : c_i = y_i\}$, $J_x = \{j \in J : d_j = x_j\}$, and $J_y = \{j \in J : d_j = y_j\}$. Since $x_i \neq y_i$ for $i = 1, ..., M$, we obtain $|I_x| + |I_y| = |J_x| + |J_y| = k$, I_x and J_x are mutually disjoint, and I_y and J_y are mutually disjoint. Therefore, if $|I_x| + |J_x| = k$, (which in turn implies that $|I_y| + |J_y| = k$), the claim holds with equality for $I' = I_x \cup J_x$, and $J' = I_y \cup J_y$. Hence, suppose without loss of generality that $|I_x| + |J_x| > k$, and $|I_y| + |J_y| < k$. Define $I'' = I_x \cup J_x$, and $J'' = I_y \cup J_y$. Let $K = I'' \cap J''$. $|I''| > k$, and $|J''| < k$. We have

$$\sum_{i \in I} c_i + \sum_{j \in J} d_j = \sum_{i \in K} (x_i + y_i) + \sum_{s \in I'' - K} x_s + \sum_{t \in J'' - K} y_t. \tag{1.18}$$

On each side of the last equation we sum exactly k components from c, ("minimum elements"), and k components from d, ("maximum elements"). Moreover, the set of components $\{x_s : s \in I'' - K\} \cup \{y_t : t \in J'' - K\}$ contains exactly $k - |K|$ minimum elements and exactly $k - |K|$ maximum elements. In particular, the set $\{x_s : s \in I'' - K\}$ contains at most $k - |K|$ maximum elements. Therefore, the set $\{x_s : s \in I'' - K\}$ contains at least $q = |I''| - |K| - (k - |K|) = |I''| - k$ minimum elements. Let $I^* \subset I'' - K$, $|I^*| = q$, denote the index set of such a subset of minimum elements. We therefore have,

$$x_i \leq y_i, i \in I^*. \tag{1.19}$$

Note that from the construction I^* and J'' are mutually disjoint. Finally define $I' = I'' - I^*$ and $J' = J'' \cup I^*$, and use (1.18) and (1.19) to observe that the claim is satisfied for this choice of sets. □

1.4 Towards Location Problems

After having presented some general properties of the ordered median functions we will now describe the use of this concept in the location context since this is the major focus of this book.

In location problems we usually are given a set $A = \{a_1, \dots, a_M\}$ of clients and we are looking for the locations of a set $X = \{x_1, \dots, x_N\}$ of new facilities. The quality of a solution is evaluated by a function on the relation between A and X, typically written as $c(A, X) = (c(a_1, X), \dots, c(a_M, X))$ or simply $c(A, X) = (c_1(X), \dots, c_M(X))$. $c(A, X)$ may express time, distance, cost, personal preferences,... Assuming that the quality of the service provided decreases with an increase of $c(a_i, X)$, the objective function to be optimized depends on the cost vector $c(A, X)$. Thus, from the server point of view, a monotonicity principle has to be required because the larger the components of the cost vector, the lower the quality of the service provided (the reader may note that this monotonicity principle is reversed when the new facilities are obnoxious or represent any risk for the users).

On the other hand, looking at the problem from the clients' point of view, it is clear that the quality of the service obtained for the group (the entire set A) does not depend on the name given to the clients in the set A. That is to say, the service is the same if the cost vector only changes the order of its components. Thus, seeing the problem from this perspective, the symmetry principle (see Milnor [138]) must hold (recall that the symmetry principle for an objective function f states that the value given by f to a point u does not depend on the order of the components of u). Therefore, for each cost vector $c_\sigma(A, X)$ whose components are a permutation of the components of $c(A, X)$, $f(c(A, X)) = f(c_\sigma(A, X))$. These two principles have been already used in the literature of location theory and their assumption is accepted (see e.g. Buhl [29], Carrizosa et al. [33], Carrizosa et. al [32] or Puerto and Fernández [169]).

We have proved that the ordered median function is compatible with these principles. By identifying x with $c(A, X)$ we can apply the concept of ordered median functions to location problems. It means that the independent variable of the general definition given in 1.1 is replaced by the cost relation among the clients and the new facilities. The components of $c(A, X)$ are related by means of the lambda vector so that different choices will generate different location models (see Table 1.1).

Of course, the main difficulty is not simply to state the general problem, but to provide structural properties and solution procedures for the respective decision spaces (continuous, network and discrete). Exactly, this will be the content of the remaining three parts of this book. Before starting these detailed parts, we want to give three illustrative examples showing the relation to classical problems in location theory.

Example 1.3
(Three points Fermat problem.) This is one of the classical problems in metric geometry and it dates back to the XVII century. Its formulation is disarmingly simple, but is really rich and still catches the attention of researchers from different areas.

According to Kuhn [126] the original formulation of this problem is credited to Pierre de Fermat (1601-1665), a French mathematician, who addressed the problem with the following challenge: *"let he who does not approve my method attempt the solution of the following problem: given three points in the plane, find a fourth point such that the sum of its distances to the three given points is a minimum".*

Denote by $x = (x_1, x_2)$ the point realizing the minimum of the Fermat problem; and assume that the coordinates of the three given points are $a = (a_1, a_2), b = (b_1, b_2), c = (c_1, c_2)$. If we take the vector $\lambda = (1, 1, 1)$, the Fermat problem consists of finding the minimum of the following ordered median function (which is in fact the classical median function):

$$f_\lambda((d(x,a), d(x,b), d(x,c))) = d(x,a) + d(x,b) + d(x,c),$$

where for instance $d(x,a) = \sqrt{(x_1 - a_1)^2 + (x_2 - a_2)^2}$ is the Euclidean distance from x to a.

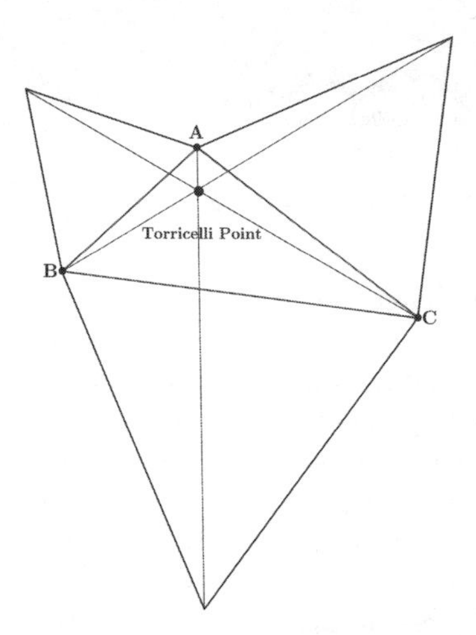

Fig. 1.2. The Torricelli point

The first geometrical solution is credited to Torricelli (1608-1647), the reader can see in Figure 1.3 the construction of the so called Torricelli point. Later in this book we will see how the same solution can be obtained using the theory of the ordered median function.

Example 1.4

In the next example we look at a realistic planning problem in the city of Kaiserslautern. We are given a part of the city map (see Figure 1.3) and we have to find a location for a take away restaurant. The model which was

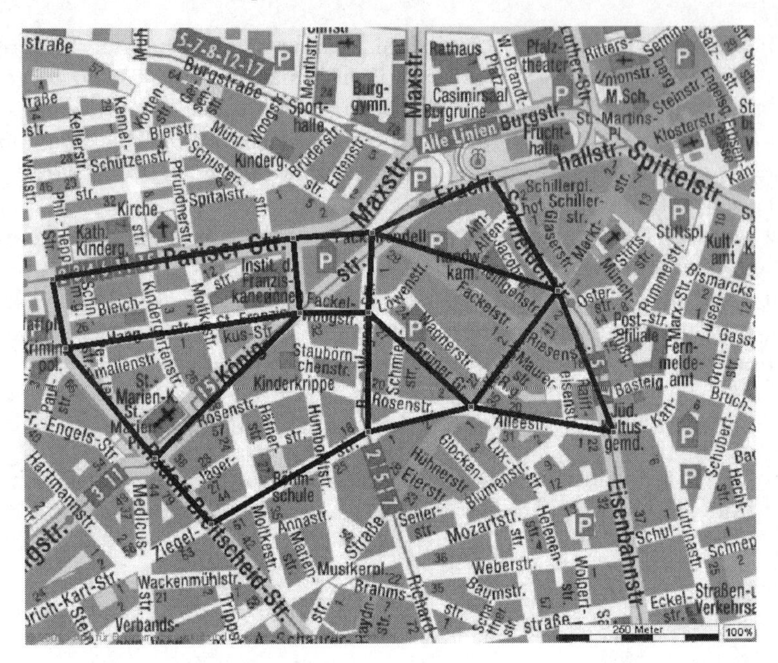

Fig. 1.3. A part of the city map of Kaiserslautern

agreed on is to use the length of the streets of the city map as measure for the distance. In addition the demand of a street is always aggregated into the next crossing. These crossings can be seen as nodes of a graph and the streets are then the corresponding edges (see Figure 1.4).

Therefore A consists of the nodes of the graph and $X = (x)$ is the location of the take away restaurant. We want to allow x to be either in a node of the graph or at any point on an edge (street). The vector $c(A, x)$ is given by the all-pair shortest path matrix (see Table 1.2).

For measuring distances among edges we simply assume that the distance to a node along an incident edge growths linear.

Moreover, it is said to be important that we do not have customers served bad. Therefore we choose a model in which we will only take into account the furthest 3 customers. All others then are remaining within a reasonable distance.

This means that we choose $\lambda = (0, \ldots, 0, 1, 1, 1)$.

As a result we get, that the optimal location would be in Node 6 with an objective value of 246. If we would have decided only to take into account the two furthest customers, we would have gotten two optimal locations: One on the edge $[6, 7]$, a quarter of the edge length away from Node 6 and another one on the edge $[6, 9]$ only a fraction of 0.113 away from Node 6. The objective value is 172. We will see later in the book how the solution procedure works.

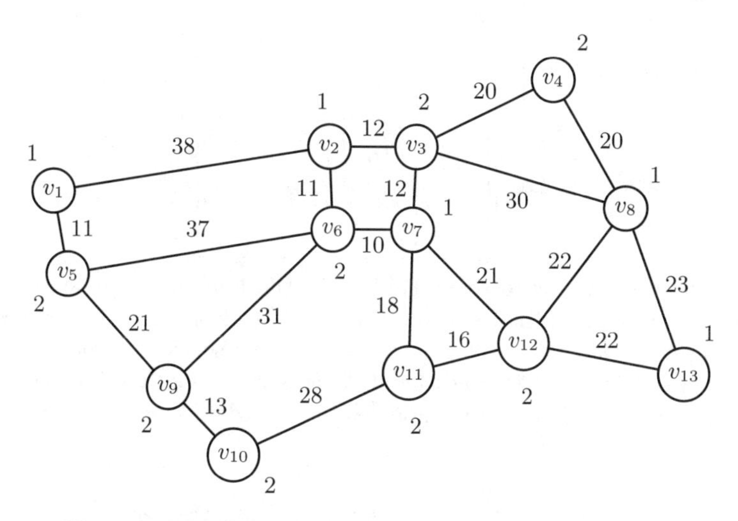

Fig. 1.4. The network model for the KL-city street map

Table 1.2. Distance matrix of Example 1.4.

	v_1	v_2	v_3	v_4	v_5	v_6	v_7	v_8	v_9	v_{10}	v_{11}	v_{12}	v_{13}
v_1	0	38	50	70	11	48	58	80	32	45	73	79	101
v_2	38	0	12	32	48	11	21	42	42	55	39	42	64
v_3	50	12	0	20	59	22	12	30	53	58	30	33	53
v_4	70	32	20	0	79	42	32	20	73	78	50	42	43
v_5	11	48	59	79	0	37	47	89	21	34	62	68	90
v_6	48	11	22	42	37	0	10	52	31	44	28	31	53
v_7	58	21	12	32	47	10	0	42	41	46	18	21	43
v_8	80	42	30	20	89	52	42	0	79	66	38	22	23
v_9	32	42	53	73	21	31	41	79	0	13	41	57	79
v_{10}	45	55	58	78	34	44	46	66	13	0	28	44	66
v_{11}	73	39	30	50	62	28	18	38	41	28	0	16	38
v_{12}	79	42	33	42	68	31	21	22	57	44	16	0	22
v_{13}	101	64	53	43	90	53	43	23	79	66	38	22	0

Example 1.5

We are coming back to the discussion between Mr. Optimistic, Mr. Pessimistic and Mrs. Realistic. The situation has changed since the local government decided to have two of these service facilities (this happened since they could not justify one). The rest of the data remains the same and is reprinted below:

	New_1	New_2	New_3	New_4
P_1	5	2	5	13
P_2	6	20	4	2
P_3	12	10	9	2
P_4	2	2	13	1
P_5	5	9	2	3

So, the task now is to select two out of the four possible locations. It is further assumed (by Mrs. Realistic) that a client always goes to the closest of the two service facilities (since one gets the same service in both locations). It is Mr. Pessimistic's turn and therefore he insists on only taking the largest costs into account. This can be modeled in the ordered median framework by setting

$$\lambda = (0,0,0,0,1) \ .$$

Moreover, $c(P_i, \{New_k, New_l\}) = \min\{c(P_i, New_k), c(P_i, New_l)\}$. We are therefore looking for a solution of the so called 2-Center problem. The possible solutions are $\{New_1, New_2\}$, $\{New_1, New_3\}$, $\{New_1, New_4\}$, $\{New_2, New_3\}$, $\{New_2, New_4\}$, $\{New_3, New_4\}$.

The computation works as follows:

$$c(A, \{New_1, New_2\}) = (2, 6, 10, 2, 5)$$

and

$$\langle \lambda, c(A, \{New_1, New_2\})_{\leq} \rangle = \langle (0,0,0,0,1), (2,2,5,6,10) \rangle = 10 \ .$$

The remaining five possibilities are computed analogously:

- $c(A, \{New_1, New_3\}) = (5, 4, 9, 2, 2)$ and the maximum is 9.
- $c(A, \{New_1, New_4\}) = (5, 2, 2, 1, 3)$ and the maximum is 5.
- $c(A, \{New_2, New_3\}) = (2, 4, 9, 2, 2)$ and the maximum is 9.
- $c(A, \{New_2, New_4\}) = (2, 2, 2, 1, 3)$ and the maximum is 3.
- $c(A, \{New_3, New_4\}) = (5, 2, 2, 1, 2)$ and the maximum is 5.

As a result Mr. Pessimistic voted for the solution $\{New_2, New_4\}$.

Part II

The Continuous Ordered Median Location Problem

2

The Continuous Ordered Median Problem

After introducing the OMf in Chapter 1, we now investigate location problems with the OMf in a continuous setting. First, we define the continuous ordered median problem (OMP). Then we take a closer look at different tools needed to work with the continuous OMP: gauges, ordered regions, elementary convex sets and bisectors.

2.1 Problem Statement

Since the early sixties much research has been done in the field of continuous location theory and a number of different models have been developed. Nowadays, continuous location has achieved an important degree of maturity. Witnesses of it are the large number of papers and research books published within this field. In addition, this development has been also recognized by the mathematical community since the AMS code 90B85 is reserved for this area of research. Continuous location problems appear very often in economic models of distribution or logistics, in statistics when one tries to find an estimator from a data set or in pure optimization problems where one looks for the optimizer of a certain function. For a comprehensive overview the reader is referred to [163] or [61].

Most of these problems share some structural properties beyond the common framework represented by the existence of a certain demand to be covered and a number of suppliers that are willing to satisfy that demand. However, despite the many coincidences found in different location problems, as far as we know an overall analysis of these common elements has never been addressed. Location theory experienced its development in a case oriented manner: Location analysts looked closer at a particular problem for which they developed results and solution procedures. This approach has provided significant advances in this field. Nevertheless, there seems to be a lack of a common knowledge, and some basic tools applicable to a large amount of problems and situations.

Our goal is to overcome this lack of not having a unified theory studying location problems with the ordered median function. In the following we will formally introduce the continuous OMP which provides a common framework for the classical continuous location problems and allows an algebraic approach to them. Moreover, this approach leads to completely new objective functions for location problems. It is worth noting that this approach emphasizes the role of the clients seen as a collectivity. From this point of view the quality of the service provided by a new facility to be located does not depend on the specific names given by the modeler to the demand points. Different ways to assign names to the demand points should not change the quality of the service, i.e. a symmetry principle must hold. This principle being important in location problems is inherent to this model. In fact, this model is much more than a simple generalization of some classical models because any location problem whose objective function is a monotone, symmetric norm of the vector of distances reduces to it (see [167], [169]). A discussion of some principles fundamental for location problems was considered in Chapter 1.

In order to formally introduce the problem we must identify the elements that constitute the model. In the continuous context the sets A and X introduced in Chapter 1 have the following meaning. A is a set of M points in $\mathbb{R}^n$ (often $\mathbb{R}^2$) and X is a set of N points in $\mathbb{R}^n$ usually the location of the new facilities. In this chapter we will mainly deal with the case $X = \{x_1\} = \{x\}$. Moreover, $c(A, X)$ will be expressed in terms of weighted distance measures from A to X.

To be more specific, we assign to each a_i, $i = 1, \ldots, M$ a respective weight w_i. The distance measure is given by a Minkowski functional defined on a compact, convex set B containing the origin in its interior. Those functions are called gauges and are written as

$$\gamma_B(x) = \inf\{r > 0 : x \in rB\}. \tag{2.1}$$

Consequently, the weighted distance measure from a_i to x is written $w_i\gamma_B(x - a_i)$. Sometimes, it is necessary to allow each $a_i \in A$ to define its own distance measure. In this case we assign each a_i a compact, convex set B_i to define $w_i\gamma_{B_i}(x - a_i)$. To simplify notation we will omit the reference to the corresponding unit ball whenever this is possible without causing confusion. In those cases, rather than γ_{B_i} we simply write γ_i. Also for the sake of brevity we will write $d_i(x) = w_i\gamma_{B_i}(x - a_i)$.

With these ingredients we can formulate the continuous OMP . For a given $\lambda = (\lambda_1, \ldots, \lambda_M)$, the general OMP can be written as

$$\inf_X f_\lambda(d(A, X)) := \langle \lambda, sort_M(d(A, X)) \rangle, \tag{2.2}$$

where $d(A, X) = (d(a_1, X), \ldots, d(a_M, X))$, and $d(a_i, X) = \inf_{y \in X} \gamma_B(y - a_i)$. For the sake of simplicity, we will write the OMf(M), $f_\lambda(d(A, X))$, in the context of location problems simply as $f_\lambda(X)$. We will denote the set of optimal solutions of Problem (2.2) as $\mathsf{M}(f_\lambda, A)$. When no explicit reference to the set A is necessary we simply write $\mathsf{M}(f_\lambda)$. It is worth noting that $\mathsf{M}(f_\lambda, A)$ might be empty. For instance, this is the case if $\lambda_i < 0$ for all $i = 1, \ldots, M$.

The reader may note that the OMP is somehow similar to the well-known Weber Problem, although it is more general because it includes as particular instances a number of other location problems; although also new useful objective functions can easily be modeled. Assume, for example, that we are not only interested in minimizing the distance to the most remote client (center objective function), but instead we would like to minimize the average distance to the 5 most remote clients (or any other number). This can easily be modeled by setting $\lambda_{M-4}, \ldots, \lambda_M = 1$ and all other lambdas equal to zero. This k-centra problem is a different way of combining average and worst-case behavior.

Also ideas from robust statistics can be implemented by only taking into account always the k_1 nearest and simultaneously the k_2 farthest.

In the field of semi-obnoxious facility location the following approach is commonly used. The weights w_i (which are assigned to existing facilities) express the attraction ($w_i > 0$) or repulsion level ($w_i < 0$) of customers at a_i with respect to a new facility. By using negative entries in the λ-vector, we are in addition able to represent the new facility point of view. This means that we can assign the existing facilities a higher repulsion level if they are closer to the new facility. Note that this is not possible with classical approaches. We will discuss this case in more detail in Section 4.4.

Example 2.1
Consider three demand points $a_1 = (1,2)$, $a_2 = (3,5)$ and $a_3 = (2,2)$ with weights $w_1 = w_2 = w_3 = 1$. Choose $\lambda_1 = \lambda_2 = \lambda_3 = 1$ to get $f_\lambda(x) = \sum_{i=1}^{3} \gamma(x - a_i)$, i.e. the Weber problem. For the second case choose $\lambda_1 = \lambda_2 = 1/2$ and $\lambda_3 = 1$ and we get: $f_\lambda(x) = 1/2 \sum_{i=1}^{3} \gamma(x - a_i) + 1/2 \max_{1 \le i \le 3} \gamma(x - a_i)$, i.e. the 1/2-centdian problem. Finally choose $\lambda_1 = \lambda_2 = 0$ and $\lambda_3 = 1$ to get: $f_\lambda(x) = \max_{1 \le i \le 3} \gamma(x - a_i)$, i.e. the center problem.

We note in passing that the objective function of OMP is non-linear. Therefore, the linear representation given in (2.2) is only region-wise defined and in general non-convex if no additional assumptions are made on λ (see Proposition 1.1 for further details).

Example 2.2
Consider two demand points $a_1 = (0,0)$ and $a_2 = (10,5)$ in the plane equipped with the l_1-norm (see (2.14) for a definition). Assume $\lambda_1 = 100$ and $\lambda_2 = 1$ and $w_1 = w_2 = 1$. We obtain only two optimal solutions to Problem (2.2), lying in each demand point. Therefore, the objective function is not convex since we have a non-convex optimal solution set.

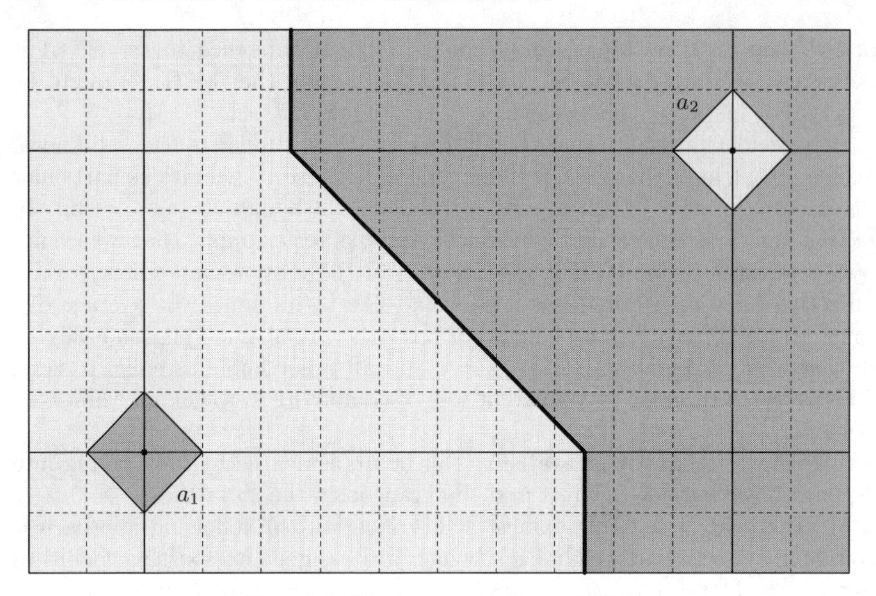

Fig. 2.1. Illustration to Example 2.2

$$
\begin{aligned}
f_\lambda(a_1) &= 100 \times 0 + 1 \times 15 = 15 \\
f_\lambda(a_2) &= 100 \times 0 + 1 \times 15 = 15 \\
f_\lambda(\frac{1}{2}(a_1 + a_2)) &= 100 \times 7.5 + 1 \times 7.5 = 757.5
\end{aligned}
$$

See Figure 2.1.

To conclude this initial overview of the OMP, we would like to address the existence of optimal solutions of Problem (2.2). The reader may note that this question is important because it may happen that this problem does not have optimal solutions. To simplify the discussion we consider the following formulation of Problem (2.2).

$$
\min_{x \in \mathbb{R}^n} f_\lambda(x) = \sum_{i=1}^{M} \lambda_i^+ d_{(i)}(x) + \sum_{i=1}^{M} \lambda_i^- d_{(i)}(x) \tag{2.3}
$$

where

$$
\lambda_i^+ = \begin{cases} \lambda_i & \text{if } \lambda_i > 0 \\ 0 & \text{otherwise} \end{cases}
$$

$$
\lambda_i^- = \begin{cases} \lambda_i & \text{if } \lambda_i < 0 \\ 0 & \text{otherwise} \end{cases}
$$

and $d_{(i)}(x)$, $i = 1, 2, \ldots, M$ was defined before.

When some lambdas are negative it may happen that the objective function $f_\lambda(x)$ has no lower bound, thus going to $-\infty$ when distances increase. In order to avoid this problem we prove a property that ensures the existence of optimal solutions of Problem (2.3).

To simplify the notation of the following theorem, set $L = \sum_{i=1}^{M}(\lambda_i^+ + \lambda_i^-)$.

Theorem 2.1 *(Necessary condition) If $L \geq 0$ then any optimal location is finite. If $L < 0$ then any optimal location is at infinity. ($\mathsf{M}(f_\lambda, A) = \emptyset$.) Moreover, if $L = 0$ then the function $f_\lambda(A)$ is bounded.*

Proof.
Recall that $d_k(x) = \gamma(x - a_k)$ for all $k = 1, \ldots, M$. Let x satisfy $\gamma(x - a_{\sigma(1)}) \leq \gamma(x - a_{\sigma(2)}) \leq \ldots \gamma(x - a_{\sigma(M)})$. By the triangle inequality, we obtain

$$\gamma(x) - \gamma(a_{\sigma(k)}) \leq \gamma(x - a_{\sigma(k)}) \leq \gamma(x) + \gamma(a_{\sigma(k)}), \text{ for } k = 1, 2, \ldots, M.$$

Since $\lambda_k^+ \geq 0$ and $\lambda_k^- \leq 0$ for all $k = 1, 2, \ldots, M$, it yields

$$\lambda_k^+(\gamma(x) - \gamma(a_{\sigma(k)})) \leq \lambda_k^+\gamma(x - a_{\sigma(k)}) \leq \lambda_k^+(\gamma(x) + \gamma(a_{\sigma(k)})), \; k = 1, \ldots, M,$$
$$\lambda_k^-(\gamma(x) + \gamma(a_{\sigma(k)})) \leq \lambda_k^-\gamma(x - a_{\sigma(k)}) \leq \lambda_k^-(\gamma(x) - \gamma(a_{\sigma(k)})), \; k = 1, \ldots, M.$$

Hence, being $\lambda_k = \lambda_k^+ + \lambda_k^-$,

$$\lambda_k\gamma(x) - |\lambda_k|\gamma(a_{\sigma(k)}) \leq \lambda_k\gamma(x - a_{\sigma(k)}) \leq \lambda_k\gamma(x) + |\lambda_k|\gamma(a_{\sigma(k)}), k = 1, 2, \ldots, M. \tag{2.4}$$

Therefore, by simple summation over the index k

$$L\gamma(x) - U \leq f_\lambda(x) \leq L\gamma(x) + U$$

where $U = \max_{\sigma \in \mathcal{P}(1\ldots M)} [\sum_{i=1}^{M} \lambda_i^+\gamma(a_{\sigma(i)}) - \sum_{j=1}^{M} \lambda_j^-\gamma(a_{\sigma(j)})]$.

First note that, when $\gamma(x) \to \infty$ and $L > 0$ then $f_\lambda(x) \to \infty$. This implies that a better finite solution must exist. On the other hand, when $L < 0$ then $f_\lambda(x) \to -\infty$, this means that the optimal solution does not exist and the value goes to infinity. Finally, if $L = 0$ the function f_λ is always bounded. □

For more details on existence issues of the OMP in general spaces the reader is referred to Chapter 6.3.

2.2 Distance Functions

In this section we will give a short description of the distance functions we will be interested in.

Distances are measured by the gauge function of B, γ_B. This function fulfills the properties of a not necessarily symmetric distance function:

$$\gamma_B(x) \geqslant 0 \quad \forall x \in \mathbb{R}^n \text{ (non-negativity)} \tag{2.5}$$

$$\gamma_B(x) = 0 \quad \Leftrightarrow \quad x = 0 \text{ (definiteness)} \tag{2.6}$$

$$\gamma_B(\mu x) = \mu\gamma_B(x) \quad \forall x \in \mathbb{R}^n, \quad \forall \mu \geqslant 0 \text{ (positive homogeneity)} \tag{2.7}$$

$$\gamma_B(x+y) \leqslant \gamma_B(x) + \gamma_B(y) \quad \forall x, y \in \mathbb{R}^n \text{ (triangle inequality)} \tag{2.8}$$

Obviously, the set B defines the unit ball of its corresponding gauge γ_B. We can observe that $\gamma_B(x) = 1$ if x lies on the boundary of B. For $x \notin B$, the value of $\gamma_B(x)$ is the factor we inflate B until we hit x. If x lies in the interior of B we shrink B until x is on the boundary of the shrunk ball. The factor by which we have to inflate or shrink B respectively gives us exactly $\gamma_B(x)$ (see the dashed ellipse in Figure reffig:gauge).

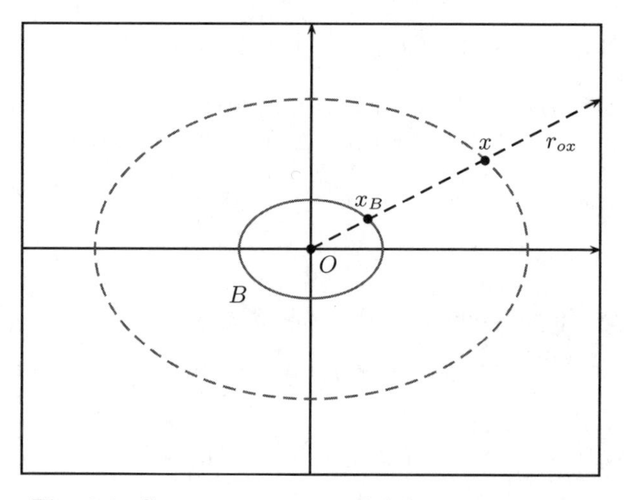

Fig. 2.2. Convex compact set B defining a gauge γ_B

Looking for a more geometrical expression of gauges, we need the definition of a foot-point.

The unique intersection point x_B of the ray starting at O and passing through x with the boundary of B is called foot-point,

$$x_B = r_{\overrightarrow{ox}} \cap bd(B). \tag{2.9}$$

This concept allows to link the analytic definition (2.1) of γ_B with the following geometric evaluation

$$\gamma_B : \mathbb{R}^n \longrightarrow \mathbb{R} \qquad , \qquad x \longmapsto \gamma_B(x) = \frac{l_2(x)}{l_2(x_C)}, \tag{2.10}$$

l_2 being the Euclidean norm (see (2.14)).

The properties (2.7) and (2.8) imply the convexity of a gauge:

$$\gamma_B(\mu x+(1-\mu)y) \leq \mu\gamma_B(x)+(1-\mu)\gamma_B(y) \quad \forall 0 \leq \mu \leq 1,\ \forall x,y \in \mathbb{R}^n. \quad (2.11)$$

Moreover, if the set B is symmetric with respect to the origin the property

$$\gamma_B(x) = \gamma_B(-x) \quad \forall x \in \mathbb{R}^n \text{ (symmetry)} \quad (2.12)$$

holds and therefore γ_B is a norm, since by (2.7) and (2.12)

$$\gamma_B(\mu x) = |\mu|\gamma_B(x) \quad \forall x \in \mathbb{R}^n, \forall \mu \in \mathbb{R} \text{ (homogeneity).} \quad (2.13)$$

γ_B being a norm then the function $d : \mathbb{R}^n \times \mathbb{R}^n \longrightarrow \mathbb{R}$ defined by $d(x,y) = \gamma_B(x-y)$ is a metric. The most well-known norms on $\mathbb{R}^n$ are the p-norms, which are defined by

$$l_p(x) := \begin{cases} \left(\sum_{i=1}^{n} |x_i|^p\right)^{\frac{1}{p}}, & 1 \leq p < \infty \\ \max_{i=1,\dots,n} \{|x_i|\}, & p = \infty \end{cases} . \quad (2.14)$$

For $p = 1, 2, \infty$ the names Manhattan norm, Euclidean norm and Chebyshev norm are in common use.

Another important concept related to the evaluation of gauge functions is polarity. The polar set B^o of B is given by

$$B^o = \{x \in \mathbb{R}^n : \langle x, p\rangle \leq 1, \quad \forall p \in B\}. \quad (2.15)$$

The polar set B^o of B induces a new gauge with unit ball B^o that is usually called dual gauge.

The next lemma connects the gauge with respect to B with its corresponding polar set. This lemma is a classical result in convex analysis and will result essential to evaluate distances in location problems. For a proof the reader is referred to [102].

Lemma 2.1 *Let B be a closed, convex set and $O \in B$. Then γ_B is the support function of the polar set B^o of B,*

$$\gamma_B(x) = \sup\{\langle x,y\rangle : y \in B^o\}. \quad (2.16)$$

Let $C \subset \mathbb{R}^n$ be a convex set. An element $p \in \mathbb{R}^n$ is said to be normal to C at $x \in C$ if $\langle p, y-x\rangle \leq 0$ for all $y \in C$. The set of all the vectors normal to C at x is called the normal cone to C at x and is denoted by $N_C(x)$.

The normal cone to B^o at x is given by

$$N_{B^o}(x) := \{p \in \mathbb{R}^n : \langle p, y-x\rangle \leq 0 \quad \forall y \in B^o\}. \quad (2.17)$$

For the sake of simplicity, we will denote the normal cone $N_{B^o}(x)$ to the unit dual ball B^o by N whenever no confusion is possible.

Table 2.1 shows the connection between the properties of the set B and the properties of the corresponding gauge γ_B.

Table 2.1. The relation between the properties of B and the properties of γ_B.

B	γ_B
bounded	definiteness
symmetric with respect to 0	symmetry
convex	triangle inequality
$0 \in \text{int}(B)$	$\gamma_B(\mathbb{R}^n) \subseteq \mathbb{R}_+$
The closure of the set B is necessary to obtain a surjective mapping γ_B.	

Now we consider what happens if we generalize the properties of a gauge-generating set S in such a way that we allow S to be non convex. First of all, notice that in this case the properties (2.5) - (2.8) may not be fulfilled anymore (see Table 2.1) and one could say that the function corresponding to a non convex unit ball is not appropriate to measure distances.

However, if S is a star-shaped compact set with respect to the origin which is contained in the interior of S, the function

$$\delta_S \quad : \quad \mathbb{R}^n \quad \to \quad \mathbb{R} \quad , \quad x \quad \mapsto \quad \inf\{r > 0 \, : \, x \in rS\} \tag{2.18}$$

fulfills the following properties:

$$\delta_S(x) \geq 0 \quad \forall\, x \in \mathbb{R}^n \;\; (\text{non-negativity}) \tag{2.19}$$

$$\delta_S(x) = 0 \quad \Leftrightarrow \quad x = 0 \;\; (\text{definiteness}) \tag{2.20}$$

$$\delta_S(\mu x) = \mu \delta_S(x) \quad \forall\, x \in \mathbb{R}^n \, , \; \forall\, \mu \geq 0 \;\; (\text{positive homogeneity}) \tag{2.21}$$

That means δ_S fulfills all the properties of a gauge except the triangular inequality. So we are not too far away from the idea of an appropriate distance function. We call the function δ_S the star-shaped gauge with respect to S. However, as a consequence of the failing of the triangular inequality δ_S is not convex.

Other extensions, as for instance the use of unbounded unit balls, have been also considered in the literature of location analysis. For further details the reader is referred to [101].

A gauge (norm) γ_C is called strictly convex if

$$\gamma_C(\mu x + (1-\mu)y) < \mu\gamma_C(x) + (1-\mu)\gamma_C(y) \quad \forall\, 0 < \mu < 1 \, , \; \forall\, x, y \in \mathbb{R}^n \, , \; x \neq y \, . \tag{2.22}$$

This property of γ_C is satisfied if and only if the corresponding unit ball C is strictly convex.

A gauge (norm) γ_C is called smooth if γ_C is differentiable for all $x \in \mathbb{R}^n \setminus \{0\}$, i. e. if $\nabla\gamma_C(x) = \left(\frac{\partial}{\partial x_i}\gamma_C(x)\right)_{i=1,\dots,n}$ exists for all $x \in \mathbb{R}^n \setminus \{0\}$.

Notice that there are strictly convex gauges which are not smooth and also smooth gauges which are not strictly convex (see Figure 2.3).

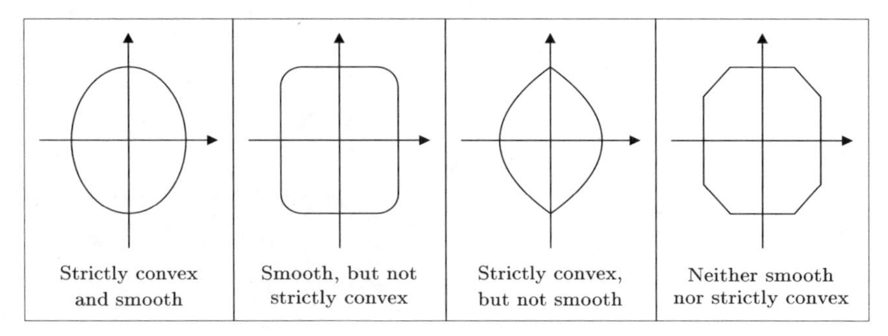

Fig. 2.3. Comparison of strictly convex and smooth gauges

Finally, notice that for a (star-shaped) gauge γ_C and a positive scalar ω the relationship

$$\omega\gamma_c = \gamma_{\frac{1}{\omega}C} \tag{2.23}$$

holds. This means that the consideration of weights can be done by changing the radius of the set C.

2.2.1 The Planar Case

We are mainly interested in problems on the plane and with polyhedral or block gauges (i.e. gauges, the unit ball of which is a polytope). Therefore, we want to describe more specifically the properties of the elements of the continuous planar formulation of Problem (2.2) which give us insights into the geometry of the model. For this reason we will particularize some of the properties that we have shown for general gauges to the case when $B \subseteq \mathbb{R}^2$ is a bounded polytope whose interior contains the zero.

The most classical examples of polyhedral norms are the rectangular or Manhattan norm and the Chebyshev or infinity norm. For the sake of readability we describe them in the following example.

Example 2.3
For an arbitrary point $x = (x_1, x_2)$ in $\mathbb{R}^2$, the l_1-norm and the l_∞-norm are:

- $l_1(x) = |x_1| + |x_2|$

  $$B_{l_1} = \{x \in \mathbb{R}^2 \, : \, l_1(x) \leq 1\} = conv(\{(-1,0),(0,-1),(1,0),(0,1)\})$$

  is the unit ball with respect to the l_1-norm (see Figure 2.4).

- $l_\infty(x) = \max\{|x_1|, |x_2|\}$

  $$B_{l_\infty} = \{x \in \mathbb{R}^2 \, : \, l_\infty(x) \leq 1\} = conv(\{(-1,-1),(1,-1),(1,1),(-1,1)\})$$

  is the unit ball with respect to the l_∞-norm (see Figure 2.4).

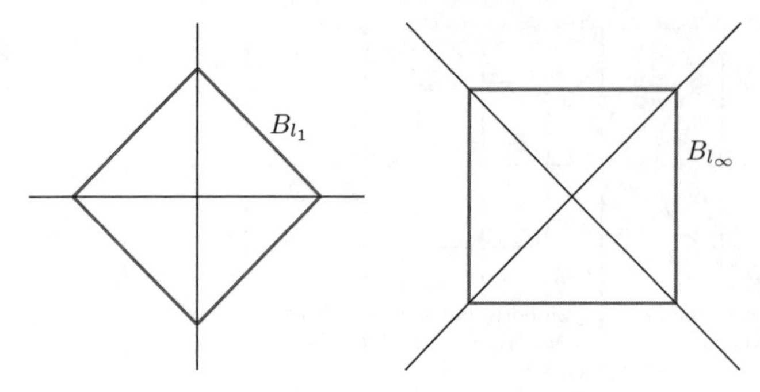

Fig. 2.4. Unit balls for l_1-norm and l_∞-norm

We denote the set of extreme points of B by $Ext(B) = \{e_g : g = 1, \ldots, G\}$. In the polyhedral case in $\mathbb{R}^2$, the polar set B^o of B is also a polytope whose extreme points are $\{e_g^o : g = 1, 2, \ldots, G\}$, see [65, 204].

The evaluation of any polyhedral gauge can be done using the following corollary of the result given in Lemma 2.1.

Corollary 2.1 *If B is a polytope in $\mathbb{R}^2$ with extreme points $Ext(B) = \{e_1, \ldots, e_G\}$ then its corresponding polyhedral gauge can be obtained by*

$$\gamma_B(x) = \max_{e_g^o \in Ext(B^o)} \langle x, e_g^o \rangle. \tag{2.24}$$

We define fundamental directions $d_1, \ldots, d_G$ as the halflines defined by 0 and $e_1, \ldots, e_G$. Further, we define Γ_g as the cone generated by d_g and d_{g+1} (fundamental directions of B) where $d_{G+1} := d_1$.

At times, it will result convenient to evaluate polyhedral gauges by means of the primal representation of the points within each fundamental cone. In this regard, we include the following lemma.

Lemma 2.2 *Let B be a polytope in $\mathbb{R}^2$ with $Ext(B) = \{e_1, \ldots, e_G\}$. Then, the value of $\gamma_B(x)$ is given as:*

$$\gamma_B(x) = \min \left\{ \sum_{g=1}^{G} \mu_g : x = \sum_{g=1}^{G} \mu_g e_g, \, \mu_g \geq 0 \right\}. \tag{2.25}$$

Moreover, if γ_B defines a block norm then:

$$\gamma_B(x) = \min \left\{ \sum_{g=1}^{\frac{G}{2}} |\mu_g| : x = \sum_{g=1}^{\frac{G}{2}} \mu_g e_g \right\}. \tag{2.26}$$

Proof.
If γ_B is a polyhedral gauge, then:

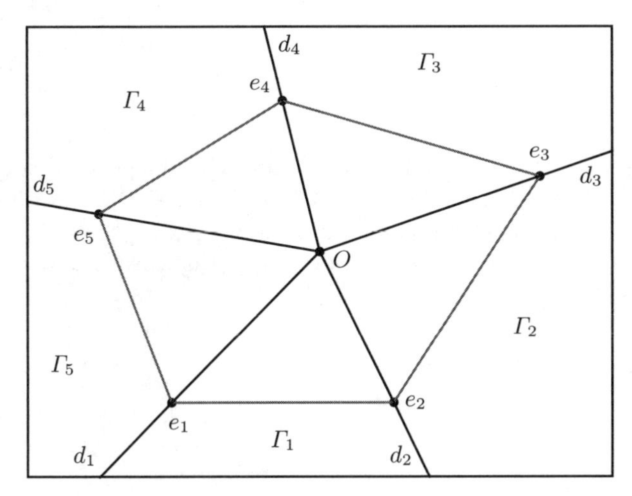

Fig. 2.5. Fundamental directions and cones generated by the extreme points of a polytope

$$\begin{aligned}
\gamma_B(x) &= \inf\left\{\mu > 0 : \frac{x}{\mu} \in B\right\} \\
&= \inf\left\{\mu > 0 : \frac{x}{\mu} = \sum_{g=1}^{G} \overline{\mu}_g e_g, \sum_{g=1}^{G} \overline{\mu}_g = 1, \overline{\mu}_g \geq 0\right\} \\
&= \inf\left\{\mu > 0 : x = \sum_{g=1}^{G} \mu_g e_g, \sum_{g=1}^{G} \mu_g = \mu, \mu_g \geq 0\right\} \\
&\quad \text{where}\, \mu_g := \mu\overline{\mu}_g, \text{ i.e., } \overline{\mu}_g := \frac{\mu_g}{\mu} \\
&= \min\left\{\sum_{g=1}^{G} \mu_g : x = \sum_{g=1}^{G} \mu_g e_g, \mu_g \geq 0\right\}.
\end{aligned}$$

Analogously if B is symmetric, i.e., $b \in B$ if and only if $-b \in B$, the set of extreme points of B can be rewritten as $Ext(B) = \{e_1, \ldots, e_{\frac{G}{2}}, -e_1, \ldots, -e_{\frac{G}{2}}\}$. This means that a point $x \in \mathbb{R}^2$ can be represented as a linear combination of $\{e_1, \ldots, e_{\frac{G}{2}}\}$ and we obtain the result. □

This result can be strengthened when we know the fundamental cone that contains the point to be evaluated. The next result shows how to compute the value of the gauge at a point x depending on which fundamental cone x lies in.

Lemma 2.3 *Let $B \subseteq \mathbb{R}^2$ be a polytope and let γ_B be its corresponding planar polyhedral gauge with $Ext(B) = \{e_1, \ldots, e_G\}$, which are numbered in counter-clockwise order. Let $x \in \Gamma_g$, i.e. $x = \alpha e_g + \beta e_{g+1}$, then $\gamma_B(x) = \alpha + \beta$.*

Proof.
Γ_g admits a representation as

$$\Gamma_g = \{x \in \mathbb{R}^2 : x = \alpha e_g + \beta e_{g+1},\ \alpha, \beta \geq 0\},\ g = 1, \dots, G \qquad (2.27)$$

where $e_{G+1} = e_1$ and e_g and e_{g+1} are linearly independent for any $g = 1, \dots, G-1$. Therefore, for any $x \in \Gamma_g$, $x = \alpha e_g + \beta e_{g+1}$ with $\alpha, \beta \geq 0$. Moreover, this representation is unique since $x \in \Gamma_g$.

Let x_o be the foot-point corresponding to x, that means, $x = \gamma_B(x) x_o$ and $x_o = \mu e_g + (1-\mu) e_{g+1}$ with $\mu \in [0,1]$. Then, $x = \alpha e_g + \beta e_{g+1}$. Since the representation of x with the generators of Γ_g is unique then $\alpha = \gamma_B(x)\mu$ and $\beta = \gamma_B(x)(1-\mu)$. Therefore,

$$\alpha + \beta = \gamma_B(x).$$

□

2.3 Ordered Regions, Elementary Convex Sets and Bisectors

2.3.1 Ordered Regions

We have already mentioned that (2.2) is a nonlinear problem because of the ordering induced by the sorting of the elements of $d(A, x)$. Therefore, it will be important to identify linearity domains of OMf, whenever possible, because it will help in solving the corresponding location problems. As a first step towards this goal we should find those regions of points where the order of the elements of the vector of weighted distances does not change. These are called ordered regions.

Definition 2.1 *Given a permutation $\sigma \in \mathcal{P}(1 \dots M)$ we denote by O_σ the ordered region given as*

$$O_\sigma = \{x \in \mathbb{R}^n : w_{\sigma(1)}\gamma(x - a_{\sigma(1)}) \leq \dots \leq w_{\sigma(M)}\gamma(x - a_{\sigma(M)})\}.$$

It is clear that these sets are regions in $\mathbb{R}^n$ where the weighted distances to the points in A ordered by the permutation σ are in non-decreasing sequence. As opposed to the case with Euclidean distance the ordered regions are in general non convex sets. Consider the following example.

Example 2.4
Let $\mathbb{R}^2$ be equipped with the rectilinear norm (l_1) and $A = \{(0,0), (2,1)\}$ with weights $w_1 = w_2 = 1$. In this case the region $O_{(1,2)}$ is not convex:

$$\begin{aligned} O_{(1,2)} &= \{x \in \mathbb{R}^2 : x_1 \leq 1/2,\ x_2 \geq 1\} \\ &\cup \{x \in \mathbb{R}^2 : x_1 \leq 3/2,\ x_2 \leq 0\} \\ &\cup \{x \in \mathbb{R}^2 : 0 \leq x_2 \leq 1,\ x_1 + x_2 \leq 3/2\} \end{aligned}$$

see Figure 2.6 where the shaded area is the set $O_{(1,2)}$.

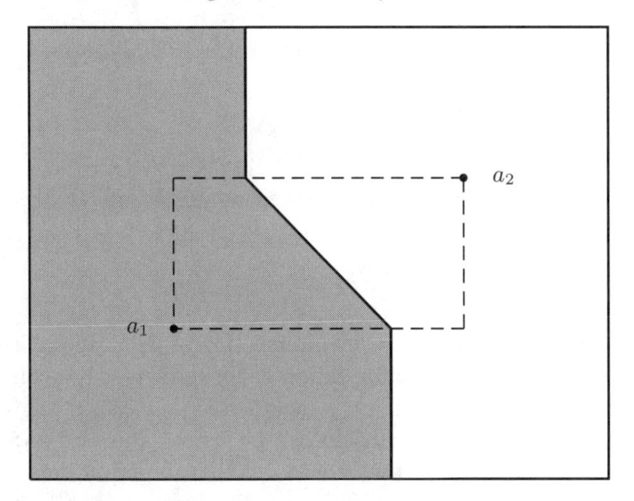

Fig. 2.6. $O_{(1,2)}$ is not convex

2.3.2 Ordered Elementary Convex Sets

Let $\mathcal{J} \neq \emptyset$ be a subset of $\mathcal{P}(1 \dots M)$; and consider the set $p = \{p_\sigma\}_{\sigma \in \mathcal{J}}$ where for each $\sigma \in \mathcal{J}$ $p_\sigma = (p_{\sigma(1)}, \dots, p_{\sigma(M)})$ with $p_{\sigma(j)} \in B^o$ for all $j = 1, \dots, M$. We let

$$OC_{\mathcal{J}}(p) = \bigcap_{\sigma \in \mathcal{J}} \Big(\bigcap_{j=1}^{M} \Big(a_{\sigma(j)} + N(p_{\sigma(j)}) \Big) \cap O_\sigma \Big). \tag{2.28}$$

Notice that when $\mathcal{J}$ reduces to a singleton, i.e. $\mathcal{J} = \{\sigma\}$, the first intersection is not considered.

We call these sets ordered elementary convex sets (o.e.c.s.). Recalling the definition of the normal cone $N(p_\sigma)$, each set $a_{\sigma(i)} + N(p_{\sigma(i)})$ is a cone with vertex at $a_{\sigma(i)}$ and generated by an exposed face of the unit ball B. The sets $C(p_\sigma) = \bigcap_{i=1}^{M} (a_{\sigma(i)} + N(p_{\sigma(i)}))$ are what Durier and Michelot call elementary convex sets (e.c.s.) (see [65] for further details). In the polyhedral case the elementary convex sets are intersections of cones generated by fundamental directions of the ball B pointed at each demand point: each elementary convex set is a polyhedron. Its vertices are called intersection points (see Figure 2.5). An upper bound of the number of elementary convex sets generated on the plane by the set of demand points A and a polyhedral norm with G fundamental directions is $O(M^2 G^2)$. For further details see [65] and [177].

It is clear from its definition that the ordered elementary convex sets are intersections of elementary convex sets with ordered regions (see (2.28)). Therefore, these sets are a refinement of the elementary convex sets. Moreover, o.e.c.s. are convex though the ordered regions O_σ are not necessarily convex.

Indeed, the intersection of any ordered region O_σ with an e.c.s. expressed as $\bigcap_{j=1}^{M}(a_j + N(p_j))$ is given by

$$OC_\sigma(p_\sigma) = \Big\{x \in \mathbb{R}^n : \langle p_{\sigma(1)}, x - a_{\sigma(1)}\rangle \leq \langle p_{\sigma(2)}, x - a_{\sigma(2)}\rangle, \dots, \langle p_{\sigma(M-1)}, x - a_{\sigma(M-1)}\rangle \leq \langle p_{\sigma(M)}, x - a_{\sigma(M)}\rangle\Big\} \cap \bigcap_{j=1}^{M}(a_j + N(p_j)).$$

This region is convex because it is the intersection of halfspaces with an e.c.s. Therefore, since $OC_\sigma(p_\sigma)$ is an intersection of convex sets it is a convex set. In fact, although each one of the o.e.c.s. is included in an e.c.s., it might coincide with no one of them, i.e. this inclusion can be strict. This situation happens because the linear behavior of the ordered median function depends on the order of the distance vector. For instance, the simplest o.e.c.s., $OC_\sigma(p)$, is given by the intersection of one classical e.c.s. $C(p)$ with one ordered region O_σ.

The following example clarifies the above discussion and shows several o.e.c.s.

Example 2.5

1. Consider $\mathbb{R}^2$ with the rectilinear (l_1) norm and $A = \{(0,0),(2,1)\}$. For $\mathcal{J} = \{(1,2)\}$ with $p = \{p_1, p_2\}$ where $p_1 = (1,1)$ and $p_2 = (-1,-1)$ then

$$OC_{(1,2)}(p) = \{x \in \mathbb{R}^2 : 0 \leq x_1,\ 0 \leq x_2 \leq 1,\ x_1 + x_2 \leq 3/2\}.$$

   Notice that this o.e.c.s. is strictly included in the e.c.s. $C(p)$ with $p = \{(1,1),(-1,-1)\}$.
2. For $\mathcal{J} = \{(1,2),\ (2,1)\}$ and $p = \Big\{p_{(1,2)}, p_{(2,1)}\Big\}$ with $p_{(1,2)} = \{(1,1),(-1,1)\}$ and $p_{(2,1)} = \{(-1,1),(1,1)\}$, we have

$$OC_{\mathcal{J}}(p) = \{x \in \mathbb{R}^2 : x_1 = 1/2,\ x_2 \geq 1\}.$$

3. Finally, for $\mathcal{J} = \{(2,1)\}$ and $p = \{p_1, p_2\}$, with $p_1 = (1,-1)$ and $p_2 = (1,1)$ we have

$$OC_{(2,1)}(p) = \{x \in \mathbb{R}^2 : x_1 \geq 2,\ 0 \leq x_2 \leq 1\}$$

   which coincides with the e.c.s. $C(p)$.

Let $\lambda \in \Lambda_M^{\leq}$ be a fixed vector. Assume that λ has $k \leq M$ different entries $\lambda_1 < \ldots < \lambda_k$ (if $k > 1$) and that each entry has r_i replications such that $\sum_{i=1}^{k} r_i = M$. Let us denote $\bar{r}_1 = 0$ and $\bar{r}_i = \sum_{j=1}^{i-1} r_j$ for $i \geq 2$. The vector λ can be written as:

$$\lambda = (\lambda_1, \overset{r_1}{\ldots}, \lambda_1, \ldots, \lambda_k, \overset{r_k}{\ldots}, \lambda_k). \tag{2.29}$$

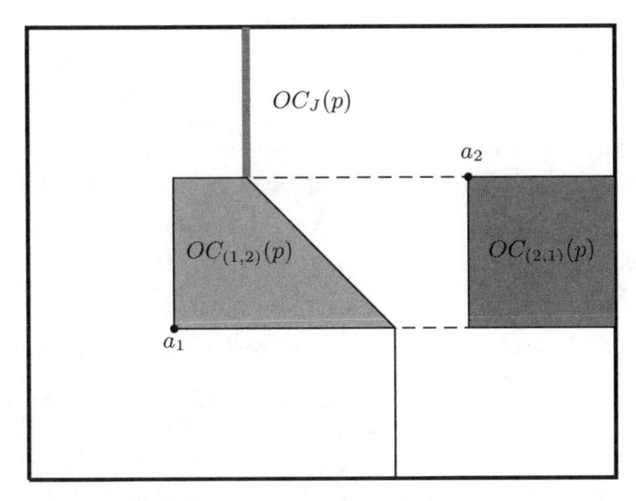

Fig. 2.7. Ordered regions of Example 2.5

Consider $\sigma = (\sigma(1), \ldots, \sigma(M)) \in \mathcal{P}(1 \ldots M)$. Let $L_\sigma(\lambda, i)$ be the set of all the permutations of $\{\sigma(\bar{r}_i + 1), \ldots, \sigma(\bar{r}_{i+1})\}$ for any $i = 1, \ldots, k$. In the same way, let $L_\sigma(\lambda)$ be the set of all the permutations $\pi \in \mathcal{P}(1 \ldots M)$ such that $\pi = (\pi_1, \ldots, \pi_k)$ with $\pi_i \in L_\sigma(\lambda, i)$ for $i = 1, \ldots, k$.

For instance, for $\sigma = (1, 2, 3, 4, 5)$ and $\lambda = (1/20, 1/10, 1/5, 1/4, 1/2)$; $L_\sigma(\lambda, i) = \{(i)\}$ for all $i = 1, \ldots, 5$. Therefore, $L_\sigma(\lambda) = \{(1, 2, 3, 4, 5)\}$.

For $\lambda = (1/4, 1/4, 1/3, 1/3, 1/2)$ we have

$$L_\sigma(\lambda, 1) = \{(1, 2), (2, 1)\},\ L_\sigma(\lambda, 2) = \{(3, 4), (4, 3)\},\ L_\sigma(\lambda, 3) = \{(5)\}.$$

Therefore,

$$L_\sigma(\lambda) = \{(1, 2, 3, 4, 5), (2, 1, 3, 4, 5), (1, 2, 4, 3, 5), (2, 1, 4, 3, 5)\};$$

and for $\lambda = (1/5, \ldots, 1/5)$

$$L_\sigma(\lambda) = \mathcal{P}(1 \ldots M).$$

In addition to the o.e.c.s. we introduce a new family of sets that will be used to describe the geometrical properties of the continuous OMP. Let $\mathcal{J}$ be a subset of the set $\mathcal{P}(1 \ldots M)$, $\mathcal{J} \neq \emptyset$, $p = \{p_\sigma\}_{\sigma \in \mathcal{J}}$ where $p_\sigma = (p_{\sigma(1)}, \ldots, p_{\sigma(M)})$ with $p_{\sigma(j)} \in B^0$ for every $\sigma \in \mathcal{J}$ and each $j = 1, \ldots, M$ and $\lambda \in S_M^{\leq}$. We let

$$OC_{\mathcal{J},\lambda}(p) = \bigcap_{\sigma \in \mathcal{J}} \Big(\bigcup_{\pi \in L_\sigma(\lambda)} \Big(\bigcap_{j=1}^{M} \big(a_{\sigma(j)} + N(p_{\sigma(j)})\big) \cap O_\pi \Big) \Big)$$

or equivalently

$$OC_{\mathcal{J},\lambda}(p) = \bigcap_{\sigma \in \mathcal{J}} \Big(\bigcap_{j=1}^{M} \Big(a_{\sigma(j)} + N(p_{\sigma(j)}) \Big) \cap \bigcup_{\pi \in L_\sigma(\lambda)} O_\pi \Big).$$

Notice that these sets are also convex. The set $\bigcup_{\pi \in L_\sigma(\lambda)} O_\pi$ can be easily described within any e.c.s. Indeed, let σ be any permutation of $\{1, \ldots, M\}$. Consider the e.c.s. $C(p) = \bigcap_{j=1}^{M} \Big(a_j + N(p_j) \Big)$ and any λ in the form given in (2.29). Since $L_\sigma(\lambda, i)$ contains all the permutations of $\{\sigma(\bar{r}_i + 1), \ldots, \sigma(\bar{r}_{i+1})\}$ this implies that within $C(p)$

$$\langle p_{\sigma(j)}, x - a_{\sigma(j)} \rangle \le \langle p_{\sigma(l)}, x - a_{\sigma(l)} \rangle \text{ or } \langle p_{\sigma(j)}, x - a_{\sigma(j)} \rangle \ge \langle p_{\sigma(l)}, x - a_{\sigma(l)} \rangle$$

for any $\bar{r}_i + 1 \le j \ne l \le \bar{r}_{i+1}$. That is to say, there is no order constraint among those positions which correspond to tied entries in λ.

On the other hand, any permutation in $L_\sigma(\lambda)$ verifies that the position of $\sigma(j)$ is always in front of the position of $\sigma(l)$ provided that $j \le \bar{r}_i < l$. Therefore, $C(p) \cap \bigcup_{\pi \in L_\sigma(\lambda)} O_\pi$ is included in

$$\{x \in \mathbb{R}^n : \langle p_{\sigma(j)}, x - a_{\sigma(j)} \rangle \le \langle p_{\sigma(l)}, x - a_{\sigma(l)} \rangle;\ \gamma(x - a_{\sigma(m)}) = \langle p_{\sigma(m)}, x - a_{\sigma(m)} \rangle,\ m = j, l\} \text{ for any } j \le \bar{r}_i < l.$$

In conclusion,

$$\begin{aligned} C(p) \cap \textstyle\bigcup_{\pi \in L_\sigma(\lambda)} O_\pi = \{x \in \mathbb{R}^n : & \gamma(x - a_{\sigma(j)}) = \langle p_{\sigma(j)}, x - a_{\sigma(j)} \rangle \ \forall j, \text{ and} \\ & \langle p_{\sigma(j)}, x - a_{\sigma(j)} \rangle \le \langle p_{\sigma(l)}, x - a_{\sigma(l)} \rangle \\ & \forall j, l;\ 1 \le j \le \bar{r}_i < l \le M \text{ and for any } 2 \le i \le k\}. \end{aligned}$$

Hence, it is a convex set. For instance, for a five demand points problem with $\sigma = (1, 2, 3, 4, 5)$ and $\lambda = (1/4, 1/4, 1/3, 1/3, 1/2)$ we have

$$\begin{aligned} & \bigcap_{j=1}^{M} \Big(a_j + N(p_j) \Big) \cap \bigcup_{\pi \in L_\sigma(\lambda)} O_\pi = \\ & \quad \{x \in \mathbb{R}^n : \langle p_{\sigma(i)}, x - a_{\sigma(i)} \rangle \le \langle p_{\sigma(j)}, x - a_{\sigma(j)} \rangle, \text{ for } i = 1, 2 \text{ and } j = 3, 4, \\ & \quad \langle p_{\sigma(k)}, x - a_{\sigma(k)} \rangle \le \langle p_{\sigma(5)}, x - a_{\sigma(5)} \rangle, \text{ for } k = 3, 4, \\ & \quad \text{and } \gamma(x - a_{\sigma(j)}) = \langle p_{\sigma(j)}, x - a_{\sigma(j)} \rangle \ \forall j\}. \end{aligned}$$

Finally, we quote for the sake of completeness a result stated in Chapter 6.3 which geometrically characterizes the solution set of the convex ordered median location problem: (Theorem 6.9) "The whole set of optimal solutions of Problem (2.2) for $\lambda \in S_M^{\le}$ always coincides with a set $OC_{\mathcal{J},\lambda}(p)$ for some particular choice of λ, J and p". This is to say, the solution set coincides with the intersection of ordered regions with elementary convex sets [169]. In the general non-convex case the optimal solution (provided that it exists) does not have to coincide with only one of these sets but there are always sets of the family $OC_{\mathcal{J},\lambda}(p)$ that are optimal solutions.

We summarize these findings in the following theorem.

Theorem 2.2 *There always exists an optimal solution of the OMP in the extreme points (vertices in the polyhedral case) of the ordered elementary convex sets.*

The set of extreme points of all ordered elementary convex sets of an OMP is denoted OIP.

2.3.3 Bisectors

As we have already seen, it does not exist a unique linear representation of the objective function of Problem (2.2) on the whole decision space.

It is easy to see, that the representation may change when $\gamma(x-a_i)-\gamma(x-a_j)$ becomes 0 for some $i,j \in \{1,\ldots,M\}$ with $i \neq j$. It is well-known that the set of points at the same distance of two given points a_1, a_2 is its bisector. When using the Euclidean distance bisectors reduce to the standard perpendicular bisectors, known by everybody from high school maths. Nevertheless, although there is a large body of literature on perpendicular bisectors, not much is known about the properties of bisectors for general gauges. Bisectors being important to identify linearity domains of the ordered median function in continuous spaces will be studied in detail later in Chapter 3.

However, in this introductory chapter we feel it is important to give an overview of the easiest properties of bisectors of polyhedral gauges on the plane. This approach will help the reader to understand concepts like ordered regions or o.e.c.s. Also some easy examples will shed light on the different structure of these sets depending on the gauge that is used for their construction.

Definition 2.2 *The weighted bisector of a_i and a_j with respect to γ is the set $Bisec(a_i,a_j) = \{x \ : w_i\gamma(x-a_i) = w_j\gamma(x-a_j), i \neq j\}$.*

As an illustration of Definition 2.2 one can see in Figure 2.1 the bisector line for the points a_1 and a_2 with the rectangular norm.

Proposition 2.1 *The weighted bisector of a_i and a_j is a set of points verifying a linear equation within each elementary convex set.*

Proof.
In an elementary convex set $\gamma(x-a_i)$ and $\gamma(x-a_j)$ can be written as $l_i(x-a_i)$ and $l_j(x-a_j)$ respectively, where l_i and l_j are linear functions. Therefore, $w_i\gamma(x-a_i) = w_j\gamma(x-a_j)$ is equivalent to $w_il_i(x-a_i) = w_jl_j(x-a_j)$ and the result follows. □

We will now give a more exact description of the complexity of a bisector.

Proposition 2.2 *The weighted bisector of a_i and a_j with respect to a polyhedral gauge γ with G extreme points has at most $O(G)$ different subsets defined by different linear equations.*

Proof.
By Proposition 2.1 weighted bisectors are set of points given by linear equations within each elementary convex set. Therefore, the unique possible breakpoints may occur on the fundamental directions.

Let us denote by $L^g_{a_i}$ the fundamental direction starting at a_i with direction e_g. On this halfline the function $w_i\gamma(x-a_i)$ is linear with constant slope and $w_j\gamma(x-a_j)$ is piecewise linear and convex. Therefore, the maximum number of zeros of $w_i\gamma(x-a_i) - w_j\gamma(x-a_j)$ when $x \in L^g_{a_i}$ is two. Hence, there are at most two breakpoints of the weighted bisector of a_i and a_j on $L^g_{a_i}$.

Repeating this argument for any fundamental direction we obtain that an upper bound for the number of breakpoints is $4G$. □

This result implies that the number of different linear expressions defining any bisector is also linear in G, the number of fundamental directions. Remark, that in some settings bisectors may have non empty interior. See for instance Figure 2.8, where we show the bisector set defined by the points $p = (0,0)$ and $q = (4,0)$ with the Chebychev norm.

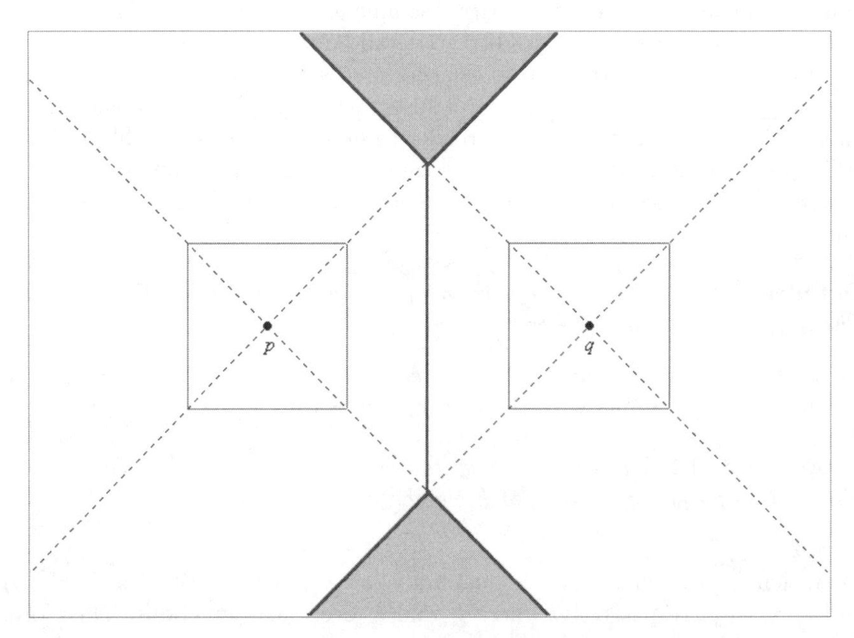

Fig. 2.8. Example of bisector with non empty interior

When at least two points are simultaneously considered the set of bisectors builds a subdivision of the plane (very similar to the well-known k−order Voronoi diagrams, see the book of Okabe et al. [155]). The cells of this subdivision coincide with the ordered regions that were formally introduced in

Definition 2.1. Notice that these regions need not to be convex sets as can be clearly seen in Figure 2.9.
The importance of these regions is that in their intersection with any elementary convex set, the OMP behaves like a Weber problem, i.e. the objective function has a unique linear representation. Recall that the intersections between ordered regions and e.c.s. are called ordered elementary convex sets (o.e.c.s). The ordered regions play a very important role in the algorithmic approach developed for solving the problem. In terms of bisectors, these regions are cells defined by at most $M-1$ bisectors of the set A.

However, the main disadvantage of dealing with these regions is their complexity. A naive analysis could lead to conclude that their number is $M!$ which would make the problem intractable. Using the underlying geometric structure we can obtain a polynomial bound which allows us to develop in Chapter 4 an efficient algorithm for solving Problem (2.2).

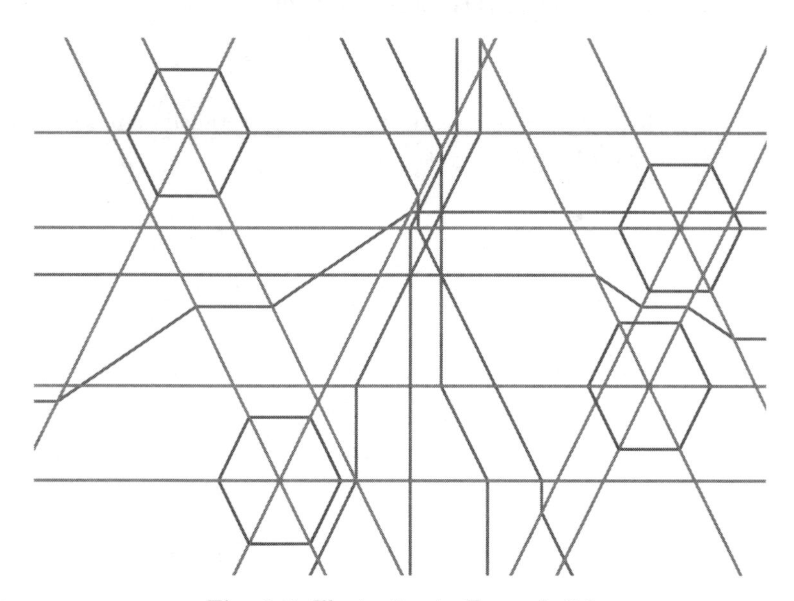

Fig. 2.9. Illustration to Example 2.6

Theorem 2.3 *An upper bound on the number of ordered regions in the plane is* $O(M^4G^2)$.

Proof.
Given two bisectors with $O(G)$ linear pieces, the maximum number of intersections is $O(G^2)$. The number of bisectors of M points is $\binom{M}{2}$, so, the maximum number of intersections between them is $O(G^2\binom{\binom{M}{2}}{2})$. By the Euler's formula,

the number of intersections has the same complexity as the number of regions. Hence, an upper bound for the number of ordered regions is $O(M^4G^2)$. □

A detailed analysis of this theorem shows that this bound is not too bad. Although, it is of order M^4G^2, it should be noted that the number of bisectors among the points in A is $\binom{M}{2}$ which is of order M^2. Therefore, even in the most favorable case of straight lines, the number of regions in worst case analysis gives $O(\binom{M}{2}^2)$ which is, in fact $O(M^4)$. Since our bisectors are polygonal with G pieces, this bound is rather tight.

Example 2.6
Figure 2.9 shows the ordered regions between the points $a_1 = (0, 11)$, $a_2 = (3, 0)$, $a_3 = (16, 8)$ and $a_4 = (15, 3)$ with the hexagonal norm whose extreme points are $Ext(B) = \{(2,0),(1,2),(-1,2),(-2,0),(-1,-2),(1,-2)\}$. For instance, the region $O_{(3,1,2)}$ is the set of points

$$\{x \in \mathbb{R}^2 \ : \ \gamma(x - a_3) \leq \gamma(x - a_1) \leq \gamma(x - a_2)\} \ .$$

In the next chapter we give a more rigorous and detailed analysis of bisectors in the plane with mixed gauges.

3

Bisectors

This chapter focuses (after a short overview of the classical concepts) on general bisectors for two points associated with different gauges. The reader may skip this chapter if he is mainly interested in the OMP. However, a deep understanding of general bisectors is one of the keys for geometrical solution methods for the OMP. Moreover, not much has yet been published on this topic. We will introduce the concept of a general bisector, show some counterintuitive properties and will end with an optimal time algorithm to compute it. In our presentation we follow [207].

3.1 Bisectors - the Classical Case

If we consider two sites (locations etc.) a and b in the plane $\mathbb{R}^2$ and a gauge γ_C, an interesting task is to find out,

> *which points of the plane can be reached faster from a and which can be reached faster from b, respectively with respect to the gauge γ_C.*

Another way to formulate the task is the following :

> *Determine the points in the plane which can be reached as fast from site a as from site b with respect to the gauge γ_C,*

i.e. determine the set

$$\mathsf{Bisec}(a,b) := \{x \in \mathbb{R}^2 \,:\, \gamma_C(x-a) = \gamma_C(x-b)\} \quad . \tag{3.1}$$

$\mathsf{Bisec}(a,b)$ is called the (classical) bisector of a and b with respect to γ_C. Furthermore, the notation

$$C_C(a,b) := \{x \in \mathbb{R}^2 \,:\, \gamma_C(x-a) \le \gamma_C(x-b)\} \tag{3.2}$$
$$D_C(a,b) := \{x \in \mathbb{R}^2 \,:\, \gamma_C(x-a) < \gamma_C(x-b)\} \tag{3.3}$$
$$C_C(b,a) := \{x \in \mathbb{R}^2 \,:\, \gamma_C(x-b) \le \gamma_C(x-a)\} \tag{3.4}$$
$$D_C(b,a) := \{x \in \mathbb{R}^2 \,:\, \gamma_C(x-b) < \gamma_C(x-a)\} \tag{3.5}$$

is used to denote the points which can not be reached faster from b, faster from a, not faster from a and faster from b, respectively.

For $x \in \mathbb{R}^2$ let x_C^a and x_C^b be the uniquely defined intersection points of $\text{bd}(C^a)$ and $\text{bd}(C^b)$ with the half-lines $d_{a,x}$ and $d_{a,x}$ respectively. According to the geometric definition (2.10) it follows that :

$$\mathsf{Bisec}(a,b) := \left\{ x \in \mathbb{R}^2 \; : \; \frac{l_2(x-a)}{l_2(x_C^a - a)} = \frac{l_2(x-b)}{l_2(x_C^b - b)} \right\}. \tag{3.6}$$

The simplest bisector one can imagine is the bisector with respect to l_2. It is the mid-perpendicular to the segment $\overline{ab}$, i.e. the line which runs through the midpoint $\frac{1}{2}(a+b)$ of the segment $\overline{ab}$ and is perpendicular to the segment $\overline{ab}$. The simple shape of the bisector is directly connected with the simple shape of the Euclidean unit ball.

[212] investigated bisectors for the so-called A-distances. The name A-distance was introduced by themselves and describes nothing but a special case of a block norm. To be more precise the corresponding unit ball of an A-distance has $2|A|$ extreme points and the Euclidean distance of all extreme points from the origin is one, i.e. all extreme points lie on the Euclidean unit sphere. For example, the Manhattan-norm l_1 and the scaled Chebyshev-norm $\sqrt{2}l_\infty$ are A-distances. [212] proved that the bisector with respect to an A-distance is a polygonal line consisting of no more than $2|A| - 1$ pieces. The bisector with respect to an A-distance can also contain degenerated parts (see Figure 3.1). However [46, 106] proved, that $\mathsf{Bisec}(a,b)$ is homeomorphic to a line, if the set C is strictly convex.

3.2 Possible Generalizations

Now there are different possibilities imaginable to generalize the concept of bisectors. The first one is the introduction of weights for the sites a and b. Here we can distinguish between additive weights $\nu_a, \nu_b \in \mathbb{R}$ and multiplicative weights $\omega_a, \omega_b \in \mathbb{R}^+$. Without loosing generality one can assign the weight 0 to a and $\nu := \nu_b - \nu_a$ to b in the additive case, whereas one can assign the weight 1 to a and $\omega := \frac{\omega_b}{\omega_a}$ to b in the multiplicative case. According to this weighted situation the analytic and geometric definition of the bisector are given by

$$\mathsf{Bisec}_{\omega,\nu}(a,b) := \{ x \in \mathbb{R}^2 \; : \; \gamma_C(x-a) = \omega\gamma_C(x-b) + \nu \} \tag{3.7}$$

and

$$\mathsf{Bisec}_{\omega,\nu}(a,b) := \left\{ x \in \mathbb{R}^2 \; : \; \frac{l_2(x-a)}{l_2(x_A^a - a)} = \frac{\omega l_2(x-b)}{l_2(x_B^b - b)} + \nu \right\} \tag{3.8}$$

respectively. In the case of the Euclidean distance the multiplicative weighted bisector is given by the classical Apollonius circle with midpoint $\frac{a-\omega^2 b}{1-\omega^2}$ and radius $\frac{\omega}{|1-\omega^2|} l_2(a-b)$ (see [8] and Figure 3.13).

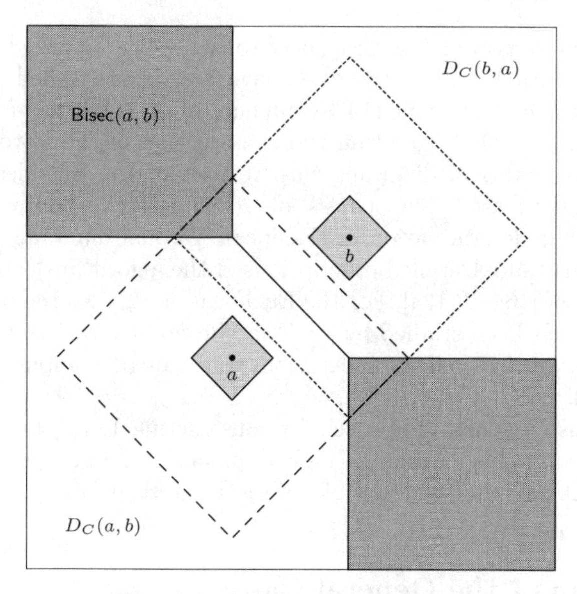

Fig. 3.1. A degenerated bisector

A more general concept for bisectors is the consideration of two different gauges γ_A and γ_B associated with the sites a and b respectively, which we will investigate in the Sections 3.3 - 3.7. This concept includes obviously the classical case of two identical gauges as well as the use of multiplicative weight. The consideration of additive weights is not included. However additive weights play a subordinate part for our purpose in location theory. Therefore we can get over this disadvantage. Nevertheless, we will have a short look on additive weights in Section 3.4.

Instead of generalizing the way of distance measuring, another generalization can also be done by increasing the number of sites. This leads to the concept of Voronoi diagrams, which generalizes the task of the preceding section, in a natural way, for a nonempty set of M points in the plane:

Let $S \neq \emptyset$ be a finite subset of $\mathbb{R}^2$, $\prec$ be a total ordering on S and γ_C be a gauge. For $a, b \in S, a \neq b$ let

$$R_C(a,b) := \begin{cases} C_C(a,b) \ , \ a \prec b \\ D_C(a,b) \ , \ b \prec a \end{cases} \quad . \tag{3.9}$$

Then

$$R_C(a,S) := \bigcap_{\substack{b \in S \\ b \neq a}} R_C(a,b) \tag{3.10}$$

is the Voronoi region of a with respect to (S, γ_C) and

$$V_C(S) := \bigcup_{a \in S} \text{bd}(R_C(a,S)) \tag{3.11}$$

is the Voronoi diagram of S with respect to γ_C.

Voronoi diagrams for arbitrary γ_C have first been studied by [39]. The Voronoi regions are star-shaped. The influence of $|S|$ on the complexity of the Voronoi diagram is $O(|S|)$ and on the construction of the Voronoi diagram is $O(|S| \log |S|)$. Voronoi diagrams with respect to the Euclidean norm are already well investigated. They consist of $O(|S|)$ many Voronoi vertices, edges and regions. The determination of Euclidean Voronoi diagrams can be done in $O(|S| \log |S|)$ time. Detailed descriptions of the algorithms can be found in the textbooks of [164, 7, 124]. For the Euclidean norm also the multiplicative weighted case has been studied by [8]. The Voronoi diagram contains $O(|S|^2)$ many Voronoi vertices, edges and regions and can be computed in $O(|S|^2)$ time (see [9]).

There is also a general approach for constructing $V_C(S)$ for arbitrary γ_C due to [69], which also works in higher dimensions, but does not directly provide insight into the structure of the resulting diagrams.

3.3 Bisectors - the General Case

In this section we will study the shape of bisectors, if we deal with two different gauges for two sites. By (2.23) we know that we can avoid the use of weights without loosing the generality of the results we develop. That means we measure the distance of a point $x \in \mathbb{R}^2$ from the site $a \in \mathbb{R}^2$ by a gauge γ_A and the distance of $x \in \mathbb{R}^2$ from the site $b \in \mathbb{R}^2$ by a different gauge γ_B. In this case the (general) bisector is defined by

$$\mathsf{Bisec}(a,b) := \{x \in \mathbb{R}^2 \, : \, \gamma_A(x-a) = \gamma_B(x-b)\} \quad . \tag{3.12}$$

Furthermore the sets

$$C_{A,B}(a,b) := \{x \in \mathbb{R}^2 \, : \, \gamma_A(x-a) \le \gamma_B(x-b)\} \tag{3.13}$$
$$D_{A,B}(a,b) := \{x \in \mathbb{R}^2 \, : \, \gamma_A(x-a) < \gamma_B(x-b)\} \tag{3.14}$$
$$C_{B,A}(b,a) := \{x \in \mathbb{R}^2 \, : \, \gamma_A(x-b) \le \gamma_B(x-a)\} \tag{3.15}$$
$$D_{B,A}(b,a) := \{x \in \mathbb{R}^2 \, : \, \gamma_A(x-b) < \gamma_B(x-a)\} \tag{3.16}$$

are defined analogously to the sets $C_C(a,b)$, $D_C(a,b)$, $C_C(b,a)$, $D_C(b,a)$ in the sense of (3.12).

Obviously

$$\mathsf{Bisec}(a,b) = \mathsf{Bisec}(b,a) \tag{3.17}$$
$$C_{A,B}(a,b) \cap C_{B,A}(b,a) = \mathsf{Bisec}(a,b) \tag{3.18}$$
$$C_{A,B}(a,b) \,\dot{\cup}\, D_{B,A}(b,a) = \mathbb{R}^2 \tag{3.19}$$
$$D_{A,B}(a,b) \,\dot{\cup}\, C_{B,A}(b,a) = \mathbb{R}^2 \tag{3.20}$$

hold.

For a function $f : \mathbb{R}^2 \to \mathbb{R}$ and a real number $z \in \mathbb{R}$ the sets

$$L_=(f,z) := \{x \in \mathbb{R}^2 : f(x) = z\} \quad (3.21)$$
$$L_\leq(f,z) := \{x \in \mathbb{R}^2 : f(x) \leq z\} \quad (3.22)$$
$$L_<(f,z) := \{x \in \mathbb{R}^2 : f(x) < z\} \quad (3.23)$$

are called the level curve, level set and the strict level set of f with respect to z, respectively.

Moreover, we define the functions

$$f_{AB} : \mathbb{R}^2 \to \mathbb{R} \quad , \quad x \mapsto \gamma_A(x-a) - \gamma_B(x-b) \quad (3.24)$$

and

$$f_{BA} : \mathbb{R}^2 \to \mathbb{R} \quad , \quad x \mapsto \gamma_B(x-b) - \gamma_A(x-a) \quad . \quad (3.25)$$

Using the concept of level curves, level sets and strict level sets we can therefore write

$$\mathsf{Bisec}(a,b) := L_=(f_{AB},0) = L_=(f_{BA},0) \quad (3.26)$$
$$C_{A,B}(a,b) := L_\leq(f_{AB},0) \quad (3.27)$$
$$D_{A,B}(a,b) := L_<(f_{AB},0) \quad (3.28)$$
$$C_{A,B}(a,b) := L_\leq(f_{BA},0) \quad (3.29)$$
$$D_{A,B}(a,b) := L_<(f_{BA},0) \quad (3.30)$$

3.3.1 Negative Results

Very often properties of classical bisectors are transmitted careless to general bisectors which leads, in fact, to mistakes. Therefore we will present for reasons of motivation some examples of general bisectors which do not have the properties of classical bisectors.

Properties of a classical bisector $\mathsf{Bisec}(a,b)$ with respect to a gauge γ_C are for instance the following:

- $\mathsf{Bisec}(a,b)$ is connected.
- If $\mathsf{Bisec}(a,b)$ contains a two-dimensional area, this area is unbounded.
- $\mathsf{Bisec}(a,b)$ is separating.
- ...

These properties are in general not valid for general bisectors $\mathsf{Bisec}(a,b)$:

- Figure 3.2 and Figure 3.3 show disconnected bisectors. The bisector in Figure 3.2 consists of three connected components which are all homeomorphic to a line. However one of the two connected components of the bisector in Figure 3.3 is a closed curve, whereas the other one is homeomorphic to a line. Therefore, Figure 3.3 contradicts also the conjecture that the bisector is either a closed curve or consists of connected components which are all homeomorphic to a line.

- The bisector in Figure 3.4 contains two bounded two-dimensional areas. Also the existence of a bounded and an unbounded two-dimensional area in one and the same bisector is possible (see Figure 3.5). As we will see later there are never more than two (un)bounded two-dimensional areas contained in a bisector $\mathsf{Bisec}(a, b)$.
- Finally Figure 3.6 presents a bisector $\mathsf{Bisec}(a, b)$ which contains a vertical half-line running through the interior of $C_{A,B}(a, b)$. That means this piece of the bisector does not separate $D_{A,B}(a, b)$ from $D_{B,A}(b, a)$.
- ...

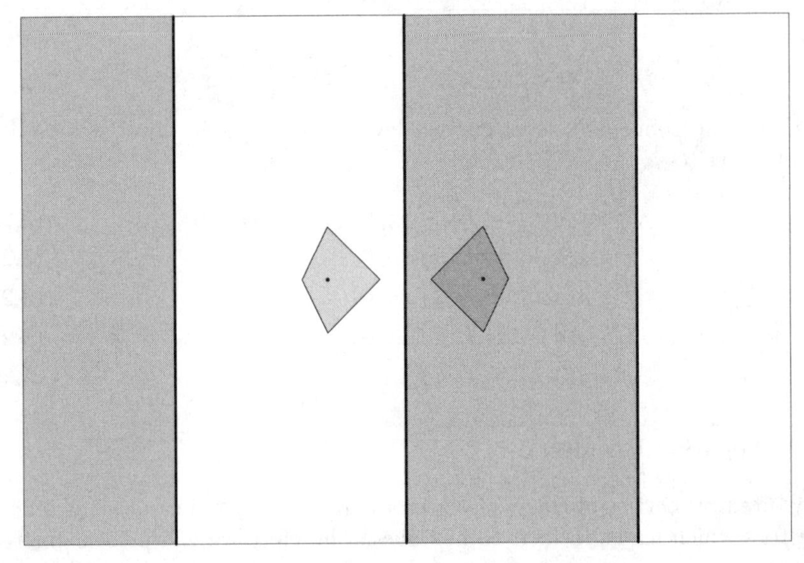

Fig. 3.2. A disconnected bisector with 3 connected components

3.3.2 Structural Properties

The first lemma gives information about the topological properties of the sets $\mathsf{Bisec}(a, b)$, $C_{A,B}(a, b)$, $C_{B,A}(b, a)$, $D_{A,B}(a, b)$ and $D_{B,A}(b, a)$.

Lemma 3.1
Let $a, b \in \mathbb{R}^2$ be two sites associated with gauges γ_A and γ_B, respectively. Then the sets $\mathsf{Bisec}(a, b), C_{A,B}(a, b), C_{B,A}(b, a)$ are closed in $\mathbb{R}^2$, and the sets $D_{A,B}(a, b), D_{B,A}(b, a)$ are open in $\mathbb{R}^2$.

Proof.
As difference of continuous functions $f_{AB} = -f_{BA}$ is also continuous and specifically lower semi-continuous. Therefore, the level sets of f_{AB} are closed

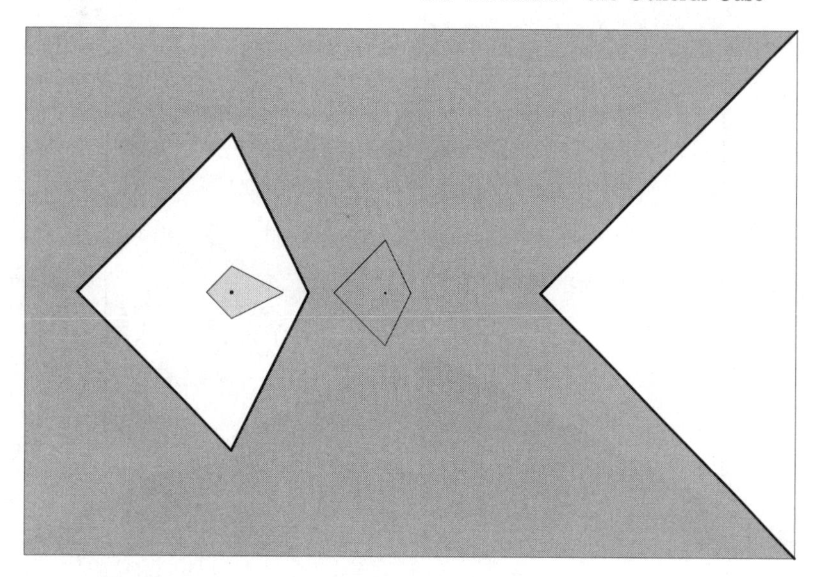

Fig. 3.3. A disconnected bisector containing a closed curve

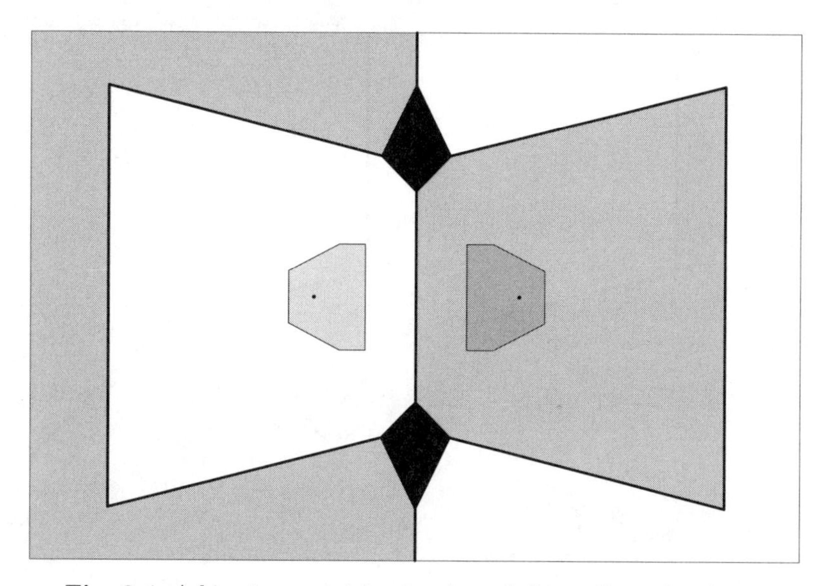

Fig. 3.4. A bisector containing two bounded two-dimensional area

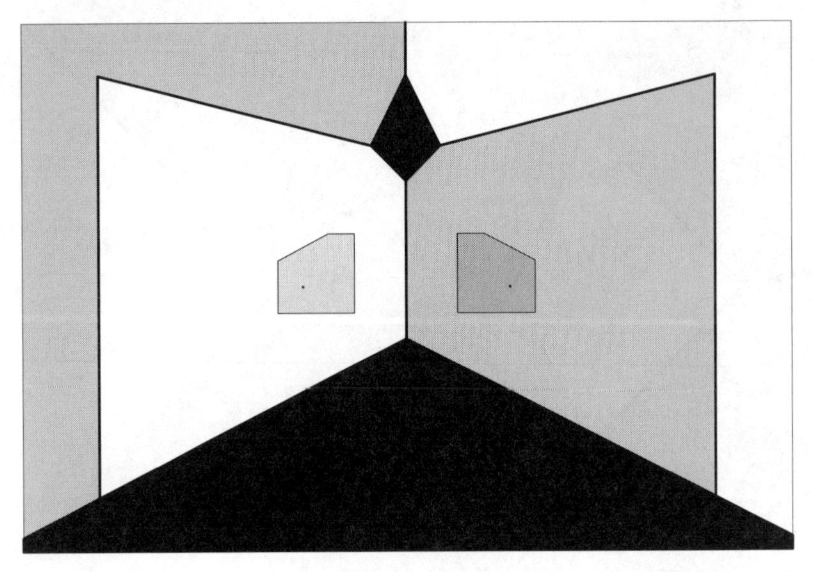

Fig. 3.5. A bisector containing a bounded and an unbounded two-dimensional area

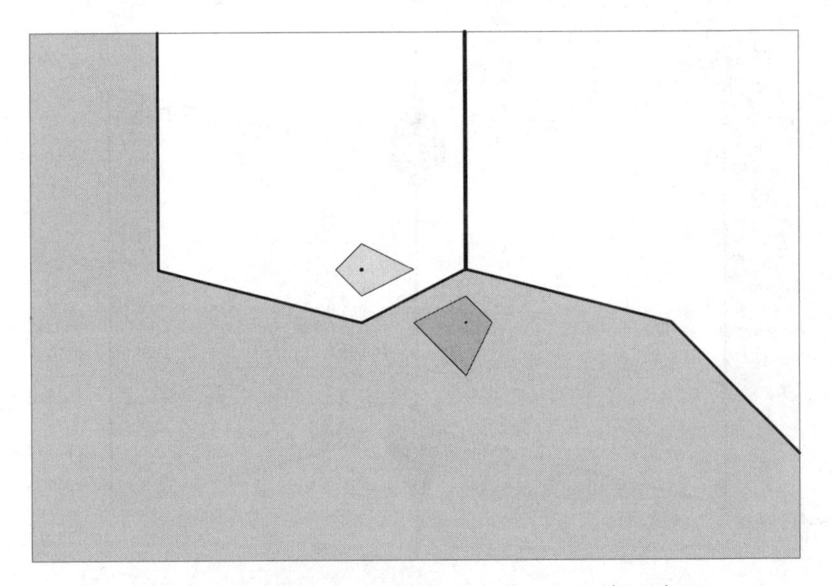

Fig. 3.6. A bisector containing a non-separating piece

in $\mathbb{R}^2$, which means $C_{A,B}(a,b) = L_{\leq}(f_{AB}, 0)$ and $C_{B,A}(b,a) = L_{\leq}(f_{BA}, 0)$ are closed. $\mathsf{Bisec}(a,b) = C_{A,B}(a,b) \cap C_{B,A}(b,a)$ is closed, since the intersection of two closed sets is closed. Finally $D_{A,B}(a,b) = \mathbb{R}^2 \setminus C_{B,A}(b,a)$ and $D_{A,B}(a,b) = \mathbb{R}^2 \setminus C_{B,A}(b,a)$ are open in $\mathbb{R}^2$, since $C_{A,B}(a,b)$ and $C_{B,A}(b,a)$ are closed in $\mathbb{R}^2$. □

However, notice that in general the following properties are *not valid* as one could see from the examples of Section 3.3.1 (see Figures 3.4, 3.5 and 3.6) :

$$\begin{aligned}
\mathsf{Bisec}(a,b) &= \mathrm{bd}(C_{A,B}(a,b)) & \mathsf{Bisec}(a,b) &= \mathrm{bd}(D_{A,B}(a,b)) \\
\mathsf{Bisec}(a,b) &= \mathrm{bd}(C_{B,A}(b,a)) & \mathsf{Bisec}(a,b) &= \mathrm{bd}(D_{B,A}(b,a)) \\
C_{A,B}(a,b) &= \mathrm{cl}(D_{A,B}(a,b)) & D_{A,B}(a,b) &= \mathrm{int}(C_{A,B}(a,b)) \\
C_{B,A}(b,a) &= \mathrm{cl}(D_{B,A}(b,a)) & D_{B,A}(b,a) &= \mathrm{int}(D_{B,A}(b,a)) \\
\mathrm{bd}(D_{B,A}(b,a)) &= \mathrm{bd}(D_{A,B}(a,b)) & \mathrm{bd}(C_{A,B}(a,b)) &= \mathrm{bd}(C_{B,A}(b,a)) \\
\mathrm{bd}(D_{A,B}(a,b)) &= \mathrm{bd}(C_{B,A}(b,a)) & \mathrm{bd}(C_{A,B}(a,b)) &= \mathrm{bd}(D_{B,A}(b,a)) \\
\mathrm{bd}(C_{A,B}(a,b)) &= \mathrm{bd}(D_{A,B}(a,b)) & \mathrm{bd}(D_{B,A}(b,a)) &= \mathrm{bd}(C_{B,A}(b,a))
\end{aligned}$$

To obtain more structural properties a case analysis is necessary.

First of all we study the trivial case $a = b$. If we additionally assume $\gamma_A = \gamma_B$, we obtain the totally trivial case $\mathsf{Bisec}(a,b) = \mathbb{R}^2$. So let γ_A and γ_B be distinct.

Lemma 3.2
Let $a \in \mathbb{R}^2$ be a site associated with gauges γ_A and γ_B. Then $\mathsf{Bisec}(a,a)$ is star-shaped with respect to a.

Proof.

Let $x \in \mathsf{Bisec}(a,a)$ and $0 \leq \lambda \leq 1$

$$\begin{aligned}
\Rightarrow \quad \gamma_A(\lambda a + (1-\lambda)x - a)) &= \gamma_A((1-\lambda)(x-a)) \\
&= (1-\lambda)\gamma_A(x-a) \\
&= (1-\lambda)\gamma_B(x-a) \\
&= \gamma_B((1-\lambda)(x-a)) \\
&= \gamma_B(\lambda a + (1-\lambda)x - a))
\end{aligned}$$

$$\Rightarrow \quad \lambda a + (1-\lambda)x \in \mathsf{Bisec}(a,a)\,.$$

□

Figure 3.7 shows an example for the trivial situation $a = b$ with polyhedral gauges γ_A and γ_B.

We assume that a and b are different from now on. Surprisingly we may have four distinct distances between the sites a and b, which result from the

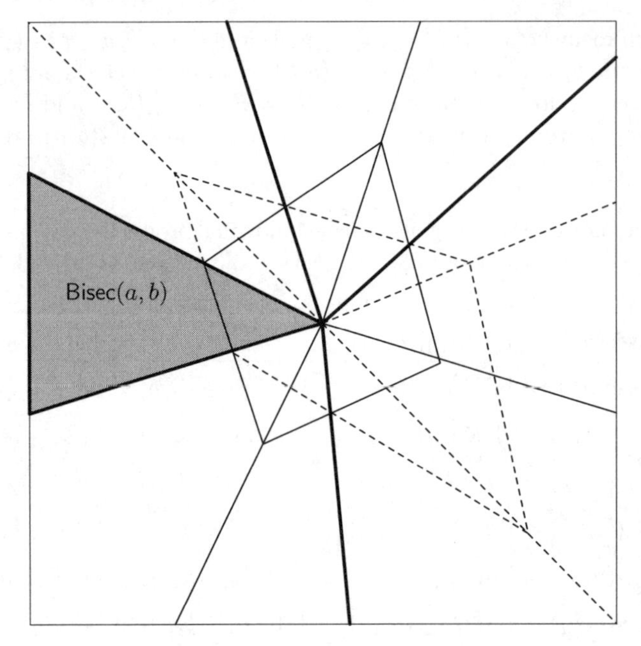

Fig. 3.7. Bisector for $a = b$

combination of the two gauges γ_A and γ_B with the two difference vectors $b-a$ and $a-b$, namely

$$\alpha^+ := \gamma_A(b-a) \tag{3.31}$$
$$\alpha^- := \gamma_A(a-b) \tag{3.32}$$
$$\beta^+ := \gamma_B(a-b) \tag{3.33}$$
$$\beta^- := \gamma_B(b-a) \quad . \tag{3.34}$$

In the following we investigate how often the bisector $\mathsf{Bisec}(a,b)$ can intersect

- the basis line

$$l_{a,b} := \{x_\lambda \in \mathbb{R}^2 \,:\, x_\lambda := (1-\lambda)a + \lambda b\,,\, \lambda \in \mathbb{R}\} \tag{3.35}$$
$$= \{x_\lambda \in \mathbb{R}^2 \,:\, x_\lambda := a + \lambda(b-a)\,,\, \lambda \in \mathbb{R}\} \tag{3.36}$$

  running through a and b,
- the half-line

$$\mathbf{d}_{a,e} := \{x_\lambda \in \mathbb{R}^2 \,:\, x_\lambda = a + \lambda e\,,\, \lambda \geq 0\} \tag{3.37}$$

  starting at a in the direction of $e \in S(0,1)$, where $S(0,1)$ is the Euclidean unit sphere.

Lemma 3.3
Let $a, b \in \mathbb{R}^2$ be two sites associated with gauges γ_A and γ_B respectively. Then

the set $\mathit{Bisec}(a,b) \cap l_{a,b} = \{x \in l_{a,b} : \gamma_A(x-a) = \gamma_B(x-b)\}$ *consists of at least one and at most three points.*

Proof.
As a consequence of (2.5) - (2.8) $\gamma_A(\,.\,-a)/_{l_{a,b}}$ and $\gamma_B(\,.\,-b)/_{l_{a,b}}$ are convex, consist of two affine-linear pieces and have a uniquely defined minimum at a and b respectively. These properties prove directly the statement (see Figure 3.8). □

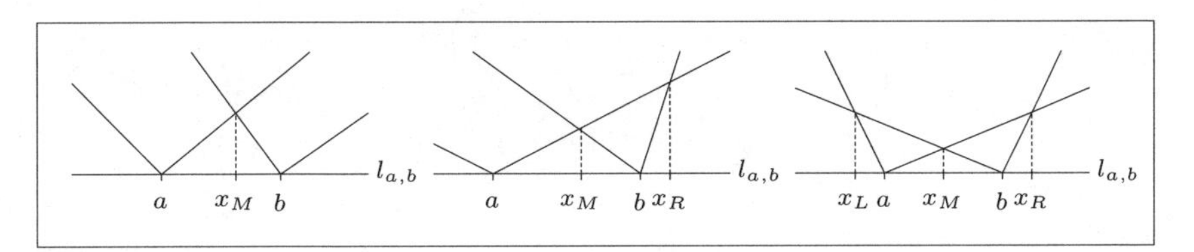

Fig. 3.8. Illustration to the proof of Lemma 3.3

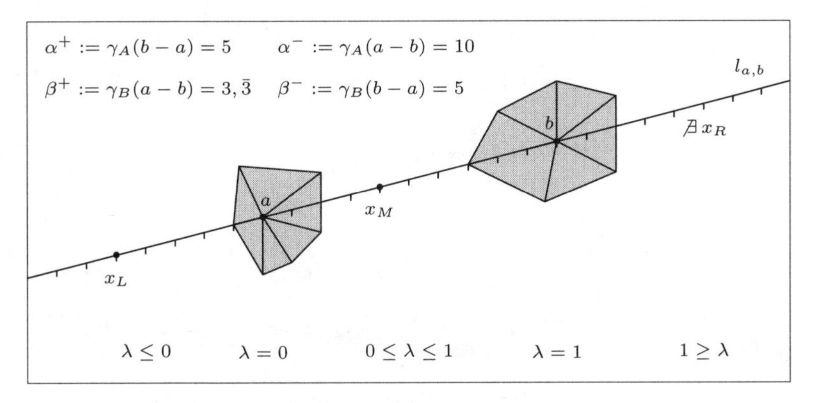

Fig. 3.9. Illustration to Lemma 3.3 and Corollary 3.1

The following corollary characterizes the points of the bisector on the basis line exactly.

Corollary 3.1
Let $a,b \in \mathbb{R}^2$ *be two sites associated with gauges* γ_A, γ_B *and for* $\lambda \in \mathbb{R}$ *define* $x_\lambda := \lambda a + (1-\lambda)b$. *Then we have:*

1. $x_M := \frac{\alpha^+}{\alpha^+ + \beta^+} a + \frac{\beta^+}{\alpha^+ + \beta^+} b \in \mathit{Bisec}(a,b)$.

2. *If* $\alpha^- > \beta^+$ *, then* $x_L := \frac{\alpha^-}{\alpha^- - \beta^+} a + \frac{-\beta^+}{\alpha^- - \beta^+} b \in \mathsf{Bisec}(a,b)$ *.*
3. *If* $\alpha^- \le \beta^+$ *, then* $\nexists \lambda \le 0$ *with* $x_\lambda \in \mathsf{Bisec}(a,b)$ *.*
4. *If* $\alpha^+ < \beta^-$ *, then* $x_R := \frac{\alpha^+}{\alpha^+ - \beta^-} a + \frac{-\beta^-}{\alpha^+ - \beta^-} b \in \mathsf{Bisec}(a,b)$ *.*
5. *If* $\alpha^+ \ge \beta^-$ *, then* $\nexists \lambda \ge 1$ *with* $x_\lambda \in \mathsf{Bisec}(a,b)$ *.*

Proof.
1.

$$\begin{aligned}
\gamma_A(x_M - a) &= \gamma_A\left(\frac{\alpha^+}{\alpha^+ + \beta^+} a + \frac{\beta^+}{\alpha^+ + \beta^+} b - a\right) \\
&= \gamma_A\left(\frac{\beta^+}{\alpha^+ + \beta^+} (b - a)\right) \\
&= \frac{\beta^+}{\alpha^+ + \beta^+} \gamma_A(b - a) \text{ , since } \frac{\beta^+}{\alpha^+ + \beta^+} \ge 0 \\
&= \frac{\alpha^+ \beta^+}{\alpha^+ + \beta^+} \\
&= \frac{\alpha^+}{\alpha^+ + \beta^+} \gamma_B(a - b) \\
&= \gamma_B\left(\frac{\alpha^+}{\alpha^+ + \beta^+} (a - b)\right) \text{ , since } \frac{\alpha^+}{\alpha^+ + \beta^+} \ge 0 \\
&= \gamma_B\left(\frac{\alpha^+}{\alpha^+ + \beta^+} a + \frac{\beta^+}{\alpha^+ + \beta^+} b - b\right) \\
&= \gamma_B(x_M - b)
\end{aligned}$$

2.

$$\begin{aligned}
\gamma_A(x_L - a) &= \gamma_A\left(\frac{\alpha^-}{\alpha^- - \beta^+} a + \frac{-\beta^+}{\alpha^- - \beta^+} b - a\right) \\
&= \gamma_A\left(\frac{\beta^+}{\alpha^- - \beta^+} (a - b)\right) \\
&= \frac{\beta^+}{\alpha^- - \beta^+} \gamma_A(a - b) \text{, since } \frac{\beta^+}{\alpha^- - \beta^+} \ge 0 \\
&= \frac{\alpha^- \beta^+}{\alpha^- - \beta^+} \\
&= \frac{\alpha^-}{\alpha^- - \beta^+} \gamma_B(a - b) \\
&= \gamma_B\left(\frac{\alpha^-}{\alpha^- - \beta^+} (a - b)\right) \text{ , since } \frac{\alpha^-}{\alpha^- - \beta^+} \ge 0 \\
&= \gamma_B\left(\frac{\alpha^-}{\alpha^- - \beta^+} a + \frac{-\beta^+}{\alpha^- - \beta^+} b - b\right) \\
&= \gamma_B(x_L - b)
\end{aligned}$$

3. For $\alpha^- \le \beta^+$ the fraction $\frac{-\beta^+}{\alpha^- - \beta^+}$ is undefined or ≥ 1, therefore there is no $\lambda \le 0$ with $x_\lambda \in \mathsf{Bisec}(a,b)$.

4.

$$\begin{aligned}
\gamma_A(x_R - a) &= \gamma_A\left(\frac{\alpha^+}{\alpha^+ - \beta^-}a + \frac{-\beta^-}{\alpha^+ - \beta^-}b - a\right) \\
&= \gamma_A\left(\frac{\beta^-}{\alpha^+ - \beta^-}(a-b)\right) \\
&= \frac{\beta^-}{\beta^- - \alpha^+}\gamma_A(b-a)\,, \text{ since } \frac{\beta^-}{\alpha^+ - \beta^-} \leq 0 \\
&= \frac{\alpha^+ \beta^-}{\beta^- - \alpha^+} \\
&= \frac{\alpha^+}{\beta^- - \alpha^+}\gamma_B(b-a) \\
&= \gamma_B\left(\frac{\alpha^+}{\alpha^+ - \beta^-}(a-b)\right)\,, \text{ since } \frac{\alpha^+}{\beta^- - \alpha^+} \leq 0 \\
&= \gamma_B\left(\frac{\alpha^+}{\alpha^+ - \beta^-}a + \frac{\beta^-}{\alpha^+ - \beta^-}b - b\right) \\
&= \gamma_B(x_R - b)
\end{aligned}$$

5. For $\alpha^+ \geq \beta^-$ the fraction $\frac{-\beta^-}{\alpha^+ - \beta^-}$ is undefined or ≤ 0, therefore there is no $\lambda \geq 1$ with $x_\lambda \in \mathsf{Bisec}(a,b)$.

□

If γ_A and γ_B are norms, the statements of Lemma 3.3 and Corollary 3.1 can be intensified to the following corollary.

Corollary 3.2

Let $a, b \in \mathbb{R}^2$ be two sites associated with norms γ_A and γ_B respectively. Then the line $l_{a,b}$ contains at least one and at most two points of the bisector $\mathsf{Bisec}(a,b)$.

Proof.

Since γ_A and γ_B are norms, the equations

$$\alpha^+ := \gamma_A(b-a) = \gamma_A(a-b) =: \alpha^- \tag{3.38}$$

$$\beta^+ := \gamma_B(a-b) = \gamma_B(b-a) =: \beta^- \tag{3.39}$$

are valid. Therefore, at most one of the conditions $\alpha^- > \beta^+$ and $\alpha^+ < \beta^-$ is fulfilled. Applying Corollary 3.1 leads to the result. □

A concrete specification of the bisector points on a half-line $\mathbf{d}_{a,e}$ is not possible. However, we can characterize the type of the set $\mathsf{Bisec}(a,b) \cap \mathbf{d}_{a,e}$.

Lemma 3.4

Let $a, b \in \mathbb{R}^2$ be two sites associated with gauges γ_A and γ_B respectively. Then the set $\mathsf{Bisec}(a,b) \cap d_{a,e} = \{x \in d_{a,e} : \gamma_A(x-a) = \gamma_B(x-b)\}$

1. *is either empty or*
2. *consists of one point or*
3. *consists of two points or*
4. *is equal to a closed segment or*
5. *is equal to a half-line.*

Moreover, if γ_B is additionally strictly convex, $\mathsf{Bisec}(a,b) \cap \mathbf{d}_{a,e}$ can not be a closed segment or a half-line.

Proof.
Since $\gamma_A(\,.\, - a)/_{\mathbf{d}_{a,e}}$ is linear by (2.7) and $\gamma_B(\,.\, - b)$ is convex by (2.11), the results follow (see Figure 3.10). □

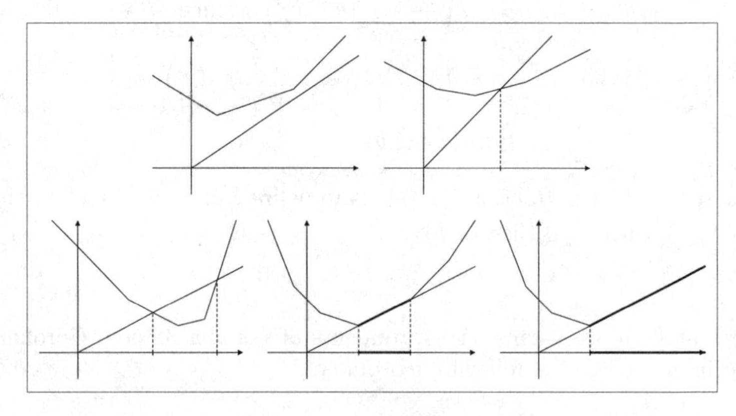

Fig. 3.10. Illustration to the proof of Lemma 3.4

Naturally the roles of a and b in Lemma 3.4 can be reversed. Therefore the statements of Lemma 3.4 are also valid for half-lines starting at b.

Now let us compare two gauges in a certain direction $e \in S(0,1)$. We say that gauge γ_A is stronger than gauge γ_B in direction $e \in S(0,1)$, if $\gamma_A(e) < \gamma_B(e)$. Note that "<" means stronger since the wider the unit ball extends in a direction the smaller factor is necessary to scale the ball until it reaches a certain point.

Lemma 3.5 ([107])
Let $a, b \in \mathbb{R}^2$ be two sites associated with gauges γ_A and γ_B, respectively. For a direction $e \in S(0,1)$ holds :

1. *$f_{AB}/_{d_{b,e}}$ and $f_{BA}/_{d_{a,e}}$ are convex.*
2. *$f_{AB}/_{d_{a,e}}$ and $f_{BA}/_{d_{b,e}}$ are concave.*
3. *$f_{AB}/_{d_{b,e}}$ is strictly decreasing, if γ_A is stronger than γ_B in direction e.*
4. *$f_{BA}/_{d_{b,e}}$ is strictly increasing, if γ_A is stronger than γ_B in direction e.*

Proof.

1. We prove the property only for $f_{AB}/_{d_{b,e}}$, since the proof for $f_{BA}/_{d_{a,e}}$ is analogous.
   Let $x, y \in d_{b,e}$ and $\mu \in [0,1]$. For x and y exist $\lambda_1, \lambda_2 \geq 0$ with $x = b + \lambda_1 e$ and $y = b + \lambda_2 e$, and therefore we have

$$\begin{aligned}
&f_{AB}(\mu x + (1-\mu)y) \\
&= f_{AB}(b + (\mu\lambda_1 + (1-\mu)\lambda_2)e) \\
&= \gamma_A(b + (\mu\lambda_1 + (1-\mu)\lambda_2)e - a) - \gamma_B((\mu\lambda_1 + (1-\mu)\lambda_2)e) \\
&= \gamma_A(\mu(b + \lambda_1 e - a) + (1-\mu)(b + \lambda_2 e - a)) - (\mu\lambda_1 + (1-\mu)\lambda_2)\gamma_B(e) \\
&\leq \mu\gamma_A(b + \lambda_1 e - a) + (1-\mu)\gamma_A(b + \lambda_2 e - a) - (\mu\gamma_B(\lambda_1 e) \\
&+ (1-\mu)\gamma_B(\lambda_2 e)) \\
&= \mu(\gamma_A(b + \lambda_1 e - a) - \gamma_B(\lambda_1 e)) + (1-\mu)(\gamma_A(b + \lambda_2 e - a) - \gamma_B(\lambda_2 e)) \\
&= \mu f_{AB}(x) + (1-\mu) f_{AB}(y) \quad ,
\end{aligned}$$

   hence $f_{AB}/_{d_{b,e}}$ is convex.
2. Follows by 1. and $f_{AB} = -f_{BA}$.
3. For $\lambda \geq 0$ we obtain with (2.7) and (2.8)

$$f_{AB}(b + \lambda e) = \gamma_A(b + \lambda e - a) - \gamma_B(\lambda e) \leq \gamma_A(b - a) + \lambda\underbrace{(\gamma_A(e) - \gamma_B(e))}_{<0} . \tag{3.40}$$

   This inequality (3.40) implies $\lim_{\lambda \to +\infty} f_{AB}(b + \lambda e) = -\infty$ and $f_{AB}(b + \lambda e) < 0$ for $\lambda > \frac{\gamma_A(b-a)}{\gamma_B(e) - \gamma_A(e)} > 0$. From (2.6) we know, that $f_{AB}(b) = \gamma_A(b - a) > 0$ and hence $f_{AB}/_{d_{b,e}}$ must be strictly decreasing as a convex function.
4. Follows by 3. and $f_{AB} = -f_{BA}$.

□

Corollary 3.3 ([107])
Let $a, b \in \mathbb{R}^2$ be two sites associated with gauges γ_A and γ_B respectively. Let $e \in S(0,1)$ and let γ_B be stronger than γ_A in direction e. Then the half-line $\mathbf{d}_{a,e}$ contains exactly one point of the bisector Bisec(a,b). Furthermore, this bisector point can be expressed as $a + \lambda e$ with $\lambda \in (0, \frac{\gamma_B(a-b)}{\gamma_A(e) - \gamma_B(e)}]$.

Proof.
Follows directly from the proof of Lemma 3.5. □

Now we continue with the geometric definition of general bisectors. For $x \in \mathbb{R}^2$ let x_A^a and x_B^b be the foot points of x with respect to a and b respectively, i.e. the uniquely defined intersection points of $\mathrm{bd}(A^a)$ and $\mathrm{bd}(B^b)$ with the respective rays $d_{a,x}$ and $d_{b,x}$. According to the geometric definition (3.6) another equivalent definition of $\mathsf{Bisec}(a,b)$ is given by:

$$\mathsf{Bisec}(a,b) := \left\{ x \in \mathbb{R}^2 \; : \; \frac{l_2(x-a)}{l_2(x_A^a - a)} = \frac{l_2(x-b)}{l_2(x_B^b - b)} \right\}. \tag{3.41}$$

In the sense of this geometric definition it can be seen, that the (translated) foot-points x_A^a and x_B^b are generating the bisector point x. Notice that the condition in Definition (3.41) describes the situation of the perspective transformation with centre x and segments $\overline{xx_A^a}$, $\overline{xa}$ and $\overline{xx_B^b}$, $\overline{xb}$, respectively. Therefore, the segment $\overline{x_A^a x_B^b}$ is parallel to the segment $\overline{ab}$ (see Figure 3.11) and we can formulate the following Lemma.

Lemma 3.6 ([107])
Let $a, b \in \mathbb{R}^2$ be two sites associated with gauges γ_A and γ_B, respectively. For a point $x \in \mathsf{Bisec}(a,b)$ the line l_{x_A, x_B} through the two translated foot points x_A^a, x_B^b is parallel to the line $l_{a,b}$. Conversely, if there are two points $u \in A^a$ and $v \in B^b$, such that $d_{a,u}$ and $d_{b,v}$ intersect in a point x and $\overline{uv}$ is parallel to $\overline{ab}$, then $x \in \mathsf{Bisec}(a,b)$.

Proof.
Follows obviously from the geometric definition of the bisector in (3.6) and the principle of the perspective transformation. □

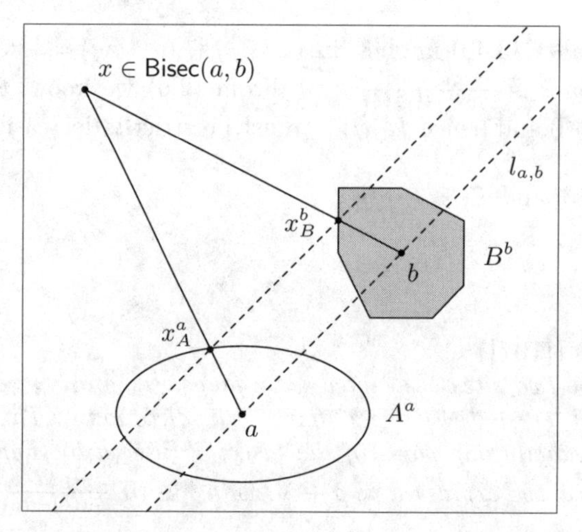

Fig. 3.11. Homothetic triangles $\Delta(x,a,b)$ and $\Delta(a,x_a,x_b)$

From this geometric property of a bisector point we can conclude another interesting fact. If we consider a point u on the boundary of A^a, we may ask, which bisector points do have the foot point u with respect to a. Of course, these bisector points lie on the half-line $d_{a,u}$. From Lemma 3.6 we know that that their foot points with respect to b must lie on a line parallel to $l_{a,b}$ and of course on $\mathrm{bd}(B^b)$. The intersection of a line with the boundary of B^b consists either of zero, one, two points or a segment, and these are the foot points on B^b which correspond with u. Therefore, the bisector restricted to the half-line starting at a through u consists either of zero, one, two points, a segment or a halfline. In this way we have found another proof for Lemma 3.4.

3.3.3 Partitioning of Bisectors

The bisector in Figure 3.2 consists of three connected components. However, only the component in the middle separates a from b, the remaining ones do not separate a from b. Therefore the middle component seems somehow to be *essential* and the other ones to be *unessential*. This observation leads to the question, whether the parts of the bisector can be distinguished in *essential* and *unessential* parts. Afterwards one could wonder, whether the plane can be partitioned in advance into two (or more) regions with respect to a and b, which contain the essential and unessential parts respectively. Finally, another interesting question is, which parts of the boundary of the unit balls A^a and B^b respectively generate the essential part in the sense of the geometric definition of $\mathsf{Bisec}(a,b)$ in (3.41). The following discussion answer the raised questions. The results we obtain are already published in [107, 108], where the interested reader can find the proofs.

Without loss of generality we assume in this section that the line $l_{a,b}$ is horizontal and that b lies on the right side of a. This situation can be produced by a rotation. Notice that the rotation concerns not only the sites, but also the unit balls.

Let t_A be the top point of A, i.e. the leftmost point of the intersection of A and the horizontal tangent to A from above, and let b_A be the bottom point of A, i.e. the corresponding point from below. The top and bottom point of B are defined analogously, i.e. "leftmost" is replaced by "rightmost". Figure 3.12 shows the situation for the translated unit balls A^a, B^b, translated top points t_A^a, t_B^b and translated bottom points b_A^a, b_B^b. For each unit ball, the plane is divided into two cones separated by two half-lines. The **face cone** $\mathsf{F}\Gamma_A$ of A is the closed cone containing $b-a$ with the half-lines d_{0,b_A}, d_{0,t_A} as boundary and the **back cone** $\mathsf{B}\Gamma_A$ of A is the open complement of $\mathsf{F}\Gamma_A$. The face and back cones of B are defined analogously. In Figure 3.12 the translated cones $\mathsf{F}\Gamma_A^a, \mathsf{B}\Gamma_A^a, \mathsf{F}\Gamma_B^b$ and $\mathsf{B}\Gamma_B^b$ are shown.

Notice that the plane $\mathbb{R}^2$ can be partitioned depending on the intersection of these cones as follows:

$$\mathbb{R}^2 := (\mathsf{F}\Gamma_A^a \cap \mathsf{F}\Gamma_B^b) \,\dot{\cup}\, (\mathsf{F}\Gamma_A^a \cap \mathsf{B}\Gamma_B^b) \,\dot{\cup}\, (\mathsf{B}\Gamma_A^a \cap \mathsf{F}\Gamma_B^b) \,\dot{\cup}\, (\mathsf{B}\Gamma_A^a \cap \mathsf{B}\Gamma_B^b) \quad . \tag{3.42}$$

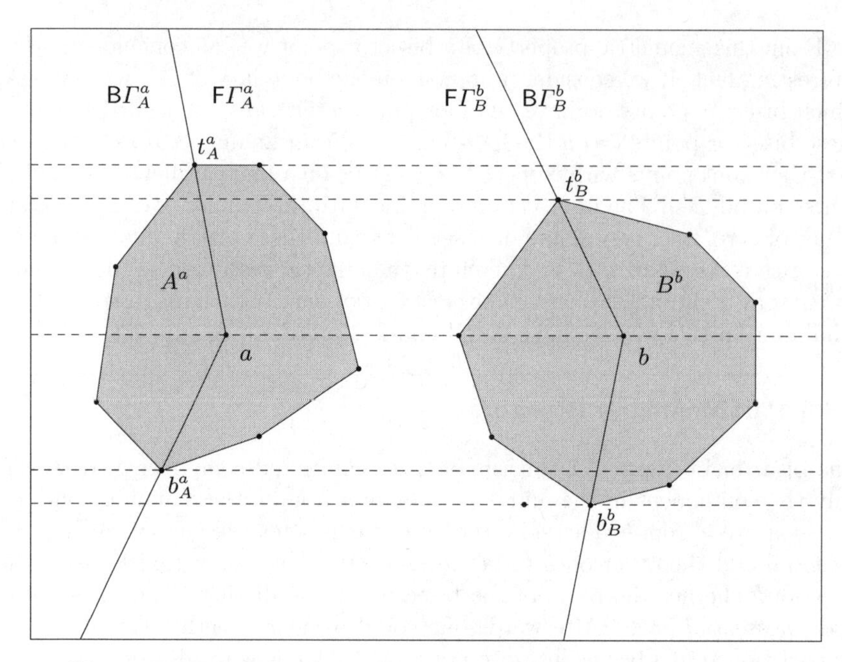

Fig. 3.12. Definition of top and bottom points, face and back cones

We denote the four created regions by

$$\mathbb{R}^2_{\mathsf{FF}} := \mathsf{F}\Gamma_A^a \cap \mathsf{F}\Gamma_B^b \text{ ("face to face" region)} \tag{3.43}$$
$$\mathbb{R}^2_{\mathsf{FB}} := \mathsf{F}\Gamma_A^a \cap \mathsf{B}\Gamma_B^b \text{ ("face to back" region)} \tag{3.44}$$
$$\mathbb{R}^2_{\mathsf{BF}} := \mathsf{B}\Gamma_A^a \cap \mathsf{F}\Gamma_B^b \text{ ("back to face" region)} \tag{3.45}$$
$$\mathbb{R}^2_{\mathsf{BB}} := \mathsf{B}\Gamma_A^a \cap \mathsf{B}\Gamma_B^b \text{ ("back to back" region)} \tag{3.46}$$

Moreover, the bisector $\mathsf{Bisec}(a, b)$ can be decomposed into four parts, also depending on the intersection of these cones :

$$\mathsf{Bisec}^{\mathsf{FF}}(a,b) := \mathsf{Bisec}(a,b) \cap \mathsf{F}\Gamma_A^a \cap \mathsf{F}\Gamma_B^b \text{ ("face to face" bisector)} \tag{3.47}$$
$$\mathsf{Bisec}^{\mathsf{FB}}(a,b) := \mathsf{Bisec}(a,b) \cap \mathsf{F}\Gamma_A^a \cap \mathsf{B}\Gamma_B^b \text{ ("face to back" bisector)} \tag{3.48}$$
$$\mathsf{Bisec}^{\mathsf{BF}}(a,b) := \mathsf{Bisec}(a,b) \cap \mathsf{B}\Gamma_A^a \cap \mathsf{F}\Gamma_B^b \text{ ("back to face" bisector)} \tag{3.49}$$
$$\mathsf{Bisec}^{\mathsf{BB}}(a,b) := \mathsf{Bisec}(a,b) \cap \mathsf{B}\Gamma_A^a \cap \mathsf{B}\Gamma_B^b \text{ ("back to back" bisector)} \tag{3.50}$$

The "face to face" bisector $\mathsf{Bisec}^{\mathsf{FF}}(a, b)$ is never empty, while there are necessary and sufficient criteria for the emptiness of the other parts, as the next lemma shows.

Lemma 3.7 ([107])
Let $a, b \in \mathbb{R}^2$ be two sites associated with gauges γ_A and γ_B respectively. The

bisector $\mathsf{Bisec}(a,b)$ *can be decomposed in four pieces* $\mathsf{Bisec}^{\mathsf{FF}}(a,b)$, $\mathsf{Bisec}^{\mathsf{FB}}(a,b)$, $\mathsf{Bisec}^{\mathsf{BF}}(a,b)$ *and* $\mathsf{Bisec}^{\mathsf{BB}}(a,b)$ *for which the following properties hold:*

1. $\mathsf{Bisec}^{\mathsf{FF}}(a,b) \neq \emptyset$.
2. $\mathsf{Bisec}^{\mathsf{BF}}(a,b) \neq \emptyset$ *iff a direction* $e \in S(0,1) \cap \mathsf{B}\Gamma_A \cap \mathsf{F}\Gamma_B$ *exists such that* γ_B *is stronger than* γ_A *in this direction.*
3. $\mathsf{Bisec}^{\mathsf{FB}}(a,b) \neq \emptyset$ *iff a direction* $e \in S(0,1) \cap \mathsf{F}\Gamma_A \cap \mathsf{B}\Gamma_B$ *exists such that* γ_A *is stronger than* γ_B *in this direction.*
4. $\mathsf{Bisec}^{\mathsf{BB}}(a,b) \neq \emptyset$ *iff* $(\partial A \cap \mathsf{B}\Gamma_A) \cap (\partial B \cap \mathsf{B}\Gamma_B) \neq \emptyset$.

In the case of two identical gauges we can strengthen the result of Lemma 3.7.

Corollary 3.4 ([107])
Let $a,b \in \mathbb{R}^2$ *be two sites associated with two identical gauges* $\gamma_A = \gamma_B$. *Then we have* $\mathsf{Bisec}(a,b) = \mathsf{Bisec}^{\mathsf{FF}}(a,b)$. *In other words: the other potential bisector parts* $\mathsf{Bisec}^{\mathsf{FB}}(a,b)$, $\mathsf{Bisec}^{\mathsf{BF}}(a,b)$ *and* $\mathsf{Bisec}^{\mathsf{BB}}(a,b)$ *are empty.*

Lemma 3.8 ([107])
Let $a,b \in \mathbb{R}^2$ *be two sites associated with gauges* γ_A *and* γ_B *respectively. If there is a bisector point* x *on the half-line* $\mathbf{d}_{a,e}$, *then* $\gamma_B(e) \leq \gamma_A(e)$. *In case of* $\gamma_B(e) = \gamma_A(e)$ *the boundary* $\mathrm{bd}(A)$ *of the unit ball* A *contains a line segment* s *parallel to the basis line* $l_{a,b}$, *and the foot point* x_A^a *lies on the supporting line of the segment* s^a.

The next lemma shows that the number of connected components of the bisector $\mathsf{Bisec}(a,b)$ essentially depends on the number of intersections of the boundaries of the two unit balls.

Lemma 3.9 ([107])
Let $a,b \in \mathbb{R}^2$ *be two sites associated with gauges* γ_A *and* γ_B *respectively. Then holds:*

1. $\mathsf{Bisec}^{\mathsf{FF}}(a,b)$ *is connected. It is separated from the rest of* $\mathsf{Bisec}(a,b)$ *if and only if* $\mathrm{bd}(A) \cap \mathrm{bd}(B) \cap \mathsf{F}\Gamma_A \cap \mathsf{F}\Gamma_B$ *is a set of exactly two points.*
2. *The number of the connected components in* $\mathsf{Bisec}^{\mathsf{BF}}(a,b)$ *is equal to the number of the connected components of the set*

$$\left\{e \in \mathbb{R}^2 : e \in \mathrm{bd}(A) \cap \mathsf{B}\Gamma_A \cap \mathsf{F}\Gamma_B \wedge \gamma_B(e) < \gamma_A(e)\right\},$$

   and analogously for $\mathsf{Bisec}^{\mathsf{FB}}(a,b)$.
3. $\mathsf{Bisec}^{\mathsf{BB}}(a,b)$ *consists of at most two connected components, but they are connected to other parts of the bisector.*

Summarizing the results of Lemma 3.9 we can conclude the following corollary.

Corollary 3.5 ([107])
Let $a,b \in \mathbb{R}^2$ *be two sites associated with gauges* γ_A *and* γ_B *respectively.*

Moreover, let K and L be the number of connected components of $\mathsf{Bisec}^{\mathsf{BF}}(a,b)$ and $\mathsf{Bisec}^{\mathsf{FB}}(a,b)$. The number of connected components of the whole bisector $\mathsf{Bisec}(a,b)$ is at least $\max\{1, K+L-3\}$ and at most $K+L+1$.

In the case of two unit balls with non-intersecting boundaries connectivity of the bisector is guaranteed.

Corollary 3.6 ([107])
Let $a,b \in \mathbb{R}^2$ be two sites associated with gauges γ_A and γ_B respectively. If γ_B is stronger than γ_A in all directions then the bisector $\mathsf{Bisec}(a,b)$ is a closed curve around the site a. Especially $\mathsf{Bisec}(a,b)$ consists of one connected component.

The result of Corollary 3.6 can be illustrated by the weighted Euclidean case where the bisector is given by the Apollonius' circle (see Figure 3.13).

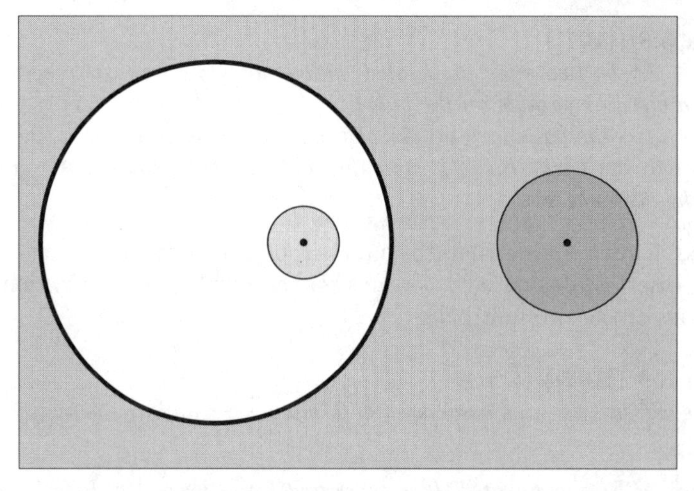

Fig. 3.13. The Apollonius' circle

Finally we present two more illustrative examples before we study bisectors of polyhedral gauges in the next section.

Example 3.1
Figure 3.14 shows a Φ-shaped bisector where all four parts $\mathsf{Bisec}^{\mathsf{FF}}(a,b)$, $\mathsf{Bisec}^{\mathsf{FB}}(a,b)$, $\mathsf{Bisec}^{\mathsf{BF}}(a,b)$ and $\mathsf{Bisec}^{\mathsf{BB}}(a,b)$ are not empty and connected to each other. This illustrates the last statement of Lemma 3.9.

Example 3.2
The bisector in Figure 3.15 illustrates the other statements of Lemma 3.9. The intersection of the unit balls A and B is shown in Figure 3.16.

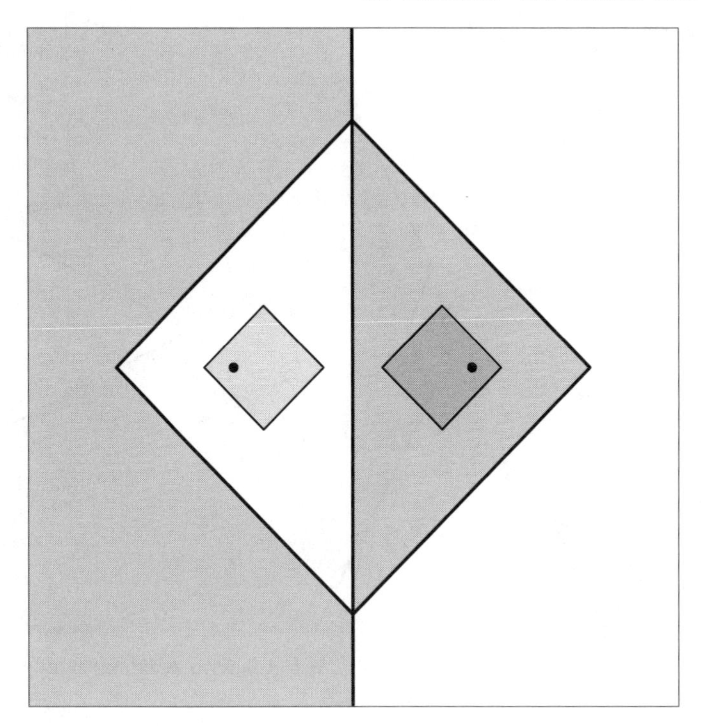

Fig. 3.14. Illustration to Example 3.1 - A Φ-shaped bisector

$\mathrm{bd}(A) \cap \mathrm{bd}(B) \cap \mathsf{F}\Gamma_A \cap \mathsf{F}\Gamma_B = \{e_2, e_7\}$ *contains exactly two points. Therefore* $\mathit{Bisec}^{\mathsf{FF}}(a,b)$ *is separated from the other components of* $\mathit{Bisec}(a,b)$.

$\{e \in \mathbb{R}^2 : e \in \mathrm{bd}(A) \cap \mathsf{B}\Gamma_A \cap \mathsf{F}\Gamma_B \wedge \gamma_B(e) < \gamma_A(e)\} = \mathrm{ri}(\overline{e_3e_4}) \,\dot{\cup}\, \mathrm{ri}(\overline{e_5e_6})$ *consists of* $K = 2$ *components and hence there are two connected components of* $\mathit{Bisec}^{\mathsf{BF}}(a,b)$.

$\{e \in \mathbb{R}^2 : e \in \mathrm{bd}(B) \cap \mathsf{F}\Gamma_A \cap \mathsf{B}\Gamma_B \wedge \gamma_A(e) < \gamma_B(e)\} = \mathrm{ri}(\overline{e_8e_1})$ *consists of* $L = 1$ *component and hence there is one connected component of* $\mathit{Bisec}^{\mathsf{BF}}(a,b)$.

Since $\mathit{Bisec}^{\mathsf{FF}}(a,b)$ *is separated from the other components of* $\mathit{Bisec}(a,b)$, *the maximal number of* $K + L + 1 = 4$ *connected components is attained.*

Naturally this example can be generalized. If A *and* B *are both regular* $2n$*-gons with the same area, centered at the origin,* A *standing on an extreme point,* B *standing on a face, then* $\mathit{Bisec}^{\mathsf{BF}}(a,b)$ *contains* n *connected components and* $\mathit{Bisec}^{\mathsf{FB}}(a,b)$ *contains* $n-1$ *connected components, whereas* $\mathit{Bisec}^{\mathsf{BB}}(a,b)$ *is empty and the one connected component of* $\mathit{Bisec}^{\mathsf{FF}}(a,b)$ *is separated from* $\mathit{Bisec}^{\mathsf{BF}}(a,b) \cup \mathit{Bisec}^{\mathsf{FB}}(a,b)$. *Therefore* $\mathit{Bisec}(a,b)$ *contains* $2n$ *components.*

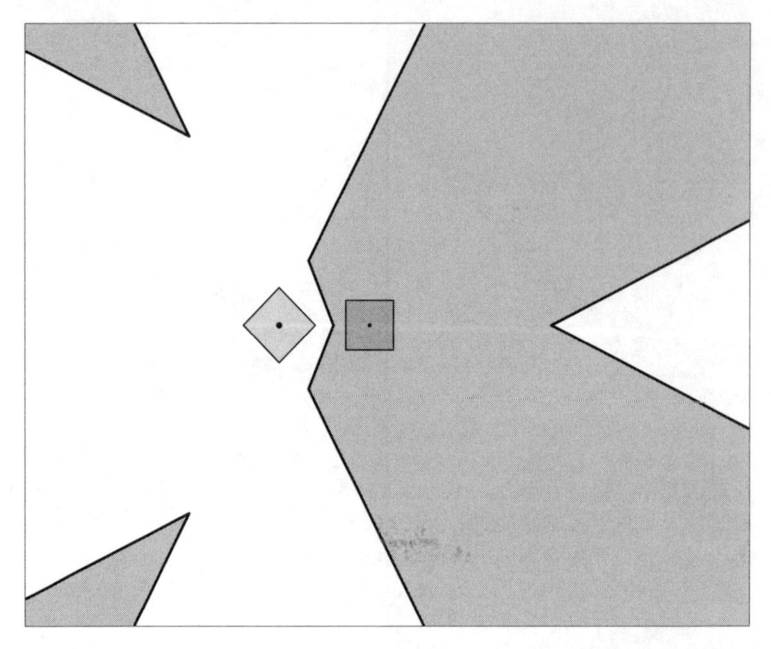

Fig. 3.15. Illustration to Example 3.2 - A bisector with maximal number of connected components

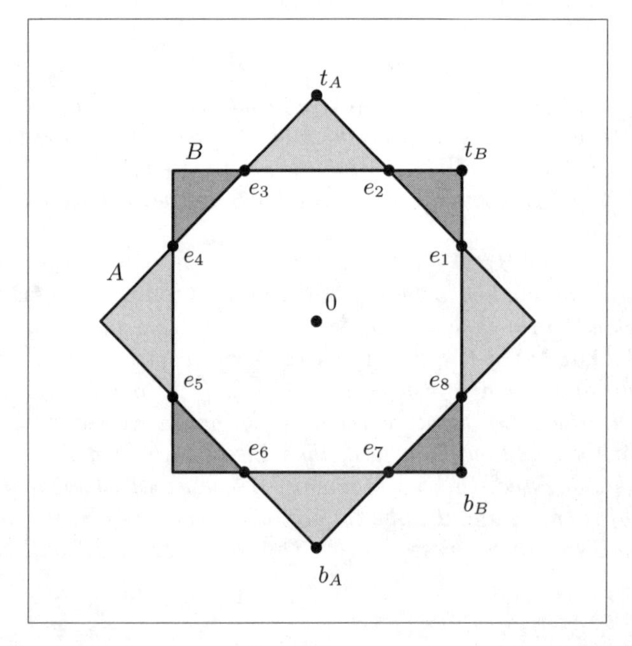

Fig. 3.16. Illustration to Example 3.2 - The intersection of the unit balls A and B

3.4 Bisectors of Polyhedral Gauges

Let the polyhedral unit balls $A, B \subseteq \mathbb{R}^2$ corresponding to the polyhedral gauges γ_A, γ_B be defined by the sets of extreme points $\{e_1^a, \ldots, e_K^a\}$ and $\{e_1^b, \ldots, e_L^b\}$ respectively. Let the corresponding fundamental directions, cones and supporting hyper-planes be denoted by d_i^a, Γ_i^a, h_i^a, $i \in \mathcal{K}$, and d_j^b, Γ_j^b, h_j^b, $j \in \mathcal{L}$, respectively.

Finally we denote the objects translated by the vector a and b respectively with the $\sim$-symbol and obtain for $i \in \mathcal{K}$ and $j \in \mathcal{L}$:

$$\tilde{A} := a + A := \{y \in \mathbb{R}^2 : y - a \in A\} \tag{3.51}$$

$$\tilde{B} := b + B := \{y \in \mathbb{R}^2 : y - b \in B\} \tag{3.52}$$

$$\tilde{e}_i^a := a + e_i^a \tag{3.53}$$

$$\tilde{e}_i^b := b + e_i^b \tag{3.54}$$

$$\tilde{d}_i^a := a + d_i^a := \{y \in \mathbb{R}^2 : y - a \in d_i^a\} \tag{3.55}$$

$$\tilde{d}_i^b := b + d_i^b := \{y \in \mathbb{R}^2 : y - b \in d_i^b\} \tag{3.56}$$

$$\tilde{\Gamma}_i^a := a + \Gamma_i^a := \{y \in \mathbb{R}^2 : y - a \in \Gamma_i^a\} \tag{3.57}$$

$$\tilde{\Gamma}_i^b := b + \Gamma_i^b := \{y \in \mathbb{R}^2 : y - b \in \Gamma_i^b\} \tag{3.58}$$

$$\tilde{h}_i^a := a + h_i^a := \{y \in \mathbb{R}^2 : \langle s_i^a, y - a\rangle \leq g_i^a\} \tag{3.59}$$

$$\tilde{h}_j^b := b + h_j^b := \{y \in \mathbb{R}^2 : \langle s_j^b, y - b\rangle \leq g_i^b\} \tag{3.60}$$

We assume $a \neq b$. Then the geometric structure generated by the translated fundamental directions is a planar subdivision and therefore we use the following notation.

For $i \in \mathcal{K}$ and $j \in \mathcal{L}$ we call $f_{ij}^{ab} := \tilde{\Gamma}_i^a \cap \tilde{\Gamma}_j^b$ a cell, if $\text{int}(f_{ij}^{ab}) \neq \emptyset$. Cells can be bounded or unbounded. For the set of cells (bounded cells, unbounded cells) we use the symbols $\mathcal{F}, \mathcal{F}_b, \mathcal{F}_u$.

$x \in \mathbb{R}^2$ is called a vertex, if $x = a$ or $x = b$ or $x = v_{ij}^{ab} := \tilde{d}_i^a \cap \tilde{d}_j^b$ for $i \in \mathcal{K}$ and $j \in \mathcal{L}$. The set of vertices is denoted by $\mathcal{V}$.

Finally we call $e \subseteq \mathbb{R}^2$ a bounded edge, if $x, y \in \mathcal{V}$ exist with $e = \overline{xy}$, $e \cap \mathcal{V} = \{x, y\}$ and $e \cap \text{int}(f) = \emptyset \ \forall \ f \in \mathcal{F}$.

If $f, \tilde{f} \in \mathcal{F}_u$, $f \neq \tilde{f}$ and $f \cap \tilde{f} \neq \emptyset$, we call $e = f \cap \tilde{f}$ an unbounded edge. For the set of edges (bounded edges, unbounded edges) the symbols $\mathrm{e}, \mathcal{E}_b, \mathcal{E}_u$ are used.

The planar subdivision is now given by $(\mathcal{V}, \mathrm{e}, \mathcal{F})$. An estimation for the number of vertices, segments and cells of a planar graph (planar subdivision) is given by the Eulerian Polyhedron Formula.

Theorem 3.1 ([26])
For a connected planar graph $G = (\mathcal{V}(G), \mathrm{e}(G))$ with $|\mathcal{V}(G)|$ vertices, $|\mathrm{e}(G)|$ edges and $|\mathcal{F}(G)|$ cells holds:

1. $|\mathrm{e}(G)| - |\mathcal{V}(G)| = |\mathcal{F}(G)| - 2$ *(Eulerian Polyhedron Formula)*
2. $|\mathrm{e}(G)| \leq 3|\mathcal{V}(G)| - 6$

3. $|\mathcal{F}(G)| \leq 2|\mathcal{V}(G)| - 4$

Applying this result to the planar subdivision $(\mathcal{V}, \mathrm{e}, \mathcal{F})$ we obtain the following estimation.

Lemma 3.10

If we denote the number of vertices by $n_{\mathcal{V}}$, the number of edges (bounded edges, unbounded edges) by n_{e} $(n_{\mathcal{E}_b}, n_{\mathcal{E}_u})$ and the number of cells (bounded cells, unbounded cells) by $n_{\mathcal{F}}$ $(n_{\mathcal{F}_b}, n_{\mathcal{F}_u})$ the following estimations for the planar subdivision $(\mathcal{V}, \mathrm{e}, \mathcal{F})$ hold:

1. $n_{\mathcal{E}_b} - n_{\mathcal{V}} = n_{\mathcal{F}_b} - 1$
2. $n_{\mathcal{E}_b} \leq 3n_{\mathcal{V}} - 6 \leq 3KL - 6$
3. $n_{\mathcal{E}_u} \leq K + L$
4. $n_{\mathcal{F}_b} \leq 2n_{\mathcal{V}} - 5 \leq 2KL - 5$
5. $n_{\mathcal{F}_u} \leq K + L$

Proof.

A vertex is either equal to a or equal to b or the intersection point of a fundamental direction of a with an fundamental direction of b. Since one fundamental direction of a can intersect at most $L-1$ fundamental directions of b, the number of vertices is bounded by $K(L-1)+2 = KL-(K-2) \leq KL$. The remaining statements follow by applying Theorem 3.1. □

The following six shapes for $f \in \mathcal{F}$ are possible (see Figure 3.17):

BT: bounded triangle
BQ: bounded quadrangle
UTS: unbounded triangle generated by two fundamental directions of the same site
UTE: unbounded triangle generated by two fundamental directions, one of each site
UQ: unbounded quadrangle
UP: unbounded pentagon

Notice that all six types of cells are convex, since they are the intersection of two convex cones. The following lemma describes the shape of a bisector inside a cell.

Lemma 3.11

Let $a, b \in \mathbb{R}^2$ be two sites associated with polyhedral gauges γ_A, γ_B. For $f \in \mathcal{F}$ we have that either

1. $f \cap \mathsf{Bisec}(a,b) = \emptyset$ *or*
2. $f \cap \mathsf{Bisec}(a,b)$ *is a single point or*
3. $f \cap \mathsf{Bisec}(a,b)$ *is a line segment or*
4. $f \cap \mathsf{Bisec}(a,b)$ *is a half-line or*
5. $f \cap \mathsf{Bisec}(a,b) = f$, *i.e.* $f \subseteq \mathsf{Bisec}(a,b)$.

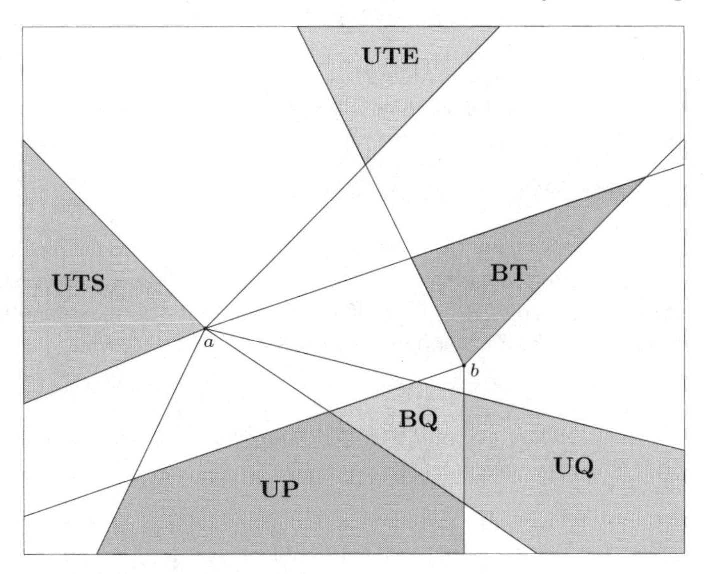

Fig. 3.17. Types of regions

Proof.
Let $f \in \mathcal{F}$. We know that $f = \tilde{\Gamma}_i^a \cap \tilde{\Gamma}_j^b$ for some $i \in \mathcal{K}$ and $j \in \mathcal{L}$. By Lemma 2.3 we know that $\gamma_A(\,.\,-a)/_{\tilde{\Gamma}_i^a}$ and $\gamma_B(\,.\,-b)/_{\tilde{\Gamma}_j^b}$ are affine-linear, that means

$$\gamma_A(x-a) = \alpha_1 x_1 + \alpha_2 x_2 + \alpha_3 \quad \forall\, x \in \tilde{\Gamma}_i^a$$

and

$$\gamma_B(x-b) = \beta_1 x_1 + \beta_2 x_2 + \beta_3 \quad \forall\, x \in \tilde{\Gamma}_j^b$$

with $\alpha_k, \beta_k \in \mathbb{R}\,,\ k \in \{1,2,3\}\,.$

The solution of the equation

$$\gamma_A(x-a) = \gamma_B(x-b) \tag{3.61}$$

which is equivalent to

$$(\alpha_1 - \beta_1)x_1 + (\alpha_2 - \beta_2)x_2 = \beta_3 - \alpha_3 \tag{3.62}$$

leads to the following case analysis:

Case 1: $\alpha_1 = \beta_1 \ \wedge\ \alpha_2 = \beta_2$
  Case 1.1: $\alpha_3 = \beta_3$
    $\Rightarrow$ Equation (3.62) is fulfilled for all $x \in \mathbb{R}^2$.
    $\Rightarrow$ $f \subseteq \mathsf{Bisec}(a,b)$.
  Case 1.2: $\alpha_3 \neq \beta_3$
    $\Rightarrow$ No $x \in \mathbb{R}^2$ fulfills equation (3.62).
    $\Rightarrow$ $f \cap \mathsf{Bisec}(a,b) = \emptyset$.

Case 2 : $\alpha_1 \neq \beta_1 \;\vee\; \alpha_2 \neq \beta_2$

$\Rightarrow$ The solution of equation (3.62) is given by a line l.

$\Rightarrow$ Since l is a line and f is convex, $l \cap f$ is the empty set, a single point, a segment or a half-line.

□

By Lemma 3.11 we know that $\mathsf{Bisec}(a, b)$ is made up of linear pieces, therefore we can speak now of bisector vertices, bisector edges and bisector cells. The following two Corollaries give information about the existence and the number of cells in $(\mathcal{V}, \mathrm{e}, \mathcal{F})$ belonging to the bisector $\mathsf{Bisec}(a, b)$.

Corollary 3.7

Let $a, b \in \mathbb{R}^2$ be two sites associated with polyhedral gauges γ_A and γ_B respectively and let $f_{ij}^{ab} \in \mathcal{F}$, then holds :

$$f_{ij}^{ab} \subseteq \mathsf{Bisec}(a,b) \Longleftrightarrow \left\{ \frac{s_i^a}{g_i^a} = \frac{s_j^b}{g_j^b} \wedge (a-b) \;\perp\; \frac{s_i^a}{g_i^a} \left(= \frac{s_j^b}{g_j^b} \right) \right\} . \tag{3.63}$$

Proof.

$\gamma_A(\,.\, - a)/_{\tilde{\Gamma}_i^a}$ is given by

$$\gamma_A(x-a) = \frac{s_{i1}^a}{g_i^a} x_1 + \frac{s_{i2}^a}{g_i^a} x_2 - \left(\frac{s_{i1}^a}{g_i^a} a_1 + \frac{s_{i2}^a}{g_i^a} a_2 \right)$$

and $\gamma_B(\,.\, - b)/_{\tilde{\Gamma}_j^b}$ is given by

$$\gamma_B(x-b) = \frac{s_{j1}^b}{g_j^b} x_1 + \frac{s_{j2}^b}{g_j^b} x_2 - \left(\frac{s_{j1}^b}{g_j^b} b_1 + \frac{s_{j2}^b}{g_j^b} b_2 \right) \quad .$$

Therefore **Case 1.1** of the proof of Lemma 3.11 occurs if and only if

$$\frac{s_{i1}^a}{g_i^a} = \frac{s_{j1}^b}{g_j^b} \wedge \frac{s_{i2}^a}{g_i^a} = \frac{s_{j2}^b}{g_j^b} \wedge \frac{s_{i1}^a}{g_i^a} a_1 + \frac{s_{i2}^a}{g_i^a} a_2 = \frac{s_{j1}^b}{g_j^b} b_1 + \frac{s_{j2}^b}{g_j^b} b_2$$

which is equivalent to

$$\frac{s_i^a}{g_i^a} = \frac{s_j^b}{g_j^b} \qquad \wedge \qquad \left\langle \frac{s_i^a}{g_i^a}, a-b \right\rangle = 0 \quad .$$

□

Corollary 3.8

Let $a, b \in \mathbb{R}^2$ be two sites associated with polyhedral gauges γ_A and γ_B respectively. Then the bisector $\mathsf{Bisec}(a, b)$ contains at most two cells of the planar subdivision $(\mathcal{V}, \mathrm{e}, \mathcal{F})$.

Proof.
There exist at most two indices of $\mathcal{K}$, which fulfill the condition $(a-b) \perp \frac{h_i^a}{g_i^a}$.
□

Now we are already able to present a naive algorithm for the determination of the bisector $\mathsf{Bisec}(a,b)$ which computes the linear pieces of $\mathsf{Bisec}(a,b)$, step by step, for each cell of $(\mathcal{V}, \mathrm{e}, \mathcal{F})$.

Algorithm 3.1: *Computing Bisec(a,b) for polyhedral gauges γ_A, γ_B*

Input: Sites $a, b \in \mathbb{R}^2$.
Polyhedral gauges $\gamma_A, \gamma_B : \mathbb{R}^2 \to \mathbb{R}$.
Output: Bisector $\mathsf{Bisec}(a,b)$.

$\mathsf{Bisec}(a,b) := \emptyset$
for $i = 1$ **TO** K **do**
  for $j = 1$ **TO** L **do**
    Compute f_{ij}^{ab}.
    if $f_{ij}^{ab} \neq \emptyset$ **then**
      Compute $l_{ij}^{ab} := \{x \in \mathbb{R}^2 : \frac{1}{g_i^a}\langle h_i^a, x-a\rangle = \frac{1}{g_j^b}\langle h_j^b, x-b\rangle\}$.
      Compute $\mathsf{Bisec}^{ij}(a,b) := f_{ij}^{ab} \cap l_{ij}^{ab}$.
      $\mathsf{Bisec}(a,b) := \mathsf{Bisec}(a,b) \cup \mathsf{Bisec}^{ij}(a,b)$.
    end
  end
end

Algorithm 3.1 has a time complexity of $O(KL)$. Notice that Lemma 3.11, Corollary 3.7 and Algorithm 3.1 are also valid for bisectors of polyhedral gauges with additive weights, since everything is only based on the affine linearity of γ_A and γ_B in the cells of $(\mathcal{V}, \mathrm{e}, \mathcal{F})$.

If we apply Lemma 3.4 to the translated fundamental directions

$$\tilde{d_i^a} := \{x_\lambda \in \mathbb{R}^2 : x_\lambda = a + \lambda e_i^a, \lambda \geq 0\}, i \in \mathcal{K}, \tag{3.64}$$

of the polyhedral gauge γ_A, we obtain the following corollary:

Corollary 3.9
Let $a, b \in \mathbb{R}^2$ be two sites associated with polyhedral gauges γ_A and γ_B respectively and let $i \in \mathcal{K}$. Then the set $\mathsf{Bisec}(a,b) \cap \tilde{d_i^a} = \{x \in \tilde{d_i^a} : \gamma_A(x-a) = \gamma_B(x-b)\}$ is either

1. *$= \emptyset$ or*
2. *consists of 1 point or*
3. *consists of 2 points or*
4. *is equal to a closed segment or*
5. *is equal to a half-line.*

Since the roles of $a(A, \gamma_A)$ and $b(B, \gamma_B)$ in Lemma 3.4 can be reversed, we obtain together with Lemma 3.11 the following corollary.

Corollary 3.10
The bisector $\mathsf{Bisec}(a, b)$ *of two sites* $a, b \in \mathbb{R}^2$ *associated with polyhedral gauges* γ_A *and* γ_B *respectively consists of at most* $2(K + L)$ *vertices, that means* $\mathsf{Bisec}(a, b)$ *runs through at most* $2(K + L)$ *cells of the planar subdivision* $(\mathcal{V}, \mathrm{e}, \mathcal{F})$.

By Corollary 3.10 the complexity of the bisector is linear in the number of extreme points of the two generating unit balls. That means the bisector $\mathsf{Bisec}(a, b)$ does in general not run through all cells of $(\mathcal{V}, \mathrm{e}, \mathcal{F})$, since the number of cells can be $O(KL)$. Nevertheless, this behavior is possible in a situation where the number of cells is $O(K + L)$ (see Figure 3.18). Moreover, we know by Corollary 3.10 that it is worthwhile to look for an algorithm with linear complexity.

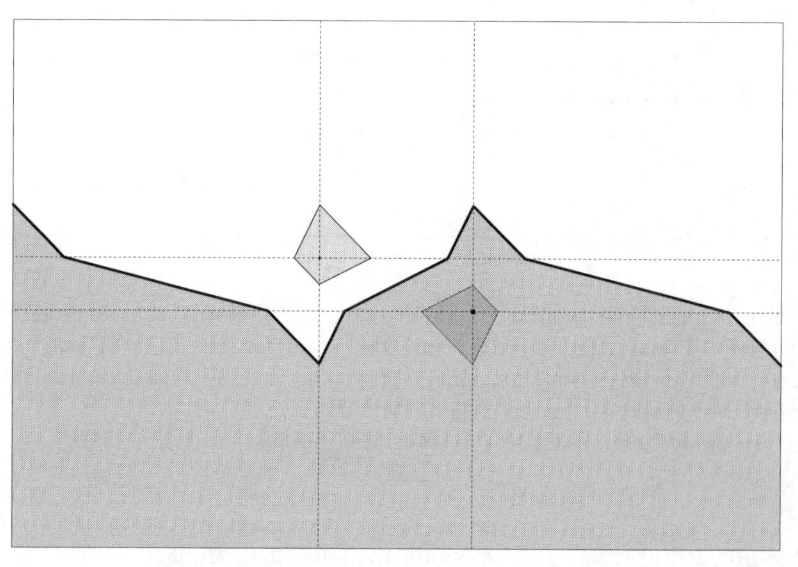

Fig. 3.18. A bisector running through all cells

The number of connected components is bounded as follows.

Lemma 3.12 ([107])
The bisector $\mathsf{Bisec}(a, b)$ *of two sites* $a, b \in \mathbb{R}^2$ *associated with polyhedral gauges* γ_A *and* γ_B *respectively consists of at most* $\min\{K, L\}$ *connected components. This bound is tight.*

Proof.
bd(A) and bd(B) intersect at most $\min\{K, L\}$ times. Therefore the sets

$$\{e \in \mathbb{R}^2 : e \in \mathrm{bd}(A) \cap \mathsf{B}\Gamma_A \cap \mathsf{F}\Gamma_B \ \wedge \ \gamma_B(e) < \gamma_A(e)\}$$

and

$$\{e \in \mathbb{R}^2 : e \in \mathrm{bd}(B) \cap \mathsf{F}\Gamma_A \cap \mathsf{B}\Gamma_B \ \wedge \ \gamma_A(e) < \gamma_B(e)\}$$

have together at most $\min\{K, L\}$ components. Now from Lemma 3.9 follows that $\mathsf{Bisec}^{\mathsf{BF}}(a,b) \cup \mathsf{Bisec}^{\mathsf{FB}}(a,b)$ has at most $\min\{K, L\}$ components. If $\mathsf{Bisec}^{\mathsf{FF}}(a,b)$ is separated from $\mathsf{Bisec}^{\mathsf{BF}}(a,b) \cup \mathsf{Bisec}^{\mathsf{FB}}(a,b)$, exactly two intersection points of bd(A) and bd(B) lie in $\mathsf{F}\Gamma_A \cup \mathsf{F}\Gamma_B$ by Lemma 3.9. This reduces the number of possible components of $\mathsf{Bisec}^{\mathsf{BF}}(a,b) \cup \mathsf{Bisec}^{\mathsf{FB}}(a,b)$ at least by one. Thus, the number of components of $\mathsf{Bisec}^{\mathsf{FF}}(a,b) \cup \mathsf{Bisec}^{\mathsf{BF}}(a,b) \cup \mathsf{Bisec}^{\mathsf{FB}}(a,b)$ is at most $\min\{K, L\}$. By Lemma 3.9 we know that $\mathsf{Bisec}^{\mathsf{BB}}(a,b)$ is always connected to other components of the bisector. Therefore $\mathsf{Bisec}(a,b) = \mathsf{Bisec}^{\mathsf{FF}}(a,b) \cup \mathsf{Bisec}^{\mathsf{BF}}(a,b) \cup \mathsf{Bisec}^{\mathsf{FB}}(a,b) \cup \mathsf{Bisec}^{\mathsf{BB}}(a,b)$ contains at most $\min\{K, L\}$ connected components. This bound is tight as Example 3.2 shows.
□

Now we present an algorithm with complexity $O(K + L)$ which makes use of the results of Subsection 3.3.3. That means we compute the bisector $\mathsf{Bisec}(a,b)$ by determining the four disjoint bisector parts $\mathsf{Bisec}^{\mathsf{FF}}(a,b)$, $\mathsf{Bisec}^{\mathsf{BF}}(a,b)$, $\mathsf{Bisec}^{\mathsf{FB}}(a,b)$ and $\mathsf{Bisec}^{\mathsf{BB}}(a,b)$ separately. Each of the four bisector components can be computed in $O(K + L)$ time while scanning the extreme points between b_A and t_A respectively b_B and t_B simultaneously in a proper way. This is possible since the linear behavior of the bisector changes only at fundamental directions (see Lemma 3.11). We describe the computation of $\mathsf{Bisec}^{\mathsf{FF}}(a,b)$ in detail. The computation of the other bisector parts can be done in a quite similar way. Therefore, only the differences will be briefly mentioned here.

Without loosing generality we assume that a and b lie on a horizontal line and b lies on the right side of a (see Subsection 3.3.3). Moreover, we assume that the extreme points $\mathrm{Ext}(A) = \{e_1^a, \ldots, e_K^a\}$ as well as the extreme points $\mathrm{Ext}(B) = \{e_1^b, \ldots, e_L^b\}$ are stored in double connected lists. Notice that the top and bottom points (see Subsection 3.3.3) fulfill $t_A, b_A \in \mathrm{Ext}(A)$ and $t_B, b_B \in \mathrm{Ext}(B)$. We denote the counterclockwise successor (predecessor) of $e_A \in \mathrm{Ext}(A)$ with s_A (p_A) and the clockwise successor (predecessor) of $e_B \in \mathrm{Ext}(B)$ with s_B (p_B). Finally $\mathrm{ycd}(x)$ returns the ordinate x_2 of the point $x = (x_1, x_2)$ and $\Gamma(x, u, v)$ denotes the cone pointed at x and bounded by the half-lines $\mathbf{d}_{x,u}$ and $\mathbf{d}_{x,v}$.

Starting from the bisector point $x_M := \frac{\alpha^+}{\alpha^+ + \beta^+} a + \frac{\beta^+}{\alpha^+ + \beta^+} b$ (see Corollary 3.1) we step through all the fundamental directions in the order of the ordinate of their defining extreme points, first upwards and then downwards. These loops are performed as long as the maximal ($\mathrm{ycd}(t_A), \mathrm{ycd}(t_B)$), respectively the minimal ($\mathrm{ycd}(b_A), \mathrm{ycd}(b_B)$), ordinate of one of the unit balls is not reached and the current line segments $\overline{e_A s_A}, \overline{e_B s_B}$, respectively $\overline{e_A p_A}, \overline{e_B p_B}$, do not intersect. If the upwards (downwards) loop terminates due to the former criterion, it remains to check if one of the special cases with horizontal boundary segments of the unit balls occur. In the latter case the upper (lower) part of $\mathsf{Bisec}^{\mathsf{FF}}(a, b)$ ends with an unbounded bisector edge and is only connected via x_M to other parts of the bisector (see Lemma 3.9).

Steps 4 and 5 can be performed in $O(K + L)$ time, since $\text{Ext}(A)$ and $\text{Ext}(B)$ are stored as double connected lists. Moreover, the loop is performed in $O(K + L)$ time, whereas all other operations take constant time. Therefore the complexity of Algorithm 3.2 is in fact $O(K + L)$.

The construction of the two potential parts of $\mathsf{Bisec}^{\mathsf{BB}}(a, b)$ can take place directly after the construction of the upper and lower part of $\mathsf{Bisec}^{\mathsf{FF}}(a, b)$, respectively. For the upper part we start with the sweep at the ordinate $y := \min\{\text{ycd}(t_A), \text{ycd}(t_B)\}$. The upper part of $\mathsf{Bisec}^{\mathsf{BB}}(a, b)$ exists if and only if the back side of A lies to the right of the back side of B in the height of y. We scan downwards the two back sides of the unit balls as in the while loop of Algorithm 3.2 until they intersect. For the lower part of $\mathsf{Bisec}^{\mathsf{BB}}(a, b)$ we proceed analogous.

For $\mathsf{Bisec}^{\mathsf{BF}}(a, b)$ and $\mathsf{Bisec}^{\mathsf{FB}}(a, b)$ we start again at the position where the construction of the upper part of $\mathsf{Bisec}^{\mathsf{FF}}(a, b)$ has stopped. For $\mathsf{Bisec}^{\mathsf{BF}}(a, b)$ we scan along the back side of A and the face side of B. Bisector pieces are generated while the face side of B lies to the left of the back side of A, i.e. γ_B is stronger than γ_A. The construction of $\mathsf{Bisec}^{\mathsf{BF}}(a, b)$ does not end at an intersection point of the back side of A with the face side of B as it was the case for $\mathsf{Bisec}^{\mathsf{FF}}(a, b)$. Here an intersection produces a half-line belonging to $\mathsf{Bisec}^{\mathsf{BF}}(a, b)$ instead. As long as γ_B is stronger than γ_A no bisector piece arises. But if another intersection point arises there will also further bisector pieces arise. Therefore, we search for the next intersection point of the boundaries of the unit balls. This is continued until we reach the position where the construction of the lower part of $\mathsf{Bisec}^{\mathsf{FF}}(a, b)$ has stopped. For the construction of $\mathsf{Bisec}^{\mathsf{FB}}(a, b)$ we proceed analogous.

Analogous to Algorithm 3.2 the Steps 2-4 of Algorithm 3.3 have a time complexity of $O(K + L)$ and we obtain the following theorem.

Theorem 3.2 ([107])
Let $a, b \in \mathbb{R}^2$ be two sites associated with polyhedral gauges γ_A and γ_B respectively. Then Algorithm 3.3 computes the bisector $\mathsf{Bisec}(a, b)$ in optimal time $O(K + L)$.

At the end we turn to the case $a = b$. In this case the planar subdivision $(\mathcal{V}, \mathrm{e}, \mathcal{F})$ consists of only one node, namely $\mathcal{V} = \{a\}$, and the plane is divided by the fundamental directions of A and B into at most $K + L$ cones ($\hat{=}$ unbounded cells of triangular shape) with apex at a (see Figure 3.7). The computation of $\mathsf{Bisec}(a, b)$ can be done in $O(K + L)$ time as in the general case.

***Algorithm 3.2**: Computing* $\mathsf{Bisec}^{\mathsf{FF}}(a,b)$ *for polyhedral gauges*

Input: Sites $a, b \in \mathbb{R}^2$
Polyhedral gauges $\gamma_A, \gamma_B : \mathbb{R}^2 \to \mathbb{R}$.
Output: Bisector $\mathsf{Bisec}^{\mathsf{FF}}(a,b)$.

$\mathsf{Bisec}^{\mathsf{FF}}(a,b) := \emptyset$.
$x_M := \frac{\alpha^+}{\alpha^+ + \beta^+} a + \frac{\beta^+}{\alpha^+ + \beta^+} b$ ($\star$ $x_M \in \mathsf{Bisec}(a,b) \cap l_{a,b}$, see Corollary 3.1. $\star$)
$x := x_M$ ($\star$ First upper part of $\mathsf{Bisec}^{\mathsf{FF}}(a,b)$ starting at (artificial) bisector vertex x_M $\star$)
Determine e_A with $\mathrm{ycd}(e_A) \leq \mathrm{ycd}(x_M) < \mathrm{ycd}(s_A)$.
Determine e_B with $\mathrm{ycd}(e_B) \leq \mathrm{ycd}(x_M) < \mathrm{ycd}(s_B)$.
while $\mathrm{ycd}(e_A) \neq \mathrm{ycd}(t_A)$ **AND** $\mathrm{ycd}(e_B) \neq \mathrm{ycd}(t_B)$ **AND** $\overline{e_A s_A} \cap \overline{e_B s_B} = \emptyset$ **do**
    if $\mathrm{ycd}(s_A) < \mathrm{ycd}(s_B)$ **then**
        Determine $v \in \overline{e_B s_B}$ with $\mathrm{ycd}(v) = \mathrm{ycd}(s_A)$. ($\star$ Partner f. p. of s_A. $\star$)
        $y := \mathbf{d}_{a,s_A} \cap \mathbf{d}_{b,v}$ ($\star$ The next bisector vertex. $\star$)
        $e_A := s_A$ ($\star$ Advance in $\mathrm{Ext}(A)$. $\star$)
        else if $\mathrm{ycd}(s_A) > \mathrm{ycd}(s_B)$ **then**
            Determine $u \in \overline{e_A s_A}$ with $\mathrm{ycd}(u) = \mathrm{ycd}(s_B)$. ($\star$ P. f. p. of s_B. $\star$)
            $y := \mathbf{d}_{a,u} \cap \mathbf{d}_{b,s_B}$ ($\star$ The next bisector vertex. $\star$)
            $e_B := s_B$ ($\star$ Advance in $\mathrm{Ext}(B)$. $\star$)
        end
        else if $\mathrm{ycd}(s_A) = \mathrm{ycd}(s_B)$ **then**
            ($\star$ s_A and s_B are partner footpoints. $\star$)
            $y := \mathbf{d}_{a,s_A} \cap \mathbf{d}_{b,s_B}$ ($\star$ The next bisector vertex. $\star$)
            $e_A := s_A$, $e_B := s_B$ ($\star$ Advance in $\mathrm{Ext}(A)$ and $\mathrm{Ext}(B)$. $\star$)
        end
    end
    $\mathsf{Bisec}^{\mathsf{FF}}(a,b) := \mathsf{Bisec}^{\mathsf{FF}}(a,b) \cup \overline{xy}$ ($\star$ A bounded bisector edge. $\star$)
        $x := y$ ($\star$ Advance in $\mathsf{Bisec}^{\mathsf{FF}}(a,b)$. $\star$)
end
if $\mathrm{ycd}(e_A) \neq \mathrm{ycd}(t_A)$ **AND** $\mathrm{ycd}(e_B) \neq \mathrm{ycd}(t_B)$ **then**
    $u := \overline{e_A s_A} \cap \overline{e_B s_B}$ ($\star$ $\overline{e_A s_A} \cap \overline{e_B s_B}$ is an unique point. $\star$)
    $\mathsf{Bisec}^{\mathsf{FF}}(a,b) := \mathsf{Bisec}^{\mathsf{FF}}(a,b) \cup \mathbf{d}_{x,u}$ ($\star$ The final unbounded bisec. edge. $\star$)
    ($\star$ Otherwise the maximal ordinate of at least one unit ball is reached and it remains to deal with horizontal boundary parts: $\star$)
    else if $\mathrm{ycd}(e_A) = \mathrm{ycd}(s_A) = \mathrm{ycd}(e_B) = \mathrm{ycd}(s_B)$ **then**
        $\mathsf{Bisec}^{\mathsf{FF}}(a,b) := \mathsf{Bisec}^{\mathsf{FF}}(a,b) \cup (\Gamma(a, e_A, s_A) \cap \Gamma(b, e_B, s_B))$ ($\star$ The final (un)bounded bisector cell. $\star$)
    end
    else if $\mathrm{ycd}(e_A) = \mathrm{ycd}(s_A)$ **then**
        Determine $v \in \overline{e_B s_B}$ with $\mathrm{ycd}(v) = \mathrm{ycd}(s_A)$.
        $\mathsf{Bisec}^{\mathsf{FF}}(a,b) := \mathsf{Bisec}^{\mathsf{FF}}(a,b) \cup (\Gamma(a, e_A, s_A) \cap \mathbf{d}_{b,v})$
    end
    else if $\mathrm{ycd}(e_B) = \mathrm{ycd}(s_B)$ **then**
        Determine $u \in \overline{e_A s_A}$ with $\mathrm{ycd}(u) = \mathrm{ycd}(s_B)$.
        $\mathsf{Bisec}^{\mathsf{FF}}(a,b) := \mathsf{Bisec}^{\mathsf{FF}}(a,b) \cup (\mathbf{d}_{a,u} \cap \Gamma(b, e_B, s_B))$
    end
end
$x := x_M$ ($\star$ Now lower part of $\mathsf{Bisec}^{\mathsf{FF}}(a,b)$ starting at x_M : analog. to upper part $\star$)

Algorithm 3.3: *Computing* $\mathsf{Bisec}(a,b)$ *for polyhedral gauges* γ_A, γ_B

Input: Sites $a, b \in \mathbb{R}^2$.
Polyhedral gauges $\gamma_A, \gamma_B : \mathbb{R}^2 \to \mathbb{R}$.
Output: Bisector $\mathsf{Bisec}(a,b)$.

Compute $\mathsf{Bisec}^{\mathsf{FF}}(a,b)$ (with Algorithm 3.2).
Compute $\mathsf{Bisec}^{\mathsf{BF}}(a,b)$ (analogous).
Compute $\mathsf{Bisec}^{\mathsf{FB}}(a,b)$ (analogous).
Compute $\mathsf{Bisec}^{\mathsf{BB}}(a,b)$ (analogous).
$\mathsf{Bisec}(a,b) := \mathsf{Bisec}^{\mathsf{FF}}(a,b) \cup \mathsf{Bisec}^{\mathsf{BF}}(a,b) \cup \mathsf{Bisec}^{\mathsf{FB}}(a,b) \cup \mathsf{Bisec}^{\mathsf{BB}}(a,b)$.

Now with the construction of the bisectors, we can solve the ordered median problem, in a geometrical way, using Algorithm 4.2 from Subsection 4.2.

It should be noted, that most examples in this chapter have been computed by an computer program containing an implementation of Algorithm 4.2. This program is available on `http://www.itwm.fhg.de/opt/LOLA/lola.html` .

3.5 Bisectors of Elliptic Gauges

In this section we study the behavior of the bisector for unit balls of elliptic shape, which is a generalization of the classical (weighted) l_2-case. Moreover, we will see at the end, that the computation of this kind of bisectors leads to an alternative way to introduce and motivate conic sections.

For positive scalars $c_1, c_2 > 0$ the set

$$C := \left\{ x \in \mathbb{R}^2 : \frac{x_1^2}{c_1^2} + \frac{x_2^2}{c_2^2} = 1 \right\} = \left\{ x \in \mathbb{R}^2 : \sqrt{\frac{x_1^2}{c_1^2} + \frac{x_2^2}{c_2^2}} = 1 \right\} \quad (3.65)$$

is called ellipse with radii c_1, c_2 (see Figure 3.19). More precisely for $c_1 > c_2$ the scalar c_1 determines half the length of the major axis, whereas the scalar c_2 determines half the length of the minor axis. In the case $c_1 < c_2$ major and minor axes are exchanged and for $c_1 = c_2$ major and minor axes are unambiguous.

The corresponding gauge is given by

$$\gamma_C \quad : \quad \mathbb{R}^2 \quad \to \quad \mathbb{R} \quad , \quad x \quad \mapsto \quad \sqrt{\frac{x_1^2}{c_1^2} + \frac{x_2^2}{c_2^2}} \quad (3.66)$$

and fulfills the properties of a norm, since C is symmetric with respect to the origin.

Notice that we can rewrite C as

$$C := \{ x \in \mathbb{R}^2 : \zeta_1 x_1^2 + \zeta_2 x_2^2 = 1 \} \quad , \quad (3.67)$$

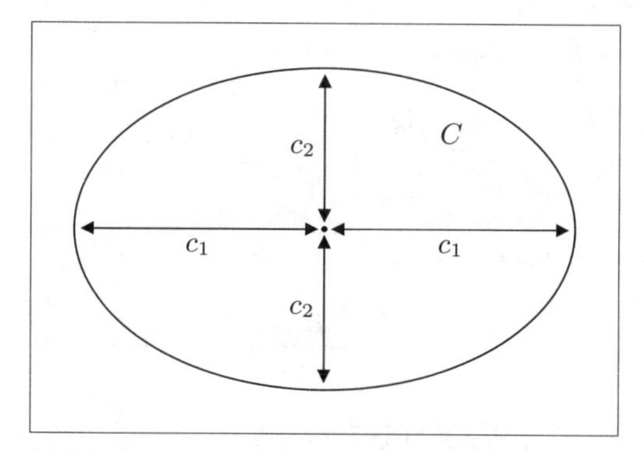

Fig. 3.19. Ellipse C with radii c_1 and c_2

if we define $\zeta_i := \frac{1}{c_i^2}$ for $i = 1, 2$ (and $c_i := \zeta_i^{-\frac{1}{2}}$ for $i = 1, 2$).

Let us assume the norms γ_A, γ_B associated with the two sites $a, b \in \mathbb{R}^2$ are defined with respect to the elliptic unit balls

$$A := \{x \in \mathbb{R}^2 \,:\, \alpha_1 x_1^2 + \alpha_2 x_2^2 = 1\} \tag{3.68}$$

and

$$B := \{x \in \mathbb{R}^2 \,:\, \beta_1 x_1^2 + \beta_2 x_2^2 = 1\} \tag{3.69}$$

with $\alpha_i, \beta_i > 0$ for $i = 1, 2$.

Since A and B intersect at most four times we can deduce, by Lemma 3.9, that $\mathsf{Bisec}(a, b)$ consists of at most two connected components. Moreover, since γ_A and γ_B are norms the line $l_{a,b}$ through a and b contains at most two points of the bisector (see Corollary 3.2).

If we want to determine the points in the plane belonging to the bisector $\mathsf{Bisec}(a, b)$, we have to solve the equation

$$\alpha_1(x_1 - a_1)^2 + \alpha_2(x_2 - a_2)^2 = \beta_1(x_1 - b_1)^2 + \beta_2(x_2 - b_2)^2 \tag{3.70}$$

which is equivalent to

$$\sum_{i=1,2} (\alpha_i - \beta_i)x_i^2 - 2(\alpha_i a_i - \beta_i b_i)x_i = \sum_{i=1,2} \beta_i b_i^2 - \alpha_i a_i^2 \tag{3.71}$$

and leads therefore to the following case analysis :

Case 1 : $\alpha_1 \neq \beta_1 \;\wedge\; \alpha_2 \neq \beta_2$

In this case Equation (3.71) is equivalent to

$$\sum_{i=1,2} (\alpha_i - \beta_i) \left(x_i^2 - 2\frac{\alpha_i a_i - \beta_i b_i}{\alpha_i - \beta_i} x_i \right) = \sum_{i=1,2} \beta_i b_i^2 - \alpha_i a_i^2 \quad . \tag{3.72}$$

By completion of the square we obtain

$$\sum_{i=1,2}(\alpha_i-\beta_i)\left(x_i-\frac{\alpha_i a_i-\beta_i b_i}{\alpha_i-\beta_i}\right)^2=\sum_{i=1,2}\beta_i b_i^2-\alpha_i a_i^2+\frac{(\alpha_i a_i-\beta_i b_i)^2}{\alpha_i-\beta_i}\quad . \tag{3.73}$$

The right side of Equation (3.73) can be rewritten as

$$\begin{aligned}&\sum_{i=1,2}\beta_i b_i^2-\alpha_i a_i^2+\frac{\alpha_i^2a_i^2-2\alpha_i\beta_i a_i b_i+\beta_i^2b_i^2}{\alpha_i-\beta_i}\\ &=\sum_{i=1,2}\frac{\alpha_i\beta_i b_i^2-\beta_i^2b_i^2-\alpha_i^2a_i^2+\alpha_i\beta_i a_i^2+\alpha_i^2a_i^2-2\alpha_i\beta_i a_i b_i+\beta_i^2b_i^2}{\alpha_i-\beta_i}\\ &=\sum_{i=1,2}\frac{\alpha_i\beta_i}{\alpha_i-\beta_i}(b_i^2-2a_ib_i+a_i^2)\\ &=\sum_{i=1,2}\frac{\alpha_i\beta_i}{\alpha_i-\beta_i}(a_i-b_i)^2\quad ,\end{aligned}$$

that means Equation (3.73) is equivalent to

$$\sum_{i=1,2}\underbrace{(\alpha_i-\beta_i)}_{=:\zeta_i\neq 0}\left(x_i-\underbrace{\frac{\alpha_i a_i-\beta_i b_i}{\alpha_i-\beta_i}}_{=:z_i}\right)^2=\underbrace{\sum_{i=1,2}\frac{\alpha_i\beta_i}{\alpha_i-\beta_i}(a_i-b_i)^2}_{=:\zeta}\quad . \tag{3.74}$$

Equation (3.74) requires a further case analysis.
Case 1.1: $\zeta:=\sum\limits_{i=1,2}\frac{\alpha_i\beta_i}{\alpha_i-\beta_i}(a_i-b_i)^2\neq 0$

Case 1.1.1: $\mathrm{sgn}(\zeta_1)=\mathrm{sgn}(\zeta_2)$ with $\zeta_i:=\alpha_i-\beta_i$ for $i=1,2$

This case also implies $\mathrm{sgn}(\zeta)=\mathrm{sgn}(\zeta_1)=\mathrm{sgn}(\zeta_2)$, that means Equation (3.74) describes an ellipse with midpoint (z_1,z_2), radii $c_i=\sqrt{\frac{\zeta}{\zeta_i}}$ for $i=1,2$ and

$$\begin{aligned}\text{major axis in } x_1\text{-direction iff } c_1>c_2 \text{ iff } \zeta_1<\zeta_2\\ \text{unambiguous major axis iff } c_1=c_2 \text{ iff } \zeta_1=\zeta_2\\ \text{major axis in } x_2\text{-direction iff } c_1<c_2 \text{ iff } \zeta_1>\zeta_2\end{aligned}$$

(see Figure 3.20).

Case 1.1.2: $\mathrm{sgn}(\zeta_1)\neq\mathrm{sgn}(\zeta_2)$ with $\zeta_i:=\alpha_i-\beta_i$ for $i=1,2$

In this case Equation (3.74) describes an hyperbola with midpoint (z_1,z_2). $c_1=\sqrt{\frac{\zeta}{\zeta_1}}$ is half the length of the transverse axis (in x_1-direction) and $c_2=\sqrt{-\frac{\zeta}{\zeta_2}}$ is half the length of the conjugate axis (in x_2-direction), if $\operatorname{sgn}(\zeta) = \operatorname{sgn}(\zeta_1)$. In the case of $\operatorname{sgn}(\zeta) = \operatorname{sgn}(\zeta_2)$ the length of the transverse axis (in x_2-direction) is given by $c_2 = \sqrt{\frac{\zeta}{\zeta_2}}$, whereas the length of the conjugate axis (in x_1-direction) is given by $c_1 = \sqrt{-\frac{\zeta}{\zeta_1}}$ (see Figure 3.21).

Case 1.2: $\zeta := \sum\limits_{i=1,2} \frac{\alpha_i \beta_i}{\alpha_i - \beta_i}(a_i - b_i)^2 = 0$

Case 1.2.1: $\operatorname{sgn}(\zeta_1) = \operatorname{sgn}(\zeta_2)$ with $\zeta_i := \alpha_i - \beta_i$ for $i = 1,2$

This case also implies $a_i = b_i$ for $i = 1,2$. Therefore, Equation (3.74) can be rewritten as

$$(\alpha_1 - \beta_1)(x_1 - a_1)^2 + (\alpha_2 - \beta_2)(x_2 - a_2)^2 = 0 \tag{3.75}$$

which is only fulfilled by the isolated point (a_1, a_2) (see Figure 3.22).

Case 1.2.2: $\operatorname{sgn}(\zeta_1) \neq \operatorname{sgn}(\zeta_2)$ with $\zeta_i := \alpha_i - \beta_i$ for $i = 1,2$

In this case Equation (3.74) is equivalent to

$$|\zeta_1|(x_1 - z_1)^2 - |\zeta_2|(x_2 - z_2)^2 = 0 \quad . \tag{3.76}$$

If we substitute $\sqrt{|\zeta_i|}(x_i - z_i)$ by $\tilde{x}_i$ for $i = 1,2$ we obtain the equation

$$\tilde{x}_1^2 - \tilde{x}_2^2 = 0 \tag{3.77}$$

which is equivalent to

$$(\tilde{x}_1 - \tilde{x}_2)(\tilde{x}_1 + \tilde{x}_2) = 0 \quad . \tag{3.78}$$

The solutions of Equation (3.78) are

$$\tilde{x}_1 = \tilde{x}_2 \ \vee \ \tilde{x}_1 = -\tilde{x}_2 \quad . \tag{3.79}$$

By re-substitution we obtain the equations

$$\sqrt{|\zeta_1|}(x_1 - z_1) = \sqrt{|\zeta_2|}(x_2 - z_2) \tag{3.80}$$

$$\sqrt{|\zeta_1|}(x_1 - z_1) = -\sqrt{|\zeta_2|}(x_2 - z_2) \tag{3.81}$$

which are equivalent to

$$x_2 = \sqrt{\frac{|\zeta_1|}{|\zeta_2|}}x_1 - \sqrt{\frac{|\zeta_1|}{|\zeta_2|}}z_1 + z_2 \tag{3.82}$$

$$x_2 = -\frac{\sqrt{|\zeta_1|}}{\sqrt{|\zeta_2|}}x_1 + \frac{\sqrt{|\zeta_1|}}{\sqrt{|\zeta_2|}}z_1 + z_2 \tag{3.83}$$

The equations (3.82) and (3.83) represent two lines with opposite slopes, which intersect at the point (z_1, z_2) (see Figure 3.23).

Case 2: $\alpha_1 \neq \beta_1 \wedge \alpha_2 = \beta_2$

In this case Equation (3.71) is equivalent to

$$(\alpha_1-\beta_1)x_1^2-2(\alpha_1 a_1-\beta_1 b_1)x_1-2\alpha_2(a_2-b_2)x_2 = \beta_1 b_1^2-\alpha_1 a_1^2+\alpha_2(b_2^2-a_2^2) \quad , \tag{3.84}$$

which requires an additional case analysis:

Case 2.1: $a_2 \neq b_2$

In this case we can solve Equation (3.84) with respect to x_2 which leads to

$$x_2 = \frac{(\alpha_1-\beta_1)x_1^2}{2\alpha_2(a_2-b_2)} - \frac{2(\alpha_1 a_1-\beta_1 b_1)x_1}{2\alpha_2(a_2-b_2)} - \frac{\beta_1 b_1^2-\alpha_1 a_1^2+\alpha_2(2b_2^2-a_2^2)}{2\alpha_2(a_2-b_2)} . \tag{3.85}$$

Using completion of the square, the right side of Equation (3.85) can be rewritten as

$$\begin{aligned}
&\frac{\alpha_1-\beta_1}{2\alpha_2(a_2-b_2)}\left(x_1^2 - 2\frac{\alpha_1 a_1-\beta_1 b_1}{\alpha_1-\beta_1}x_1 + \left(\frac{\alpha_1 a_1-\beta_1 b_1}{\alpha_1-\beta_1}\right)^2\right) \\
&-\frac{\beta_1 b_1^2-\alpha_1 a_1^2+\alpha_2(b_2^2-a_2^2)}{2\alpha_2(a_2-b_2)} - \frac{\alpha_1-\beta_1}{2\alpha_2(a_2-b_2)} \cdot \frac{(\alpha_1 a_1-\beta_1 b_1)^2}{(\alpha_1-\beta_1)^2} \\
=\ &\frac{\alpha_1-\beta_1}{2\alpha_2(a_2-b_2)}\left(x_1 - \frac{\alpha_1 a_1-\beta_1 b_1}{\alpha_1-\beta_1}\right)^2 \\
&-\frac{\alpha_1\beta_1(a_1-b_1)^2-(\alpha_1-\beta_1)\alpha_2(a_2^2-b_2^2)}{2\alpha_2(\alpha_1-\beta_1)(a_2-b_2)} \\
=\ &\frac{\alpha_1-\beta_1}{2\alpha_2(a_2-b_2)}\left(x_1 - \frac{\alpha_1 a_1-\beta_1 b_1}{\alpha_1-\beta_1}\right)^2 \\
&-\frac{\alpha_1\beta_1(a_1-b_1)^2}{2\alpha_2(\alpha_1-\beta_1)(a_2-b_2)} + \frac{a_2+b_2}{2} \quad ,
\end{aligned}$$

that means Equation (3.85) is equivalent to

$$\begin{aligned}
x_2 = &\underbrace{\frac{\alpha_1-\beta_1}{2\alpha_2(a_2-b_2)}}_{=:\zeta}\left(x_1 - \underbrace{\frac{\alpha_1 a_1-\beta_1 b_1}{\alpha_1-\beta_1}}_{=:z_1}\right)^2 \\
&\underbrace{-\frac{\alpha_1\beta_1(a_1-b_1)^2}{2\alpha_2(\alpha_1-\beta_1)(a_2-b_2)} + \frac{a_2+b_2}{2}}_{=:z_2} \quad .
\end{aligned} \tag{3.86}$$

Equation (3.86) describes a parabola with horizontal directrix (focal line) and apex (z_1, z_2), which is stretched (or compressed) by the factor ζ (see Figure 3.24).

Case 2.2 : $a_2 = b_2$

In this case Equation (3.84) is equivalent to the following quadratic equation in x_1 :

$$\underbrace{(\alpha_1 - \beta_1)}_{=:a} x_1^2 + \underbrace{2(\beta_1 b_1 - \alpha_1 a_1)}_{=:b} x_1 + \underbrace{(\alpha_1 a_1^2 - \beta_1 b_1^2)}_{=:c} = 0 \tag{3.87}$$

Since

$$\begin{aligned} b^2 - 4ac &= (2(\beta_1 b_1 - \alpha_1 a_1))^2 - 4(\alpha_1 - \beta_1)(\alpha_1 a_1^2 - \beta_1 b_1^2) \\ &= 4\alpha_1 \beta_1 (a_1 - b_1)^2 \\ &\geq 0 \end{aligned}$$

Equation (3.87) has at least one solution and we can distinguish two cases :

Case 2.2.1 : $a_1 \neq b_1$

In this case we have $b^2 - 4ac = 4\alpha_1\beta_1(a_1 - b_1)^2 > 0$, which means Equation (3.87) has the two solutions

$$\begin{aligned} x_1 &= \frac{-b + \sqrt{b^2 - 4ac}}{2a} = \frac{-2(\beta_1 b_1 - \alpha_1 a_1) + \sqrt{4\alpha_1\beta_1(a_1 - b_1)^2}}{2(\alpha_1 - \beta_1)} \\ &= \frac{\alpha_1 a_1 - \beta_1 b_1 + \sqrt{\alpha_1\beta_1}|a_1 - b_1|}{\alpha_1 - \beta_1} \end{aligned} \tag{3.88}$$

and

$$\begin{aligned} \tilde{x}_1 &= \frac{-b - \sqrt{b^2 - 4ac}}{2a} = \frac{-2(\beta_1 b_1 - \alpha_1 a_1) - \sqrt{4\alpha_1\beta_1(a_1 - b_1)^2}}{2(\alpha_1 - \beta_1)} \\ &= \frac{\alpha_1 a_1 - \beta_1 b_1 - \sqrt{\alpha_1\beta_1}|a_1 - b_1|}{\alpha_1 - \beta_1} \end{aligned} \tag{3.89}$$

which represent vertical lines (see Figure 3.25).

Case 2.2.2 : $a_1 = b_1$

In this case we have $b^2 - 4ac = 4\alpha_1\beta_1(a_1 - b_1)^2 = 0$, which means

$$x_1 = \frac{-b}{2a} = \frac{-2(\beta_1 b_1 - \alpha_1 a_1)}{2(\alpha_1 - \beta_1)} = \frac{(\alpha_1 - \beta_1)a_1}{\alpha_1 - \beta_1} = a_1 \tag{3.90}$$

is the uniquely defined solution of Equation (3.87) and represents a vertical line (see Figure 3.26).

Case 3 : $\alpha_1 = \beta_1 \wedge \alpha_2 \neq \beta_2$

Analogously to **Case 2** with reversed roles of the indices.

Case 4: $\alpha_1 = \beta_1 \ \wedge \ \alpha_2 = \beta_2$

In this case Equation (3.71) is equivalent to

$$-2 \sum_{i=1,2} (\alpha_i a_i - \beta_i b_i) x_i = \sum_{i=1,2} \beta_i b_i^2 - \alpha_i a_i^2 \tag{3.91}$$

or

$$2\alpha_1(a_1 - b_1)x_1 + 2\alpha_2(a_2 - b_2)x_2 = \alpha_1(a_1^2 - b_1^2) + \alpha_2(a_2^2 - b_2^2) \quad . \tag{3.92}$$

Depending on a_1, a_2, b_1, b_2 we can distinguish the following four cases:
Case 4.1: $a_1 \neq b_1 \ \wedge \ a_2 \neq b_2$

This case implies

$$x_2 = -\frac{\alpha_1(a_1 - b_1)}{\alpha_2(a_2 - b_2)} x_1 + \frac{\alpha_1(a_1^2 - b_1^2) + \alpha_2(a_2^2 - b_2^2)}{2\alpha_2(a_2 - b_2)} \quad . \tag{3.93}$$

Equation (3.93) describes a line, which is neither horizontal nor vertical (see Figure 3.27).
Case 4.2: $a_1 \neq b_1 \ \wedge \ a_2 = b_2$

This case implies

$$x_1 = \frac{a_1 + b_1}{2} \quad . \tag{3.94}$$

Equation (3.94) describes a vertical line (see Figure 3.28).
Case 4.3: $a_1 = b_1 \ \wedge \ a_2 \neq b_2$

This case implies

$$x_2 = \frac{a_2 + b_2}{2} \quad . \tag{3.95}$$

Equation (3.95) describes a horizontal line (see Figure 3.29).
Case 4.4: $a_1 = b_1 \ \wedge \ a_2 = b_2$

This case implies

$$0x_1 + 0x_2 = 0 \tag{3.96}$$

which is fulfilled for all $x \in \mathbb{R}^2$ (see Figure 3.30).

Table 3.1 summarizes the case analysis.

Finally, to illustrate the above case analysis we classify the well-known situation of two unit balls given by circles (Euclidean case). The consideration of two Euclidean circles means $\alpha_1 = \alpha_2$, $\beta_1 = \beta_2$ (or $\gamma_A = \sqrt{\alpha_1} l_2$ and $\gamma_A = \sqrt{\beta_1} l_2$) and therefore $\zeta_1 = \zeta_2$, i.e. **Case 1.1.2**, **Case 1.2.2**, **Case 2** and **Case 3** can not occur. Because of $\zeta_1 = \zeta_2$ the ellipse of **Case 1.1.1** describes the Apollonius' circle. Since $\alpha_1 = \alpha_2$ Equation (3.92) can be written as $\langle a - b, x - \frac{a+b}{2} \rangle$ and describes the perpendicular line to the segment $\overline{ab}$ at

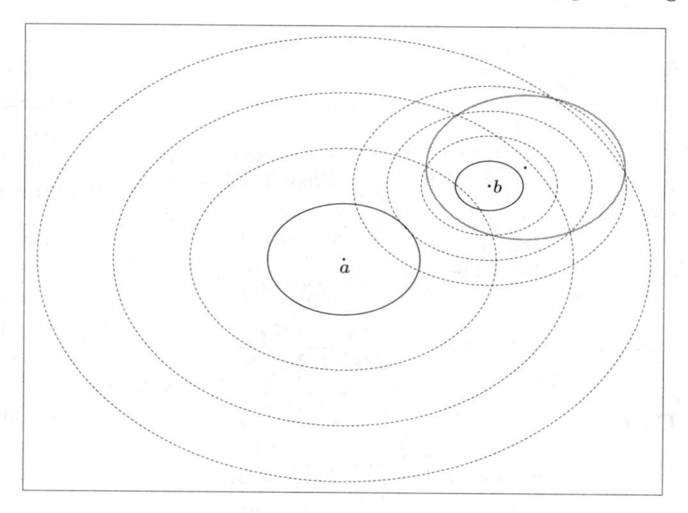

Fig. 3.20. Illustration to Case 1.1.1

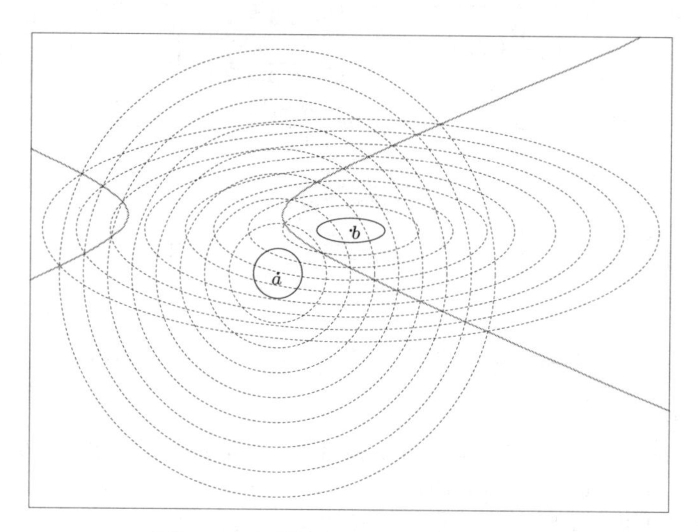

Fig. 3.21. Illustration to Case 1.1.2

Table 3.1. Summary of the case analysis.

<table>
<tr><th colspan="3">Case analysis</th><th>Bisector</th></tr>
<tr><td rowspan="4">Case 1
$\alpha_1 \neq \beta_1 \ \wedge \ \alpha_2 \neq \beta_2$</td><td rowspan="2">Case 1.1
$\zeta \neq 0$</td><td>Case 1.1.1
$\mathrm{sgn}(\zeta_1) = \mathrm{sgn}(\zeta_2)$</td><td>Ellipse</td></tr>
<tr><td>Case 1.1.2
$\mathrm{sgn}(\zeta_1) \neq \mathrm{sgn}(\zeta_2)$</td><td>Hyperbola</td></tr>
<tr><td rowspan="2">Case 1.2
$\zeta = 0$</td><td>Case 1.2.1
$\mathrm{sgn}(\zeta_1) = \mathrm{sgn}(\zeta_2)$</td><td>Isolated Point</td></tr>
<tr><td>Case 1.2.2
$\mathrm{sgn}(\zeta_1) \neq \mathrm{sgn}(\zeta_2)$</td><td>2 lines with opposite slopes</td></tr>
<tr><td rowspan="3">Case 2
$\alpha_1 \neq \beta_1 \ \wedge \ \alpha_2 = \beta_2$</td><td colspan="2">Case 2.1
$a_2 \neq b_2$</td><td>Parabola with horizontal directrix</td></tr>
<tr><td rowspan="2">Case 2.2
$a_2 = b_2$</td><td>Case 2.2.1
$a_1 \neq b_1$</td><td>2 vertical lines</td></tr>
<tr><td>Case 2.2.2
$a_1 = b_1$</td><td>1 vertical line</td></tr>
<tr><td rowspan="3">Case 3
$\alpha_1 = \beta_1 \ \wedge \ \alpha_2 \neq \beta_2$</td><td colspan="2">Case 3.1
$a_1 \neq b_1$</td><td>Parabola with vertical directrix</td></tr>
<tr><td rowspan="2">Case 3.2
$a_1 = b_1$</td><td>Case 3.2.1
$a_2 \neq b_2$</td><td>2 horizontal lines</td></tr>
<tr><td>Case 3.2.2
$a_2 = b_2$</td><td>1 horizontal line</td></tr>
<tr><td rowspan="4">Case 4
$\alpha_1 = \beta_1 \ \wedge \ \alpha_2 = \beta_2$</td><td colspan="2">Case 4.1
$a_1 \neq b_1 \ \wedge \ a_2 \neq b_2$</td><td>1 line (neither horizontal nor vertical)</td></tr>
<tr><td colspan="2">Case 4.2
$a_1 \neq b_1 \ \wedge \ a_2 = b_2$</td><td>1 vertical line</td></tr>
<tr><td colspan="2">Case 4.3
$a_1 = b_1 \ \wedge \ a_2 \neq b_2$</td><td>1 horizontal line</td></tr>
<tr><td colspan="2">Case 4.4
$a_1 = b_1 \ \wedge \ a_2 = b_2$</td><td>Plane $\mathbb{R}^2$</td></tr>
</table>

the midpoint $\frac{a+b}{2}$ in **Case 4.1**, **Case 4.2** and **Case 4.3**. **Case 1.2.1** and **Case 4.4** include obviously the situation of two Euclidean circles with the same midpoint.

Moreover, we can see that the right-most column of Table 3.1 contains all regular (or non-degenerate) conic sections, i. e.

ellipse, hyperbola, parabola,

as well as all singular (or degenerate) conic sections, i. e.

two lines, pair of parallel lines (or double line), isolated point.

Hence the computation of bisectors for elliptic gauges can be used as an alternative way to introduce in the theory of conic sections. The classical way

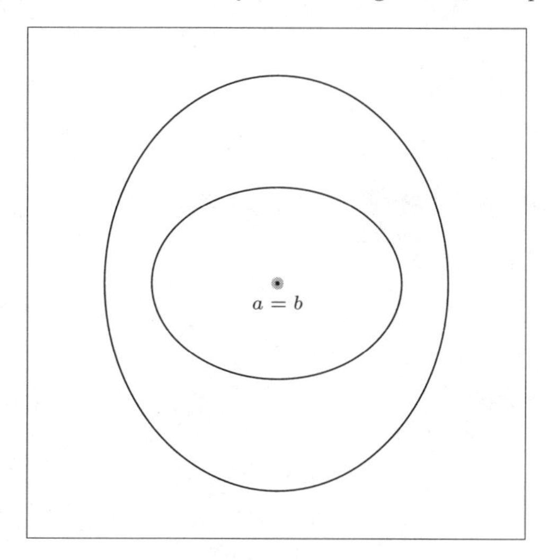

Fig. 3.22. Illustration to Case 1.2.1

to introduce the conic sections is the intersection of a double cone K with a hyper-plane H (see [125, 24]). Depending on

1. the vertex angle δ of the double cone K,
2. the gradient angle ϱ of the hyper-plane H,
3. the location of the vertex S with respect to the hyper-plane H

the different conic sections can be classified (see Figure 3.31). As we have seen here, computing the bisector for elliptic gauges, we obtain the different conic sections depending on

1. (half) the length of the major and minor axis of the ellipse A,
2. (half) the length of the major and minor axis of the ellipse B,
3. the location of the sites a and b with respect to each other.

Notice that the epigraphs of the *two* gauges are "cones" with ellipses (instead of circles) as horizontal cross-sections (see Figure 3.32). Therefore we have in some sense the relationship to the classical *double* cone (with intersecting hyper-plane).

3.6 Bisectors of a Polyhedral Gauge and an Elliptic Gauge

In this section we study the behavior of a bisector generated by a polyhedral gauge and an elliptic gauge.

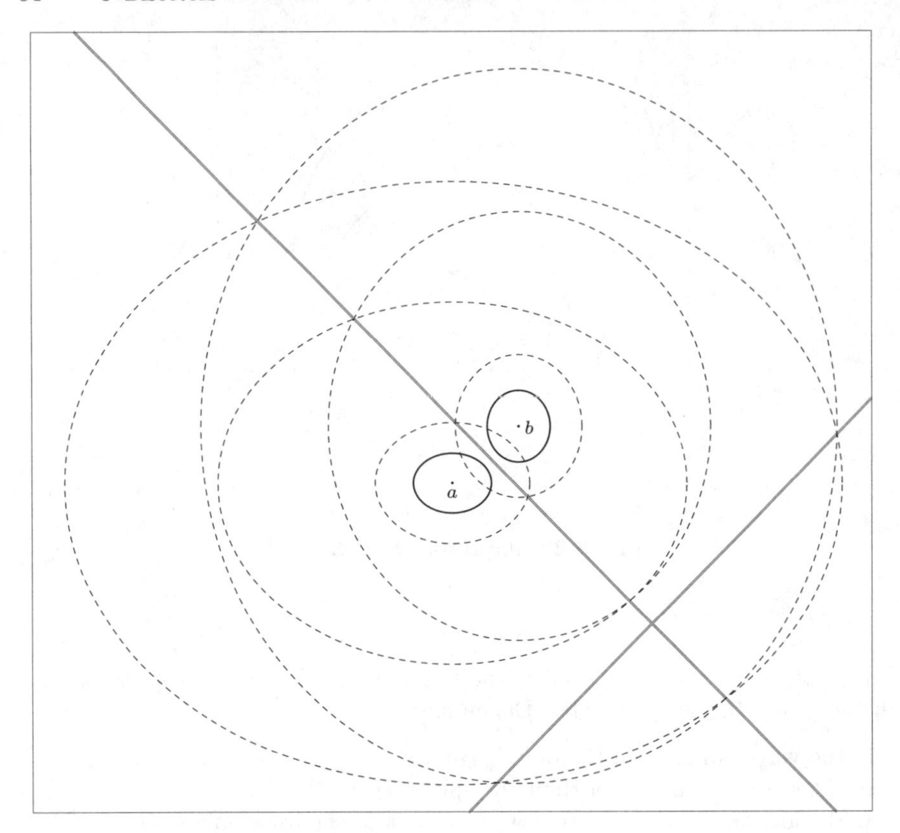

Fig. 3.23. Illustration to Case 1.2.2

Let site $a \in \mathbb{R}^2$ be associated with a polyhedral gauge γ_A. The translated extreme points of unit ball A define a partition of the plane in convex cones $\tilde{\Gamma}_i^a, i \in \mathcal{K}$, pointed at a as described in Sections 2.2 and 3.3. γ_A behaves linearly in each cone $\tilde{\Gamma}_i^a, i \in \mathcal{K}$. Hence we can assume that γ_A in a fixed cone $\tilde{\Gamma}^a \in \left\{ \tilde{\Gamma}_i^a : i \in \mathcal{K} \right\}$ is given by

$$\gamma_A(x) = s_1 x_1 + s_2 x_2 \tag{3.97}$$

where s_1, s_2 are real numbers with $s_1 \neq 0$ or $s_2 \neq 0$.

Let the elliptic gauge γ_B associated with site $b \in \mathbb{R}^2$ be given by

$$\gamma_B(x) = \sqrt{\xi_1 x_1^2 + \xi_2 x_2^2} \tag{3.98}$$

where ξ_1, ξ_2 are positive real numbers (see page 75).

The boundary of the polyhedral unit ball A and the elliptic unit ball B intersect at most $2K$ times. Therefore, the bisector $\mathsf{Bisec}(a, b)$ will contain not more than K connected components by Lemma 3.9.

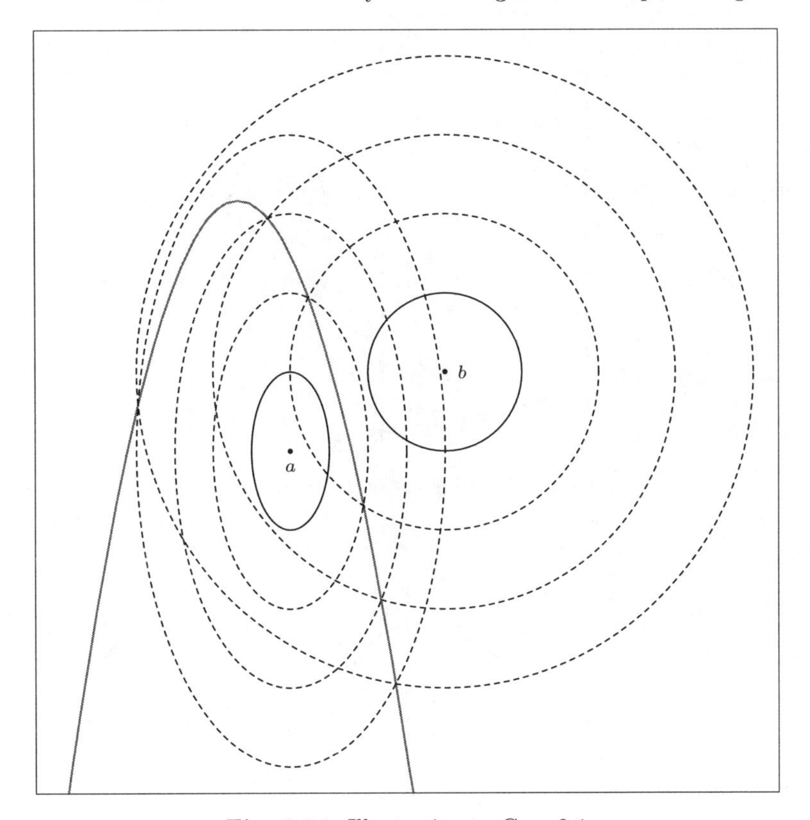

Fig. 3.24. Illustration to Case 2.1

If we can determine the bisector $\mathsf{Bisec}(a,b)$ in cone $\tilde{\Gamma}^a$, we can determine the whole bisector $\mathsf{Bisec}(a,b)$ in K steps.

To determine the bisector in cone $\tilde{\Gamma}^a$ we have to solve the equation

$$\gamma_A(x-a) = \gamma_B(x-b) \tag{3.99}$$

under the side constraint $x \in \tilde{\Gamma}^a$, which is equivalent to

$$s_1(x_1-a_1)+s_2(x_2-a_2) \;=\; \sqrt{\xi_1(x_1-b_1)^2+\xi_2(x_2-b_2)^2}\quad . \tag{3.100}$$

Squaring both sides of Equation (3.100) leads to

$$\begin{aligned} s_1^2(x_1-a_1)^2+2s_1s_2(x_1-a_1)(x_2-a_2)+s_2^2(x_2-a_2)^2 \\ = \xi_1(x_1-b_1)^2+\xi_2(x_2-b_2)^2 \end{aligned} \tag{3.101}$$

which is equivalent to

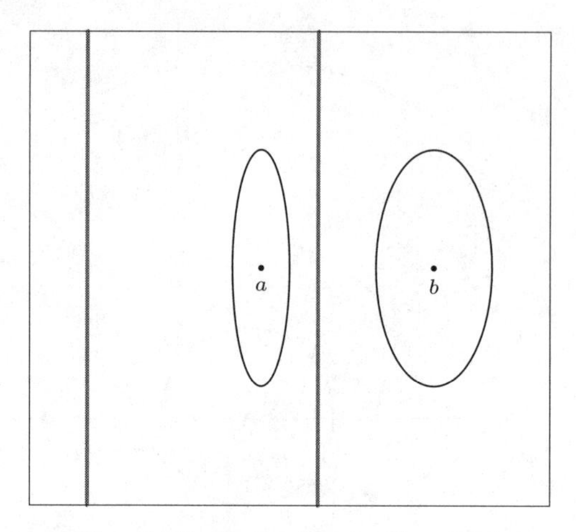

Fig. 3.25. Illustration to Case 2.2.1

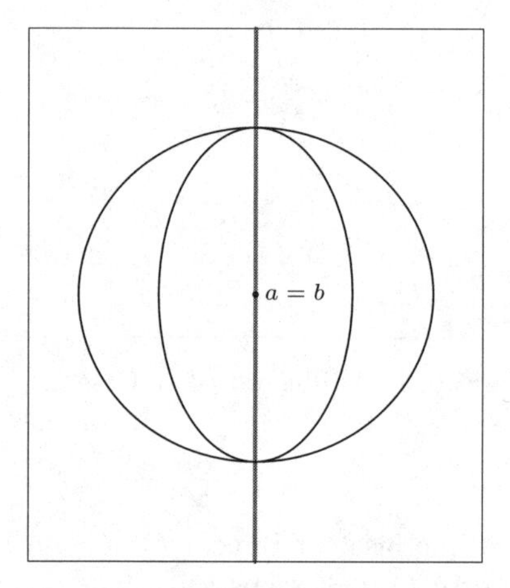

Fig. 3.26. Illustration to Case 2.2.2

$$\sum_{i=1,2} s_i^2(x_i^2 - 2a_ix_i + a_i^2) \; + \; 2s_1s_2(x_1x_2 - a_1x_2 - a_2x_1 + a_1a_2)$$
$$= \sum_{i=1,2} \xi_i(x_i^2 - 2b_ix_i + b_i^2) \, . \quad (3.102)$$

Sorting Equation (3.102) with respect to the variables x_1 and x_2 leads to

$$\sum_{i=1,2} (\xi_i - s_i^2)x_i^2 + 2(s_i^2a_i + s_1s_2a_j - \xi_ib_i)x_i \; - \; 2s_1s_2x_1x_2$$
$$= \sum_{i=1,2} s_i^2a_i^2 - \xi_ib_i^2 \; + \; 2s_1s_2a_1a_2 \quad (3.103)$$

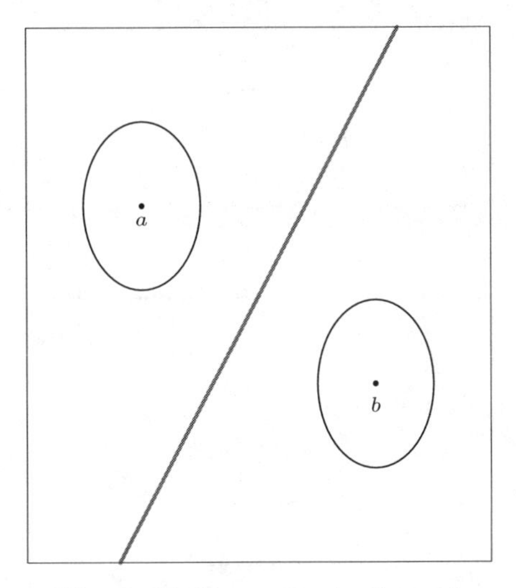

Fig. 3.27. Illustration to Case 4.1

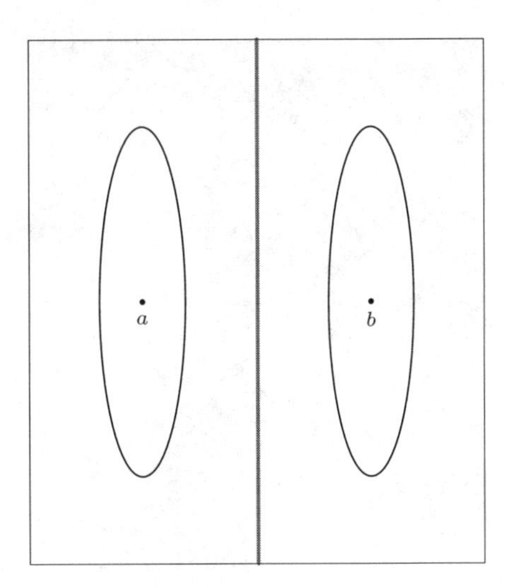

Fig. 3.28. Illustration to Case 4.2

with $j := 1 + (i \bmod 2)$, which is equivalent to

$$\sum_{i=1,2} \eta_i x_i^2 + 2(s_i^2 a_i + s_1 s_2 a_j - \xi_i b_i) x_i \; - \; 2 s_1 s_2 x_1 x_2$$
$$+ \sum_{i=1,2} \xi_i b_i^2 \; - \; (s_1 a_1 + s_2 a_2)^2 \; = \; 0 \quad (3.104)$$

with $\eta_i := \xi_i - s_i^2$ for $i = 1, 2$. Equation (3.104) can be rewritten as

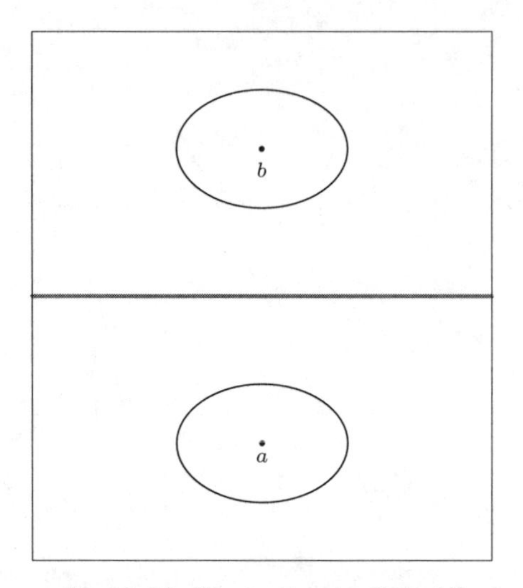

Fig. 3.29. Illustration to Case 4.3

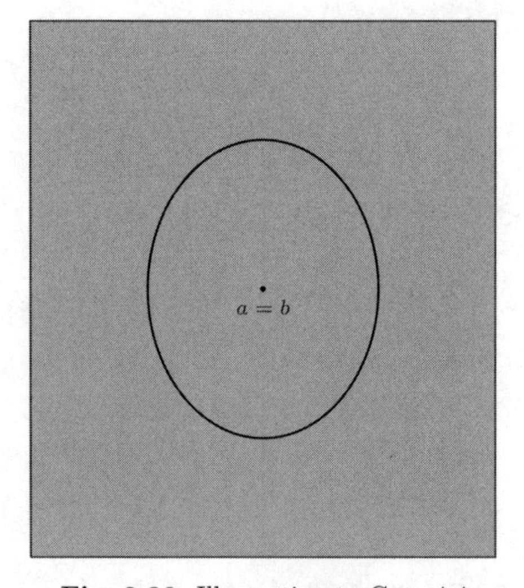

Fig. 3.30. Illustration to Case 4.4

$$x^\top Sx + 2r^\top x + q = 0 \tag{3.105}$$

with

$$S = \begin{pmatrix} \eta_1 & -s_1 s_2 \\ -s_1 s_2 & \eta_2 \end{pmatrix}, \ r = \begin{pmatrix} s_1^2 a_1 + s_1 s_2 a_2 - \xi_1 b_1 \\ s_2^2 a_2 + s_1 s_2 a_1 - \xi_2 b_2 \end{pmatrix},$$
$$q = \sum_{i=1,2} \xi_i b_i^2 - (s_1 a_1 + s_2 a_2)^2 . \tag{3.106}$$

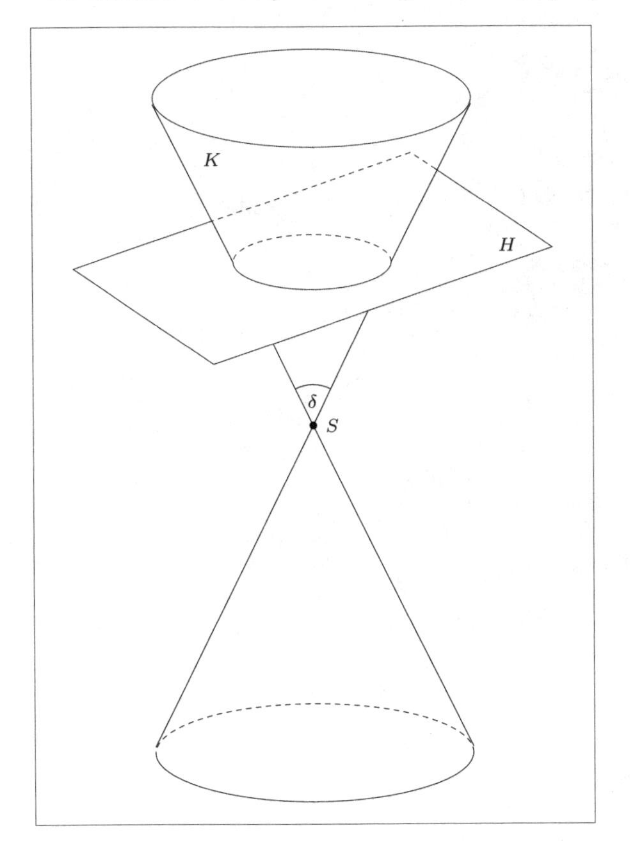

Fig. 3.31. Intersection of a *double* cone with a hyperplane

To avoid the mixed quadratic term $-2s_1s_2x_1x_2$ in Equation (3.105) a reduction to normal form (principal-axis transformation) is necessary. The principal-axis transformation is based on the following well-known theorem of linear algebra (see for example [75]).

Theorem 3.3
For a symmetric matrix S exists a orthogonal matrix P, such that $P^\top SP =:$ D is a diagonal matrix. The diagonal elements of D are the (always real) eigenvalues λ_i of S (with the corresponding multiplicity) and the columns of P are the corresponding normed eigenvectors v_i.

By setting $x := Py$ we obtain the normal form

$$\begin{aligned} 0 &= (Py)^\top SPy + 2r^\top Py + q \\ &= yP^\top SPy + 2(P^\top r)^\top y + q \end{aligned}$$

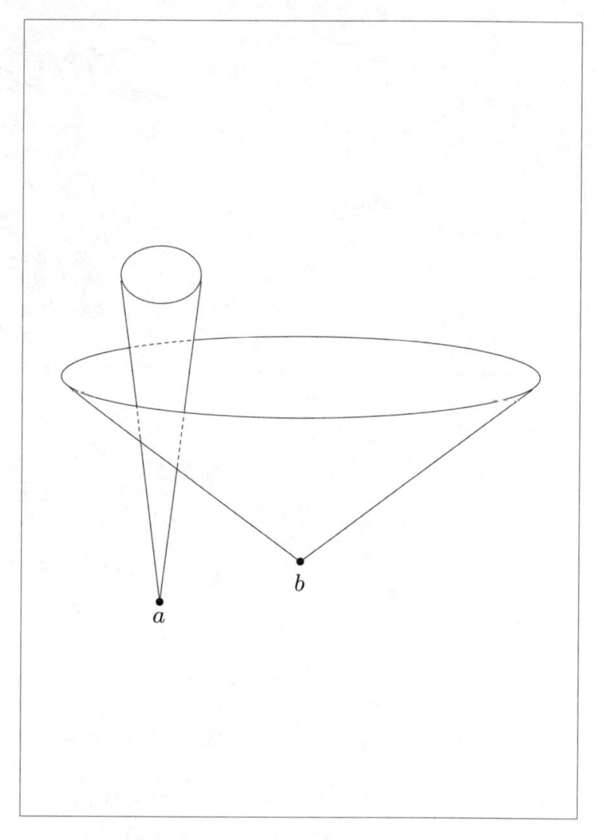

Fig. 3.32. Intersection of the epigraphs of *two* elliptic gauges

$$
\begin{aligned}
&= y^\top D y + 2 \begin{pmatrix} v_1^\top r \\ v_2^\top r \end{pmatrix}^\top y + q \\
&= \lambda_1 y_1^2 + \lambda_2 y_2^2 + 2 v_1^\top r y_1 + 2 v_2^\top r y_2 + q \quad ,
\end{aligned}
\tag{3.107}
$$

which is an algebraic equation of second degree. Consequently the bisector $\mathsf{Bisec}(a,b) \cap \tilde{\Gamma}^a$ will be part of a conic section in the $y_1 y_2$-coordinate system (compare with Equation (3.71) in Section 3.5). We will investigate later how to decide in advance which conic section, depending on s_1, s_2, ξ_1, ξ_2, is created.

To determine the eigenvalues of D we have to solve the quadratic equation

$$
\begin{aligned}
0 &= \det(S - \lambda I) \\
&= \det \begin{pmatrix} \eta_1 - \lambda & -s_1 s_2 \\ -s_1 s_2 & \eta_2 - \lambda \end{pmatrix} \\
&= (\eta_1 - \lambda)(\eta_2 - \lambda) - s_1^2 s_2^2
\end{aligned}
$$

$$= \lambda^2 - (\eta_1 + \eta_2)\lambda + \eta_1\eta_2 - s_1^2 s_2^2 \quad . \tag{3.108}$$

The discriminant is given by

$$\begin{aligned} d &= (\eta_1 + \eta_2)^2 - 4(\eta_1\eta_2 - s_1^2 s_2^2) \\ &= \eta_1^2 + 2\eta_1\eta_2 + \eta_2^2 - 4\eta_1\eta_2 + 4s_1^2 s_2^2 \\ &= (\eta_1 - \eta_2)^2 + 4s_1^2 s_2^2 \ \geq \ 0 \end{aligned} \tag{3.109}$$

and hence the eigenvalues by

$$\lambda_1 = \frac{\eta_1 + \eta_2 - \sqrt{d}}{2} \text{ and } \lambda_2 = \frac{\eta_1 + \eta_2 + \sqrt{d}}{2} \quad . \tag{3.110}$$

For the determination of the corresponding normed eigenvectors, respectively P, a case analysis depending on s_1, s_2, η_1, η_2 is necessary.

Case 1: $s_1 s_2 = 0$, i.e. $s_1 = 0 \ \vee \ s_2 = 0$

This case implies $d = (\eta_1 - \eta_2)^2$ and therefore

$\lambda_1 = \frac{\eta_1+\eta_2-|\eta_1-\eta_2|}{2} = \min\{\eta_1, \eta_2\}$ and $\lambda_2 = \frac{\eta_1+\eta_2+|\eta_1-\eta_2|}{2} = \max\{\eta_1, \eta_2\}$.

Case 1.1: $\eta_1 \leq \eta_2$

That means $\lambda_1 = \eta_1$ and $\lambda_2 = \eta_2$. The (identity) matrix $P = \left(\begin{smallmatrix} 1 & 0 \\ 0 & 1 \end{smallmatrix}\right)$ is orthogonal and satisfies $P^\top SP = S = \left(\begin{smallmatrix} \eta_1 & 0 \\ 0 & \eta_2 \end{smallmatrix}\right) = \left(\begin{smallmatrix} \lambda_1 & 0 \\ 0 & \lambda_2 \end{smallmatrix}\right) = D$. Notice that this case does not describe a real principal-axis transformation, since P is the identity matrix.

Case 1.2: $\eta_1 > \eta_2$

That means $\lambda_1 = \eta_2$ and $\lambda_2 = \eta_1$. The matrix $P = \left(\begin{smallmatrix} 0 & 1 \\ 1 & 0 \end{smallmatrix}\right)$ is orthogonal and satisfies $P^\top SP = \left(\begin{smallmatrix} \eta_2 & 0 \\ 0 & \eta_1 \end{smallmatrix}\right) = \left(\begin{smallmatrix} \lambda_1 & 0 \\ 0 & \lambda_2 \end{smallmatrix}\right) = D$. Notice that this case only describes an exchange of the coordinate axes, since $P = \left(\begin{smallmatrix} 0 & 1 \\ 1 & 0 \end{smallmatrix}\right)$.

Case 2: $s_1 s_2 \neq 0$, i.e. $s_1 \neq 0 \ \wedge \ s_2 \neq 0$

This case implies $\sqrt{d} = \sqrt{(\eta_1 - \eta_2)^2 + 4s_1^2 s_2^2} > |\eta_1 - \eta_2|$ and therefore $d \pm \sqrt{d}(\eta_1 - \eta_2) > 0$. Hence the following normed eigenvectors

$$v_1 = \frac{1}{\sqrt{2(d+\sqrt{d}(\eta_1-\eta_2))}} \begin{pmatrix} 2s_1 s_2 \\ \eta_1 - \eta_2 + \sqrt{d} \end{pmatrix} \quad ,$$

$$v_2 = \frac{1}{\sqrt{2(d-\sqrt{d}(\eta_1-\eta_2))}} \begin{pmatrix} 2s_1 s_2 \\ \eta_1 - \eta_2 - \sqrt{d} \end{pmatrix} \quad ,$$

are well defined. The verification of the orthogonality of $P = \left(\begin{smallmatrix} v_{11} & v_{12} \\ v_{21} & v_{22} \end{smallmatrix}\right)$ and of $P^\top SP = D$ is quite lengthy, but straightforward. Therefore, we drop it here.

Now we investigate, which conic section describes Equation (3.107). This depends on the slope of γ_B in comparison to the slope of γ_A in direction of the steepest ascent of γ_B. The steepest ascent of the linear function γ_A is

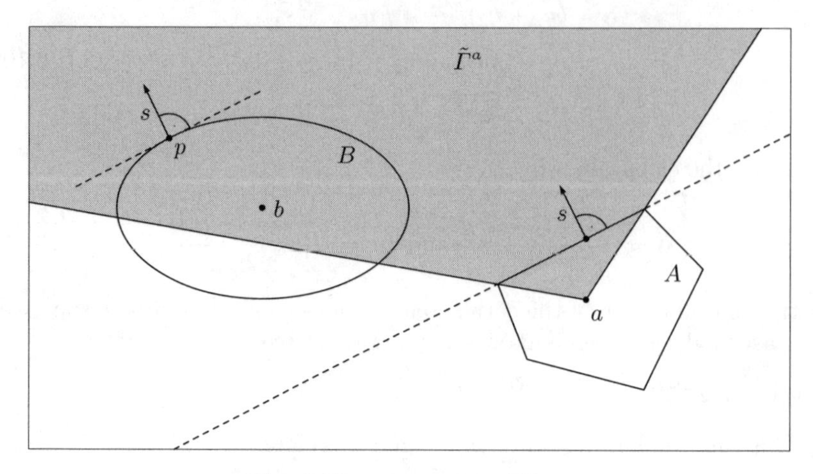

Fig. 3.33. s tangent to B in p

in the direction of its gradient $\nabla\gamma_A = (s_1, s_2)$ and the slope is given by $\frac{\partial}{\partial_s}\gamma_A = \langle s, \frac{s}{l_2(s)}\rangle = l_2(s)$. The gradient s of γ_A is normal to

$$B = \left\{x \in \mathbb{R}^2 \,:\, \sqrt{\xi_1 x_1^2 + \xi_2 x_2^2} = 1\right\} \quad \text{at} \quad p = \frac{1}{\sqrt{\frac{s_1^2}{\xi_1} + \frac{s_2^2}{\xi_2}}}\left(\frac{s_1}{\xi_1}, \frac{s_2}{\xi_2}\right), \tag{3.111}$$

since the tangential line to B at a point $p \in \mathrm{bd}(B)$ is given by

$$p_1\xi_1 x_1 + p_2\xi_2 x_2 = 1\,. \tag{3.112}$$

(see Figure 3.33). The gradient of γ_B is given by

$$\nabla\gamma_B(x) = \left(\frac{\xi_1 x_1}{\gamma_B(x)}, \frac{\xi_2 x_2}{\gamma_B(x)}\right) \quad \text{for} \quad x \in \mathbb{R}^2 \setminus \{0\} \tag{3.113}$$

and therefore the directional derivative of γ_B at p in the direction of s is given by

$$\frac{\partial}{\partial_s}\gamma_B(p) = \langle\nabla\gamma_B(p), \frac{s}{l_2(s)}\rangle = \frac{l_2(s)}{\sqrt{\frac{s_1^2}{\xi_1} + \frac{s_2^2}{\xi_2}}} \tag{3.114}$$

To decide on the kind of the conic section, we compare $\frac{\partial}{\partial_s}\gamma_A$ with $\frac{\partial}{\partial_s}\gamma_B(p)$. We have that

$$\frac{\partial}{\partial_s}\gamma_A > \frac{\partial}{\partial_s}\gamma_B(p) \quad\Leftrightarrow\quad {}_2(s) > \frac{l_2(s)}{\sqrt{\frac{s_1^2}{\xi_1}+\frac{s_2^2}{\xi_2}}} \quad\Leftrightarrow\quad \sqrt{\frac{s_1^2}{\xi_1}+\frac{s_2^2}{\xi_2}} > 1$$

$$\Leftrightarrow\quad \frac{s_1^2}{\xi_1}+\frac{s_2^2}{\xi_2} > 1 \quad . \tag{3.115}$$

Moreover, the location of the sites a and b with respect to s determines whether the conic section is degenerated or not. Hence we can distinguish depending on $\varphi(s_1,s_2,\xi_1,\xi_2) := \frac{s_1^2}{\xi_1}+\frac{s_2^2}{\xi_2}$ and $\psi(a,b,s) := \langle a-b, s\rangle$ the following cases:

Case 1: $\varphi(s_1,s_2,\xi_1,\xi_2) < 1$
  Case 1.1: $\psi(a,b,s) \neq 0$: $\mathsf{Bisec}(a,b) \cap \tilde{\Gamma}^a \subseteq$ Ellipse.
  Case 1.2: $\psi(a,b,s) = 0$: $\mathsf{Bisec}(a,b) \cap \tilde{\Gamma}^a \subseteq$ Isolated point.
Case 2: $\varphi(s_1,s_2,\xi_1,\xi_2) = 1$
  Case 2.1: $\psi(a,b,s) \neq 0$: $\mathsf{Bisec}(a,b) \cap \tilde{\Gamma}^a \subseteq$ Parabola.
  Case 2.2: $\psi(a,b,s) = 0$: $\mathsf{Bisec}(a,b) \cap \tilde{\Gamma}^a \subseteq$ Line.
Case 3: $\varphi(s_1,s_2,\xi_1,\xi_2) > 1$
  Case 3.1: $\psi(a,b,s) \neq 0$: $\mathsf{Bisec}(a,b) \cap \tilde{\Gamma}^a \subseteq$ Hyperbola.
  Case 3.2: $\psi(a,b,s) = 0$: $\mathsf{Bisec}(a,b) \cap \tilde{\Gamma}^a \subseteq$ Two lines.

We illustrate the above discussion with the following example.

Example 3.3
Consider the sites $a = (0,0)$ and $b = (6,0)$ in the plane. Assume the unit ball of site a is defined by the extreme points $(2,-2)$, $(2,1)$, $(\frac{1}{2},1)$, $(-\frac{1}{2},0)$ and $(-\frac{1}{2},-2)$, whereas site b is associated with the Euclidean unit ball (see Figure 3.34). The input data implies $\xi_1 = 1$, $\xi_2 = 1$, $\eta_1 = 1-s_1^2$, $\eta_2 = 1-s_2^2$, $S = \begin{pmatrix} 1-s_1^2 & -s_1s_2 \\ -s_1s_2 & 1-s_2^2\end{pmatrix}$, $r = \begin{pmatrix} -6 \\ 0\end{pmatrix}$ and $q = 36$. Moreover we obtain $\varphi(s_1,s_2,\xi_1,\xi_2) = s_1^2+s_2^2$ and $\psi(a,b,c) = -6s_1$.
The bisector is computed in five steps.

1. *$\tilde{\Gamma}_1^a$ defined by $\tilde{e}_1^a = (2,-2)$ and $\tilde{e}_2^a = (2,1)$:*

   $s_1 = \frac{1}{2}$ and $s_2 = 0$ imply $\eta_1 = \frac{3}{4}$ and $\eta_2 = 1$. Therefore we obtain $d = \frac{1}{16}$ and the eigenvalues $\lambda_1 = \frac{3}{4}$, $\lambda_2 = 1$. Since $s_2 = 0$ and $\eta_1 < \eta_2$, we have **Case 1.1** *with $P = \begin{pmatrix} 1 & 0 \\ 0 & 1\end{pmatrix}$ and obtain (without a principal-axis transformation)*

$$\tfrac{3}{4}x_1^2 + x_2^2 - 12x_1 + 36 = 0 \quad ,$$

   which is equivalent to the ellipse (Notice $\varphi(s_1,s_2,\xi_1,\xi_2) = \frac{1}{4} < 1$, $\psi(a,b,s) = -3 \neq 0$.)

$$\frac{1}{4^2}(x_1-8)^2 + \frac{1}{(2\sqrt{3})^2}x_2^2 = 1 \quad .$$

   The side constraints are $x_2 \geq -x_1 \wedge x_2 \leq \frac{1}{2}x_1$ or equivalently $-x_1 \leq x_2 \leq \frac{1}{2}x_1$.

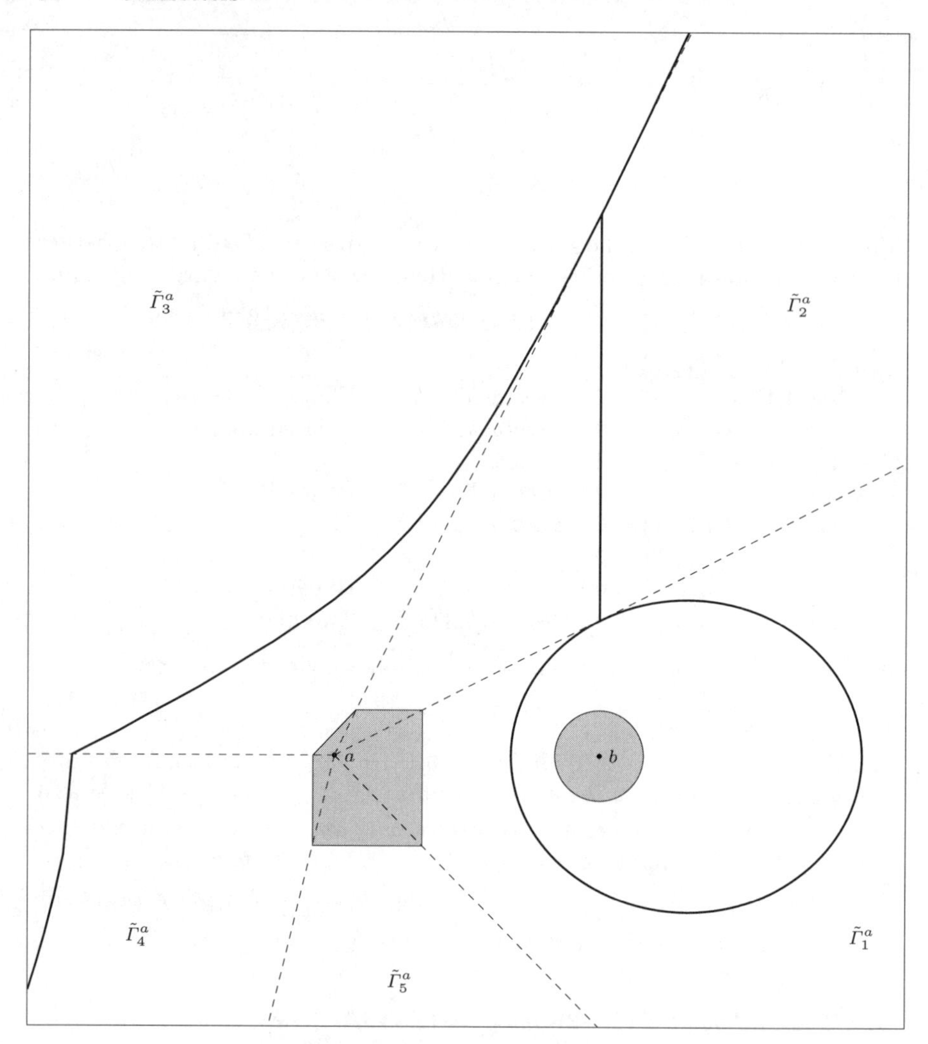

Fig. 3.34. Illustration to Example 3.3

A simple computation shows that the line $x_2 = -x_1$ *does not intersect the ellipse, whereas the line* $x_2 = \frac{1}{2}x_1$ *touches the ellipse in* $(6,3)$. *Moreover the point* $(4,0) \in \tilde{\Gamma}_1^a$ *is also a point of the ellipse, therefore the ellipse is totally included in* $\tilde{\Gamma}_1^a$.

2. $\tilde{\Gamma}_2^a$ *defined by* $\tilde{e}_2^a = (2,1)$ *and* $\tilde{e}_3^a = \left(\frac{1}{2},1\right)$:

$s_1 = 0$ *and* $s_2 = 1$ *imply* $\eta_1 = 1$ *and* $\eta_2 = 0$. *Therefore we obtain* $d = 1$ *and the eigenvalues* $\lambda_1 = 0$, $\lambda_2 = 1$. *Since* $s_1 = 0$ *and* $\eta_1 > \eta_2$, *we have* **Case 1.2** *with* $P = \left(\begin{smallmatrix} 0 & 1 \\ 1 & 0 \end{smallmatrix}\right)$ *and obtain*

$$y_2^2 - 12y_2 + 36 = 0 \quad ,$$

which is equivalent to

$$(y_2 - 6)^2 = 0$$

and therefore only fulfilled by the line (Notice $\varphi(s_1, s_2, \xi_1, \xi_2) = 1$, $\psi(a, b, s) = 0$.*)*

$$y_2 = 6 \qquad or \qquad x_1 = 6 \quad .$$

The side constraints $x_2 \geq \frac{1}{2}x_1 \wedge x_2 \leq 2x_1$ *imply* $3 \leq x_2 \leq 12$. *Therefore, in* $\tilde{\Gamma}_2^a$ *the bisector is given by a vertical line segment from* $(6, 3)$ *to* $(6, 12)$ *in the* x_1x_2*-coordinate system.*

3. $\tilde{\Gamma}_3^a$ *defined by* $\tilde{e}_3^a = \left(\frac{1}{2}, 1\right)$ *and* $\tilde{e}_4^a = \left(-\frac{1}{2}, 0\right)$ *:*

   $s_1 = -2$ *and* $s_2 = 2$ *imply* $\eta_1 = -3$ *and* $\eta_2 = -3$. *Therefore we obtain* $d = 64$ *and the eigenvalues* $\lambda_1 = -7$, $\lambda_2 = 1$. *Since* $s_1 \neq 0$ *and* $s_2 \neq 0$, *we have* **Case 2** *with* $P = \frac{\sqrt{2}}{2}\begin{pmatrix} +1 & +1 \\ -1 & +1 \end{pmatrix}$ *and obtain*

$$-7y_1^2 + y_2^2 - 6\sqrt{2}y_1 - 6\sqrt{2}y_2 + 36 = 0 \quad ,$$

   which is equivalent to the hyperbola (Notice $\varphi(s_1, s_2, \xi_1, \xi_2) = 8 > 1$, $\psi(a, b, s) = 12 \neq 0$.*)*

$$\frac{1}{(\frac{12}{7})^2}(y_1 + \tfrac{3\sqrt{2}}{7})^2 - \frac{1}{(\frac{12\sqrt{7}}{7})^2}(y_2 + 3\sqrt{2})^2 = 1.$$

   This hyperbola is tangent to the line $y_2 = -3y_1$ *in* $\left(-3\sqrt{2}, 9\sqrt{2}\right)$ *(or* $x_2 = 2x_1$ *in* $(6, 12)$ *the* x_1x_2*-coordinate system) and intersects the line* $y_2 = -y_1$ *in* $\left(-3\sqrt{2}, -3\sqrt{2}\right)$ *(or* $x_2 = 0$ *in* $(-6, 0)$ *in the* x_1x_2*-coordinate system).*

4. $\tilde{\Gamma}_4^a$ *defined by* $\tilde{e}_4^a = \left(-\frac{1}{2}, 0\right)$ *and* $\tilde{e}_5^a = \left(-\frac{1}{2}, -2\right)$ *:*

   $s_1 = -2$ *and* $s_2 = 0$ *imply* $\eta_1 = -3$ *and* $\eta_2 = 1$. *Therefore we obtain* $d = 16$ *and the eigenvalues* $\lambda_1 = -3$, $\lambda_2 = 1$. *Since* $s_2 = 0$ *and* $\eta_1 < \eta_2$, *we have* **Case 1.1** *with* $P = \left(\begin{smallmatrix} 1 & 0 \\ 0 & 1 \end{smallmatrix}\right)$ *and obtain (without principal-axis transformation)*

$$-3x_1^2 + x_2^2 - 12x_1 + 36 = 0 \quad ,$$

   which is equivalent to the hyperbola (Notice $\varphi(s_1, s_2, \xi_1, \xi_2) = 4 > 1$, $\psi(a, b, s) = 12 \neq 0$.*)*

$$\frac{1}{4^2}(x_1 + 2)^2 - \frac{1}{(4\sqrt{3})^2}x_2^2 = 1 \quad .$$

   The side constraints are $x_1 \leq 0 \wedge x_1 \geq 4x_2$ *or equivalently* $4x_2 \leq x_1 \leq 0$.

5. $\tilde{\Gamma}_5^a$ *defined by* $\tilde{e}_5^a = \left(-\frac{1}{2}, -2\right)$ *and* $\tilde{e}_1^a = (2, -2)$ *:*

   $s_1 = 0$ *and* $s_2 = \frac{1}{2}$ *imply* $\eta_1 = 1$ *and* $\eta_2 = \frac{3}{4}$. *Therefore, we obtain* $d = \frac{1}{16}$ *and the eigenvalues* $\lambda_1 = \frac{3}{4}$, $\lambda_2 = 1$. *Since* $s_1 = 0$ *and* $\eta_1 > \eta_2$, *we have* **Case 1.2** *with* $P = \left(\begin{smallmatrix} 0 & 1 \\ 1 & 0 \end{smallmatrix}\right)$ *and obtain*

$$\tfrac{3}{4}y_1^2 + y_2^2 - 12y_2 + 36 = 0 \quad ,$$

which is equivalent to

$$\tfrac{3}{4}y_1^2 + (y_2 - 6)^2 = 0$$

and therefore only fulfilled by the isolated point $y = (0,6)$ *respectively* $x = (6,0)$ *(Notice* $\varphi(s_1, s_2, \xi_1, \xi_2) = \frac{1}{4} < 1$, $\psi(a,b,s) = 0.$ *). Since* $x = (6,0)$ *does not fulfill the side constraints* $x_2 \leq 4x_1 \ \wedge \ x_2 \leq -x_1$ *, the bisector does not run through* $\tilde{\Gamma}_5^a$*.*

Figure 3.34 shows the bisector Bisec(a,b)*. Notice that the vertical segment from* $(6,3)$ *to* $(6,12)$*, which is part of the bisector in* $\tilde{\Gamma}_2^a$*, is not separating.*

3.7 Approximation of Bisectors

In this section we will investigate the approximation of bisectors. The approximation of bisectors is of interest in those cases where the determination of the exact bisector Bisec(a,b) is too difficult. Difficulties will occur if the curves describing the boundaries $\mathrm{bd}(A)$ and $\mathrm{bd}(B)$ of the two generating unit balls are analytical not easy to handle. In this case one should choose two unit balls $\tilde{A}, \tilde{B}$ which are similar to A and B, respectively, and easy to handle, for instance polyhedral approximations of A and B. Good approximations of the unit balls will lead to a good approximation of the bisector. Therefore, the approximation of a bisector Bisec(a,b) is strongly related to the approximation of the generating unit balls A and B. Hence we consider first the "similarity" of unit balls before we consider the "similarity" of bisectors.

We measure the similarity of two unit balls $C, \tilde{C} \subseteq \mathbb{R}^n$ by the expressions

$$\Delta_{C,\tilde{C}} := \sup_{x \in \mathbb{R}^n \setminus \{0\}} \left| \frac{\gamma_C(x) - \gamma_{\tilde{C}}(x)}{l_2(x)} \right| = \sup_{x \in S(0,1)} |\gamma_C(x) - \gamma_{\tilde{C}}(x)| \qquad (3.116)$$

or

$$\nabla_{C,\tilde{C}} := \sup_{x \in \mathbb{R}^n \setminus \{0\}} \left| \frac{\gamma_C(x) - \gamma_{\tilde{C}}(x)}{\gamma_{\tilde{C}}(x)} \right| = \sup_{x \in \mathrm{bd}(\tilde{C})} |\gamma_C(x) - 1| \qquad (3.117)$$

which compare the maximal difference between $\gamma_C(x)$ and $\gamma_{\tilde{C}}(x)$ with respect to the Euclidean unit ball and to $\tilde{C}$ respectively. The smaller $\Delta_{C,\tilde{C}}$ and $\nabla_{C,\tilde{C}}$ are, the more similar are C and $\tilde{C}$ or γ_C and $\gamma_{\tilde{C}}$, respectively (see Example 3.4).

Notice that $\Delta_{C,\tilde{C}} = \Delta_{\tilde{C},C}$, but $\nabla_{C,\tilde{C}} \neq \nabla_{\tilde{C}.C}$ in general.

Example 3.4
We consider the Manhattan-norm l_1*, the Euclidean norm* l_2*, the Chebyshev-*

norm l_∞ *and their corresponding unit balls and obtain:*

$\Delta_{C,\tilde{C}} = \Delta_{\tilde{C},C}$			
p	1	2	∞
1	0	$\sqrt{2}-1$	$\frac{\sqrt{2}}{2}$
2	$\sqrt{2}-1$	0	$1-\frac{\sqrt{2}}{2}$
∞	$\frac{\sqrt{2}}{2}$	$1-\frac{\sqrt{2}}{2}$	0

$\nabla_{C,\tilde{C}} \neq \nabla_{\tilde{C},C}$			
p	1	2	∞
1	0	$\sqrt{2}-1$	1
2	$1-\frac{\sqrt{2}}{2}$	0	$\sqrt{2}-1$
∞	$\frac{1}{2}$	$1-\frac{\sqrt{2}}{2}$	0

How do we measure the goodness of a point x of an approximated bisector? For that purpose we consider the absolute error

$$d^{\text{abs}}_{A,B}(x) := |\gamma_A(x-a) - \gamma_B(x-b)| \tag{3.118}$$

and the relative error

$$d^{\text{rel}}_{A,B}(x) := \left| \frac{\gamma_A(x-a) - \gamma_B(x-b)}{\gamma_A(x-a) + \gamma_B(x-b)} \right| \tag{3.119}$$

of the point x. For motivation see the following example.

Example 3.5
Let us consider the sites $a = (0,0)$ *and* $b = (10,0)$*, both associated with the* 1*-norm. Then the exact bisector is given by the vertical line* $\mathsf{Bisec}(a,b) = \{x \in \mathbb{R}^2 : x_1 = 5\}$*.*

Consider the points $u = (4.99, 0)$, $v = (5.05, 95)$, $w = (4, 995)$ *and* $y = (6, 0)$*. Which one would you accept as a point of an approximated bisector?*

On the first sight all those points whose distance to a *differs not too much from the distance to* b*. The absolute differences of the distances are given by*

$$\begin{aligned} d^{\text{abs}}_{A,B}(u) &= |4.99 - 5.01| = 0.02 \\ d^{\text{abs}}_{A,B}(v) &= |100.05 - 99.95| = 0.1 \\ d^{\text{abs}}_{A,B}(w) &= |999 - 1001| = 2 \\ d^{\text{abs}}_{A,B}(y) &= |6 - 4| = 2 \end{aligned} \quad ,$$

therefore the only acceptable point seems to be u*.*

However, from the objective point of view it seems to be more suitable to consider the relative difference of the distances. For instance, in the case of point y *the absolute difference is* $d^{\text{abs}}_{A,B}(y) = 2$*, whereas* $l_1(y-a) = 6$ *and* $l_1(y-b) = 4$*, i.e. the distance of* y *to* a *is* 50 % *larger than the distance of* y *to* b*. For the point* w *the absolute error is also* $d^{\text{abs}}_{A,B}(w) = 2$*, but* $l_1(w-a) = 999$ *and* $l_1(w-b) = 1001$*, i.e. the distance of* w *to* b *is only* 0.200... % *larger than the distance of* w *to* a*, which is neglectable. Therefore, we should consider the absolute difference of the distances in relation to the covered distances:*

$$
\begin{aligned}
d^{\text{rel}}_{A,B}(u) &= \left|\frac{4.99-5.01}{4.99+5.01}\right| = \frac{0.02}{10} = 0.002 \\
d^{\text{rel}}_{A,B}(v) &= \left|\frac{100.05-99.95}{100.05+99.95}\right| = \frac{0.1}{200} = 0.0005 \\
d^{\text{rel}}_{A,B}(w) &= \left|\frac{999-1001}{999+1001}\right| = \frac{2}{2000} = 0.001\ldots \\
d^{\text{rel}}_{A,B}(y) &= \left|\frac{6-4}{6+4}\right| = \frac{2}{10} = 0.2
\end{aligned}
$$

Now, from the objective point of view, we can accept u, v and w as points of an approximated bisector.

The following lemma gives information about the goodness of bisector points of the approximated bisector depending on the similarity of the unit balls. Example 3.6 illustrates the result of the lemma.

Lemma 3.13
Let $a, b \in \mathbb{R}^2$ be two sites associated with gauges γ_A, γ_B respectively $\gamma_{\tilde{A}}, \gamma_{\tilde{B}}$ and let K, K_a, K_b be real numbers with $K > \frac{l_2(a-b)}{2}$, $K_a > \frac{\gamma_{\tilde{A}}(b-a)}{2}$, $K_b > \frac{\gamma_{\tilde{B}}(a-b)}{2}$. Then the following (absolute and relative) error bounds hold:

1. $d^{\text{abs}}_{A,B}(x) \le \Delta_{A,\tilde{A}} l_2(x-a) + \Delta_{B,\tilde{B}} l_2(x-b) \quad \forall\, x \in \mathsf{Bisec}_{\tilde{A},\tilde{B}}(a,b)$.
2. $d^{\text{abs}}_{A,B}(x) \le \nabla_{A,\tilde{A}} \gamma_{\tilde{A}}(x-a) + \nabla_{B,\tilde{B}} \gamma_{\tilde{B}}(x-b) \quad \forall\, x \in \mathsf{Bisec}_{\tilde{A},\tilde{B}}(a,b)$.
3. $d^{\text{abs}}_{A,B}(x) \le K(\Delta_{A,\tilde{A}} + \Delta_{B,\tilde{B}}) \quad \forall\, x \in \mathsf{Bisec}_{\tilde{A},\tilde{B}}(a,b) \cap B\left(\frac{a+b}{2}, K - \frac{l_2(a-b)}{2}\right)$.
4. $d^{\text{abs}}_{A,B}(x) \le K_a \nabla_{A,\tilde{A}} + K_b \nabla_{B,\tilde{B}}$
$\forall\, x \in \mathsf{Bisec}_{\tilde{A},\tilde{B}}(a,b) \cap B\left(\frac{a+b}{2}, K_a - \frac{\gamma_{\tilde{A}}(b-a)}{2}\right) \cap B\left(\frac{a+b}{2}, K_b - \frac{\gamma_{\tilde{B}}(a-b)}{2}\right)$.
5. $d^{\text{rel}}_{A,B}(x) \le \frac{\nabla_{A,\tilde{A}} + \nabla_{B,\tilde{B}}}{2-(\nabla_{A,\tilde{A}} + \nabla_{B,\tilde{B}})} \quad \forall\, x \in \mathsf{Bisec}_{\tilde{A},\tilde{B}}(a,b)$.
6. $d^{\text{rel}}_{A,B}(x) \le \frac{\nabla_{A,\tilde{A}} + \nabla_{B,\tilde{B}}}{2} \quad \forall\, x \in \mathsf{Bisec}_{\tilde{A},\tilde{B}}(a,b)$, *if $A \subseteq \tilde{A}$ and $B \subseteq \tilde{B}$.*

Proof.

1. Let $x \in \mathsf{Bisec}_{\tilde{A},\tilde{B}}(a,b) \quad \Rightarrow \quad d^{\text{abs}}_{A,B}(x) = |\gamma_A(x-a) - \gamma_B(x-b)| = \ldots$
Case 1: $\gamma_A(x-a) \ge \gamma_B(x-b)$
$$
\begin{aligned}
\ldots &= \gamma_A(x-a) - \gamma_B(x-b) \\
&\le \gamma_{\tilde{A}}(x-a) + \Delta_{A,\tilde{A}} l_2(x-a) - (\gamma_{\tilde{B}}(x-b) - \Delta_{B,\tilde{B}} l_2(x-b)) \\
&= \Delta_{A,\tilde{A}} l_2(x-a) + \Delta_{B,\tilde{B}} l_2(x-b).
\end{aligned}
$$
Case 2: $\gamma_A(x-a) \le \gamma_B(x-b)$
$$
\begin{aligned}
\ldots &= \gamma_B(x-b) - \gamma_A(x-a) \\
&\le \gamma_{\tilde{B}}(x-b) + \Delta_{B,\tilde{B}} l_2(x-b) - (\gamma_{\tilde{A}}(x-a) - \Delta_{A,\tilde{A}} l_2(x-a)) \\
&= \Delta_{A,\tilde{A}} l_2(x-a) + \Delta_{B,\tilde{B}} l_2(x-b).
\end{aligned}
$$
2. Analogously to 1.

3. Let $x \in \mathsf{Bisec}_{\tilde{A},\tilde{B}}(a,b) \cap B\left(\frac{a+b}{2}, K - \frac{l_2(a-b)}{2}\right) \Rightarrow$

$$\begin{aligned} d^{\text{abs}}_{A,B}(x) &\leq \Delta_{A,\tilde{A}} l_2(x-a) + \Delta_{B,\tilde{B}} l_2(x-b) \text{ by 1.} \\ &= \Delta_{A,\tilde{A}} l_2\left(x - \tfrac{a+b}{2} + \tfrac{b-a}{2}\right) + \Delta_{B,\tilde{B}} l_2\left(x - \tfrac{a+b}{2} + \tfrac{a-b}{2}\right) \\ &\leq \Delta_{A,\tilde{A}}\left(l_2\left(x - \tfrac{a+b}{2}\right) + l_2\left(\tfrac{b-a}{2}\right)\right) + \Delta_{B,\tilde{B}}\left(l_2\left(x - \tfrac{a+b}{2}\right) + l_2\left(\tfrac{a-b}{2}\right)\right) \\ &\leq \Delta_{A,\tilde{A}} K + \Delta_{B,\tilde{B}} K = K(\Delta_{A,\tilde{A}} + \Delta_{B,\tilde{B}}) \,. \end{aligned}$$

4. Analogously to 3.
5. Let $x \in \mathsf{Bisec}_{\tilde{A},\tilde{B}}(a,b)$. Then 2. implies

$$\begin{aligned} d^{\text{rel}}_{A,B}(x) &= \left| \frac{\gamma_A(x-a) - \gamma_B(x-b)}{\gamma_A(x-a) + \gamma_B(x-b)} \right| \\ &\leq \frac{\nabla_{A,\tilde{A}} \gamma_{\tilde{A}}(x-a) + \nabla_{B,\tilde{B}} \gamma_{\tilde{B}}(x-b)}{\gamma_A(x-a) + \gamma_B(x-b)} \\ &\leq \frac{\nabla_{A,\tilde{A}} \gamma_{\tilde{A}}(x-a) + \nabla_{B,\tilde{B}} \gamma_{\tilde{B}}(x-b)}{(1-\nabla_{A,\tilde{A}})\gamma_{\tilde{A}}(x-a) + (1-\nabla_{B,\tilde{B}})\gamma_{\tilde{B}}(x-b)} \\ &= \frac{\nabla_{A,\tilde{A}} + \nabla_{B,\tilde{B}}}{(1-\nabla_{A,\tilde{A}}) + (1-\nabla_{B,\tilde{B}})} \\ &= \frac{\nabla_{A,\tilde{A}} + \nabla_{B,\tilde{B}}}{2 - (\nabla_{A,\tilde{A}} + \nabla_{B,\tilde{B}})} \quad . \end{aligned}$$

6. Let $x \in \mathsf{Bisec}_{\tilde{A},\tilde{B}}(a,b)$. $\gamma_A(x-a) \geq \gamma_{\tilde{A}}(x-a)$, $\gamma_B(x-b) \geq \gamma_{\tilde{B}}(x-b)$ and 2. imply

$$\begin{aligned} d^{\text{rel}}_{A,B}(x) = \left| \frac{\gamma_A(x-a) - \gamma_B(x-b)}{\gamma_A(x-a) + \gamma_B(x-b)} \right| &\leq \frac{\nabla_{A,\tilde{A}} \gamma_{\tilde{A}}(x-a) + \nabla_{B,\tilde{B}} \gamma_{\tilde{B}}(x-b)}{\gamma_{\tilde{A}}(x-a) + \gamma_{\tilde{B}}(x-b)} \\ &= \frac{\nabla_{A,\tilde{A}} + \nabla_{B,\tilde{B}}}{2} \,. \end{aligned}$$

□

Example 3.6
Consider the sites $a = (0,10)$ *and* $b = (15,5)$ *associated with the gauges* $\gamma_A = l_2, \gamma_{\tilde{A}} = l_\infty$ *and* $\gamma_B = l_2, \gamma_{\tilde{B}} = l_1$. *For* $x = (10,0)$ *we have*

$$\gamma_{\tilde{A}}(x-a) = \max\{|10-0|, |0-10|\} = 10 = |10-15| + |0-5| = \gamma_{\tilde{B}}(x-b)\,,$$

i.e. $x \in \mathit{Bisec}_{\tilde{A},\tilde{B}}(a,b)$, *and*

$$\begin{aligned} l_2(x-a) &= \sqrt{(10-0)^2 + (0-10)^2} = 10\sqrt{2}, \\ l_2(x-b) &= \sqrt{(10-15)^2 + (0-5)^2} = 5\sqrt{2} \end{aligned}$$

which imply the absolute error bounds

$$\Delta_{A,\tilde{A}} l_2(x-a) + \Delta_{B,\tilde{B}} l_2(x-b) = (1-\tfrac{\sqrt{2}}{2})\, 10\sqrt{2} + \left(\sqrt{2}-1\right) 5\sqrt{2} = 5\sqrt{2}$$

and

$$\nabla_{A,\tilde{A}} \gamma_{\tilde{A}}(x-a) + \nabla_{B,\tilde{B}} \gamma_{\tilde{B}}(x-b) = (1-\tfrac{\sqrt{2}}{2})\, 10 + \left(\sqrt{2}-1\right) 10 = 5\sqrt{2}\,.$$

Since $d^{\text{abs}}_{A,B}(x) = \left|10\sqrt{2} - 5\sqrt{2}\right| = 5\sqrt{2}$, *both absolute error bounds are sharp for* x. *The relative error bounds are given by*

$$\frac{\nabla_{A,\tilde{A}} + \nabla_{B,\tilde{B}}}{2 - (\nabla_{A,\tilde{A}} + \nabla_{B,\tilde{B}})} = \frac{1 - \frac{\sqrt{2}}{2} + \sqrt{2} - 1}{2 - (1 - \frac{\sqrt{2}}{2} + \sqrt{2} - 1)} = 0,5469\ldots$$

and

$$\frac{\nabla_{A,\tilde{A}} + \nabla_{B,\tilde{B}}}{2} = \frac{1 - \frac{\sqrt{2}}{2} + \sqrt{2} - 1}{2} = \frac{\sqrt{2}}{4} = 0.3535\ldots \quad ,$$

both are not sharp for x, *since* $d^{\text{rel}}_{A,B}(x) = \left|\frac{10\sqrt{2}-5\sqrt{2}}{10\sqrt{2}+5\sqrt{2}}\right| = \frac{5\sqrt{2}}{15\sqrt{2}} = \frac{1}{3}$.

The approximation of a convex set $C \subseteq \mathbb{R}^2$ can be done by a set $\underline{C} \subseteq \mathbb{R}^2$ with $\underline{C} \subseteq C$, i.e. C is approximated by $\underline{C}$ from the interior, or by a set $\overline{C} \subseteq \mathbb{R}^2$ with $C \subseteq \overline{C}$, i.e. C is approximated by $\overline{C}$ from the exterior. Taking into account this fact for the two generating unit balls A and B of a bisector we can bound the region which contains $\mathsf{Bisec}(a,b)$ by considering two bisectors $\mathsf{Bisec}_{\underline{A},\overline{B}}(a,b)$ and $\mathsf{Bisec}_{\overline{A},\underline{B}}(a,b)$ with $\underline{A} \subseteq A \subseteq \overline{A}$ and $\underline{B} \subseteq B \subseteq \overline{B}$.

Lemma 3.14
Let $a, b \in \mathbb{R}^2$ *be two sites associated with gauges* $\gamma_{\underline{A}}, \gamma_A, \gamma_{\overline{A}}$ *respectively* $\gamma_{\underline{B}}, \gamma_B, \gamma_{\overline{B}}$ *such that* $\underline{A} \subseteq A \subseteq \overline{A}$ *and* $\underline{B} \subseteq B \subseteq \overline{B}$. *Then the following expressions describe the same set:*

1. $C_{\overline{A},\underline{B}}(a,b) \cap C_{\overline{B},\underline{A}}(b,a)$
2. $\mathbb{R}^2 \setminus \left(D_{\underline{A},\overline{B}}(a,b) \cup D_{\underline{B},\overline{A}}(b,a) \right)$

Proof.

$$C_{\overline{A},\underline{B}}(a,b) \cap C_{\overline{B},\underline{A}}(b,a)$$

$$= \left(\mathbb{R}^2 \setminus D_{\underline{B},\overline{A}}(b,a) \right) \bigcap \left(\mathbb{R}^2 \setminus D_{\underline{A},\overline{B}}(a,b) \right)$$

$$= \mathbb{R}^2 \setminus \left(D_{\underline{A},\overline{B}}(a,b) \cup D_{\underline{B},\overline{A}}(b,a) \right)$$

□

Let us denote the set described by the expressions of Lemma 3.14 by $E_{\underline{A},\overline{A},\underline{B},\overline{B}}(a,b)$. Obviously

$$E_{\underline{A},\overline{A},\underline{B},\overline{B}}(a,b) = E_{\underline{B},\overline{B},\underline{A},\overline{A}}(b,a) \tag{3.120}$$

holds since the definition is symmetric in $\underline{A}, \overline{A}$ and $\underline{B}, \overline{B}$.

Lemma 3.15
Let $a, b \in \mathbb{R}^2$ be two sites associated with gauges $\gamma_{\underline{A}}, \gamma_A, \gamma_{\overline{A}}$ respectively $\gamma_{\underline{B}}, \gamma_B, \gamma_{\overline{B}}$ such that $\underline{A} \subseteq A \subseteq \overline{A}$ and $\underline{B} \subseteq B \subseteq \overline{B}$. Then the bisector $\mathsf{Bisec}(a,b)$ is included in $E_{\underline{A},\overline{A},\underline{B},\overline{B}}(a,b)$, i.e. $\mathsf{Bisec}(a,b) \subseteq E_{\underline{A},\overline{A},\underline{B},\overline{B}}(a,b)$.

Proof.
For $x \in \mathsf{Bisec}(a,b)$ and $\underline{A} \subseteq A \subseteq \overline{A}$, $\underline{B} \subseteq B \subseteq \overline{B}$ holds

$$\gamma_{\overline{A}}(x-a) \le \gamma_A(x-a) = \gamma_B(x-b) \le \gamma_{\underline{B}}(x-b)$$

and

$$\gamma_{\overline{B}}(x-b) \le \gamma_B(x-b) = \gamma_A(x-a) \le \gamma_{\underline{A}}(x-a)$$

which implies

$$x \in C_{\overline{A},\underline{B}}(a,b) \quad \wedge \quad x \in C_{\overline{B},\underline{A}}(b,a)$$

and therefore

$$x \in C_{\overline{A},\underline{B}}(a,b) \cap C_{\overline{B},\underline{A}}(b,a) = E_{\underline{A},\overline{A},\underline{B},\overline{B}}(a,b) \quad .$$

□

To make the results of Lemma 3.15 clearer we present two examples.

Example 3.7
Figure 3.35 shows the situation of Lemma 3.15 for the sites $a = (0,0)$, $b = (8,4)$ and the gauges $\gamma_{\underline{A}} = \gamma_{\underline{B}} = l_1$, $\gamma_A = \gamma_B = l_2$, $\gamma_{\overline{A}} = \gamma_{\overline{B}} = l_\infty$. Notice that $\mathsf{Bisec}(a,b)$, $\mathsf{Bisec}_{\underline{A},\overline{B}}(a,b)$ as well as $\mathsf{Bisec}_{\overline{A},\underline{B}}(a,b)$ consist of one connected component. This property will be violated in the following example.

Example 3.8
Figure 3.36 shows the situation of Lemma 3.15 for the sites $a = (0,0)$, $b = (6,3)$ and the gauges $\gamma_{\underline{A}} = \frac{1}{2}l_1$, $\gamma_A = \frac{1}{2}l_2$, $\gamma_{\overline{A}} = \frac{1}{2}l_\infty$, $\gamma_{\underline{B}} = l_1$, $= \gamma_B = l_2$, $\gamma_{\overline{B}} = l_\infty$. Here $\mathsf{Bisec}(a,b)$ and $\mathsf{Bisec}_{\overline{A},\underline{B}}(a,b)$ are closed curves, while $\mathsf{Bisec}_{\underline{A},\overline{B}}(a,b)$ is disconnected (with two connected components). Obviously, an approximated bisector should keep the character (number and type of connected components) of the exact bisector. Therefore this example illustrates not only Lemma 3.15, but also demonstrates that the approximation of $\gamma_A = \frac{1}{2}l_2$ by $\gamma_{\underline{A}} = \frac{1}{2}l_1$ and $\gamma_B = l_2$ by $\gamma_{\overline{B}} = l_\infty$ is not precise enough. However, the approximation of $\gamma_A = \frac{1}{2}l_2$ by $\gamma_{\overline{A}} = \frac{1}{2}l_\infty$ and $\gamma_B = l_2$ by $\gamma_{\underline{B}} = l_1$ keeps already the character of the bisector. This is due to the fact that $\overline{A}$ and $\underline{B}$ share the same properties as A and B with respect to Lemma 3.7 - Corollary 3.6.

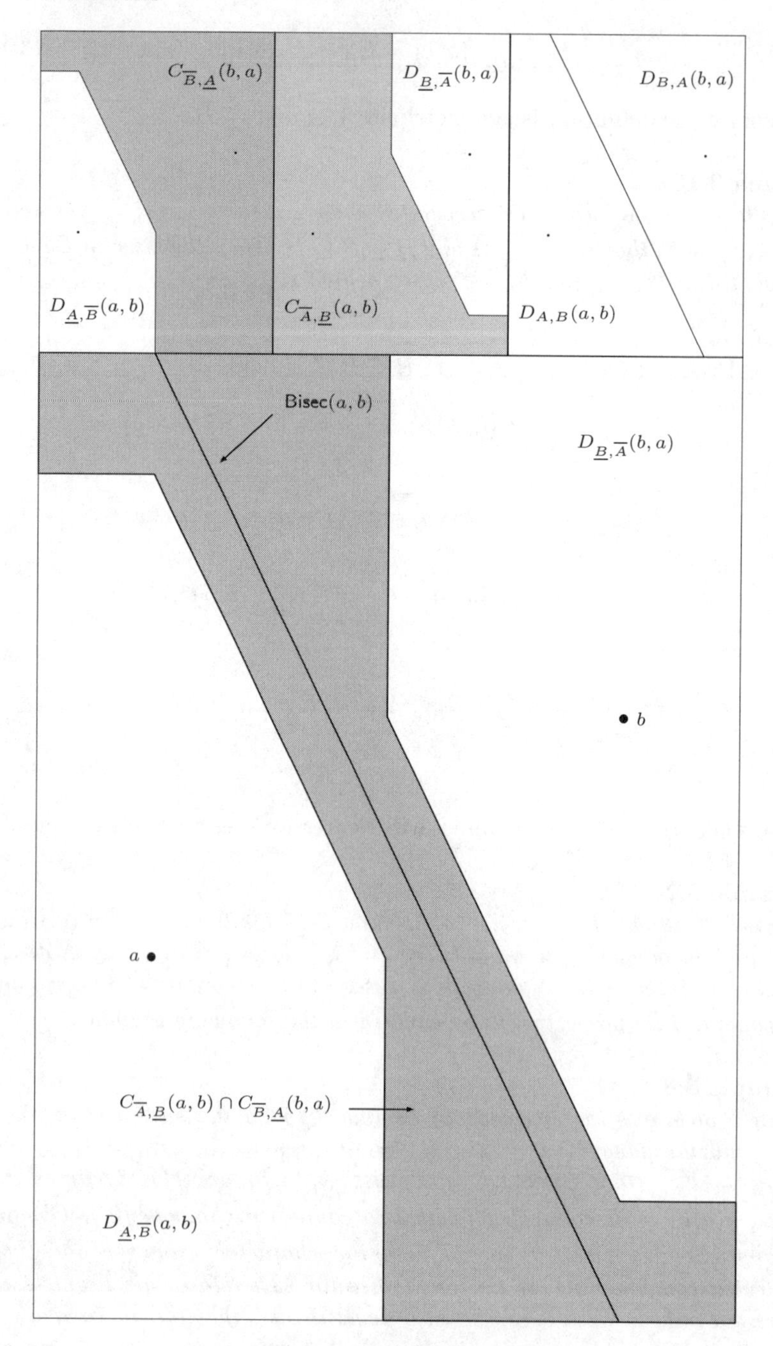

Fig. 3.35. Illustration to Example 3.7

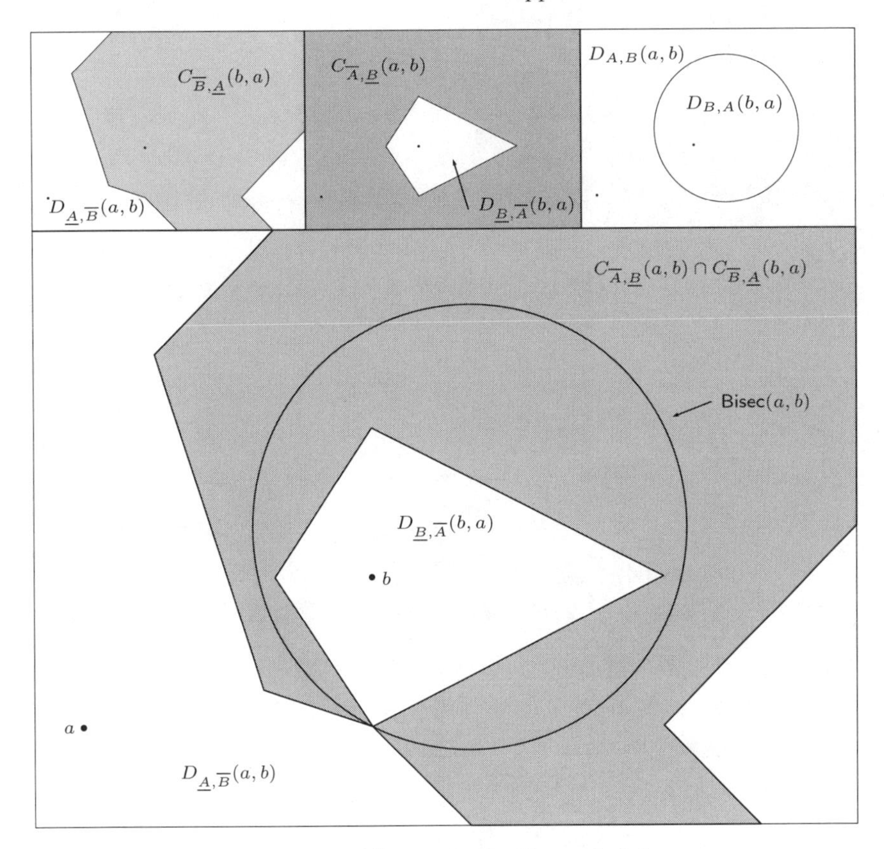

Fig. 3.36. Illustration to Example 3.8

Nevertheless, Bisec$_{\overline{A},\underline{B}}(a,b)$ *is still far away from a good approximation. This observations can be summarized as follows: An exact bisector* Bisec(a,b) *and its approximation* $B_{\tilde{A},\tilde{B}}(a,b)$ *have the same character (number and type of connected components), if* $\tilde{A}$ *and* $\tilde{B}$ *share the same properties as* A *and* B *with respect to Lemma 3.7 - Corollary 3.6. Hence the approximations of the unit balls should be chosen carefully.*

4

The Single Facility Ordered Median Problem

In Chapter 2 we have seen many useful structural properties of the continuous OMP. In this chapter we will concentrate on solution algorithms for the single facility case using as basis the results of Chapter 2. First we will present an approach for the convex case (λ non-decreasing) based on linear programming (see also [177] and [54]) which is applicable for the polyhedral case in $\mathbb{R}^n$. However, as will be explained, the case with general λ will lead to a highly inefficient search procedure. Therefore, we introduce also a geometric approach which is limited to low dimensions, but has the advantage that it is valid for any type of λ. Then we proceed to show how the approaches with polyhedral gauges can be used to solve also the non-polyhedral case. The following section is devoted to the OMP with positive and negative weights whereas in the final section we present the best algorithms known up-to-date for the polyhedral version of the OMP, following the results in [117]. All results are accompanied by a detailed complexity analysis.

4.1 Solving the Single Facility OMP by Linear Programming

Let γ_i be the gauge associated with the facility a_i, $i = 1, \dots, M$, which is defined by the extreme points of a polytope B_i, $Ext(B_i) = \{e_g^i : g = 1, \dots, G_i\}$. Moreover, we denote by B_i^o the polar polytope of B_i and its set of extreme points as $Ext(B_i^o) = \{e_g^{i^o} : g = 1, \dots, G_i\}$.

Throughout this section we will assume $\lambda \in S_M^{\leq}$. Therefore, using Theorem 1.3 we can write the ordered median function as:

$$f_\lambda(x) = \sum_{i=1}^{M} \lambda_i d_{(i)}(x) = \max_{\sigma \in \mathcal{P}(1 \dots M)} \sum_{i=1}^{M} \lambda_i w_{\sigma(i)} \gamma_{\sigma(i)}(x - a_{\sigma(i)}).$$

Restricting ourselves to polyhedral gauges, a straightforward approach, using linear programming, can be given to solve the OMP.
For a fixed $\sigma \in \mathcal{P}(1 \dots M)$, consider the following linear program:

$$\begin{aligned}(P_\sigma) \min \sum_{i=1}^{M} \lambda_i z_{\sigma(i)} & \\ \text{s.t. } w_i \langle e_g^{i^o}, x - a_i \rangle \le z_i \ \forall e_g^{i^o} \in Ext(B_i^o), & \ i = 1, 2, \dots, M \\ z_{\sigma(i)} \le z_{\sigma(i+1)} \qquad & \ i = 1, 2, \dots, M-1.\end{aligned}$$

In the next lemma we will show that this linear program provides optimal solutions to the OMP in the case where an optimal solution of P_σ is in O_σ.

Lemma 4.1 *For any $x \in O_\sigma$ we have:*

1. *x can be extended to a feasible solution $(x, z_1, \dots, z_M)$ of P_σ.*
2. *The optimal objective value of P_σ with x fixed equals $f_\lambda(x)$.*

Proof.

1. Since $\gamma_i(x - a_i) = \max_{e_g^{i^o} \in Ext(B_i^o)} \langle e_g^{i^o}, x - a_i \rangle$ (see Corollary 2.1), we can fix $z_i := w_i \gamma_i(x - a_i)$ which satisfies the first set of inequalities in P_σ. From $x \in O_\sigma$ we have
$w_{\sigma(i)} \gamma_{\sigma(i)}(x - a_{\sigma(i)}) \le w_{\sigma(i+1)} \gamma_{\sigma(i+1)}(x - a_{\sigma(i+1)})$ for $i = 1, \dots, M-1$. Therefore, $z_{\sigma(i)} \le z_{\sigma(i+1)}$ and also the second set of inequalities is fulfilled.
2. In the first set of inequalities in P_σ we have
$z_i \ge w_i \max_{e_g^{i^o} \in Ext(B_i^o)} \langle e_g^{i^o}, x - a_i \rangle$. Therefore, in an optimal solution (x, z^*) of P_σ for fixed x we have $z_i^* = w_i \max_{e_g^{i^o} \in Ext(B_i^o)} \langle e_g^{i^o}, x - a_i \rangle = w_i \gamma_i(x - a_i)$. Hence, we have for the objective function value of P_σ

$$\sum_{i=1}^{M} \lambda_i z_{\sigma(i)}^* = \sum_{i=1}^{M} \lambda_i w_{\sigma(i)} \gamma_{\sigma(i)}(x - a_{\sigma(i)}) = f_\lambda(x).$$

□

Corollary 4.1 *If for an optimal solution (x^*, z^*) of P_σ we have $x^* \in O_\sigma$ then x^* is also an optimal solution to the OMP restricted to O_σ.*

Nevertheless, it is not always the case that the optimal solutions to P_σ belong to O_σ. We show in the next example one of these situations.

Example 4.1
Consider the 1-dimensional situation with three existing facilities $a_1 = 0$, $a_2 = 2$ and $a_3 = 4$ and corresponding weights $w_1 = w_2 = w_3 = 1$. Moreover, $\lambda_1 = 1$, $\lambda_2 = 2$, and $\lambda_3 = 4$.

This input data generates the ordered regions $O_{123} = (-\infty, 1]$, $O_{213} = [1, 2]$, $O_{231} = [2, 3]$ and $O_{321} = [3, \infty)$ as shown in Figure 4.1.

The linear program P_{123} is given by

$$\begin{aligned} \min \quad & 1z_1 + 2z_2 + 4z_3 \\ \text{s.t.} \quad & \\ & \pm(x-0) \leq z_1 \\ & \pm(x-2) \leq z_2 \\ & \pm(x-4) \leq z_3 \\ & z_1 \leq z_2 \\ & z_2 \leq z_3 \end{aligned}$$

$(x, z_1, z_2, z_3) = (2, 2, 2, 2)$ is feasible for P_{123} with objective value $1 \cdot 2 + 2 \cdot 2 + 4 \cdot 2 = 14$, but $x = 2 \notin O_{123}$. Moreover, the optimal solution restricted to O_{123} is given by $x = 1$ with $f_\lambda(1) = 1 \cdot 1 + 2 \cdot 1 + 4 \cdot 3 = 15 > 14$, whereas the global optimal solution is given by $x = 2$ with $f_\lambda(2) = 1 \cdot 0 + 2 \cdot 2 + 4 \cdot 2 = 12$.

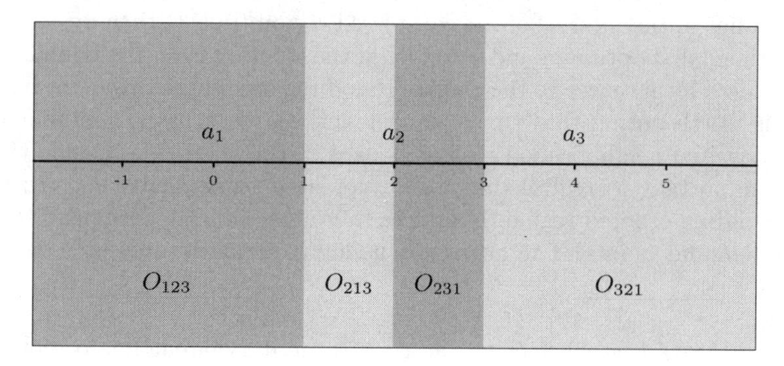

Fig. 4.1. Illustration to Example 4.1

The following result shows that in the case outlined in Example 4.1 we can determine another ordered region where we can continue the search for an optimal solution.

Lemma 4.2 *If for an optimal solution (x^*, z^*) of P_σ we have $x^* \in O_{\sigma'}$ and $x \notin O_\sigma$ with $\sigma' \neq \sigma$ then*

$$\min_{x \in O_{\sigma'}} f_\lambda(x) < \min_{x \in O_\sigma} f_\lambda(x).$$

Proof.
Since $\sigma' \neq \sigma$ there exist at least two indices i and j such that for $x \in O_\sigma$ we have $w_i\gamma_i(x - a_i) \leq w_j\gamma_j(x - a_j)$ and for $x^* \in O_{\sigma'}$ we have $w_i\gamma_i(x - a_i) > w_j\gamma_j(x - a_j)$. But (x^*, z^*) is feasible for P_σ, which means $z_i^* \leq z_j^*$ and

$$z_i^* \geq w_i \max_{e_g^{i^o} \in Ext(B_i^o)} \langle e_g^{i^o}, x^* - a_i \rangle = w_i\gamma_i(x^* - a_i).$$

So,

$$z_j^* \geq w_i\gamma_i(x^* - a_i).$$

Together we get for $x^* \in O_{\sigma'}$ and (x^*, z^*) being feasible for P_σ that $z_j^* \geq w_i\gamma_i(x^* - a_i) > w_j\gamma_j(x^* - a_j)$ in P_σ. This implies, that the optimal objective value for P_σ which is $\sum_{i=1}^{M} \lambda_i z_{\sigma(i)}^*$ is greater than $f_\lambda(x^*)$. But from Lemma 4.1 we know that since $x^* \in O_{\sigma'}$ the optimal objective value of $P_{\sigma'}$ equals $f_\lambda(x^*)$. $\square$

Our initial assumption, $\lambda \in S_M^{\leq}$, implies the convexity of the OMf. Using additionally Lemma 4.2, we can develop a descent algorithm for this problem. For each ordered region the problem is solved as a linear program which geometrically means either finding the locally best solution in this ordered region or detecting that this region does not contain the global optimum by using Lemma 4.2. In the former case two situations may occur. First, if the solution lies in the interior of the considered region (in $\mathbb{R}^n$) then by convexity this is the global optimum and secondly, if the solution is on the boundary we have to do a local search in the neighborhood regions where this point belongs to. It is worth noting that to accomplish this search a list $\mathcal{L}$ containing the already visited neighborhood regions is used in the algorithm. Beside this, it is also important to realize that we do not need to explicitly construct the corresponding ordered region. It suffices to evaluate and to sort the distances to the demand points. The above discussion is formally presented in Algorithm 4.1.

Algorithm 4.1 is efficient in the sense that it is polynomially bounded. Once the dimension of the problem is fixed, its complexity is dominated by the complexity of solving a linear program for each ordered region. Since for any fixed dimension the number of ordered regions is polynomially bounded and the interior point method solves linear programs in polynomial time, Algorithm 4.1 is polynomial in the number of cells.

Example 4.2

Consider three facilities $a_1 = (0, 2.5)$, $a_2 = (5.5, 0)$ and $a_3 = (5.5, 6)$ with the same gauge γ corresponding to l_1-norm, $\lambda = (1, 2, 3)$ and all weights equal to one.

The following problem is formulated:

$$\min_{x \in \mathbb{R}^2} \gamma_{(1)}(x - A) + 2\gamma_{(1)}(x - A) + 3\gamma_{(1)}(x - A).$$

We show in Figure 4.2 the ordered elementary convex sets for this problem. Notice that the thick lines represent the bisectors between the points in A, while the thin ones are the fundamental directions of the norm.

Algorithm 4.1: *Solving the convex single facility OMP via linear programming*

Choose x^o as an appropriate starting point.
Initialize $\mathcal{L} := \emptyset$, $y^* = x^o$.
Look for the ordered region, O_{σ^o} which y^* belongs to, where σ^o determines the order.
1 Solve the linear program P_{σ^o}.
Let $u^o = (x_1^o, x_2^o, z_\sigma^o)$ be an optimal solution.
if $x^0 = (x_1^o, x_2^o) \notin O_{\sigma^o}$ **then**
| determine a new ordered region O_{σ^o}, where x^o belongs to and go to 1
end
Let $y^o = (x_1^o, x_2^o)$.
if *y^o belongs to the interior of O_{σ^o}* **then**
| set $y^* = y^o$ and go to 2.
end
if $f_\lambda(y^o) \neq f_\lambda(y^*)$ **then**
| $\mathcal{L} := \{\sigma^o\}$.
end
if *there exist i and j verifying*

$$w_{\sigma^o(i)}\gamma_{\sigma^o(i)}(y^o - a_{\sigma^o(i)}) = w_{\sigma^o(j)}\gamma_{\sigma^o(j)}(y^o - a_{\sigma^o(j)}) \quad i < j$$

such that

$$(\sigma^o(1), \dots, \sigma^o(j), \dots, \sigma^o(i), \dots, \sigma^o(M)) \notin \mathcal{L}$$

then
| $y^* := y^o$, $\sigma^o := (\sigma^o(1), \dots, \sigma^o(j), \dots, \sigma^o(i), \dots, \sigma^o(M))$, $\mathcal{L} := \mathcal{L} \cup \{\sigma^o\}$
| go to 1
else
| go to 2 (Optimum found).
end
2 Output y^*.

We solve the problem using Algorithm 4.1. Starting with $x^o = a_3 = (5.5, 6) \in O_{\sigma^o}$ for $\sigma^o = (3, 2, 1)$, we get an optimal solution in two iterations. In the first one, we have to solve the linear program, with respect to σ^o, given by:

$$\begin{aligned} (P_{\sigma^o}) \ \min\ & z_3 + 2z_2 + 3z_1 \\ \text{s.t. } & \langle e_g^{i^o}, x - a_i \rangle \le z_i \ \ e_g^{i^o} \in Ext(B_i^o) = Ext(B_{l_1}^o) = Ext(B_{l_\infty}), \\ & \qquad\qquad i = 1, 2, 3 \\ & z_3 \le z_2 \\ & z_2 \le z_1 . \end{aligned}$$

The optimal solution is $u^o = (5.5, 3, 6, 3, 3)$ with objective value 27. Thus, we get the point $x^1 = b_9 = (5.5, 3)$ that lies on the boundary of the ordered regions with respect to $\sigma^o = (3, 2, 1)$ and $\sigma^1 = (2, 3, 1)$. Following the algorithm, we have to solve P_{σ^1} to verify that in this new region there is no

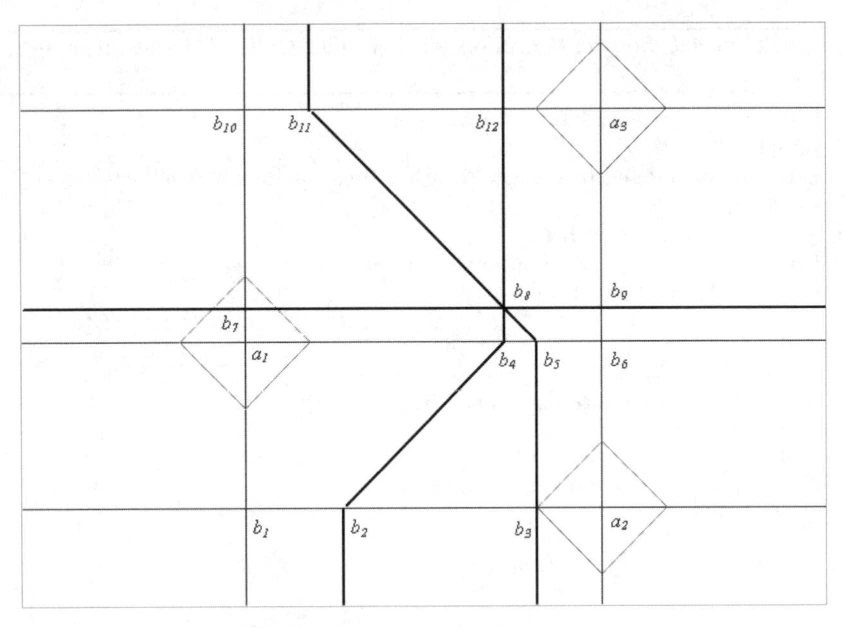

Fig. 4.2. Illustration to Example 4.2

better solution. In the second iteration we obtain $u^1 = (5.5, 2.5, 5.5, 2.5, 3.5)$ with objective value 26. From u^1 we get the point $x^2 = b_6 = (5.5, 2.5)$ that lies in the ordered region with respect to $\sigma = (2, 3, 1)$. This point cannot be improved in its neighborhood, therefore it is an optimal solution.

In principle, we may use the linear programming approach also in the non-convex case. However, this only can be done by enumerating all ordered regions, since otherwise we might get stuck in a local minimum. Many of the ordered regions might even be empty. Therefore, we will present in Section 4.2 an approach which explicitly uses the underlying geometry to be able to tackle the non-convex case in the plane.

4.2 Solving the Planar Ordered Median Problem Geometrically

In Section 2.3.2 we have seen (Theorem 2.2) that there exists an optimal solution of the OMP on the vertices of the ordered elementary convex sets, usually called ordered intersection points (OIP). Consequently, the geometrical method will be based on this result.

For the sake of simplicity, in the presentation we restrict ourselves to the planar case. Furthermore, to solve the problem geometrically, we will only consider the full dimensional ordered elementary convex sets. If we construct the rays based on the fundamental directions and the bisectors, we will have a planar subdivision, where the regions are called **cells** (see Figure 4.3).

The set of all cells will be denoted by $\mathcal{C}$.

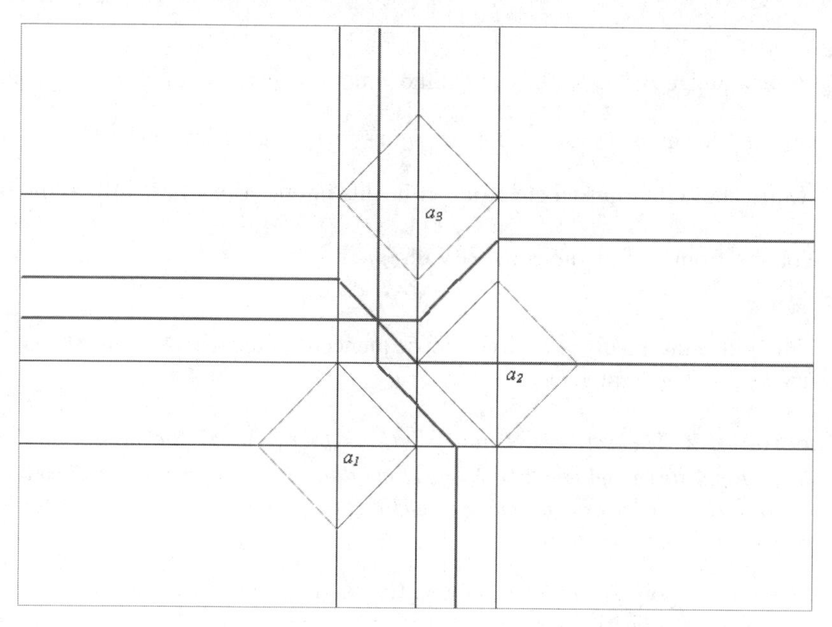

Fig. 4.3. Bisectors, fundamental directions and corresponding cells

These regions are convex. Recall the discussion on the o.e.c.s. after (2.28). Therefore, if we construct all the fundamental directions and the bisectors we divide the plane into convex regions. In addition, we have seen that the ordered median function f_λ is convex for $0 \leq \lambda_1 \leq \cdots \leq \lambda_M$. Therefore, we know that all local minima are global minima. Furthermore, we have:

Theorem 4.1 *The ordered median function fulfills the following properties:*

1. *f_λ is affine linear in each cell $C \in \mathcal{C}$.*
2. *If $\lambda \in S_M^{\leq}$ the level curves of f_λ are convex polygons in $\mathbb{R}^2$, which are affine linear in each cell $C \in \mathcal{C}$.*

Proof.

1. For $C \in \mathcal{C}$ there exists a permutation $\pi \in \mathcal{P}(1 \ldots M)$ such that $C \subseteq O_\pi$. Therefore, for $x \in C \subseteq O_\pi$ we have

$$f_\lambda(x) = \sum_{i=1}^{M} \lambda_i w_{\pi(i)} \gamma_{\pi(i)}(x - a_{\pi(i)}) = \sum_{i=1}^{M} \underbrace{\lambda_{\pi^{-1}(i)} w_i}_{=:\overline{w}_i} \gamma_i(x - a_i)$$

$$= \sum_{i=1}^{M} \overline{w}_i \gamma_i(x - a_i).$$

Every gauge $\gamma_i(x - a_i)$ is an affine function in each fundamental cone $a_i + \Gamma_i^j$ for all $j \in \{1, \ldots, G_i\}$, so, $\gamma_i(x - a_i)$ is also affine in $\bigcap_{i=1}^{M} (a_i + \Gamma_i^j)$. Since the cell C must be contained in this intersection, f_λ is also affine in C.

2. Follows from 1. and the convexity of f_λ.

□

The following result is a direct consequence of Theorem 4.1 and the convexity of f_λ (Theorem 1.3).

Theorem 4.2 *The set of optimal solutions $M(f_\lambda, A)$ of the single facility OMP with polyhedral gauges and $\lambda \in S_M^{\leq}$ is either a vertex or a bounded convex union of cells or edges in the planar subdivision generated by the fundamental directions and the bisectors.*

Even in the non-convex case, solving the single facility OMP can be reduced to evaluate the objective function at each vertex of every cell as shown in the next corollary.

Corollary 4.2 *There is at least one optimal solution for the planar OMP with polyhedral gauges in a vertex of a cell, i.e., in the set of the ordered intersection points OIP. Furthermore*

$$M(f_\lambda, A) = \bigcup_{C \in \mathcal{C}} conv\{(\arg \min_{x \in OIP} f_\lambda(x)) \bigcap C\} .$$

Proof.
There exists an optimal solution in a vertex of a cell because the objective function is affine linear in every cell by Theorem 4.1. If there are several optimal vertices in a cell, the convex hull of these optima is also optimal, since the ordered median function is affine linear in each cell. □

This means that for solving the non-convex case we have to compute all the intersection points. The following lemma shows that we only have to look for intersection points that are inside the Euclidean ball centered at the origin and with radius R, as shown in Figure 4.4. As a side-effect we can replace half-lines by line segments.

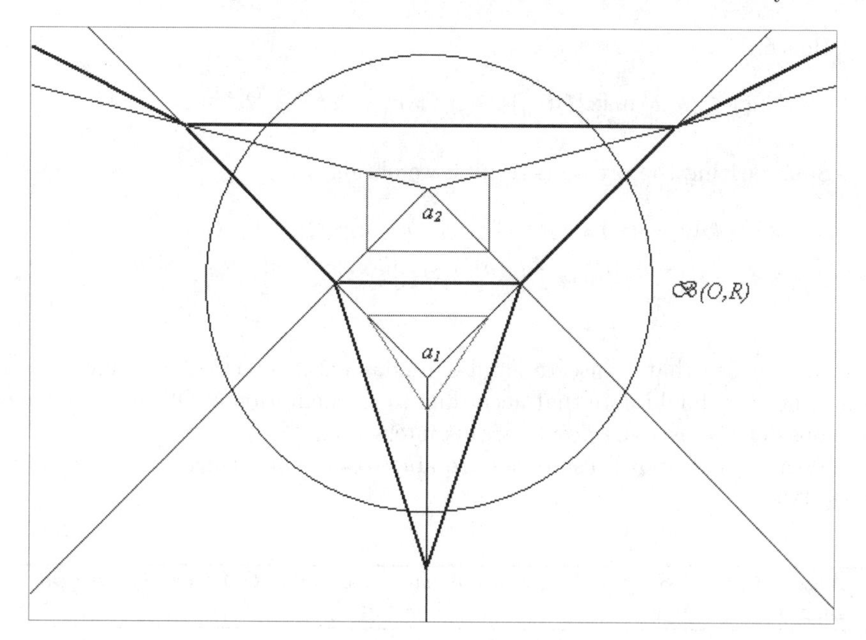

Fig. 4.4. Euclidean ball that contains optimal solutions

Lemma 4.3 *For* $x \in M(f_\lambda, A)$ *we obtain*

$$l_2(x) \leq \max_{1 \leq i \leq M} \{\gamma_i(-a_i) r^i_{max} + l_2(a_i)\} := R, \tag{4.1}$$

where $r^i_{max} = \max_{g \in G_i} \{l_2(e^i_g)\}$.

Proof.
By [153] we know that there exists an index $1 \leq i_x \leq M$ with

$$\gamma_{i_x}(x - a_{i_x}) \leq \gamma_{i_x}(-a_{i_x}).$$

Moreover, let $x \in \mathbb{R}^2$ and x_o be the corresponding foot-point, then

$$\gamma_B(x) = \frac{l_2(x)}{l_2(x_o)} \text{ (see (2.10))}.$$

Moreover, we know that there exists an index $g_o \in G$ such that $x \in \Gamma_{g_o}$, therefore x_o can be written as $x_o = \alpha e_{g_o} + (1-\alpha) e_{g_o+1}$, where $\alpha \in (0,1)$.

$$\begin{aligned} l_2(x_o) &= l_2(\alpha e_{g_o} + (1-\alpha) e_{g_o+1}) \\ &\leq \alpha l_2(e_{g_o}) + (1-\alpha) l_2(e_{g_o+1}) \\ &\leq \alpha \max_{g \in G}\{l_2(e_g)\} + (1-\alpha) \max_{g \in G}\{l_2(e_g)\} \\ &= \max_{g \in G}\{l_2(e_g)\}. \end{aligned}$$

Hence,

$$l_2(x) \leq \gamma_i(x) \max_{g \in G_i}\{l_2(e_g^i)\} = \gamma_i(x) r_{max}^i \; \forall x \in \mathbb{R}^2 \, \forall i = 1, \ldots, M.$$

Summarizing the previous results, we obtain

$$\begin{aligned} l_2(x) &\leq l_2(x - a_{i_x}) + l_2(a_{i_x}) \leq \gamma_{i_x}(x - a_{i_x}) r_{max}^{i_x} + l_2(a_{i_x}) \\ &\leq \gamma_{i_x}(-a_{i_x}) r_{max}^{i_x} + l_2(a_{i_x}) \leq \max_{1 \leq i \leq M} \{\gamma_i(-a_i) \cdot r_{max}^i + l_2(a_i)\}. \end{aligned}$$

□

In the case that a bisector and a fundamental direction coincide over a segment, one should note that according to the definition of OIP we only have to consider the extreme points of this intersection.

With all the above results we are able to present Algorithm 4.2 to solve the OMP.

***Algorithm 4.2**: Solving the general planar single facility OMP with polyhedral gauges*

---

1. Construct all fundamental directions and bisectors.
2. Transform them in segments selecting only the part of them that lies in the Euclidean ball $\mathcal{B}(O, R)$.
3. Compute the set of all the intersection points, that lie inside the Euclidean ball $\mathcal{B}(O, R)$.
4. Evaluate the objective function f_λ at each intersection point.
5. Choose the point(s) with minimal objective value.
6. Output $\mathsf{M}(f_\lambda, A) = \bigcup_{C \in \mathcal{C}} conv\{(\arg\min_{x \in \mathsf{OIP}} f_\lambda(x)) \bigcap C\}$.

To study the complexity of this algorithm, we proceed as follows. Let $G = \max_{1 \leq i \leq M} \{G_i\}$ denote the maximal number of fundamental directions of the different gauges in the problem. The number of fundamental directions is in the worst case $O(GM)$. The maximal number of linear pieces of every bisector is $O(G)$ and there are at most $O(M^2)$ bisectors, so, an upper bound of bisector linear pieces is $O(GM^2)$. Therefore, we have to compute the arrangement of the lines induced by these $O(GM^2)$ segments. The complexity of this step is $O(G^2M^4)$ (see [68]). Since evaluating the function f_λ takes $O(M \log(GM))$ solving the OMP requires $O(G^4M^5 \log(GM))$ time. The last step will be to construct the convex hull of the optimal vertices that lie in the same cell in the non-convex case and otherwise, the convex hull of all the optimal vertices. This task can be performed without increasing the overall complexity using the information of the arrangement of lines previously computed.

The difficulty of this algorithm is the computation of the intersection points, since we have to construct the fundamental directions and the bisectors. The computation of the fundamental directions is fairly easy because of their definition, but the construction of the bisectors is not a trivial task as we have seen in Chapter 3.

Nevertheless, this method has some advantages compared to Algorithm 4.1 as we can see with the next example.

Example 4.3
We want to solve the OMP with respect to $\lambda = (1, 2, 3)$ and considering $A = \{a_1 = (0, 2.5), a_2 = (5.5, 0), a_3 = (5.5, 6)\}$ with l_1-norm.

When we compute the cells as shown in Figure 4.2 we obtain the vertices b_i, $i = 1, \dots, 12$. In Table 4.1 we have the values of f_λ at the intersection points. Therefore, the optimal solution is the segment $\overline{b_5 b_6}$, since in this case the ordered median function is convex, with objective value 26, see Figure 4.5.

Table 4.1. Data of Example 4.3 and Example 4.4.

x	(x_1, x_2)	d			$\lambda = (1, 2, 3)$	$\lambda = (1, 1, 0)$	$\lambda = (1, 0, 0)$
a_1	$(0, 2.5)$	0	8	9	43	8	**0**
a_2	$(5.5, 0)$	8	0	6	36	**6**	**0**
a_3	$(5.5, 6)$	9	6	0	39	**6**	**0**
b_1	$(0, 0)$	2.5	5.5	11.5	48	8	2.5
b_2	$(1.5, 0)$	4	4	10	42	8	4
b_3	$(4.5, 0)$	7	1	7	36	8	1
b_4	$(4, 2.5)$	4	4	5	27	8	4
b_5	$(4.5, 2.5)$	4.5	3.5	4.5	**26**	8	3.5
b_6	$(5.5, 2.5)$	5.5	2.5	3.5	**26**	**6**	2.5
b_7	$(0, 3)$	0.5	8.5	8.5	43	9	0.5
b_8	$(4, 3)$	4.5	4.5	4.5	27	9	4.5
b_9	$(5.5, 3)$	6	3	3	27	**6**	3
b_{10}	$(0, 6)$	3.5	11.5	5.5	49	9	3.5
b_{11}	$(1, 6)$	4.5	10.5	4.5	45	9	4.5
b_{12}	$(4, 6)$	7.5	7.5	1.5	39	9	1.5

Notice that Algorithm 4.1 provides a single optimal solution, since this algorithm stops when it finds a solution in the interior of an ordered region.

On the contrary, with the geometrical approach we get all minima. Moreover, computing once the set of ordered intersection points OIP, it is possible to solve different instances of OMP for different choices of λ (as shown in Table 4.1). Notice that using Algorithm 4.1 we have to fix λ in advance. In

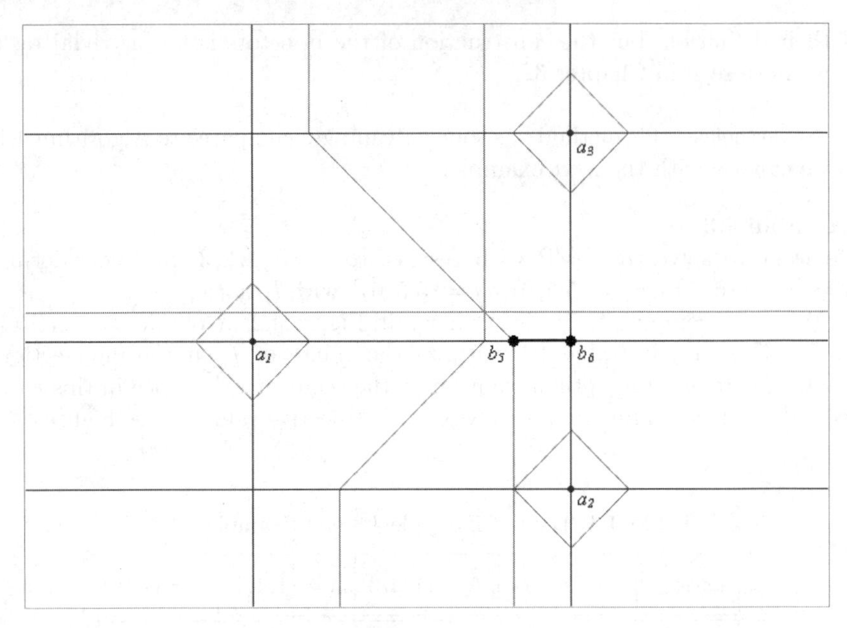

Fig. 4.5. Illustration to Example 4.3 with $\lambda = (1, 2, 3)$

addition, Algorithm 4.2 is also valid for the non-convex case (see Example 4.4), whereas Algorithm 4.1 is only applicable in the convex one.

On the other hand, the application of the geometrical method to $\mathbb{R}^3$ (in general, higher dimensions) is more difficult due to the complexity of the computation of the bisectors and arrangements.

Example 4.4
Continuing with the data from Example 4.3, we consider the following cases:

a) $\lambda = (1, 1, 0)$

The optimal value of f_λ is 6 and it is attained at a_2, a_3, b_6 and b_9 (see Table 4.1). Moreover, a_3 and b_9 are on the same cell, so the segment $\overline{a_3 b_9}$ is also optimal. Likewise, the segments $\overline{b_9 b_6}$ and $\overline{b_6 a_2}$ are optimal and then the complete segment $\overline{a_3 a_2}$ contains all the optimal solutions.

We have obtained a set of optimal solutions that is convex, see Figure 4.6. Nevertheless, the objective function is not convex, since

$$b_7 = \frac{1}{2} b_1 + \frac{1}{2} b_{10}$$

but

$$f_\lambda(b_7) = 9 > \frac{1}{2} f_\lambda(b_1) + \frac{1}{2} f_\lambda(b_{10}) = 8.5.$$

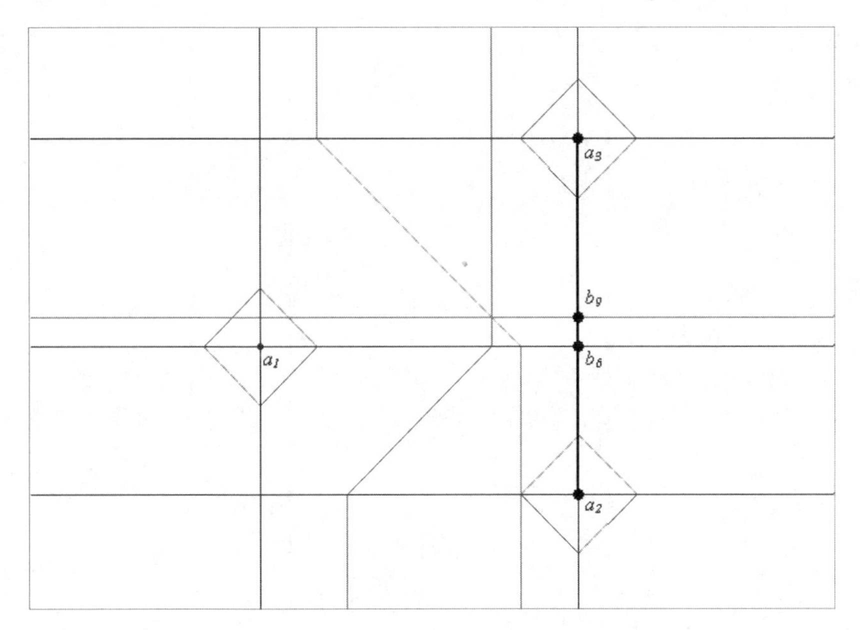

Fig. 4.6. Illustration to Example 4.4 with $\lambda = (1,1,0)$

b) $\lambda = (1,0,0)$

In this case, the optimal value 0 is attained at any of the existing facilities (see Table 4.1), but they do not lie in the same cell. Hence, the solution is not a convex set as shown in Figure 4.7.

4.3 Non Polyhedral Case

In the previous sections we considered polyhedral gauges. For some special cases the results presented so far can be adapted also to non polyhedral gauges. In the case of the Euclidean norm, for example, a number of algorithms exist to determine the bisectors needed for constructing the ordered regions (see [155]). Then, instead of linear programs, convex programs have to be solved in these ordered regions. However, this is not possible for general gauges since no general method is known to compute their bisectors. Therefore we will use the results of the polyhedral case to develop a general scheme for solving the considered problems under general gauges (not necessarily polyhedral).

Example 4.5
We address the resolution of the following OMP with three points $a_1 = (5,0), a_2 = (0,5)$, $a_3 = (10,10)$ and different non polyhedral norms at each

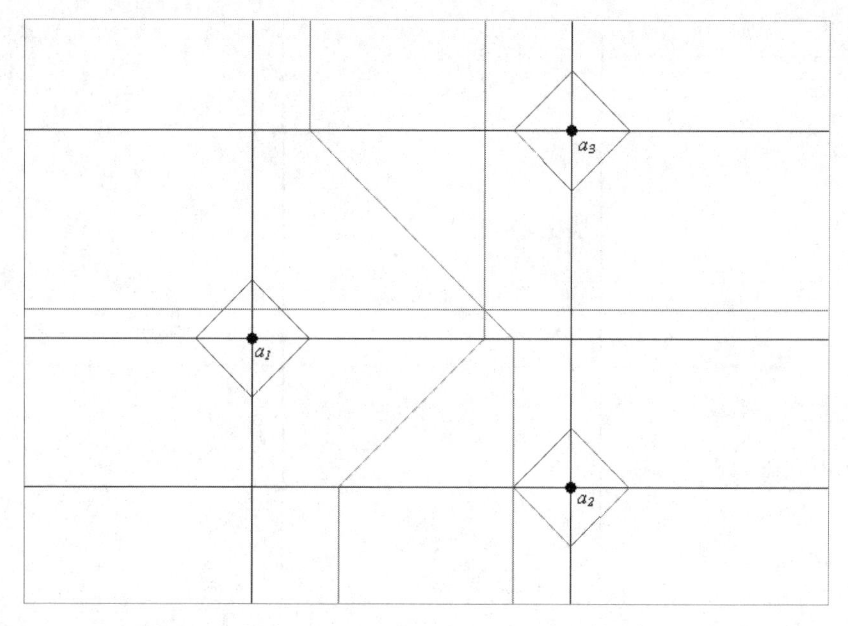

Fig. 4.7. Illustration to Example 4.4 with $\lambda = (1, 0, 0)$

point:

$$\begin{cases} d_1(x) = \|x - a_1\| = \sqrt{\frac{1}{4}(x_1 - 5)^2 + \frac{1}{9}(x_2 - 0)^2} \\ d_2(x) = \|x - a_2\|_2 = \sqrt{(x_1 - 0)^2 + (x_2 - 5)^2} \\ d_3(x) = \|x - a_3\|_2 = \sqrt{(x_1 - 10)^2 + (x_2 - 10)^2} \end{cases}$$

The problem to be solved is:

$$\min_{x \in \mathbb{R}^2} f_\lambda(x) := \sum_{i=1}^{3} \frac{1}{2^i} d_{(i)}(x).$$

In order to solve the problem we compute the different ordered regions. Let $g_{ij}(x) = d_i^2(x) - d_j^2(x)$ for all $i, j = 1, 2, 3$, this results in

$$g_{12}(x) = -3/4x_1^2 - 5/2x_1 - 75/4 - 8/9x_2^2 + 10x_2,$$
$$g_{13}(x) = 5/36x_1^2 - 5/18x_1 - 1075/36 - 5/36x_2^2 + 5x_2,$$
$$g_{23}(x) = 8/9x_1^2 - 100/9 + 3/4x_2^2 - 5x_2 + 20/9x_1.$$

Therefore, the bisectors of the three demand points are:

$$\mathsf{Bisec}(a_1, a_2) = \{(x_1, x_2) : g_{12}(x) = 0\}$$

$$\text{Bisec}(a_1, a_3) = \{(x_1, x_2) : g_{13}(x) = 0\}$$

$$\text{Bisec}(a_2, a_3) = \{(x_1, x_2) : g_{23}(x) = 0\}$$

Figure 4.8 shows the shape of the bisectors among the three points.

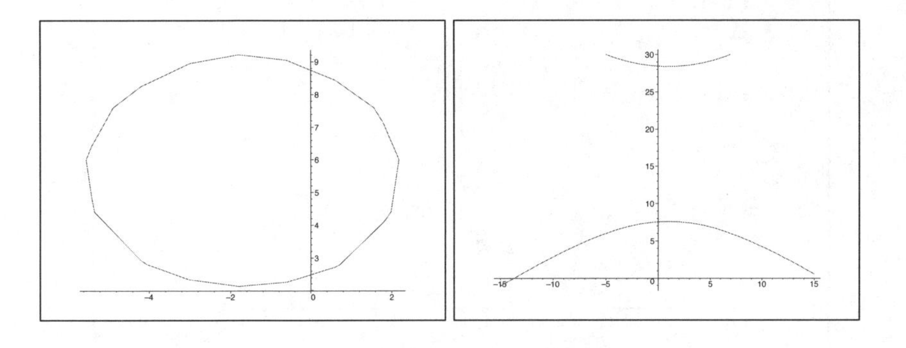

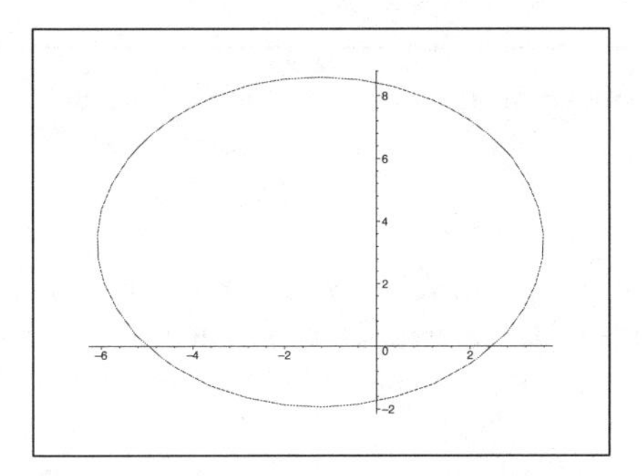

Fig. 4.8. Bisec(a_1,a_2), Bisec(a_1,a_3) and Bisec(a_2,a_3) in Example 4.5

To solve the problem the different ordered regions were approximated by convex regions. In total we distinguish 16 regions which give rise to 16 non-linear problems with constant order of the distance functions. A map of some of the ordered regions is shown in Figure 4.9.

The problems were solved with a nonlinear programming routine, *Primal-DualLogBarrier*, implemented in Maple. The objective functions, feasible regions and optimal solutions of each one of the problems are shown in Table 4.2. The optimal solution is the one achieving the minimum among the values in that table:

$$f_\lambda(x[13]) = \min_{1 \leq i \leq 16} f_\lambda(x[i]) = 2.142.$$

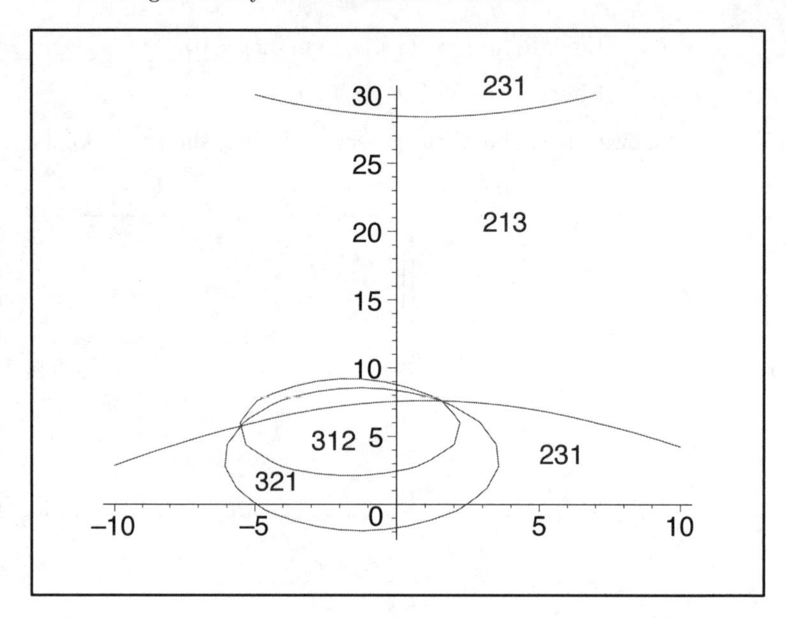

Fig. 4.9. Bisectors among all the demand points and some ordered regions in Example 4.5

The rest of this section is devoted to show that under mild hypothesis the optimal solutions of OMP with non polyhedral gauges can be arbitrarily approximated by a sequence of optimal solutions of OMP with polyhedral gauges.

Let B be the unit ball of the gauge $\gamma_B(\cdot)$, $\{C^k\}_{k\in\mathbb{N}}$ an increasing sequence of polyhedra included in B and $\{D^k\}_{k\in\mathbb{N}}$ a decreasing sequence of polyhedra including B, that is,

$$C^k \subset C^{k+1} \subset B \subset D^{k+1} \subset D^n \quad \text{for all } k = 1, 2, \ldots$$

Let $\gamma_{C^k}(\cdot)$ and $\gamma_{D^k}(\cdot)$ be the gauges whose unit balls are C^k and D^k respectively.

Proposition 4.1 *If $C^k \subset B \subset D^k$ we have that*

$$\gamma_{C^k}(x) \geq \gamma_B(x) \geq \gamma_{D^k}(x) \quad \forall x \in \mathbb{R}^n.$$

The proof follows directly from the definition of gauges.

Recall that given two compact sets A, B the Hausdorff distance between A and B is

$$d_H(A, B) = \max(\max_{x\in A} d_2(x, B), \max_{y\in B} d_2(A, y))$$

where $d_2(x, B) = \min_{y\in B} d_2(x, y)$ being d_2 the Euclidean distance.

Table 4.2. Approximated ordered regions in Example 4.5

Number	Objective	Region	$x = (x_1, x_2)$	$f_\lambda(x)$
(P1)	$\frac{d1}{4} + \frac{d2}{2} + \frac{d3}{8}$	$g_{13}(x) \geq 0,\ x_2 \geq 9.2$	(0.997,9.178)	3.44
(P2)	$\frac{d1}{8} + \frac{d2}{2} + \frac{d3}{4}$	$g_{13}(x) \leq 0,\ x_1 \geq 3.55,\ x_2 \leq 9.2$	(3.55,5.378)	2.816
(P3)	$\frac{d1}{8} + \frac{d2}{2} + \frac{d3}{4}$	$g_{13}(x) \leq 0,\ x_1 \leq -6,\ x_2 \leq 9.2$	(6.00,5.432)	5.182
(P4)	$\frac{d1}{8} + \frac{d2}{2} + \frac{d3}{4}$	$g_{13}(x) \leq 0,\ x_1 \geq -6,$ $x_1 \leq 3.55,\ x_2 \leq -2$	(1.352,-2.000)	5.472
(P5)	$\frac{d1}{8} + \frac{d2}{4} + \frac{d3}{2}$	$g_{23}(x) \leq 0,\ x_1 \geq -6,$ $x_1 \leq 3.55,\ x_2 \leq 2$	(2.189,1.999)	3.509
(P6)	$\frac{d1}{4} + \frac{d2}{8} + \frac{d3}{2}$	$g_{12}(x) \geq 0,\ g_{13}(x) \leq 0$	(2.189, 6.215)	2.548
(P7)	$\frac{d1}{8} + \frac{d2}{4} + \frac{d3}{2}$	$g_{23}(x) \leq 0,\ x_2 \geq 2,$ $7.5{*}x_1{-}4.3{*}x_2{+}3.75 \geq 0$	(2.921,5.966)	2.602
(P8)	$\frac{d1}{8} + \frac{d2}{4} + \frac{d3}{2}$	$g_{23}(x) \leq 0,\ x_2 \geq 2,$ $-5.5{*}x_1{-}4.5{*}x_2{-}11 \geq 0$	(-3.688,2.064)	4.752
(P9)	$\frac{d1}{2} + \frac{d2}{8} + \frac{d3}{4}$	$g_{12}(x) \geq 0,\ g_{23}(x) \leq 0,$ $3.2{*}x_1{-}5.5{*}x_2{+}49.5 \leq 0$	(-.7328,8.574)	3.392
(P10)	$\frac{d1}{2} + \frac{d2}{4} + \frac{d3}{8}$	$g_{12}(x) \geq 0,\ g_{23}(x) \geq 0,$ $x_1 - 2.1 * x_2 + 20 \leq 0$	(-.823, 9.131)	3.614
(P11)	$\frac{d1}{4} + \frac{d2}{2} + \frac{d3}{8}$	$g_{13}(x) \geq 0,\ x_2 \leq 9.2,$ $3.2{*}x_1{+}1.5{*}x_2{-}16.5 \geq 0$	(1.598, 7.591)	2.665
(P12)	$\frac{d1}{4} + \frac{d2}{2} + \frac{d3}{8}$	$g_{13}(x) \geq 0,\ x_2 \leq 9.2,$ $-15.5 * x_1 + 10.3 * x_2 - 155 \leq 0$	(.5136,7.596)	2.592
(P13)	$\frac{d1}{8} + \frac{d2}{2} + \frac{d3}{4}$	$g_{13}(x) \leq 0,\ x_2 \geq 28.2$	(1.0,6.510)	2.142
(P14)	$\frac{d1}{8} + \frac{d2}{2} + \frac{d3}{4}$	$g_{13}(x) \leq 0,\ x_1 \leq 3.55,\ x_2 \geq -2,$ $2.5 * x_1 - 2.5 * x_2 - 6.25 \geq 0$	(3.550, 1.050)	4.00
(P15)	$\frac{d1}{8} + \frac{d2}{2} + \frac{d3}{4}$	$g_{13}(x) \leq 0,\ x_1 \geq -6,\ x_2 \geq -2,$ $2 * x_1 + x_2 + 10 \leq 0$	(-6.0,2.0)	5.713
(P16)	$\frac{d1}{2} + \frac{d2}{8} + \frac{d3}{4}$	$g_{12}(x) \geq 0,\ g_{23}(x) \leq 0,$ $3.2 * x_1 - 5.5 * x_2 + 49.5 \geq 0,$ $x_2 \geq 7.6$	(1.592,7.60)	2.670

Proposition 4.2 *Let K be a compact set. If C^k converges to B and D^k converges to B under the Hausdorff metric then for all $\varepsilon > 0$ there exists k_0 such that for all $k \geq k_0$*

$$\max_{x \in K} |f_\lambda^{C^k}(x) - f_\lambda(x)| < \varepsilon$$

$$\max_{x \in K} |f_\lambda^{D^k}(x) - f_\lambda(x)| < \varepsilon$$

being $f_\lambda^{C^k}(x) := \sum_{i=1}^M \lambda_i d_{(i)}^{C^k}(x)$ *and* $f_\lambda^{D^k}(x) := \sum_{i=1}^M \lambda_i d_{(i)}^{D^k}(x)$.

Proof.
We only prove the first inequality. The second being analogue.

Since C^k converges to B under the Hausdorff metric verifying $C^k \subset C^{k+1}$ for all $k \in \mathbb{N}$, and K is a compact set then given $\varepsilon > 0$ there exists k_a for every $a \in A$ such that if $k > k_A := \max_{a \in A} k_a$ then $\left|\gamma_B(x-a) - \gamma_{C^k}(x-a)\right| < \dfrac{\varepsilon}{\sum_{i=1}^M w_i \sum_{i=1}^M \lambda_i} \quad \forall x \in K$.

By continuity we have that for any i, j and any $x \in K$ verifying that $w_i\gamma_B(x - a_i) < w_j\gamma_B(x - a_j)$ there exists k_0 such that for all $k > k_0$

$$w_i\gamma_{C^k}(x - a_i) < w_j\gamma_{C^k}(x - a_j)$$

On the other hand, if there exists j, l and $x \in K$ such that $w_j\gamma_B(x - a_j) = w_l\gamma_B(x - a_l)$ then there also exists k_0 and a permutation σ^{k_0} such that for all $k > k_0$ would satisfy: 1) $w_{\sigma^{k_0}(j)}\gamma_{C^k}(x - a_{\sigma^{k_0}(j)}) = d_{(j)}^{C^k}(x)$, and 2) $w_{\sigma^{k_0}(j)}\gamma_B(x - a_{\sigma^{k_0}(j)}) = d_{(j)}^B(x)$. Hence, we have for any $x \in K$ and $k > \max\{k_A, k_0\}$ that

$$d_{(j)}^B(x) = w_{\sigma^{k_0}(j)}\gamma_B(x - a_{\sigma^{k_0}(j)})$$

$$d_{(j)}^{C^k}(x) = w_{\sigma^{k_0}(j)}\gamma_{C^k}(x - a_{\sigma^{k_0}(j)}).$$

Therefore, for any $x \in K$ and $k > \max\{k_A, k_0\}$ we obtain that

$$\begin{aligned} |f_\lambda^{C^k}(x) - f_\lambda(x)| &= \textstyle\sum_{i=1}^M \lambda_i |d_{(i)}^B(x) - d_{(i)}^{C^k}(x)| \\ &= \textstyle\sum_{i=1}^M \lambda_i w_{\sigma^{k_0}(i)} |\gamma_B(x - a_{\sigma^{k_0}(i)}) - \gamma_{C^k}(x - a_{\sigma^{k_0}(i)})| < \varepsilon. \end{aligned}$$

□

Corollary 4.3 *i) If C^k converges to B under the Hausdorff metric, then $f_\lambda^{C^k}(x)$ converges to $f_\lambda(x)$, and the sequence $\{f_\lambda^{C^k}(x)\}_{k \in \mathbb{N}}$ is decreasing.*
ii) If D^k converges to B under the Hausdorff metric, then $f_\lambda^{D^k}(x)$ converges to $f_\lambda(x)$, and the sequence $\{f_\lambda^{D^k}(x)\}_{k \in \mathbb{N}}$ is increasing.

In the following, we use another type of convergence, called epi-convergence (see Definition 1.9 in the book by Attouch [6]). Let $\{g; g^\nu, \nu = 1, 2, \ldots\}$ be a collection of extended-values functions. We say that g^ν epi-converges to g if for all x,

$$\inf_{x^\nu \to x} \liminf_{\nu \to \infty} g^\nu(x^\nu) \geq g(x)$$

$$\inf_{x^\nu \to x} \limsup_{\nu \to \infty} g^\nu(x^\nu) \leq g(x)\ ,$$

where the infima are with respect to all subsequences converging to x. The epi-convergence is very important because it establishes a relationship between the convergence of functionals and the convergence of the sequence of their minima. Further details can be found in the book by Attouch [6].

Our next result states the theoretical convergence of the proposed scheme.

Theorem 4.3 *i) Let $\{x_k\}_{k\in\mathbb{N}}$ be a sequence such that $x_k \in \mathcal{M}(f_\lambda^{C^k}, A)$ then any accumulation point of $\{x_k\}_{k\in\mathbb{N}}$ belongs to $\mathcal{M}(f_\lambda, A)$.*

ii) Let $\{x^k\}_{k\in\mathbb{N}}$ be a sequence such that $x^k \in \mathcal{M}(f_\lambda^{D^k}, A)(x)$ then any accumulation point of $\{x^k\}_{k\in\mathbb{N}}$ belongs to $\mathcal{M}(f_\lambda, A)$.

Proof.

We only prove the first part, because the proof of the second one is built on the same pattern using Proposition 2.41. in [6] instead of Proposition 2.48.

First of all, since the sequence $\{f_\lambda^{C^k}\}_{k\in\mathbb{N}}$ is a decreasing sequence applying Theorem 2.46 in [6] we obtain that the sequence $\{f_\lambda^{C^k}(x)\}_{k\in\mathbb{N}}$ is epi-convergent.

In addition, we get from Proposition 2.48 in [6] that

$$\lim_{k\to\infty} \inf_{x\in\mathbb{R}^n} f_\lambda^{C^k}(x) = \inf_{x\in\mathbb{R}^n} \lim_{k\to\infty} f_\lambda^{C^k}(x) = \inf_{x\in\mathbb{R}^n} f_\lambda(x) \tag{4.2}$$

Since $\{f_\lambda^{C^k}\}_{k\in\mathbb{N}}$ is epi-convergent, we get from Theorem 2.12 in [6] that any accumulation point of the sequence $\{x_k\}_{k\in\mathbb{N}}$ is an optimal solution of the problem with objective function f_λ. □

4.4 Continuous OMPs with Positive and Negative Lambdas

During the last years, a new version of the Weber problem has been considered dealing simultaneously with positive and negative weights, see for instance [45, 58, 194, 195]. This formulation allows to locate a facility which may be attractive for some demand points and repulsive for others. Indeed, a nuclear plant, a garbage dump or a sewage plant may both be desirable to the main users, who want to be close to their location in order to minimize transportation cost, and obnoxious to residents who want to have the facility located far from their vicinity. Another example could be the location of a chemical plant. This plant can be undesirable for the closest cities because of the pollution it emits, but at the same time it could be desirable for other cities due to the jobs it may offer.

Different techniques have been developed for solving these kinds of problems, and a large number of them are based on the theory of difference of convex (*d.c.*) programming, see for instance [104, 37, 38]. The *d.c.* programming solves the problem reducing it to an equivalent problem dealing with a concave function restricted to a convex region.

In this section, we consider the ordered median problem with positive and negative lambdas. Our purpose is to present an approach to solve this problem with polyhedral norms which can also be adapted to solve the ordered median problem with positive and negative lambdas under general norms.

In this section, we present a necessary condition for the existence of optimal solution. In addition, we give a finite dominating set for the case with polyhedral norms. For this particular case, two exact algorithms are developed. These algorithms find an optimal solution of the problem in finite time. Finally, we outline how all the results proved for polyhedral norms can be extended to general norms using the same methodology presented in Section 4.3. For further details on the topic covered in this section the reader is referred to [146, 171].

The problem which we are going to analyze was already introduced in (2.3). We repeat it for the sake of readability:

$$\min_{x \in \mathbb{R}^n} f_\lambda(x) = \sum_{i=1}^{M} \lambda_i^+ d_{(i)}(x) + \sum_{i=1}^{M} \lambda_i^- d_{(i)}(x) \tag{4.3}$$

where as usual distances $d_i(x) = \gamma(x - a_i)$ for $i = 1, \ldots, M$, and

$$\lambda_i^+ = \begin{cases} \lambda_i & \text{if } \lambda_i > 0 \\ 0 & \text{otherwise,} \end{cases} \qquad \lambda_i^- = \begin{cases} \lambda_i & \text{if } \lambda_i < 0 \\ 0 & \text{otherwise.} \end{cases}$$

These models are quite general and allow the formulation of different interesting situations. For instance, if we consider that the absolute value of all the negative lambdas is less than the positive ones, then the model would represent the location of an obnoxious facility (nuclear plant, garbage dump ...) which is very dangerous for the nearest centers but economical reasons force to locate not too far away, so that the transportation cost is not too expensive.

The relationship among the positive and negative values of the lambda weights plays an important role in the existence of optimal solution of Problem (4.3). Let $L = \sum_{i=1}^{M}(\lambda_i^+ + \lambda_i^-)$. When some lambdas are negative it may happen that the objective function f_λ has no lower bound, thus f_λ is going to $-\infty$ when distances increase. In order to avoid this problem, we proved in Theorem 2.1 (see Chapter 2) the following result.

Theorem 4.4 *(Necessary condition) If $L \geq 0$ then any optimal location is finite. If $L < 0$ then any optimal location is at infinity. Moreover, if $L = 0$ then the function f_λ is bounded.*

We start with a localization results for the set of optimal solutions of Problem (4.3).

Theorem 4.5 *If $L > 0$ Problem (4.3) has always an optimal solution on an ordered intersection point.*

Proof.
The function f_λ is linear within each ordered elementary convex set. Therefore, in each ordered elementary convex set f_λ must attain at least one of its minima at some extreme point of this region. Since the global minimum is given by the minimum of the minima on each ordered elementary convex set, the result follows. □

Using Theorem 4.5 we can solve the problem be enumeration. However, we will show in the following, that more structural properties can be obtained.

Without loss of generality we assume that optimal solutions exist. Before we proceed to show the equivalent formulations of our original problem we show that the objective function of the ordered median problem with negative and positive lambdas is neither convex nor *d.c.*. Therefore, we cannot apply either the usual techniques of convex analysis or the usual tools of *d.c.* programming. Note that in our model we do not assume that the weights are in increasing order. Thus, neither the positive summation of the distances nor the negative one are convex functions. This implies that the objective function cannot be decomposed as a difference of two convex functions, i.e., it is not *d.c.*. Indeed, given the objective function

$$f_\lambda(x) = \sum_{i=1}^{M} \lambda_i^+ d_{(i)}(x) - \sum_{i=1}^{M} |\lambda_i^-| d_{(i)}(x)$$

it can be seen that the functions $\sum_{i=1}^{M} \lambda_i^+ d_{(i)}(x)$ and $\sum_{i=1}^{M} |\lambda_i^-| d_{(i)}(x)$ are convex functions if and only if the sequence of non zero coefficients are in non-decreasing order [48], that is,

$$\lambda_i \leq \lambda_{i+1} \quad \forall i = 1, \ldots, M$$

and

$$|\lambda_j| \leq |\lambda_{j+1}| \quad \forall j = 1, \ldots, M.$$

This fact implies that in the general case we deal with an objective function which is not *d.c.*. However, despite the fact that the objective function is not *d.c.*, this problem has some special properties that allow us to solve it using equivalent formulations.

Lemma 4.4 *An optimal solution of Problem (4.3) can be obtained solving the following problem:*

$$\min\ g(x,t) := t + \textstyle\sum_{j=1}^{M} \lambda_j^- d_{(j)}(x) \tag{4.4}$$
$$\begin{aligned} s.t.: \quad & \textstyle\sum_{i=1}^{M} \lambda_i^+ d_{(i)}(x) \le t, \\ & (x,t) \in \mathbb{R}^n \times \mathbb{R}. \end{aligned}$$

Proof.
Consider the set $D = \{(x,t) \in \mathbb{R}^n \times \mathbb{R} : \sum_{j=1}^{M} \lambda_j^+ d_{(j)}(x) \le t\}$, then

$$f_\lambda(x) \le g(x,t) \quad \forall (x,t) \in D, \tag{4.5}$$

and

$$f_\lambda(x) = g(x, \sum_{j=1}^{M} \lambda_j^+ d_{(j)}(x)). \tag{4.6}$$

So, if $g(y^*,t^*) = \min_{(y,t)\in D} g(y,t)$, we obtain that

$$t^* = \sum_{j=1}^{M} \lambda_j^+ d_{(j)}(y^*)$$

because $g(x,t) \ge g(x, \sum_{j=1}^{M} \lambda_j^+ d_{(j)}(x))$, $\forall\, x \in D$. Therefore, we have that y^* is an optimal solution of Problem (4.3).

Note that, if x^* is an optimal solution of Problem (4.3), then by applying (4.5), $(x^*, \sum_{j=1}^{M} \lambda_j^+ d_{(j)}(x^*))$ is also an optimal solution of Problem (4.4).

□

It should be noted, that if we restrict Problem (4.4) to any convex domain where the order of distances is fixed, the objective function of this problem is concave. However, the special difficulty of Problem (4.4) is that its feasible region is not necessarily convex.

Nevertheless, it is possible to transform Problem (4.4) into a linear program within each ordered elementary convex set. The discussion above allows us to establish a reformulation of Problem (4.4) which is adequate to our algorithmic approach.

Let us consider an ordered region O_σ and the elementary convex set $C = \bigcap_{j=1}^{M}(a_{\sigma(j)} + N(p_{\sigma(j)}))$. Then, an optimal solution of Problem (4.3) can be obtained by solving Problem (4.7) in each ordered elementary convex set

$$(\mathrm{P}(C \cap O_\sigma))\ \min\ t + \sum_{j=1}^{M} \lambda_j^- z_{\sigma(j)} \tag{4.7}$$

$$\text{s.t.:} \quad \sum_{i=1}^{M} \lambda_i^+ z_{\sigma(i)} \le t,$$

$$\langle p_{\sigma(j)}, x - a_{\sigma(j)} \rangle = z_{\sigma(j)} \quad j = 1, \ldots, M, \tag{4.8}$$

$$x - a_{\sigma(j)} \in N(p_{\sigma(j)}) \quad j = 1, \ldots, M, \tag{4.9}$$

$$z_{\sigma(k)} \le z_{\sigma(k+1)} \quad k = 1, 2, \ldots, M.$$

Indeed, $x - a_{\sigma(j)} \in N(p_{\sigma(j)})$ is equivalent to $\gamma(x - a_{\sigma(j)}) = \langle p_{\sigma(j)}, x - a_{\sigma(j)} \rangle$. Therefore, $z_{\sigma(j)} = \gamma(x - a_{\sigma(j)})$ for any $x \in C$.
Let us define the sets

$$D = \{(x,t) \in (C \cap O_\sigma) \times \mathbb{R} \ : \ \sum_{i=1}^{M} \lambda_i^+ \gamma(x - a_{\sigma(i)}) \le t\};$$

$$D' = \{(x,t,z) \in \mathbb{R}^n \times \mathbb{R} \times \mathbb{R}^M \ : \ \sum_{j=1}^{M} \lambda_j^+ z_{\sigma(j)} \le t, \langle p_{\sigma(j)}, x - a_{\sigma(j)} \rangle = z_{\sigma(j)},$$
$$x \in a_{\sigma(j)} + N(p_{\sigma(j)}),\ j = 1, \dots, M,\ \ z_{\sigma(k)} \le z_{\sigma(k+1)}\ ,$$
$$k = 1, 2, \dots, M-1\}.$$

The projection of the first $M+1$ entries of these two sets coincide. Additionally, D' is the feasible set of (4.7). Thus, combining both results and the fact that the whole family of ordered elementary convex sets induces a subdivision of $\mathbb{R}^n$, we obtain that if an optimal solution of Problem (4.4) exists within $O_\sigma \cap C$, it can be obtained by solving Problem (4.7).

4.4.1 The Algorithms

In order to solve Problem (4.3), at least, two different approaches can be followed. First, the problem can be solved by direct application of Theorem 4.5. This implies to carry out a global search in the set of ordered intersection points (OIP). The second approach consists of solving the problem in each ordered elementary convex set where we have the equivalent linear programming formulation (4.7).

The first approach (Algorithm 4.3) is conceptually very simple but implies to compute the entire set of ordered intersection points. This method is useful when it is combined with efficient algorithms to maintain arrangements of hyperplanes. The second approach, using the equivalent formulations presented above, takes advantage of the linear structure of the objective function inside each ordered elementary convex set to avoid the whole enumeration of the ordered intersection points. Anyway, as the considered problem is neither convex nor *d.c.*, we have to solve a linear program in each ordered elementary convex set. Basically, the second approach is a backtracking search in the set of ordered elementary convex sets.

To simplify the presentation of Algorithm 4.4, let $\mathsf{val}(\mathrm{P}(\mathrm{C}_1))$ denote the optimal value of Problem (4.7) in the ordered elementary convex set C_1. (See Algorithm 4.4.)

The algorithms 4.3 and 4.4 are efficient because we can find in both cases an optimal solution of the original problem in polynomial time. Let γ^0 be the dual norm of the norm γ, and let B^0 be its unit ball. Assume that B^0 has G extreme points that we denote by e_g^0, $g = 1 \dots, G$. Let $k > n$ denote the number of fundamental directions of the norm γ. Consider the following

Algorithm 4.3: *Evaluation of all ordered intersection points*

1 Compute the subdivision induced by the arrangement of hyperplanes generated by all the gauges and the bisectors between each pair of demand points.
2 Compute the objective value in each ordered intersection point.
3 Output $x^* = \arg\min_{x \in \mathsf{OIP}} f_\lambda(x)$ and $f_\lambda(x^*)$.

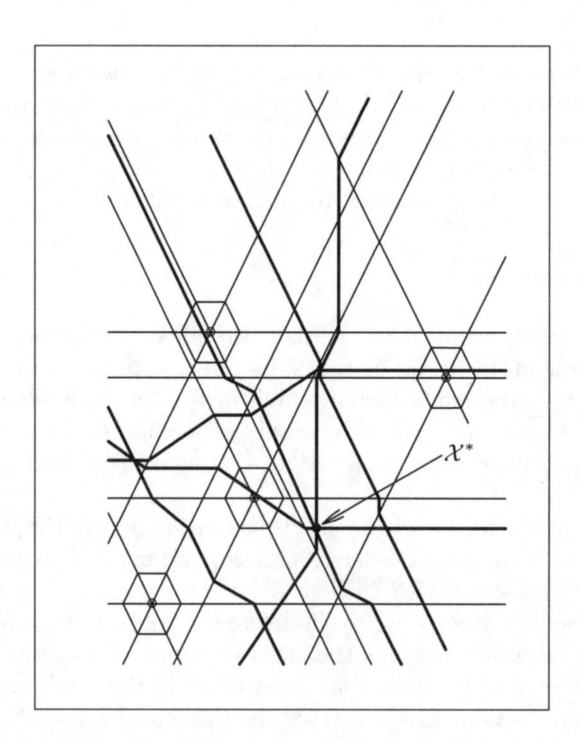

Fig. 4.10. Example with attractive and repulsive facilities

arrangements of hyperplanes. For each demand point a_i, $i = 1, \ldots, M$ we consider G hyperplanes which normal vectors are e_g^0, $g = 1 \ldots, G$. The entire collection of bisectors $\{w_j \gamma(x, a_j) - w_i \gamma(x, a_i) = 0\}$ (piecewise linear forms) is defined by at most $O((MG)^2)$ hyperplanes. These hyperplanes induce a cell partition of $\mathbb{R}^n$ that can be computed in $O((MG)^{2n})$ time for any fixed $n \geq 2$, (see Edelsbrunner [68]). In the first algorithm we have to evaluate the objective function in each intersection point which needs $O(Mlog(Mk))$. Therefore, the overall complexity is $O((MG)^{2n} M \log(kM))$.

Algorithm 4.4: *Solving the single facility OMP with positive and negative lambdas via linear programming*

(Initialization)
Let $v^* = +\infty$
$X^* = \emptyset$.
$\mathcal{L}_R = \mathcal{L}_A = \emptyset$.
Choose C_0 being any ordered elementary convex set.
Find $A(C_0)$ the set of ordered elementary convex sets adjacent to C_0. (Use the data structure by Edelsbrunner [68])
Do $\mathcal{L}_A = \mathcal{L}_A \cup A(C_0)$.
(Main)
repeat
  Choose C_1 in $\mathcal{L}_A$.
  Solve Problem (4.7) in C_1.
  if $\mathsf{val}(\mathrm{P(C_1)}) < \mathrm{v}^*$ **then**
    $v^* = \mathsf{val}(\mathrm{P(C_1)})$ and $\mathrm{X}^* = \arg\min \mathrm{P(C_1)}$
  end
  $\mathcal{L}_A = (\mathcal{L}_A \setminus C_1) \cup (A(C_1) \setminus \mathcal{L}_R)$
  $\mathcal{L}_R = \mathcal{L}_R \cup C_1$
until $\mathcal{L}_A = \emptyset$

In the second case we have to solve a linear program in each ordered elementary convex set. Each one of these linear problems can be solved in polynomial time T, and we obtain an overall complexity of $O(T(MG)^{2n})$.

In conclusion, Algorithm 4.3 and 4.4 can find an optimal solution to Problem (4.3) in finite time.

Example 4.6
Consider the location problem

$$\min_{x \in \mathbb{R}} -d_{(1)}(x) - 1.25d_{(2)}(x) + 1.5d_{(3)}(x) + 1.75d_{(4)}(x)$$

where $A = \{(3,0),(0,11),(16,8),(-4,-7)\}$, and γ_B is the hexagonal gauge with $Ext(B) = \{(2,0),(1,2),(-1,2),(-2,0),(-1,-2),(1,-2)\}$.

We obtain that the optimal solution is the point $\mathsf{M}(f_\lambda) = \{(7.25,-2)\}$ with objective value 11.125. (See Figure 4.10.) Table 4.3 shows the optimal objective value and the x-coordinates of Problem (4.7) within the ordered elementary convex sets of this example.

An Illustrative Example

Example 4.7
As a final example we consider:

Table 4.3. Evaluations given by Algorithm 4.4

It.	$x = (x_1, x_2)$	**Objective**	**It.**	$x = (x_1, x_2)$	**Objective**
1	(-7.50,0.00)	23.75	16	(6.50,8.00)	13.87
2	(-6.50,-2.00)	24.70	17	(7.00,8.00)	13.62
3	(-5.25,2.50)	17.43	18	(7.25,-3.50)	12.44
4	(-4.00,-7.00)	31.00	19	(7.25,-2.00)	11.12
5	(-1.25,8.50)	20.37	20	(7.25,0.00)	12.37
6	(0.00,-0.50)	20.06	21	(7.25,8.00)	13.12
7	(0.50,2.00)	15.56	22	(7.25,25.5)	29.37
8	(0.50,5.50)	22.12	23	(8.75,11.5)	17.7
9	(2.50,6.00)	16.19	24	(8.75,18.5)	22.12
10	(3.00,0.00)	16.56	25	(9.12,-5.75)	12.53
11	(3.50,0.00)	15.56	26	(9.12,4.50)	12.53
12	(3.50, 8.00)	15.00	27	(12.0,-2.00)	18.82
13	(4.25,-9.50)	22.62	28	(14.5,-7.00)	17.56
14	(4.25,2.50)	15.87	29	(16.0,8.00)	26.56
15	(5.75,12.5)	15.37	30	(16.2,8.5)	26.43

i	a_i	w_i	γ_i
1	(2,6.5)	1	l_1
2	(5,9.5)	1	l_∞
3	(6.5,2)	1	l_∞
4	(11,9.5)	1	l_1

We will present the solution of the 1-facility OMP for some choices of λ.

For the classical Weber function ($\lambda = (1, 1, 1, 1)$), the set of optimal solutions is given by the elementary convex set $conv\{a_1, a_2, b_1\}$ with $b_1 = (8, 6.5)$, and the optimal value of the corresponding objective function is 19.5.

The output of the program to solve this problem is shown in Figure 4.11. One can see here, that in the Classical Weber Problem the bisectors do not play a role, since the three points are intersections of fundamental directions.

In the Center case with $\lambda = (0, 0, 0, 1)$, the segment $\overline{b_1 b_2}$ with $b_1 = (8, 6.5)$ and $b_2 = (6.5, 8)$ is optimal and the objective function value is 6. The existing facilities a_1 and a_4 are the bottleneck points, since $\overline{b_1 b_2} \subset \mathsf{Bisec}(a_1, a_4)$.

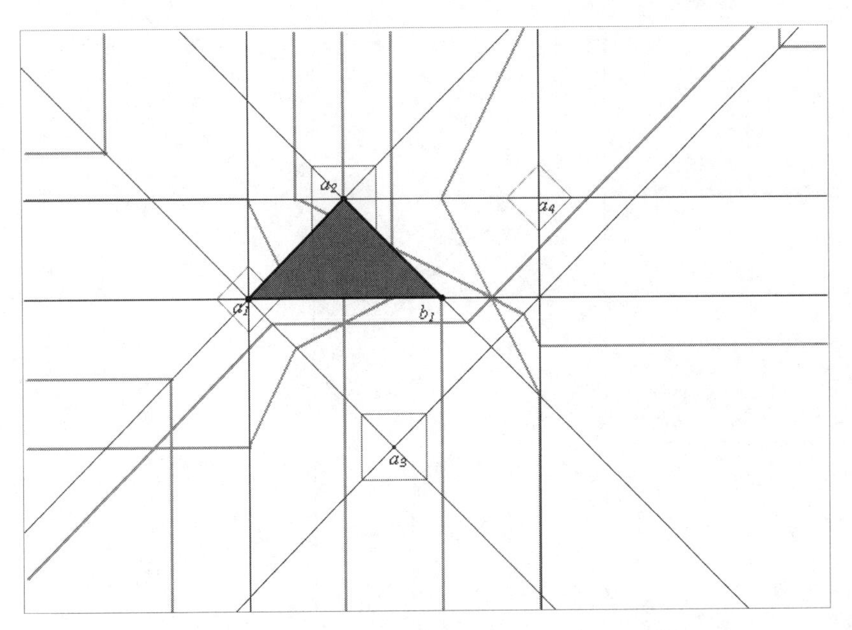

Fig. 4.11. Optimal solutions for $\lambda = (1, 1, 1, 1)$

The output for this case is shown in Figure 4.12. Observe that b_2 cannot be determined only with fundamental directions.

For the ordered median function with $\lambda = (0, 1, 2, 3)$ the optimal solution reduces to the single point $b_1 = (8, 6.5)$, as shown in Figure 4.13, with objective value 34.5.

The ordered median function corresponding to the above cases is convex, but we can also compute the solution in the non-convex case. The most simple example is to choose $\lambda = (1, 0, 0, 0)$, and the optimal solutions coincide with any existing facility with objective value 0. Hence, we would not obtain a convex solution set. Furthermore, it would not be a connected solution either.

If $\lambda = (1, 1, 0, 2)$ the solution is given by the union of two segments, $\overline{b_1 b_2}$ and $\overline{b_2 b_3}$ with $b_1 = (8, 6.5)$, $b_2 = (6.5, 8)$, and $b_3 = (4.5, 9)$. In this case the set of optimal solutions is connected but not convex and the objective value is 19.5.

The output is shown in Figure 4.14, and we can see that the set of optimal solutions is not convex, since the convex hull of the points b_1, b_2, and b_3 is the triangle whose vertices are these three points. Moreover, we can see as before with b_2, that b_3 cannot be determined only by intersecting fundamental directions.

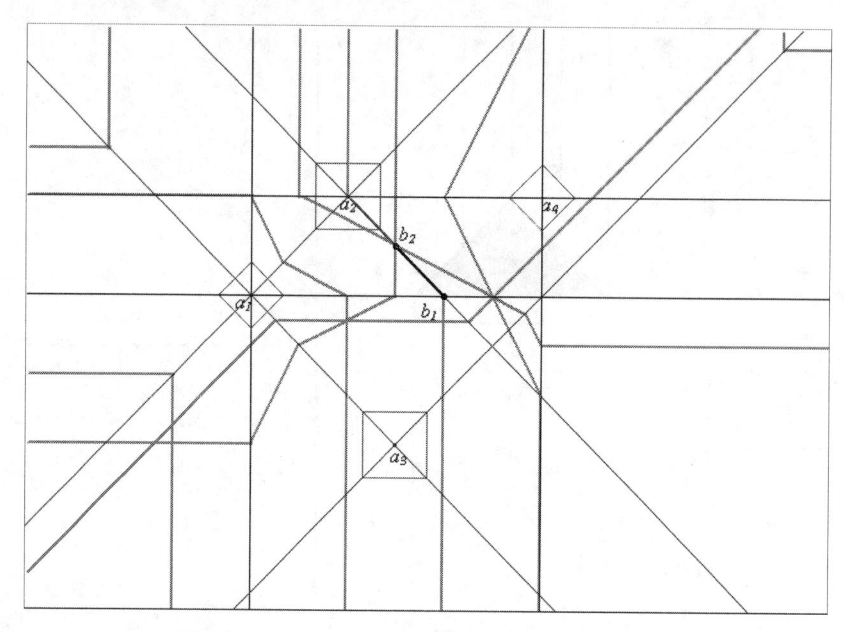

Fig. 4.12. Optimal solutions for $\lambda = (0, 0, 0, 1)$

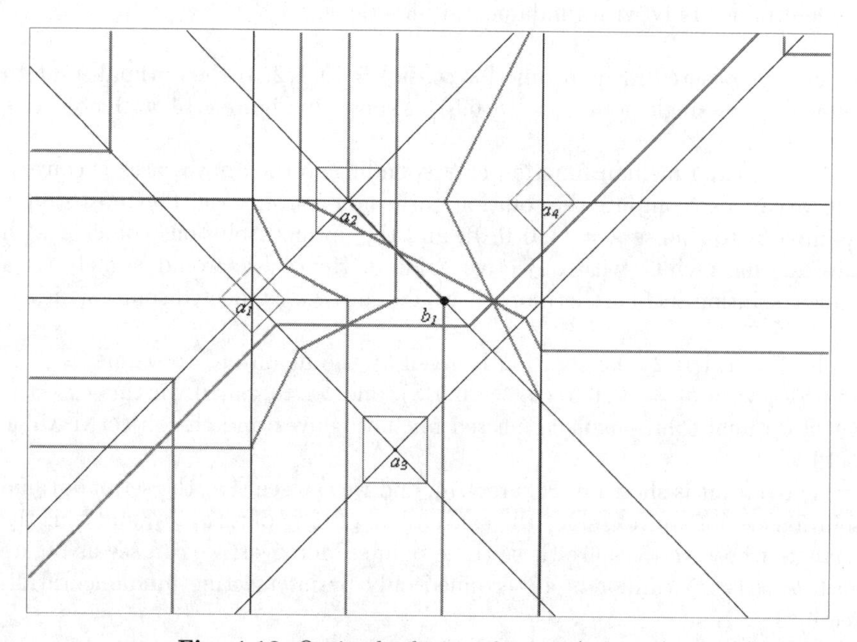

Fig. 4.13. Optimal solutions for $\lambda = (0, 1, 2, 3)$

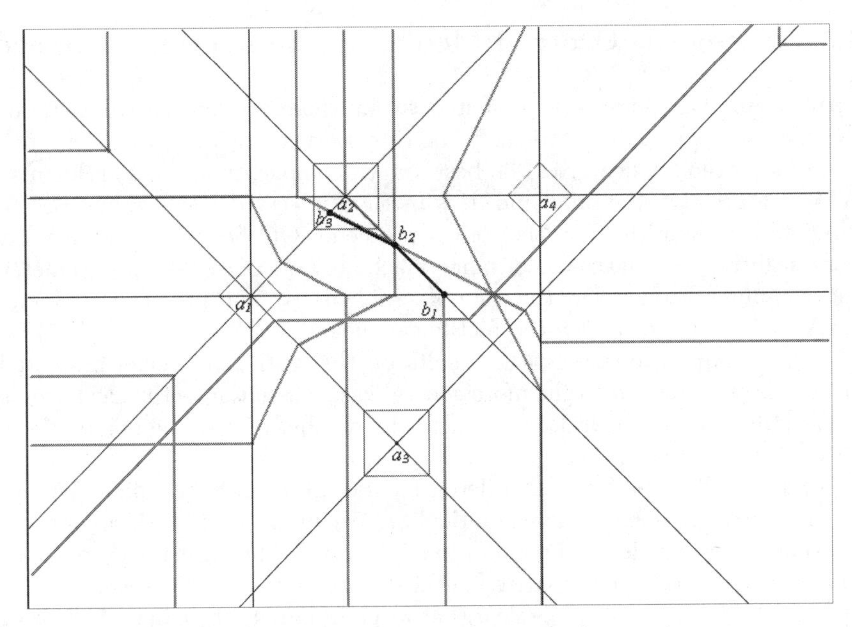

Fig. 4.14. Optimal solutions for $\lambda = (1, 1, 0, 2)$

The last example shows the behavior of the OMP with some negative weights. We choose $\lambda = (2, 2, -3, 1)$. The intersection points between the segments are

Intersection Points					
(18.5, 14)	(18.5, 15.5)	(18.5, 23)	(14, 9.5)	(12.5, 9.5)	(11, 33.26)
(11, 15.50)	(11, 8)	(11, 6.5)	(11, 5)	(11, 3.5)	(11, −2.5)
(11, −33.26)	(10.1, 9.5)	(10.5, 6)	(10, 5.5)	(9.5, 5)	(9.5, 6.5)
(9.27, 6.62)	(8.75, 5.75)	(8, 5.75)	(8, 3.5)	(8, 0.5)	(8, 9.5)
(8, 6.5)	(7.25, 7.25)	(6.5, 11)	(6.5, 9.5)	(6.5, 8)	(6.5, 6.5)
(6.50, 2.00)	(5.75, 5.75)	(5, 6.5)	(5, 5.75)	(5, 3.5)	(5, 0.5)
(4.99, 9.50)	(4.5, 9)	(3.5, 11)	(3.5, 5)	(3.5, 9.5)	(3, 7.5)
(2.75, 5.75)	(2, 12.5)	(2, 9.5)	(2, 5)	(2, −2.5)	(2, 2)
(1.99, 6.5)	(1.01, 9.5)	(−0.41, −4.91)	(−0.41, 2.59)	(−0.41, 2)	(−1, 9.5)
(−0.41, 4.09)	(−1, 2)	(−2.41, 16.91)	(−2.41, 10.91)	(−2.5, 2)	

The minimum value of the objective function is −3. The optimal solution is attained at the points (8, 3.5), (8, 0.5), (6.5, 11), (6.5, 9.5), (4.5, 9), (3.5, 11), (3.5, 9.5).

4.5 Finding the Ordered Median in the Rectilinear Space

Apart from the approaches presented so far there are alternative methods to solve ordered median problems. In this section, we want to focus on a powerful method which gives the best known complexity for this problem so far (see [117]). On the one hand, this method is very efficient giving rise to subquadratic algorithms for the convex problems. On the other hand, it is far from intuitive and makes use of fancy machinery of computational geometry and computer science. For the sake of simplicity we will present the results for $X = \mathbb{R}^n$ equipped with the rectilinear metric, i.e. $d(x, y) = \gamma(x - y) := \sum_{i=1}^{n} |x_i - y_i|$. Nevertheless, the results on the rectilinear model presented below can be extended to the more general case where the rectilinear norm is replaced by any polyhedral norm where the number of extreme points of the unit ball is constant.

First we solve the general ordered median problem on the line. We are given M demand points ordered on the line. We know that there is an optimal solution either on a demand point or in a bisector. In this case, bisectors correspond to the $O(M^2)$ points satisfying the equations $\{w_j d(x, a_j) - w_i d(x, a_i) = 0\}$ for any $i, j = 1, \ldots, M$. It is clear that we can compute them in $O(M^2)$ time. Then, since the evaluation of the ordered median function takes $O(M \log M)$, we can find an optimal solution of the problem by complete enumeration in $O(M^3 \log M)$.

If we consider the convex case, i.e. lambdas arranged in non-decreasing sequence, we can improve the above complexity up to $O(M \log^2 M)$ time. Using the convexity of the objective function, we first apply a binary search to identify a pair of adjacent demand points, $[a_i, a_{i+1}]$, containing the ordered median x^*. Since it takes $O(M \log M)$ time to evaluate the objective at any point a_j, the segment $[a_i, a_{i+1}]$ is identified in $O(M \log^2 M)$ time. Restricting ourselves to this segment we note that for $j = 1, \ldots, M$, $w_j|x - a_j|$ is a linear function of the parameter x.

To find the optimum, x^*, we use the general parametric procedure of Megiddo [132], with the modification in Cole [43]. The reader is referred to these references for a detailed discussion of the parametric approach. We only note that the master program that we apply is the sorting of the M linear functions, $\{w_j|x - a_j|\}$, $j = 1, ..., M$ (using x as the parameter). The test for determining the location of a given point x' w.r.t. x^* is based on calculating the objective $f_\lambda(x)$ and determining its one-sided derivatives at x'. This can clearly be done in $O(M \log M)$ time. We now conclude that with the above test the parametric approach in [132, 43] will find x^* in $O(M \log^2 M)$ time.

Now, we proceed with the general case. Suppose that n is fixed, and consider first the general case of the ordered median function. The collection consisting of the $O(M^2)$ (piecewise linear) bisectors $\{w_j d(x, v_j) - w_i d(x, v_i) = 0\}$, $i, j = 1, ..., M$, and the $O(M)$ hyperplanes, which are parallel to the axes and pass through $\{a_1, ..., a_M\}$, induces a cell partition of $\mathbb{R}^n$. This partition can be computed in $O(M^{2n})$ time for any fixed $n \geq 2$, (see Edelsbrunner [68]). If there is a finite ordered median, then at least one of the $O(M^{2n})$ vertices of the partition is an ordered median (notice that only if there are some negative weights finite ordered medians may not exist). Hence, by evaluating the objective at each vertex and each infinite ray of the partition we solve the problem in $O(M^{2n+1}\log M)$ time.

In the convex case in $\mathbb{R}^n$ we can directly use the approach of Cohen and Megiddo [42] to get a complexity of $O(M\log^{2n+1} M)$. This approach relies only on the fact that the objective function can be evaluated at any point just using additions, multiplications by a scalar and comparisons. Clearly, the ordered median objective function in this case falls into this class. More precisely, the complexity analysis involves several components which we will discuss now. First, we have to give a bound T on the number of operations needed to evaluate the objective function at a given point. In the case of ordered median functions this is $O(M\log M)$. The number of comparisons to be performed is $O(M\log M)$, and this can be done in $r = O(\log M)$ parallel phases using $C_i = O(M)$ effort in each phase [2]. Then the result by [42] states that the bound to find an optimal solution is $O(n^3T(\sum_{i=1}^{r}\lceil\log C_i\rceil)^n)$ for a fixed dimension n. In our case we achieve the bound $O(M\log^{2n+1} M)$ (the same bound can also be achieved by using the results in Tokuyama [196]). The bound in [42] is achieved by (recursively) solving $O(\log^2 M)$ recursive calls to instances of lower dimension. For $n = 1$ the above general bound (applied to this problem) gives $O(M\log^3 M)$. However, note that for the OMP we actually solve the case $n = 1$ in $O(M\log^2 M)$ time, an improvement by a factor of $\log M$. (See also Section 9.3.2.) Therefore, for any fixed $n \geq 2$, the bound will be reduced by a factor of $\log M$, and we have the following result.

Theorem 4.6 *Suppose that $w_j \geq 0$, for all a_j, and $0 \leq \lambda_1 \leq \ldots \leq \lambda_M$. Then for any fixed n, the (convex) rectilinear ordered median problem can be solved in $O(M\log^{2n} M)$ time.*

For comparison purposes, we note that the important special case of the k-centrum functions ($\lambda = (0, ..., 0, 1, ..., 1)$) has been recently solved in $O(M)$ time, for any fixed n, in Ogryczak and Tamir [154]. A detailed presentation is also included in Section 9.2.

5

Multicriteria Ordered Median Problems

5.1 Introduction

In the process of locating a new facility usually more than one decision maker is involved. This is due to the fact that typically the cost connected to the decision is relatively high. Of course, different persons may (or will) have different (conflicting) objectives. On other occasions, different scenarios must be compared in order to be implemented, or simply uncertainty in the parameters leads to consider different replications of the objective function. If only one objective has to be taken into account a broad range of models is available in the literature (see Chapter 11 in [59] and [61]). In contrast to that only a few papers looked at (more realistic) models for facility location, where more than one objective is involved (see [91, 168, 74, 178]). Even in those papers, one of the main deficiencies of the existing approaches is that only a few number (in most papers 1) of different types of objectives can be considered and in addition, solution approaches depend very much on the specific chosen metric. Also a detailed complexity analysis is missing in most of the papers.

On the other hand, there is a clear need for flexible models where the complexity status is known. These are prerequisites for a successful implementation of a decision support system for location planning which can really be used by decision-makers. Therefore, in this chapter we present a model for continuous multicriteria OMP (see Chapter 1) which fulfills the requirement of flexibility with respect to the choice of objective functions and allows the use of polyhedral gauges as measure of distances. Consequently, we will build on the notation and the results of the previous chapters.

The outline of the rest of the chapter is as follows. In Section 5.2 multicriteria problems and level sets are presented. Section 5.3 is devoted to the bicriteria case in the plane, while Sections 5.4 and 5.5 extend these results to the general planar Q-criteria case. The chapter ends with some concluding remarks where the generalization to the non-convex case is discussed. Throughout the chapter we keep track of the complexity of the algorithms.

5.2 Multicriteria Problems and Level Sets

Suppose we are in the situation of Chapter 1, where in the motivating example each decision maker wants to have a different objective function. Assume that there are Q different decision makers. This means that we are given Q functions, $F^1, \dots, F^Q$ from $\mathbb{R}^2$ to $\mathbb{R}$, that we want to optimize simultaneously. Evaluating these functions we get points in a Q-dimensional objective space where we do not have the canonical order of $\mathbb{R}$. Thus, it is clear that standard single objective optimization does not apply to this situation. Hence, we have to look for a different solution concept which in this chapter is the set of Pareto solutions. Recall that a point $x \in \mathbb{R}^2$ is called a Pareto location (or Pareto optimal) if there exists no $y \in \mathbb{R}^2$ such that

$$F^q(y) \leq F^q(x) \quad \forall q = 1, \dots, Q \qquad \text{and } F^p(y) < F^p(x) \text{ for some } p = 1, \dots, Q.$$

We denote the set of Pareto solutions by $\mathsf{M}_{\mathrm{Par}}\left(\mathrm{F}^1, \dots, \mathrm{F}^Q\right)$ or simply by $\mathsf{M}_{\mathrm{Par}}$ if this is possible without causing confusion.

For technical reasons we will also use the concepts of weak Pareto optimality and strict Pareto optimality. A point $x \in \mathbb{R}^2$ is called a weak Pareto location (or weakly Pareto optimal) if there exists no $y \in \mathbb{R}^2$ such that

$$F^q(y) < F^q(x) \quad \forall\, q = 1, \dots, Q\,.$$

We denote the set of weak Pareto solutions by $\mathsf{M}_{\mathrm{w-Par}}\left(\mathrm{F}^1, \dots, \mathrm{F}^Q\right)$ or simply by $\mathsf{M}_{\mathrm{w-Par}}$ when no confusion is possible. A point $x \in \mathbb{R}^2$ is called a strict Pareto location (or strictly Pareto optimal) if there exists no $y \in \mathbb{R}^2$ such that

$$F^q(y) \leq F^q(x) \quad \forall\, q = 1, \dots, Q\,.$$

We denote the set of strict Pareto solutions by $\mathsf{M}_{\mathrm{s-Par}}\left(\mathrm{F}^1, \dots, \mathrm{F}^Q\right)$ or simply by $\mathsf{M}_{\mathrm{s-Par}}$ when it is possible without confusion. Note that $\mathsf{M}_{\mathrm{s-Par}} \subseteq \mathsf{M}_{\mathrm{Par}} \subseteq \mathsf{M}_{\mathrm{w-Par}}$.

In order to obtain a geometrical characterization for a point to be a Pareto solution, we use the concept of level sets.

For a function $F : \mathbb{R}^2 \to \mathbb{R}$ the level set for a value $\rho \in \mathbb{R}$ is given by

$$L_{\leq}(F, \rho) := \{x \in \mathbb{R}^2 \,:\, F(x) \leq \rho\}$$

and the level curve for a value $\rho \in \mathbb{R}$ is given by

$$L_{=}(F, \rho) := \{x \in \mathbb{R}^2 \,:\, F(x) = \rho\}.$$

Using the level sets and level curves it is easy to prove that (see e.g. [91]):

$$x \in \mathsf{M}_{\text{w-Par}}\left(\mathrm{F}^1, \ldots, \mathrm{F}^Q\right) \Leftrightarrow \bigcap_{q=1}^{Q} L_<(F^q, F^q(x)) = \emptyset \tag{5.1}$$

$$x \in \mathsf{M}_{\text{Par}}\left(\mathrm{F}^1, \ldots, \mathrm{F}^Q\right) \Leftrightarrow \bigcap_{q=1}^{Q} L_\leq(F^q, F^q(x)) = \bigcap_{q=1}^{Q} L_=(F^q, F^q(x)) \tag{5.2}$$

$$x \in \mathsf{M}_{\text{s-Par}}\left(\mathrm{F}^1, \ldots, \mathrm{F}^Q\right) \Leftrightarrow \bigcap_{q=1}^{Q} L_\leq(F^q, F^q(x)) = \{x\} \tag{5.3}$$

To conclude the preliminaries, we recall a result by Warburton [203] that ensures that the set M_{Par} is connected, provided that the objective functions are quasi-convex.

5.3 Bicriteria Ordered Median Problems

In this section we restrict ourselves to the bicriteria case, since - as will be seen later - it is the basis for solving the Q-criteria case.

To this end, in this section we are looking for the Pareto solutions of the following vector optimization problem in $\mathbb{R}^2$

$$\min_{x \in \mathbb{R}^2} \begin{pmatrix} f^1_\lambda(x) := \sum_{i=1}^{M} \lambda^1_i d^1_{(i)}(x) \\ f^2_\lambda(x) := \sum_{i=1}^{M} \lambda^2_i d^2_{(i)}(x) \end{pmatrix}$$

where the weights λ^q_i are in non-decreasing order with respect to the index i for each $q = 1, 2$, that is,

$$0 \leq \lambda^q_1 \leq \lambda^q_2 \leq \ldots \leq \lambda^q_M\,, \quad q = 1, 2\,, \text{ i.e. } \lambda^q \in S^{\leq}_M$$

and $d^q_{(i)}(x)$ depends on the set W^q of importance given to the existing facilities by the q-th criterion, $q = 1, 2$. Therefore, the previous vector optimization problem is convex, as was discussed in Section 5.2.

Note that in a multicriteria setting each objective function f^q_λ, $q = 1, \ldots, Q$, generates its own set of bisector lines. Therefore, in the multicriteria case the ordered elementary convex sets are generated by all the fundamental directions $\overrightarrow{Oe^i_g}$, $i = 1, \ldots, M$, $g = 1, \ldots, G_i$, and the bisector lines $\mathsf{Bisec}^q(a_i, a_j)$, $q = 1, \ldots, Q$.

We are able to give a geometrical characterization of the set M_{Par} by the following theorem.

Theorem 5.1 *$\mathsf{M}_{\text{Par}}\left(\mathrm{f}^1_\lambda, \mathrm{f}^2_\lambda\right)$ is a connected chain from $\mathsf{M}(\mathrm{f}^1_\lambda)$ to $\mathsf{M}(\mathrm{f}^2_\lambda)$ consisting of facets or vertices of cells or complete cells.*

Proof.
First of all, we know that $\mathsf{M}_{\mathrm{Par}} \neq \emptyset$, so we can choose $x \in \mathsf{M}_{\mathrm{Par}}$. There exists at least one cell $\mathrm{C} \in \mathcal{C}$ with $x \in \mathrm{C}$. Hence, three cases can occur:

1. $x \in int(\mathrm{C})$: Since $x \in \mathsf{M}_{\mathrm{Par}}$ we obtain

$$\bigcap_{q=1}^{Q} L_{\leq}(f_\lambda^q, f_\lambda^q(x)) = \bigcap_{q=1}^{Q} L_{=}(f_\lambda^q, f_\lambda^q(x))$$

and by linearity of the OMP in each cell we have

$$\bigcap_{q=1}^{Q} L_{\leq}(f_\lambda^q, f_\lambda^q(y)) = \bigcap_{q=1}^{Q} L_{=}(f_\lambda^q, f_\lambda^q(y)) \quad \forall y \in \mathrm{C}$$

which means $y \in \mathsf{M}_{\mathrm{Par}}\ \forall \mathrm{y} \in \mathrm{C}$, hence $\mathrm{C} \subseteq \mathsf{M}_{\mathrm{Par}}$.

2. $x \in \overline{ab} := conv\{a, b\} \subset bd(\mathrm{C})$ and $a, b \in Ext(\mathrm{C})$. We can choose $y \in int(\mathrm{C})$ and 2 cases can occur:
   a) $y \in \mathsf{M}_{\mathrm{Par}}$. Hence we can continue as in Case 1.
   b) $y \notin \mathsf{M}_{\mathrm{Par}}$. Therefore using the linearity we obtain first

$$\bigcap_{q=1}^{Q} L_{\leq}(f_\lambda^q, f_\lambda^q(z)) \neq \bigcap_{q=1}^{Q} L_{=}(f_\lambda^q, f_\lambda^q(z)) \quad \forall z \in int(\mathrm{C})$$

   and second, we have

$$\bigcap_{q=1}^{Q} L_{\leq}(f_\lambda^q, f_\lambda^q(z)) = \bigcap_{q=1}^{Q} L_{=}(f_\lambda^q, f_\lambda^q(z)) \quad \forall z \in \overline{ab}$$

   since $x \in \mathsf{M}_{\mathrm{Par}}$. Therefore we have that $\mathrm{C} \not\subseteq \mathsf{M}_{\mathrm{Par}}$ and $\overline{ab} \subseteq \mathsf{M}_{\mathrm{Par}}$.

3. $x \in Ext(\mathrm{C})$. We can choose $y \in int(\mathrm{C})$ and two cases can occur:
   a) If $y \in \mathsf{M}_{\mathrm{Par}}$, we can continue as in Case 1.
   b) If $y \notin \mathsf{M}_{\mathrm{Par}}$, we choose $z_1, z_2 \in Ext(\mathrm{C})$ such that $\overline{xz_1}, \overline{xz_2}$ are faces of C,
      i. If z_1 or z_2 are in $\mathsf{M}_{\mathrm{Par}}$, we can continue as in Case 2.
      ii. If z_1 and z_2 are not in $\mathsf{M}_{\mathrm{Par}}$, then using the linearity in the same way as before we obtain that $(\mathrm{C} \setminus \{x\}) \cap \mathsf{M}_{\mathrm{Par}} = \emptyset$.

Hence, we obtain that the set of Pareto solutions consists of complete cells, complete faces and vertices of these cells. Since we know that the set $\mathsf{M}_{\mathrm{Par}}$ is connected, the proof is completed. □

In the following we develop an algorithm to solve the bicriteria OMP. The idea of this algorithm is to start at a vertex x of the cell structure which belongs to $\mathsf{M}_{\mathrm{Par}}$, say $x \in \mathsf{M}_{1,2} := \arg\min_{\mathrm{x} \in \mathsf{M}(\mathrm{f}_\lambda^1)} \mathrm{f}_\lambda^2(\mathrm{x})$ (set of optimal lexicographical locations, see [148]). Then using the connectivity of $\mathsf{M}_{\mathrm{Par}}$ the algorithm proceeds moving from vertex x to another Pareto optimal vertex y of the cell structure which is connected with the previous one by an elementary convex set. This procedure is repeated until the end of the chain in $\mathsf{M}_{2,1} := \arg\min_{x \in \mathsf{M}(f_\lambda^2)} f_\lambda^1(x)$ is reached.

Let $x \in \mathsf{M}_{\mathrm{Par}} \cap \mathrm{C}$ then using the notation of the proof of Theorem 5.1 (see Figure 5.1) and the linearity of the level sets in each cell, we can distinguish the following disjoint cases:

A : $\mathrm{C} \subseteq \mathsf{M}_{\mathrm{Par}}$, i.e. C is contained in the chain.
B : $\overline{xy}$ and $\overline{xz}$ are candidates for $\mathsf{M}_{\mathrm{Par}}$ and $int(C) \not\subset \mathsf{M}_{\mathrm{Par}}$.
C : $\overline{xy}$ is candidate for $\mathsf{M}_{\mathrm{Par}}$ and $\overline{xz}$ is not contained in $\mathsf{M}_{\mathrm{Par}}$.
D : $\overline{xz}$ is candidate for $\mathsf{M}_{\mathrm{Par}}$ and $\overline{xy}$ is not contained in $\mathsf{M}_{\mathrm{Par}}$.
E : Neither $\overline{xy}$ nor $\overline{xz}$ are contained in $\mathsf{M}_{\mathrm{Par}}$.

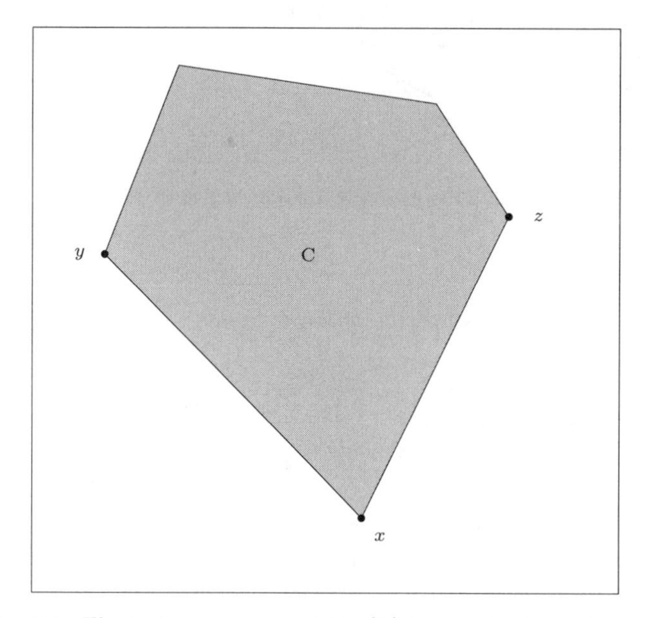

Fig. 5.1. Illustration to $y, x, z \in Ext(\mathrm{C})$ in counterclockwise order

We denote by $\mathbf{sit}(\mathrm{C}, x)$ the situation appearing in cell C according to the extreme point x of C. The following lemma states when a given segment belongs to the Pareto-set in terms of the $\mathbf{sit}(\cdot, \cdot)$ function.

Lemma 5.1 *Let* $\mathrm{C}_1, \ldots, \mathrm{C}_{P_x}$ *be the cells containing the intersection point* x*, considered in counterclockwise order, and* $y_1, \ldots, y_{P_x}$ *the intersection points adjacent to* x*, considered in counterclockwise order (see Figure 5.2). If* $x \in$

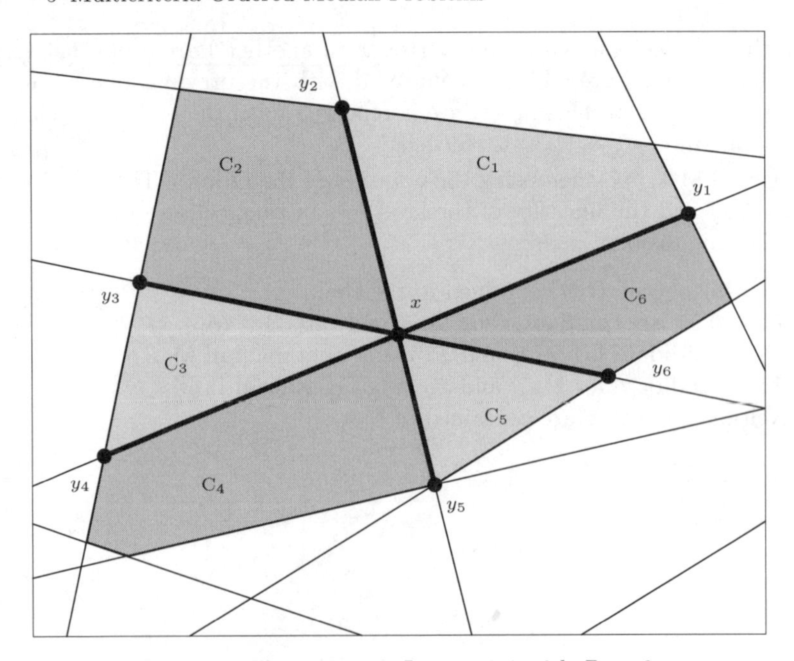

Fig. 5.2. Illustration to Lemma 5.1 with $P_x = 6$

$\mathcal{M}_{\text{Par}}$ *and* $i \in \{1, \dots, P_x\}$*, then the following holds:*

$$\overline{xy_{i+1}} \subseteq \mathcal{M}_{\text{Par}} \iff \left\{ \begin{array}{ll} & \mathbf{sit}(\mathrm{C}_i, x) = \mathbf{A} \\ or & \mathbf{sit}(\mathrm{C}_{i+1}, x) = \mathbf{A} \\ or & \left\{ \begin{array}{c} \mathbf{sit}(\mathrm{C}_i, x) \in \{\mathbf{B}, \mathbf{C}\} \\ \textit{and } \mathbf{sit}(\mathrm{C}_{i+1}, x) \in \{\mathbf{B}, \mathbf{D}\} \end{array} \right\}. \end{array} \right\}$$

Proof.
$x^\perp := (-x_2, x_1)$ is perpendicular to $x = (x_1, x_2)$. For $q = 1, 2$ let s_q be the direction vector of the lines, which describe the level curves $L_=(f_\lambda^q, .)$ in $int(C)$. Let H_C be the half-space with respect to the line

$$l_C := \left\{ y \in \mathbb{R}^2 \, : \, y = x + \eta(x_R - x) \, , \, \eta \in \mathbb{R} \right\}$$

which contains C (see Figure 5.3). Without loss of generality assume that s_q points into H_C, i.e. $\langle (x_R - x)^\perp, s_q \rangle \geq 0$ (If $\langle (x_R - x)^\perp, s_q \rangle < 0$, set $s_q := -s_q$.).

The following case analysis shows how to determine which of the five situations (A,B,C,D or E) occurs.

Case 1 :

$$x \in \operatorname{argmin}_{y \in C}\{f_\lambda^1(y)\} \qquad \wedge \qquad x \in \operatorname{argmin}_{y \in C}\{f_\lambda^2(y)\}$$

$$\iff (f_\lambda^1(x) \leq f_\lambda^1(x_L) \wedge f_\lambda^1(x) \leq f_\lambda^1(x_R)) \wedge (f_\lambda^2(x) \leq f_\lambda^2(x_L) \wedge f_\lambda^2(x) \leq f_\lambda^2(x_R))$$

Case 1.1 :

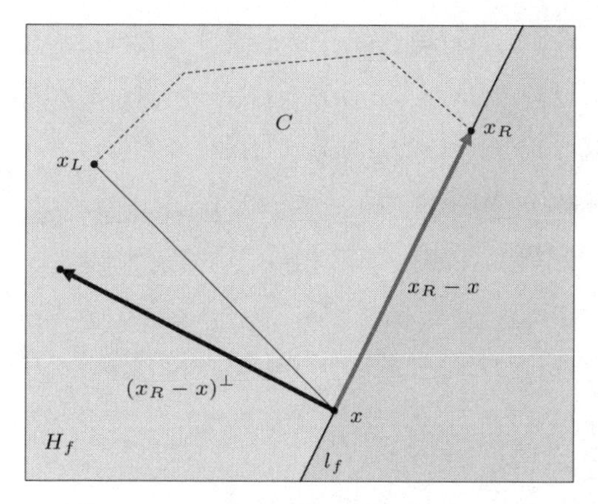

Fig. 5.3. Halfspace H_C with $C \subseteq H_C$

At least 3 out of 4 inequalities are strict

$$\Longrightarrow \big((f_\lambda^1(x) < f_\lambda^1(x_L) \wedge f_\lambda^1(x) < f_\lambda^1(x_R)) \wedge (f_\lambda^2(x) < f_\lambda^2(x_L) \wedge f_\lambda^2(x) < f_\lambda^2(x_R))\big)$$
$$\vee \ \big((f_\lambda^1(x) = f_\lambda^1(x_L) \wedge f_\lambda^1(x) < f_\lambda^1(x_R)) \wedge (f_\lambda^2(x) < f_\lambda^2(x_L) \wedge f_\lambda^2(x) < f_\lambda^2(x_R))\big)$$
$$\vee \ \big((f_\lambda^1(x) < f_\lambda^1(x_L) \wedge f_\lambda^1(x) = f_\lambda^1(x_R)) \wedge (f_\lambda^2(x) < f_\lambda^2(x_L) \wedge f_\lambda^2(x) < f_\lambda^2(x_R))\big)$$
$$\vee \ \big((f_\lambda^1(x) < f_\lambda^1(x_L) \wedge f_\lambda^1(x) < f_\lambda^1(x_R)) \wedge (f_\lambda^2(x) = f_\lambda^2(x_L) \wedge f_\lambda^2(x) < f_\lambda^2(x_R))\big)$$
$$\vee \ \big((f_\lambda^1(x) < f_\lambda^1(x_L) \wedge f_\lambda^1(x) < f_\lambda^1(x_R)) \wedge (f_\lambda^2(x) < f_\lambda^2(x_L) \wedge f_\lambda^2(x) = f_\lambda^2(x_R))\big)$$

$\Longrightarrow$ x dominates x_L as well as x_R

$\Longrightarrow$ Situation **E** occurs.

Case 1.2 :

Exactly 2 of 4 inequalities are strict.

Case 1.2.1 :

$$(f_\lambda^1(x) < f_\lambda^1(x_L) \ \wedge \ f_\lambda^1(x) < f_\lambda^1(x_R)) \wedge (f_\lambda^2(x) = f_\lambda^2(x_L) \ \wedge \ f_\lambda^2(x) = f_\lambda^2(x_R))$$

$\Longrightarrow$ x dominates x_L as well as x_R

$\Longrightarrow$ Situation **E** occurs.

Moreover

$\Longrightarrow \{x\} = \text{argmin}_{y \in C}\{f_\lambda^1(y)\}$ and f_λ^2 is constant on C

$\Longrightarrow \{x\} = \mathsf{M}_{2,1}$, since f_λ^2 convex and $int(C) \neq \emptyset$

$\Longrightarrow$ The end of the chain is achieved.

Case 1.2.2 :

$$\big((f_\lambda^1(x) = f_\lambda^1(x_L) \wedge f_\lambda^1(x) = f_\lambda^1(x_R)) \wedge (f_\lambda^2(x) < f_\lambda^2(x_L) \wedge f_\lambda^2(x) < f_\lambda^2(x_R))\big)$$
$$\vee \; \big((f_\lambda^1(x) < f_\lambda^1(x_L) \wedge f_\lambda^1(x) = f_\lambda^1(x_R)) \wedge (f_\lambda^2(x) = f_\lambda^2(x_L) \wedge f_\lambda^2(x) < f_\lambda^2(x_R))\big)$$
$$\vee \; \big((f_\lambda^1(x) = f_\lambda^1(x_L) \wedge f_\lambda^1(x) < f_\lambda^1(x_R)) \wedge (f_\lambda^2(x) < f_\lambda^2(x_L) \wedge f_\lambda^2(x) = f_\lambda^2(x_R))\big)$$

$\Longrightarrow$ x dominates x_L as well as x_R

$\Longrightarrow$ Situation **E** occurs.

Case 1.2.3 :

$$\big((f_\lambda^1(x) < f_\lambda^1(x_L) \wedge f_\lambda^1(x) = f_\lambda^1(x_R)) \wedge (f_\lambda^2(x) < f_\lambda^2(x_L) \wedge f_\lambda^2(x) = f_\lambda^2(x_R))\big)$$

$\Longrightarrow$ x dominates x_L

$\Longrightarrow$ Situation **D** occurs.

Case 1.2.4 :

$$\big((f_\lambda^1(x) = f_\lambda^1(x_L) \wedge f_\lambda^1(x) < f_\lambda^1(x_R)) \wedge (f_\lambda^2(x) = f_\lambda^2(x_L) \wedge f_\lambda^2(x) < f_\lambda^2(x_R))\big)$$

$\Longrightarrow$ x dominates x_R

$\Longrightarrow$ Situation **C** occurs.

Case 1.3 :

Exactly 1 of 4 inequalities is strict.

Case 1.3.1 :

$$\big((f_\lambda^1(x) < f_\lambda^1(x_L) \wedge f_\lambda^1(x) = f_\lambda^1(x_R)) \wedge (f_\lambda^2(x) = f_\lambda^2(x_L) \wedge f_\lambda^2(x) = f_\lambda^2(x_R))\big)$$
$$\vee \; \big((f_\lambda^1(x) = f_\lambda^1(x_L) \wedge f_\lambda^1(x) < f_\lambda^1(x_R)) \wedge (f_\lambda^2(x) = f_\lambda^2(x_L) \wedge f_\lambda^2(x) = f_\lambda^2(x_R))\big)$$

$\Longrightarrow$ $\{x\} \in \text{argmin}_{y \in C}\{f_\lambda^1(y)\}$ and f_λ^2 is constant on C

$\Longrightarrow$ $\{x\} \in \mathsf{M}_{2,1}$, since f_λ^2 convex and $int(C) \neq \emptyset$

$\Longrightarrow$ The end of the chain is achieved.

Case 1.3.2 :

$$(f_\lambda^1(x) = f_\lambda^1(x_L) \wedge f_\lambda^1(x) = f_\lambda^1(x_R)) \wedge (f_\lambda^2(x) < f_\lambda^2(x_L) \wedge f_\lambda^2(x) = f_\lambda^2(x_R))$$

$\Longrightarrow$ f_λ^1 is constant on C and $\overline{xx_R} = \text{argmin}_{y \in C}\{f_\lambda^2(y)\}$

$\Longrightarrow$ $\overline{xx_R} = \mathsf{M}_{1,2}$

$\Longrightarrow$ Situation **D** occurs.

Case 1.3.3 :

$$(f_\lambda^1(x) = f_\lambda^1(x_L) \wedge f_\lambda^1(x) = f_\lambda^1(x_R)) \wedge (f_\lambda^2(x) = f_\lambda^2(x_L) \wedge f_\lambda^2(x) < f_\lambda^2(x_R))$$

$\Longrightarrow$ f_λ^1 is constant on C and $\overline{xx_L} = \text{argmin}_{y \in C}\{f_\lambda^2(y)\}$

$\Longrightarrow$ $\overline{xx_L} = \mathsf{M}_{1,2}$

$\Longrightarrow$ Situation **C** occurs.

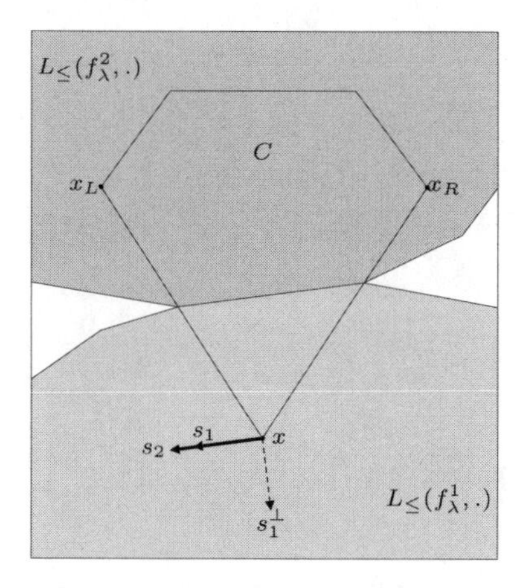

Fig. 5.4. Illustration to Case 2.1

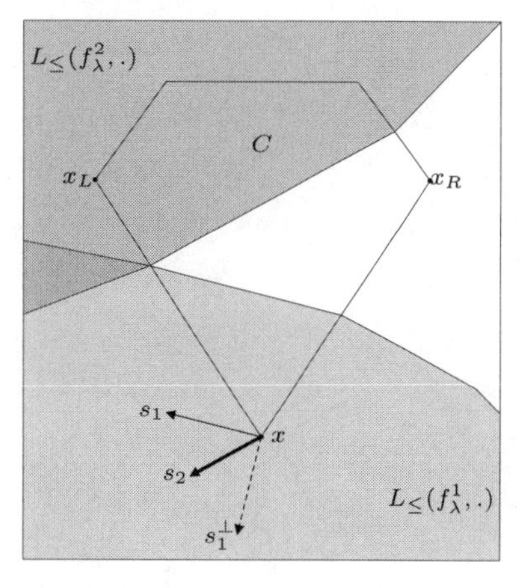

Fig. 5.5. Illustration to Case 2.2

Case 1.4 :

None of the 4 inequalities is strict.

$\Longrightarrow (f^1_\lambda(x) = f^1_\lambda(x_L) \wedge f^1_\lambda(x) = f^1_\lambda(x_R)) \wedge (f^2_\lambda(x) = f^2_\lambda(x_L) \wedge f^2_\lambda(x) = f^2_\lambda(x_R))$

$\Longrightarrow f^1_\lambda$ and f^2_λ are constant on C

$\Longrightarrow C \subseteq \mathsf{M}(f^1_\lambda)$ and $C \subseteq \mathsf{M}(f^2_\lambda)$, since f^1_λ, f^2_λ convex and $int(C) \neq \emptyset$

$\Longrightarrow$ Contradiction to $\mathsf{M}(f^1_\lambda) \cap \mathsf{M}(f^2_\lambda) = \emptyset$.

Case 2 :

$$x \in \operatorname{argmin}_{y \in C}\{f^1_\lambda(y)\} \qquad \wedge \qquad x \notin \operatorname{argmin}_{y \in C}\{f^2_\lambda(y)\}$$

$\Longleftrightarrow (f^1_\lambda(x) \leq f^1_\lambda(x_L) \wedge f^1_\lambda(x) \leq f^1_\lambda(x_R)) \wedge (f^2_\lambda(x) > f^2_\lambda(x_L) \vee f^2_\lambda(x) > f^2_\lambda(x_R))$

Case 2.1 :

$\langle s_1^\perp, s_2 \rangle = 0$

$\Longrightarrow$ Situation **A** occurs (see Figure 5.4).

Case 2.2 :

$\langle s_1^\perp, s_2 \rangle > 0$

$\Longrightarrow$ Situation **C** occurs (see Figure 5.5).

Case 2.3 :

$\langle s_1^\perp, s_2 \rangle < 0$

Case 2.3.1 :

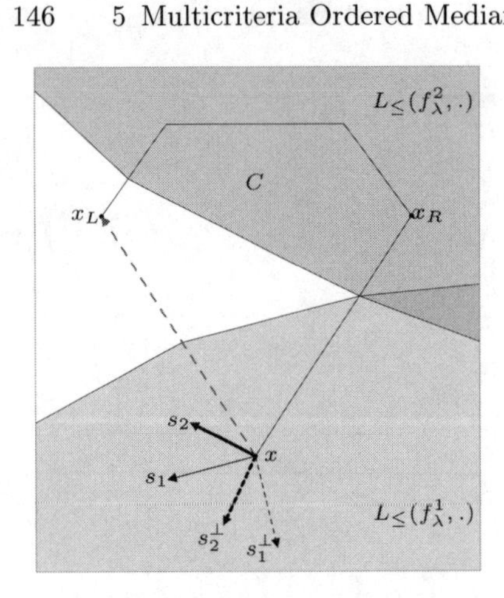

Fig. 5.6. Illustration to Case 2.3.1

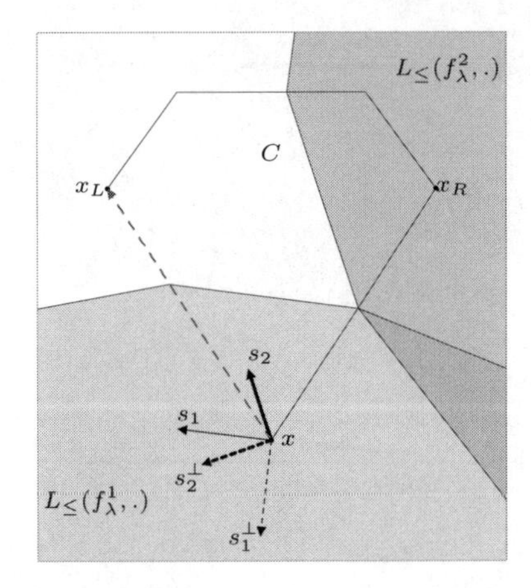

Fig. 5.7. Illustration to Case 2.3.2.2

$$\langle s_2^\perp, x_L - x\rangle \leq 0$$

$\Longrightarrow$ Situation **D** occurs (see Figure 5.6).

Case 2.3.2 :

$\langle s_2^\perp, x_L - x\rangle > 0\,.$

Case 2.3.2.1 :

$$f_\lambda^2(x_L) = f_\lambda^2(x_R)$$

$\Longrightarrow L_=(f_\lambda^2,.)$ runs through x_L and x_R

$\Longrightarrow \langle s_2^\perp, x_L - x\rangle \leq 0$

$\Longrightarrow$ Contradiction to **Case 2.3.2**.

Case 2.3.2.2 :

$$f_\lambda^2(x_L) > f_\lambda^2(x_R)$$

$\Longrightarrow$ Situation **D** occurs (see Figure 5.7).

Case 2.3.2.3 :

$$f_\lambda^2(x_L) < f_\lambda^2(x_R)$$

$\Longrightarrow$ Situation **C** occurs (see Figure 5.8).

Case 3 :

$$x \notin \operatorname{argmin}_{y\in C}\{f_\lambda^1(y)\} \qquad \wedge \qquad x \in \operatorname{argmin}_{y\in C}\{f_\lambda^2(y)\}$$

$$\Longleftrightarrow (f_\lambda^1(x) > f_\lambda^1(x_L) \vee f_\lambda^1(x) > f_\lambda^1(x_R)) \wedge (f_\lambda^2(x) \leq f_\lambda^2(x_L) \wedge f_\lambda^2(x) \leq f_\lambda^2(x_R))$$

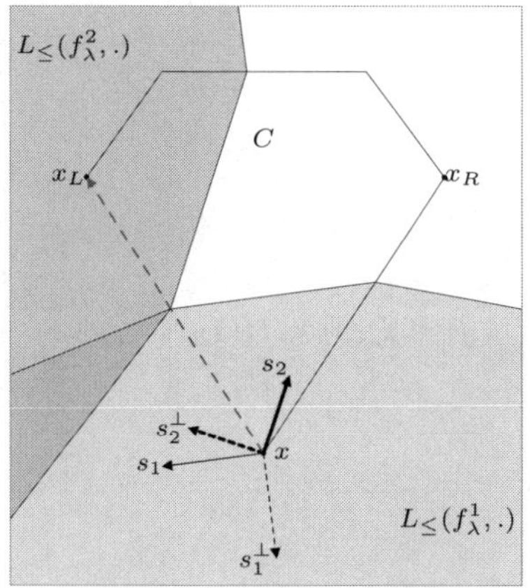

Fig. 5.8. Illustration to Case 2.3.2.3

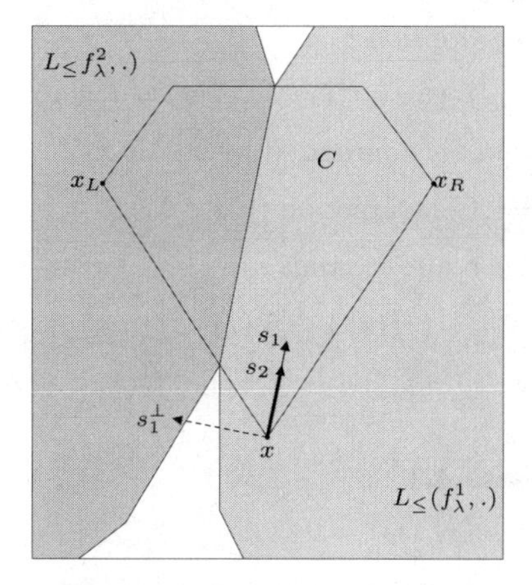

Fig. 5.9. Illustration to Case 4.1

The case analysis corresponds to the case analysis of **Case 2**, with reversed roles of f_λ^1 and f_λ^2.

Case 4 :

$$x \notin \operatorname{argmin}_{y \in C}\{f_\lambda^1(y)\} \qquad \wedge \qquad x \notin \operatorname{argmin}_{y \in C}\{f_\lambda^2(y)\}$$

$$\iff (f_\lambda^1(x) > f_\lambda^1(x_L) \vee f_\lambda^1(x) > f_\lambda^1(x_R)) \wedge (f_\lambda^2(x) > f_\lambda^2(x_L) \vee f_\lambda^2(x) > f_\lambda^2(x_R))$$

Assumption :

$$f_\lambda^1(x) > f_\lambda^1(x_L) \wedge f_\lambda^2(x) > f_\lambda^2(x_L)$$

$\implies x_L$ dominates x

$\implies$ Contradiction to $x \in \mathsf{M}_{\text{Par}}$.

Assumption :

$$f_\lambda^1(x) > f_\lambda^1(x_R) \wedge f_\lambda^2(x) > f_\lambda^2(x_R)$$

$\implies x_R$ dominates x

$\implies$ Contradiction to $x \in \mathsf{M}_{\text{Par}}$.

Therefore it holds :

$$\begin{aligned}
& \big(f_\lambda^1(x) > f_\lambda^1(x_L) \wedge f_\lambda^1(x) \le f_\lambda^1(x_R) \ \wedge\ f_\lambda^2(x) \le f_\lambda^2(x_L) \wedge f_\lambda^2(x) > f_\lambda^2(x_R)\big) \\
\vee\ & \big(f_\lambda^1(x) \le f_\lambda^1(x_L) \wedge f_\lambda^1(x) > f_\lambda^1(x_R) \ \wedge\ f_\lambda^2(x) > f_\lambda^2(x_L) \wedge f_\lambda^2(x) \le f_\lambda^2(x_R)\big) \\
\iff & \quad \big(f_\lambda^1(x_L) < f_\lambda^1(x) \le f_\lambda^1(x_R) \quad \wedge \quad f_\lambda^2(x_R) < f_\lambda^2(x) \le f_\lambda^2(x_L)\big) \\
\vee & \quad \big(f_\lambda^1(x_R) < f_\lambda^1(x) \le f_\lambda^1(x_L) \quad \wedge \quad f_\lambda^2(x_L) < f_\lambda^2(x) \le f_\lambda^2(x_R)\big)
\end{aligned}$$

Assumption :

$$f_\lambda^1(x) = f_\lambda^1(x_L) \vee f_\lambda^1(x) = f_\lambda^1(x_R) \vee f_\lambda^2(x) = f_\lambda^2(x_L) \vee f_\lambda^2(x) = f_\lambda^2(x_R)$$

$\Longrightarrow x$ is dominated by x_R or x_L

$\Longrightarrow$ Contradiction to $x \in \mathsf{M}_{\mathrm{Par}}$.

Therefore it holds :

$$\begin{array}{lccc} & \left(f_\lambda^1(x_L) < f_\lambda^1(x) < f_\lambda^1(x_R)\right. & \wedge & \left.f_\lambda^2(x_R) < f_\lambda^2(x) < f_\lambda^2(x_L)\right) \\ \vee & \left(f_\lambda^1(x_R) < f_\lambda^1(x) < f_\lambda^1(x_L)\right. & \wedge & \left.f_\lambda^2(x_L) < f_\lambda^2(x) < f_\lambda^2(x_R)\right) \end{array}$$

Case 4.1 :

$\langle s_1^\perp, s_2 \rangle = 0$

$\Longrightarrow$ Situation **A** occurs (see Figure 5.9) .

Case 4.2 :

$\langle s_1^\perp, s_2 \rangle \neq 0$.

Case 4.2.1 :

$$\begin{array}{ll} & \left(\langle s_1^\perp, s_2\rangle > 0 \wedge \left(f_\lambda^1(x_L) < f_\lambda^1(x) < f_\lambda^1(x_R) \wedge f_\lambda^2(x_R) < f_\lambda^2(x) < f_\lambda^2(x_L)\right)\right) \\ \vee & \left(\langle s_1^\perp, s_2\rangle < 0 \wedge \left(f_\lambda^1(x_R) < f_\lambda^1(x) < f_\lambda^1(x_L) \wedge f_\lambda^2(x_L) < f_\lambda^2(x) < f_\lambda^2(x_R)\right)\right) \end{array}$$

$\Longrightarrow \exists y \in C$, which dominates x (see Figure 5.10)

$\Longrightarrow$ Contradiction to $x \in \mathsf{M}_{\mathrm{Par}}$.

Case 4.2.2 :

$$\begin{array}{ll} & \left(\langle s_1^\perp, s_2\rangle > 0 \wedge \left(f_\lambda^1(x_R) < f_\lambda^1(x) < f_\lambda^1(x_L) \wedge f_\lambda^2(x_L) < f_\lambda^2(x) < f_\lambda^2(x_R)\right)\right) \\ \vee & \left(\langle s_1^\perp, s_2\rangle < 0 \wedge \left(f_\lambda^1(x_L) < f_\lambda^1(x) < f_\lambda^1(x_R) \wedge f_\lambda^2(x_R) < f_\lambda^2(x) < f_\lambda^2(x_L)\right)\right) \end{array}$$

$\Longrightarrow$ Situation **B** occurs (see Figure 5.11) .

□

These results allow to describe Algorithm 5.1 for computing $\mathsf{M}_{\mathrm{Par}}\left(\mathrm{f}_\lambda^1, \mathrm{f}_\lambda^2\right)$.

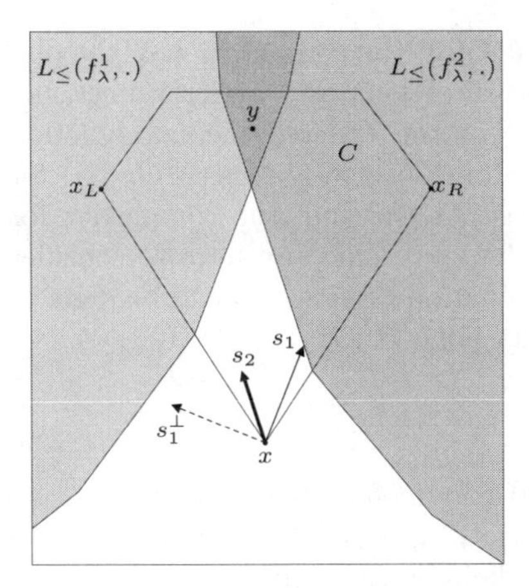

Fig. 5.10. Illustration to Case 4.2.1

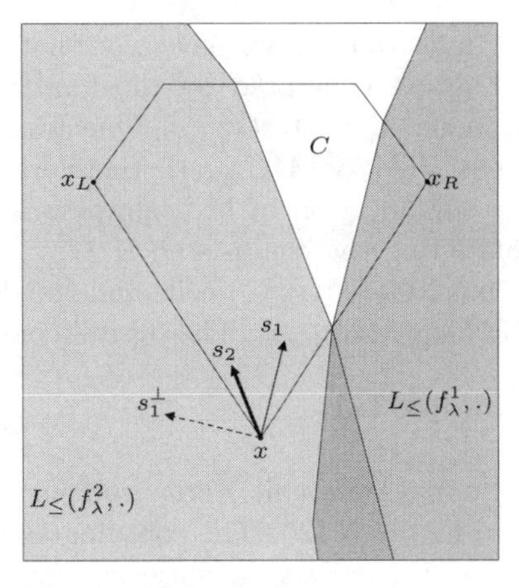

Fig. 5.11. Illustration to Case 4.2.2

Algorithm 5.1: *Computing* $M_{\mathrm{Par}}\left(f_\lambda^1, f_\lambda^2\right)$.

Compute the planar graph generated by the cells.
Compute the two sets of lexicographical locations $M_{1,2}$, $M_{2,1}$.
if $M_{1,2} \cap M_{2,1} \neq \emptyset$ **then**
  | ($\star$ trivial case : $M(f_\lambda^1) \cap M(f_\lambda^2) \neq \emptyset$ $\star$) $M_{\mathrm{Par}} := \mathrm{co}\{M_{1,2}\}$
else
  | ($\star$ non trivial case : $M(f_\lambda^1) \cap M(f_\lambda^2) = \emptyset$ $\star$)
  | $M_{\mathrm{Par}} := M_{1,2} \cup M_{2,1}$
  | Choose $x \in M_{1,2} \cap \mathcal{GIP}$ and $i := 0$
  | **while** $x \notin M_{2,1}$ **do**
  | | **repeat**
  | | | $i := i + 1$
  | | | **if** $i > P_x$ **then** $i := i - P_x$
  | | **until** $\mathbf{sit}(C_i, x) = \mathbf{A}$ *OR* ($\mathbf{sit}(C_i, x) \in \{\mathbf{B},\mathbf{C}\}$ *AND* $\mathbf{sit}(C_{i+1}, x) \in \{\mathbf{B},\mathbf{D}\}$)
  | | **if** $\mathbf{sit}(C_i, x) = \mathbf{A}$ **then**
  | | | ($\star$ We have found a bounded cell. $\star$) $M_{\mathrm{Par}} := M_{\mathrm{Par}} \cup C_i$
  | | **else**
  | | | ($\star$ We have found a bounded face. $\star$) $M_{\mathrm{Par}} := M_{\mathrm{Par}} \cup \overline{xy_i}$
  | | **end**
  | | $temp := x$
  | | $x := y_i$
  | | $i := i_x(temp) - 1$ ($\star$ Where $i_x(temp)$ is the index of temp in the list of generalized intersection points adjacent to x. $\star$)
  | **end**
end
Output: $M_{\mathrm{Par}}\left(f_\lambda^1, f_\lambda^2\right)$.

In Section 4.2 we have seen that for the OMP the computation of the planar graph generated by the fundamental directions and bisector lines can be done in $O(M^4G^2_{\max})$. One evaluation of the ordered median function needs $O(M\log(MG_{\max}))$, therefore we obtain $O(M^5G^2_{\max}\log(MG_{\max}))$ for the computation of lexicographic solutions. At the end, the complexity for computing the chain is $O(M^5G^2_{\max}\log(MG_{\max}))$, since we have to consider at most $O(M^4G^2_{max})$ cells and the determination of **sit**$(. , .)$ can be done in $O(M\log(MG_{max}))$. The overall complexity is $O(M^5G^2_{\max}\log(MG_{\max}))$.

Example 5.1

Let us consider a bicriteria problem with 20 existing facilities $A = \{a_1, \dots, a_{20}\}$ (see Figure 5.12). The coordinates $a_m = (x_m, y_m)$ and the weights $w^q_m, q = 1, 2$, of the existing facilities are given by the following table:

m	1	2	3	4	5	6	7	8	9	10
a_m	(1, 7)	(2, 19)	(7, 14)	(7, 44)	(8, 6)	(9, 23)	(10, 33)	(11, 48)	(14, 1)	(14, 13)
w^1_m	10	5	5	3	15	3	1	1	10	7
w^2_m	3	4	1	5	1	2	6	10	0	3

m	11	12	13	14	15	16	17	18	19	20
a_m	(16, 36)	(17, 43)	(19, 9)	(22, 20)	(24, 34)	(25, 45)	(27, 4)	(28, 49)	(29, 28)	(31, 37)
w^1_m	1	1	5	3	0	0	7	2	2	2
w^2_m	5	6	2	2	5	10	2	15	10	7

All existing facilities are associated with the 1-norm. Moreover, we assume that f^1_λ is a median function, while f^2_λ is a center function, i. e.

$$\lambda^q_m := \begin{cases} 1\ , & q = 1 \lor m = M \\ 0\ , & \text{otherwise} \end{cases} .$$

The optimal solution of the median function f^1_λ is unique and given by $\mathsf{M}(f^1_\lambda) = \{(10, 7)\}$ with the (optimal) objective value $z^*_1 = 1344$. Therefore we have $\mathsf{M}_{1,2} = \mathsf{M}(\mathsf{f}^1_\lambda)$.

However, the optimal solution of the center function f^2_λ is given by the segment $\mathsf{M}(f^2_\lambda) = \overline{(23\frac{1}{6}, 41\frac{1}{6})\,(25\frac{1}{4}, 43\frac{1}{4})}$ with (optimal) objective value $z^*_2 = 190$.

The center solution $\mathsf{M}(f^2_\lambda)$ lies on the bisector generated by (a_8, w^2_8) and (a_{19}, w^2_{19}). Moreover, we have $\mathsf{M}_{2,1} = \{(23\frac{1}{6}, 41\frac{1}{6})\}$. The bicriteria chain (consisting of 5 cells and 14 edges with respect to the fundamental directions and bisectors drawn in Figure 5.12) is given by

$$
\begin{aligned}
\mathsf{M}_{\mathrm{Par}} = \quad & \mathrm{conv}\{\,(10,7)\,,(11,7)\,,(11,9)\,,(10,9)\} \\
\cup \; & \overline{(11,9)\,(14,9)} \cup \overline{(14,9)\,(14,19)} \\
\cup \; & \overline{(14,19)\,(19,19)} \cup \overline{(19,19)\,(19,20)} \\
\cup \; & \mathrm{conv}\{\,(19,20)\,,(22,20)\,,(22,23)\,,(19,23)\} \\
\cup \; & \mathrm{conv}\{\,(22,23)\,,(26\tfrac{4}{5},23)\,,(25\tfrac{4}{5},28)\,,(22,28)\} \\
\cup \; & \overline{(25\tfrac{4}{5},28)\,(23\tfrac{1}{6},41\tfrac{1}{6})} \quad .
\end{aligned}
$$

Notice that the face $\overline{(26\frac{4}{5},23)\,(25\frac{4}{5},28)}$ of the last polygon and the segment $\overline{(25\frac{4}{5},28)\,(23\frac{1}{6},41\frac{1}{6})}$ are part of the bisector generated by the existing facilities $a_8 = (11,48)$ and $a_{18} = (28,49)$ with respect to $w_8^2 = 10$ and $w_{18}^2 = 15$.

Figure 5.12 shows the existing facilities, the fundamental directions, the bicriteria chain, two (out of $2\binom{M}{2} = 380$) bisectors and the four inflated unit balls determining the center solution.

5.4 The 3-Criteria Case

In this section we turn to the 3-criteria case and develop an efficient algorithm for computing $\mathsf{M}_{\mathrm{Par}}\left(\mathrm{f}_\lambda^1, \mathrm{f}_\lambda^2, \mathrm{f}_\lambda^3\right)$ using the results of the bicriteria case.

We denote by

$$
C_\infty(\mathbb{R}_0^+, \mathbb{R}^2) := \left\{\varphi \,|\, \varphi : \mathbb{R}_0^+ \to \mathbb{R}^2, \varphi \,\text{continuous}, \lim_{t\to\infty} l_2(\varphi(t)) = +\infty\right\}, \quad (5.4)
$$

where $l_2(x)$ is the Euclidean norm of the point x. $C_\infty(\mathbb{R}_0^+, \mathbb{R}^2)$ is the set of continuous curves, which map the set of nonnegative numbers $\mathbb{R}_0^+ := [0,\infty)$ into the two-dimensional space $\mathbb{R}^2$ and whose image $\varphi(\mathbb{R}_0^+)$ is unbounded in $\mathbb{R}^2$.

For a set $C \subseteq \mathbb{R}^2$ we define the enclosure of C by

$$
\begin{aligned}
\mathrm{encl}\,(C) := \big\{x \in \mathbb{R}^2 \;:\; & \exists \varepsilon > 0 \text{ with } B(x,\varepsilon) \cap C = \emptyset\,,\, \exists\, t_\varphi \in [0,\infty) \,\text{with} \\
& \varphi(t_\varphi) \in C \,\text{for all}\, \varphi \in C_\infty(\mathbb{R}_0^+, \mathbb{R}^2) \,\text{and}\, \varphi(0) = x\,\big\}\,, \quad (5.5)
\end{aligned}
$$

where $B(x,\varepsilon) = \{y \in \mathbb{R}^2 \;:\; l_2(y-x) \le \varepsilon\}$. Notice that $C \cap \mathrm{encl}\,(C) = \emptyset$. Informally spoken $\mathrm{encl}\,(C)$ contains all the points which are surrounded by C, but do not belong to C themselves.

Lemma 5.2
If $x \in \mathbb{R}^2$ is dominated by $y \in \mathbb{R}^2$ with respect to strict Pareto optimality, then $z_\lambda := x + \lambda(x-y) \in \mathbb{R}^2$ with $\lambda \ge 0$ is dominated by x with respect to strict Pareto optimality.

Proof.
From $z_\lambda := x + \lambda(x-y), \lambda \ge 0$ it follows that $x = \frac{1}{1+\lambda} z_\lambda + \frac{\lambda}{1+\lambda} y$ with $\lambda \ge 0$.

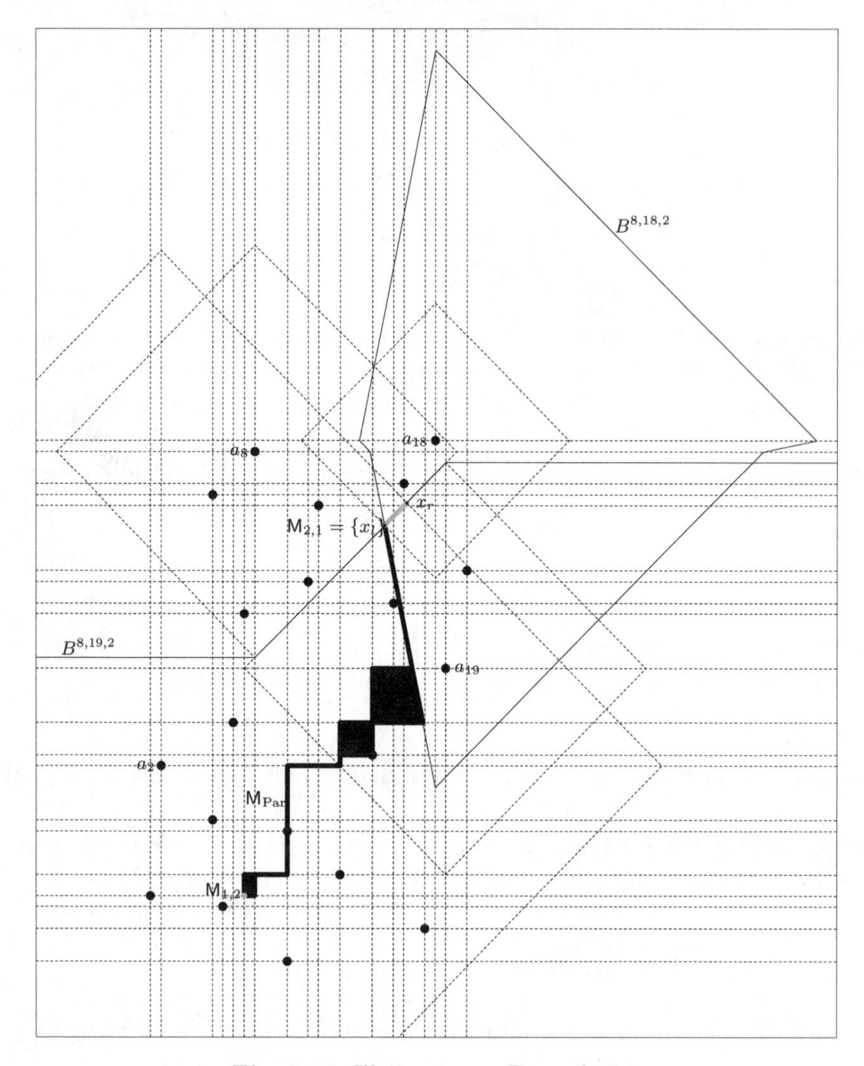

Fig. 5.12. Illustration to Example 5.1

Using that y dominates x with respect to strict Pareto optimality and the convexity of $f_\lambda^1, \ldots, f_\lambda^Q$, we obtain

$$f_\lambda^q(x) \leq \frac{1}{1+\lambda} f_\lambda^q(z_\lambda) + \frac{\lambda}{1+\lambda} f_\lambda^q(y) < \frac{1}{1+\lambda} f_\lambda^q(z_\lambda) + \frac{\lambda}{1+\lambda} f_\lambda^q(x)$$

for $q = 1 \ldots Q$, which implies

$$(1+\lambda) f_\lambda^q(x) < f_\lambda^q(z_\lambda) + \lambda f_\lambda^q(x) \quad \text{for all } q = 1, \ldots, Q.$$

Hence, $f_\lambda^q(x) < f_\lambda^q(z_\lambda)$ for all $q = 1, \ldots, Q$, i.e. x dominates z_λ with respect to strict Pareto optimality. $\square$

We denote the union of the bicriteria chains including the 1-criterion solutions by

$$\mathsf{M}^{\text{gen}}_{\text{Par}}\left(\mathrm{f}^1_\lambda, \mathrm{f}^2_\lambda, \mathrm{f}^3_\lambda\right) := \bigcup_{q=1}^{3} \mathsf{M}(\mathrm{f}^q_\lambda) \cup \bigcup_{q=1}^{2} \bigcup_{p=q+1}^{3} \mathsf{M}_{\text{Par}}\left(\mathrm{f}^p_\lambda, \mathrm{f}^q_\lambda\right) \quad . \tag{5.6}$$

We use the abbreviation *gen* since this set will *generate* the set $\mathsf{M}_{\text{Par}}\left(\mathrm{f}^1_\lambda, \mathrm{f}^2_\lambda, \mathrm{f}^3_\lambda\right)$.

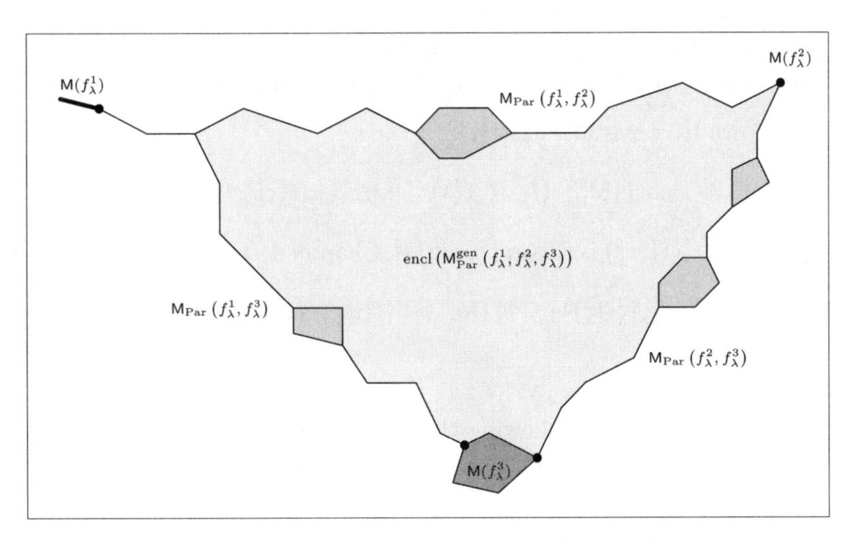

Fig. 5.13. The enclosure of $\mathsf{M}^{\text{gen}}_{\text{Par}}\left(\mathrm{f}^1_\lambda, \mathrm{f}^2_\lambda, \mathrm{f}^3_\lambda\right)$

The next lemma gives a detailed geometric description of parts of the Pareto solution set which are needed to build up $\mathsf{M}_{\text{Par}}\left(\mathrm{f}^1_\lambda, \mathrm{f}^2_\lambda, \mathrm{f}^3_\lambda\right)$. We will also learn more about the part of the plane which is crossed by the Pareto chains. For a set A, let $cl(A)$ denote the topological closure of A.

Lemma 5.3

$$\mathrm{cl}\left(\mathrm{encl}\left(M^{\text{gen}}_{\text{Par}}\left(\mathrm{f}^1_\lambda, \mathrm{f}^2_\lambda, \mathrm{f}^3_\lambda\right)\right)\right) \subseteq M_{\text{s-Par}}\left(\mathrm{f}^1_\lambda, \mathrm{f}^2_\lambda, \mathrm{f}^3_\lambda\right).$$

Proof.
Let $x \in \mathrm{encl}\left(\mathsf{M}^{\text{gen}}_{\text{Par}}\left(\mathrm{f}^1_\lambda, \mathrm{f}^2_\lambda, \mathrm{f}^3_\lambda\right)\right)$. Assume $x \notin \mathsf{M}_{\text{s-Par}}\left(\mathrm{f}^1_\lambda, \mathrm{f}^2_\lambda, \mathrm{f}^3_\lambda\right)$. Then there exists a point $y \in \mathbb{R}^2$ which dominates x with respect to strict Pareto optimality. Consider the curve

$$\begin{array}{rcl} \varphi : \mathbb{R}^+_0 & \rightarrow & \mathbb{R}^2 , \\ t & \mapsto & x + t(x-y) \end{array} \quad .$$

Obviously φ is continuous and fulfills

$$\lim_{t\to\infty} l_2((\varphi(t)) = +\infty, \text{ i.e. } \varphi \in C_\infty(\mathbb{R}_0^+, \mathbb{R}^2) .$$

Moreover, $\varphi(0) = x$. Since $x \in \operatorname{encl}\left(\mathsf{M}^{\mathrm{gen}}_{\mathrm{Par}}\left(\mathrm{f}^1_\lambda, \mathrm{f}^2_\lambda, \mathrm{f}^3_\lambda\right)\right)$, there exists $t \in [0, \infty)$ with

$$z_t := \varphi(t) \in \mathsf{M}^{\mathrm{gen}}_{\mathrm{Par}}\left(\mathrm{f}^1_\lambda, \mathrm{f}^2_\lambda, \mathrm{f}^3_\lambda\right) \quad . \tag{5.7}$$

By Lemma 5.2 we have $f^q_\lambda(x) \le f^q_\lambda(z_t)$ for all $q = 1, \ldots, 3$. Hence, we can continue with the following case analysis with respect to (5.7):

Case 1: $z_t \in \mathsf{M}(\mathrm{f}^q_\lambda)$ for some $q \in \{1, 2, 3\}$:
$\Rightarrow\ x \in \mathsf{M}(\mathrm{f}^q_\lambda)\ \Rightarrow\ \mathrm{x} \in \mathsf{M}^{\mathrm{gen}}_{\mathrm{Par}}\left(\mathrm{f}^1_\lambda, \mathrm{f}^2_\lambda, \mathrm{f}^3_\lambda\right)$. This is a contradiction.
Case 2: $z_t \in \mathsf{M}_{\mathrm{Par}}\left(\mathrm{f}^p_\lambda, \mathrm{f}^q_\lambda\right)$ for some $p, q \in \{1, 2, 3\}, p < q$:
$\Rightarrow\ x \in \mathsf{M}_{\mathrm{Par}}\left(\mathrm{f}^p_\lambda, \mathrm{f}^q_\lambda\right)\ \Rightarrow\ \mathrm{x} \in \mathsf{M}^{\mathrm{gen}}_{\mathrm{Par}}\left(\mathrm{f}^1_\lambda, \mathrm{f}^2_\lambda, \mathrm{f}^3_\lambda\right)$. This is a contradiction.

Therefore, we have $x \in \mathsf{M}_{\mathrm{s-Par}}\left(\mathrm{f}^1_\lambda, \mathrm{f}^2_\lambda, \mathrm{f}^3_\lambda\right)$, i.e.

$$\operatorname{encl}\left(\mathsf{M}^{\mathrm{gen}}_{\mathrm{Par}}\left(\mathrm{f}^1_\lambda, \mathrm{f}^2_\lambda, \mathrm{f}^3_\lambda\right)\right) \subseteq \mathsf{M}_{\mathrm{s-Par}}\left(\mathrm{f}^1_\lambda, \mathrm{f}^2_\lambda, \mathrm{f}^3_\lambda\right).$$

Since $\mathsf{M}_{\mathrm{s-Par}}\left(\mathrm{f}^1_\lambda, \mathrm{f}^2_\lambda, \mathrm{f}^3_\lambda\right)$ is closed (see [211], Chapter 4, Theorem 27) we obtain

$$\operatorname{cl}\left(\operatorname{encl}\left(\mathsf{M}^{\mathrm{gen}}_{\mathrm{Par}}\left(\mathrm{f}^1_\lambda, \mathrm{f}^2_\lambda, \mathrm{f}^3_\lambda\right)\right)\right) \subseteq \operatorname{cl}\left(\mathsf{M}_{\mathrm{s-Par}}\left(\mathrm{f}^1_\lambda, \mathrm{f}^2_\lambda, \mathrm{f}^3_\lambda\right)\right) = \mathsf{M}_{\mathrm{s-Par}}\left(\mathrm{f}^1_\lambda, \mathrm{f}^2_\lambda, \mathrm{f}^3_\lambda\right) \ .$$

□

Finally we obtain the following theorem which provides a subset as well as a superset of $\mathsf{M}_{\mathrm{Par}}\left(\mathrm{f}^1_\lambda, \mathrm{f}^2_\lambda, \mathrm{f}^3_\lambda\right)$.

Theorem 5.2

$$\begin{aligned}\operatorname{encl}\left(\mathsf{M}^{\mathrm{gen}}_{\mathrm{Par}}\left(\mathrm{f}^1_\lambda, \mathrm{f}^2_\lambda, \mathrm{f}^3_\lambda\right)\right) &\subseteq \mathsf{M}_{\mathrm{Par}}\left(\mathrm{f}^1_\lambda, \mathrm{f}^2_\lambda, \mathrm{f}^3_\lambda\right)\\ &\subseteq \mathsf{M}^{\mathrm{gen}}_{\mathrm{Par}}\left(\mathrm{f}^1_\lambda, \mathrm{f}^2_\lambda, \mathrm{f}^3_\lambda\right) \cup \operatorname{encl}\left(\mathsf{M}^{\mathrm{gen}}_{\mathrm{Par}}\left(\mathrm{f}^1_\lambda, \mathrm{f}^2_\lambda, \mathrm{f}^3_\lambda\right)\right).\end{aligned}$$

Proof.

$$\begin{aligned}\operatorname{encl}\left(\mathsf{M}^{\mathrm{gen}}_{\mathrm{Par}}\left(\mathrm{f}^1_\lambda, \mathrm{f}^2_\lambda, \mathrm{f}^3_\lambda\right)\right) &\subseteq \mathsf{M}_{\mathrm{s-Par}}\left(\mathrm{f}^1_\lambda, \mathrm{f}^2_\lambda, \mathrm{f}^3_\lambda\right) \ \text{(by Lemma 5.3)}\\ &\subseteq \mathsf{M}_{\mathrm{Par}}\left(\mathrm{f}^1_\lambda, \mathrm{f}^2_\lambda, \mathrm{f}^3_\lambda\right)\\ &\subseteq \mathsf{M}_{\mathrm{w-Par}}\left(\mathrm{f}^1_\lambda, \mathrm{f}^2_\lambda, \mathrm{f}^3_\lambda\right)\\ &\subseteq \mathsf{M}^{\mathrm{gen}}_{\mathrm{Par}}\left(\mathrm{f}^1_\lambda, \mathrm{f}^2_\lambda, \mathrm{f}^3_\lambda\right) \cup \operatorname{encl}\left(\mathsf{M}^{\mathrm{gen}}_{\mathrm{Par}}\left(\mathrm{f}^1_\lambda, \mathrm{f}^2_\lambda, \mathrm{f}^3_\lambda\right)\right)\\ &\quad \text{(by Theorem 3.2 in [178]).}\end{aligned}$$

□

Now it remains to consider the Pareto optimality of the set $\mathsf{M}^{\mathrm{gen}}_{\mathrm{Par}}\left(\mathrm{f}^1_\lambda, \mathrm{f}^2_\lambda, \mathrm{f}^3_\lambda\right)$ with respect to the three objective functions $f^1_\lambda, f^2_\lambda, f^3_\lambda$. For a cell $C \in \mathcal{C}$ we define the collapsing and the remaining parts of C with respect to Q-criteria optimality by

$$\operatorname{col}_Q(C) := \left\{x \in C \,:\, x \notin \mathsf{M}_{\mathrm{Par}}\left(\mathrm{f}^1_\lambda, \ldots, \mathrm{f}^Q_\lambda\right)\right\} \tag{5.8}$$

and

$$\mathrm{rem}_Q(C) := \left\{x \in C \, : \, x \in \mathsf{M}_{\mathrm{Par}}\left(\mathrm{f}^1_\lambda, \ldots, \mathrm{f}^Q_\lambda\right)\right\} \quad . \tag{5.9}$$

Using the differentiability of the objective functions in the interior of the cells we obtain the following lemma.

Lemma 5.4
Let $C \in \mathcal{C}$ be a cell. The following statements hold.

1. $\mathrm{col}_Q(C) \dot{\cup} \mathrm{rem}_Q(C) = C$.
2. *Either* $\mathrm{rem}_Q(C) = C$ *or* $\mathrm{rem}_Q(C) \subseteq \mathrm{bd}(C)$. *In the latter case* $\mathrm{rem}_Q(C)$ *is either empty or consists of complete faces and/or extreme points of* C.
3. *For* $C \subseteq \mathsf{M}(\mathrm{f}^p_\lambda)$ *with* $p \in \{1,2,3\}$ *and* $x \in \mathrm{int}(C)$*, we have:*

$$x \in \mathrm{rem}_3(C) \;\Leftrightarrow\; \left\{\begin{array}{l} \exists \xi \in \mathbb{R} \textit{ with } \nabla f^q_\lambda(x) = \xi \nabla f^r_\lambda(x) \\ \textit{and } \xi < 0 \textit{ for } q,r \in \{1,2,3\} \setminus \{p\}, q < r \end{array}\right\}. \tag{5.10}$$

For $C \subseteq \mathsf{M}_{\mathrm{Par}}\left(\mathrm{f}^p_\lambda, \mathrm{f}^q_\lambda\right)$ *with* $p,q \in \{1,2,3\}, p < q$*, and* $x \in \mathrm{int}(C)$*, we have:*

$$x \in \mathrm{rem}_3(C) \;\Leftrightarrow\; \left\{\begin{array}{l} \exists \xi^p, \xi^q \in \mathbb{R} \textit{ with } \nabla f^r_\lambda(x) = \xi^p \nabla f^p_\lambda(x)\,, \\ \nabla f^r_\lambda(x) = \xi^q \nabla f^q_\lambda(x) \\ \textit{and } \xi^p \xi^q \le 0 \textit{ for } r \in \{1,2,3\} \setminus \{p,q\} \end{array}\right\}. \tag{5.11}$$

Proof.

1. Follows directly from the definition of $\mathrm{col}_Q(C)$ and $\mathrm{rem}_Q(C)$.
2. If $\mathrm{int}(C) \cap \mathsf{M}_{\mathrm{Par}}\left(\mathrm{f}^1_\lambda, \mathrm{f}^2_\lambda, \mathrm{f}^3_\lambda\right) \neq \emptyset$ we have $C \subseteq \mathsf{M}_{\mathrm{Par}}\left(\mathrm{f}^1_\lambda, \mathrm{f}^2_\lambda, \mathrm{f}^3_\lambda\right)$ and hence $\mathrm{rem}_Q(C) = C$. This follows analogously to the proof of Theorem 5.1.
If $\mathrm{int}(C) \cap \mathsf{M}_{\mathrm{Par}}\left(\mathrm{f}^1_\lambda, \mathrm{f}^2_\lambda, \mathrm{f}^3_\lambda\right) = \emptyset$, then $\mathrm{rem}_Q(C) \subseteq \mathrm{bd}(C)$. The rest follows analogously to the proof of Theorem 5.1.
3. If $C \subseteq \mathsf{M}(\mathrm{f}^p_\lambda)$ for $p \in \{1,2,3\}$ and $x \in \mathrm{int}(C)$, we have $L_=(f^p_\lambda, f^p_\lambda(x)) = L_\le(f^p_\lambda, f^p_\lambda(x))$ and therefore there exist $q, r \in \{1,2,3\} \setminus \{p\}$, $q < r$ such that

$$\begin{aligned} x \in \mathrm{rem}_3(f) &\Leftrightarrow L_=(f^q_\lambda, f^q_\lambda(x)) \cap L_=(f^r_\lambda, f^r_\lambda(x)) \\ &\quad= L_\le(f^q_\lambda, f^q_\lambda(x)) \cap L_\le(f^r_\lambda, f^r_\lambda(x)) \\ &\Leftrightarrow \nabla f^q_\lambda(x) = \xi \nabla f^r_\lambda(x) \text{ with } \xi < 0. \end{aligned}$$

If $C \subseteq \mathsf{M}_{\mathrm{Par}}\left(\mathrm{f}^p_\lambda, \mathrm{f}^q_\lambda\right)$ for $p, q \in \{1,2,3\}$ and $x \in \mathrm{int}(C)$, there exists $\xi \in \mathbb{R}$ with

$$\nabla f^p_\lambda(x) = \xi \nabla f^q_\lambda(x) \text{ with } \xi < 0\,.$$

(Notice that the trivial case $\mathsf{M}(\mathrm{f}^1_\lambda) \cap \mathsf{M}(\mathrm{f}^2_\lambda) \neq \emptyset$, i.e. $\nabla f^p_\lambda(x) = 0 = \nabla f^q_\lambda(x)$ is included.)
The Pareto optimality condition

$$\bigcap_{q=1}^{3} L_{=}(f_\lambda^q, f_\lambda^q(x)) = \bigcap_{q=1}^{3} L_{\leq}(f_\lambda^q, f_\lambda^q(x))$$

for the 3 criteria is fulfilled if and only if the level curve $L_{=}^{r}(x)$, $r \in \{1,2,3\} \setminus \{p,q\}$, has the same slope as $L_{=}(f_\lambda^p, f_\lambda^p(x))$ and $L_{=}(f_\lambda^q, f_\lambda^q(x))$, i.e. if and only if $\xi^p, \xi^q \in \mathbb{R}$ exist with

$$\nabla f_\lambda^r(x) = \xi^p \nabla f_\lambda^p(x)\,,\ \nabla f_\lambda^r(x) = \xi^q \nabla f_\lambda^q(x) \text{ and } \xi^p \xi^q \leq 0\,.$$

□

Summing up the preceding results we get a complete geometric characterization of the set of Pareto solutions for three criteria.

Theorem 5.3

$$\begin{aligned} \mathsf{M}_{\text{Par}}\left(\mathrm{f}_\lambda^1, \mathrm{f}_\lambda^2, \mathrm{f}_\lambda^3\right) = &\left(\mathsf{M}_{\text{Par}}^{\text{gen}}\left(\mathrm{f}_\lambda^1, \mathrm{f}_\lambda^2, \mathrm{f}_\lambda^3\right) \cup \operatorname{encl}\left(\mathsf{M}_{\text{Par}}^{\text{gen}}\left(\mathrm{f}_\lambda^1, \mathrm{f}_\lambda^2, \mathrm{f}_\lambda^3\right)\right)\right) \\ &\setminus \{x \in \mathbb{R}^2 : \exists C \in \mathcal{C}, C \subseteq \mathsf{M}_{\text{Par}}^{\text{gen}}\left(\mathrm{f}_\lambda^1, \mathrm{f}_\lambda^2, \mathrm{f}_\lambda^3\right), \mathrm{x} \in \mathrm{col}_3(\mathrm{C})\}. \end{aligned}$$

Proof.
Let $y \in \mathsf{M}_{\text{Par}}\left(\mathrm{f}_\lambda^1, \mathrm{f}_\lambda^2, \mathrm{f}_\lambda^3\right)$. Then we have by Theorem 5.2

$$y \in \mathsf{M}_{\text{Par}}^{\text{gen}}\left(\mathrm{f}_\lambda^1, \mathrm{f}_\lambda^2, \mathrm{f}_\lambda^3\right) \cup \operatorname{encl}\left(\mathsf{M}_{\text{Par}}^{\text{gen}}\left(\mathrm{f}_\lambda^1, \mathrm{f}_\lambda^2, \mathrm{f}_\lambda^3\right)\right)\,.$$

Moreover for $C \in \mathcal{C}$ with $y \in C$ we have $y \in \mathrm{rem}_3(C)$, i.e. $y \notin \mathrm{col}_3(C)$. This implies

$$\begin{aligned} y \in &\left(\mathsf{M}_{\text{Par}}^{\text{gen}}\left(\mathrm{f}_\lambda^1, \mathrm{f}_\lambda^2, \mathrm{f}_\lambda^3\right) \cup \operatorname{encl}\left(\mathsf{M}_{\text{Par}}^{\text{gen}}\left(\mathrm{f}_\lambda^1, \mathrm{f}_\lambda^2, \mathrm{f}_\lambda^3\right)\right)\right) \\ &\setminus \{x \in \mathbb{R}^2 : \exists C \in \mathcal{C}, C \subseteq \mathsf{M}_{\text{Par}}^{\text{gen}}\left(\mathrm{f}_\lambda^1, \mathrm{f}_\lambda^2, \mathrm{f}_\lambda^3\right), \mathrm{x} \in \mathrm{col}_3(\mathrm{C})\}\quad. \end{aligned}$$

Now let

$$\begin{aligned} y \in &\left(\mathsf{M}_{\text{Par}}^{\text{gen}}\left(\mathrm{f}_\lambda^1, \mathrm{f}_\lambda^2, \mathrm{f}_\lambda^3\right) \cup \operatorname{encl}\left(\mathsf{M}_{\text{Par}}^{\text{gen}}\left(\mathrm{f}_\lambda^1, \mathrm{f}_\lambda^2, \mathrm{f}_\lambda^3\right)\right)\right) \\ &\setminus \{x \in \mathbb{R}^2 : \exists C \in \mathcal{C}, C \subseteq \mathsf{M}_{\text{Par}}^{\text{gen}}\left(\mathrm{f}_\lambda^1, \mathrm{f}_\lambda^2, \mathrm{f}_\lambda^3\right), \mathrm{x} \in \mathrm{col}_3(\mathrm{C})\}\quad. \end{aligned}$$

We distinguish the following cases:

Case 1: $y \in \operatorname{encl}\left(\mathsf{M}_{\text{Par}}^{\text{gen}}\left(\mathrm{f}_\lambda^1, \mathrm{f}_\lambda^2, \mathrm{f}_\lambda^3\right)\right)$. Then $y \in \mathsf{M}_{\text{Par}}\left(\mathrm{f}_\lambda^1, \mathrm{f}_\lambda^2, \mathrm{f}_\lambda^3\right)$ by Theorem 5.2.
Case 2: $y \in \mathsf{M}_{\text{Par}}^{\text{gen}}\left(\mathrm{f}_\lambda^1, \mathrm{f}_\lambda^2, \mathrm{f}_\lambda^3\right)$
Case 2.1: $\exists C \in \mathcal{C}, C \subseteq \mathsf{M}_{\text{Par}}^{\text{gen}}\left(\mathrm{f}_\lambda^1, \mathrm{f}_\lambda^2, \mathrm{f}_\lambda^3\right)$ with $y \in C$

$$\Rightarrow\ y \notin \mathrm{col}_3(C)\ \Rightarrow\ y \in \mathrm{rem}_3(C)\ \Rightarrow\ y \in \mathsf{M}_{\text{Par}}\left(\mathrm{f}_\lambda^1, \mathrm{f}_\lambda^2, \mathrm{f}_\lambda^3\right).$$

Case 2.2: $\nexists\, C \in \mathcal{C}, C \subseteq \mathsf{M}_{\text{Par}}^{\text{gen}}\left(\mathrm{f}_\lambda^1, \mathrm{f}_\lambda^2, \mathrm{f}_\lambda^3\right)$ with $y \in C$

$$\begin{aligned} &\Rightarrow\ L_{\leq}(f_\lambda^p, f_\lambda^p(y)) \cap L_{\leq}(f_\lambda^q, f_\lambda^q(y)) = \{y\} \text{ for some } p,q \in \{1,2,3\}, p<q \\ &\Rightarrow\ \textstyle\bigcap_{q=1}^{3} L_{\leq}(f_\lambda^q, f_\lambda^q(y)) = \{y\} \\ &\Rightarrow\ y \in \mathsf{M}_{\text{s-Par}}\left(\mathrm{f}_\lambda^1, \mathrm{f}_\lambda^2, \mathrm{f}_\lambda^3\right) \subseteq \mathsf{M}_{\text{Par}}\left(\mathrm{f}_\lambda^1, \mathrm{f}_\lambda^2, \mathrm{f}_\lambda^3\right). \end{aligned}$$

□

In the OMP the gradients $\nabla f^q_\lambda(x), q \in \{1,2,3\}$, (in those points where they are well-defined) can be computed in $O(M \log(MG_{\max}))$ time (analogous to the evaluation of the function). Therefore, we can test, with (5.10) and (5.11), in $O(M \log(MG_{\max}))$ time if a cell $C \in \mathcal{C}, C \subseteq \mathsf{M}^{\text{gen}}_{\text{Par}}(\mathrm{f}^1_\lambda, \mathrm{f}^2_\lambda, \mathrm{f}^3_\lambda)$ collapses. We obtain the following algorithm for the 3-criteria OMP with time complexity $O(M^5 G^2_{\max} \log(MG_{\max}))$.

***Algorithm 5.2**: Computing* $\mathsf{M}_{\text{Par}}(\mathrm{f}^1_\lambda, \mathrm{f}^2_\lambda, \mathrm{f}^3_\lambda)$.

Compute the planar graph generated by the cells $\mathcal{C}$.
Compute $\mathsf{M}_{\text{w-Par}}(\mathrm{f}^1_\lambda, \mathrm{f}^2_\lambda), \mathsf{M}_{\text{w-Par}}(\mathrm{f}^1_\lambda, \mathrm{f}^3_\lambda), \mathsf{M}_{\text{w-Par}}(\mathrm{f}^2_\lambda, \mathrm{f}^3_\lambda)$ using Algorithm 5.1.
$\mathsf{M}^{\text{gen}}_{\text{Par}}(\mathrm{f}^1_\lambda, \mathrm{f}^2_\lambda, \mathrm{f}^3_\lambda) := \mathsf{M}_{\text{w-Par}}(\mathrm{f}^1_\lambda, \mathrm{f}^2_\lambda) \cup \mathsf{M}_{\text{w-Par}}(\mathrm{f}^1_\lambda, \mathrm{f}^3_\lambda)$
$\qquad \cup \mathsf{M}_{\text{w-Par}}(\mathrm{f}^2_\lambda, \mathrm{f}^3_\lambda)$.
$\mathsf{M}_{\text{Par}}(\mathrm{f}^1_\lambda, \mathrm{f}^2_\lambda, \mathrm{f}^3_\lambda) := \mathsf{M}^{\text{gen}}_{\text{Par}}(\mathrm{f}^1_\lambda, \mathrm{f}^2_\lambda, \mathrm{f}^3_\lambda) \cup \text{encl}\left(\mathsf{M}^{\text{gen}}_{\text{Par}}(\mathrm{f}^1_\lambda, \mathrm{f}^2_\lambda, \mathrm{f}^3_\lambda)\right)$.
foreach $C \in \mathcal{C}$ *with* $C \subseteq \mathsf{M}^{\text{gen}}_{\text{Par}}(\mathrm{f}^1_\lambda, \mathrm{f}^2_\lambda, \mathrm{f}^3_\lambda)$ **do**
  Compute $\text{col}_3(C)$ using Lemma 5.4.
  $\mathsf{M}_{\text{Par}}(\mathrm{f}^1_\lambda, \mathrm{f}^2_\lambda, \mathrm{f}^3_\lambda) := \mathsf{M}_{\text{Par}}(\mathrm{f}^1_\lambda, \mathrm{f}^2_\lambda, \mathrm{f}^3_\lambda) \setminus \text{col}_3(\mathrm{C})$.
end
Output: $\mathsf{M}_{\text{Par}}(\mathrm{f}^1_\lambda, \mathrm{f}^2_\lambda, \mathrm{f}^3_\lambda)$.

Finally we present an example to illustrate the preceding results.

Example 5.2
We consider the four existing facilities $a_1 = (2, 6.5), a_2 = (5, 9.5), a_3 = (6.5, 2)$ and $a_4 = (11, 9.5)$ (see Figure 5.14 where the bisector $\mathsf{Bisec}(a_i, a_j)$ has been denoted by B_{ij}). a_1 and a_4 are associated with the l_1-norm, whereas a_2 and a_3 are associated with the l_∞-norm. We consider three OMf f^q_λ for $q = 1, 2, 3$ defined by the following weights:

q	w^q_1	w^q_2	w^q_3	w^q_4	λ^q_1	λ^q_2	λ^q_3	λ^q_4
1	1	1	0	1	1	1	1	1
2	1	1	1	0	1	1	1	1
3	1	1	1	1	0	0	0	1

We obtain the optimal solutions $\mathsf{M}(\mathrm{f}^1_\lambda) = \{\mathrm{a}_2\}, \mathsf{M}(\mathrm{f}^2_\lambda) = \{\mathrm{a}_1\}$ and $\mathsf{M}(\mathrm{f}^3_\lambda) = \overline{(6.5, 8), (8, 6.5)}$. The sets $\mathsf{M}^{\text{gen}}_{\text{Par}}(\mathrm{f}^1_\lambda, \mathrm{f}^2_\lambda, \mathrm{f}^3_\lambda)$ and $\mathsf{M}_{\text{Par}}(\mathrm{f}^1_\lambda, \mathrm{f}^2_\lambda, \mathrm{f}^3_\lambda)$ are drawn in Figure 5.15 and Figure 5.16, respectively. Both figures show a part of the whole situation presented in Figure 5.14.

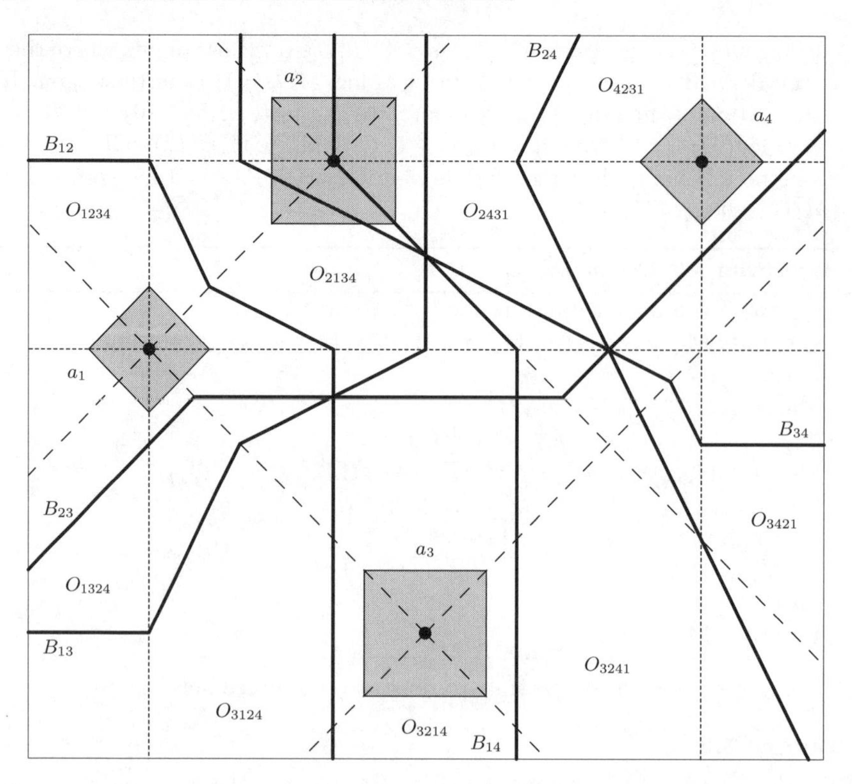

Fig. 5.14. Bisector lines and ordered regions generated by 4 existing facilities $a_1, \dots, a_4$ associated with the l_1-norm respectively the l_∞-norm for $W := \{1, 1, 1, 1\}$

5.5 The Case $Q > 3$

Now we turn to the Q-Criteria case ($Q > 3$). The results in this section are based on similar ones corresponding to the 3-criteria case. First of all, we obtain a chain of set-inclusions for the solution set $\mathsf{M}_{\text{Par}}\left(\mathrm{f}_\lambda^1, \dots, \mathrm{f}_\lambda^Q\right)$.

Theorem 5.4

1.
$$\bigcup_{\substack{p,q,r=1,\dots,Q\\ p<q<r}} \mathrm{cl}\left(\mathrm{encl}\left(\mathsf{M}_{\text{Par}}^{\text{gen}}\left(\mathrm{f}_\lambda^p, \mathrm{f}_\lambda^q, \mathrm{f}_\lambda^r\right)\right)\right) \subseteq \mathsf{M}_{\text{Par}}\left(\mathrm{f}_\lambda^1, \dots, \mathrm{f}_\lambda^Q\right)$$

2.
$$\mathsf{M}_{\text{Par}}\left(\mathrm{f}_\lambda^1, \dots, \mathrm{f}_\lambda^Q\right) \subseteq$$
$$\bigcup_{\substack{p,q,r=1,\dots,Q\\ p<q<r}} \mathsf{M}_{\text{Par}}^{\text{gen}}\left(\mathrm{f}_\lambda^p, \mathrm{f}_\lambda^q, \mathrm{f}_\lambda^r\right) \cup \bigcup_{\substack{p,q,r=1,\dots,Q\\ p<q<r}} \mathrm{encl}\left(\mathsf{M}_{\text{Par}}^{\text{gen}}\left(\mathrm{f}_\lambda^p, \mathrm{f}_\lambda^q, \mathrm{f}_\lambda^r\right)\right)$$

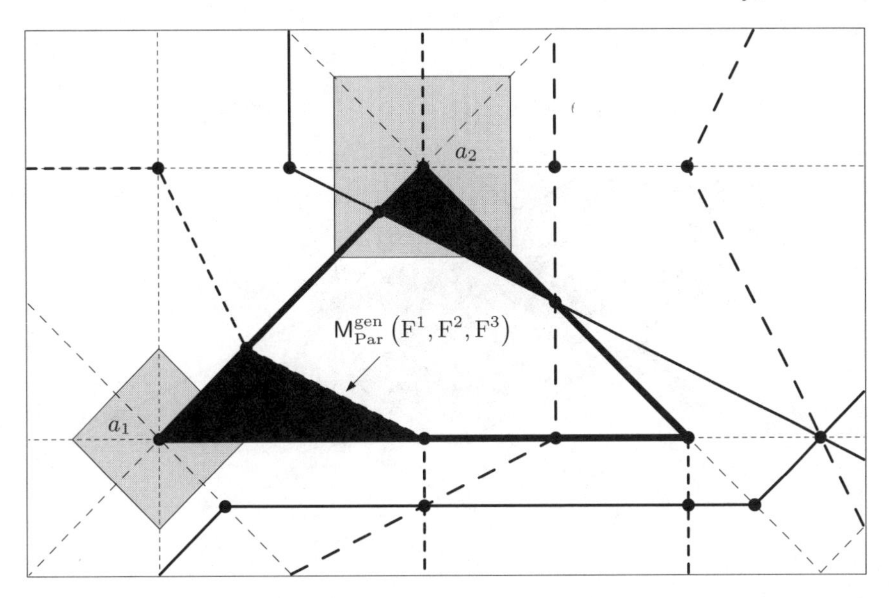

Fig. 5.15. Illustration to Example 5.2

Proof.

1. $$x \in \bigcup_{\substack{p,q,r=1,\ldots,Q \\ p<q<r}} \mathrm{cl}\left(\mathrm{encl}\left(\mathsf{M}^{\mathrm{gen}}_{\mathrm{Par}}\left(\mathrm{f}^{\mathrm{p}}_{\lambda}, \mathrm{f}^{\mathrm{q}}_{\lambda}, \mathrm{f}^{\mathrm{r}}_{\lambda}\right)\right)\right)$$

$\Leftrightarrow x \in \mathrm{cl}\left(\mathrm{encl}\left(\mathsf{M}^{\mathrm{gen}}_{\mathrm{Par}}\left(\mathrm{f}^{\mathrm{p}}_{\lambda}, \mathrm{f}^{\mathrm{q}}_{\lambda}, \mathrm{f}^{\mathrm{r}}_{\lambda}\right)\right)\right)$ for some $p, q, r = 1, \ldots, Q, p < q < r$

$\Rightarrow x \in \mathsf{M}_{\mathrm{s-Par}}\left(\mathrm{f}^{\mathrm{p}}_{\lambda}, \mathrm{f}^{\mathrm{q}}_{\lambda}, \mathrm{f}^{\mathrm{r}}_{\lambda}\right)$
for some $p, q, r = 1, \ldots, Q, p < q < r$, by Lemma 5.3

$\Leftrightarrow L_{\leq}(f^p_\lambda, f^p_\lambda(x)) \cap L_{\leq}(f^q_\lambda, f^q_\lambda(x)) \cap L_{\leq}(f^r_\lambda, f^r_\lambda(x)) = \{x\}$
for some $p, q, r = 1, \ldots, Q, p < q < r$, by (5.3)

$\Rightarrow \bigcap_{q=1}^{Q} L_{\leq}(f^q_\lambda, f^q_\lambda(x)) = \{x\}$, since $x \in L_{\leq}(f^q_\lambda, f^q_\lambda(x))$ for all $q = 1, \ldots, Q$

$\Leftrightarrow x \in \mathsf{M}_{\mathrm{s-Par}}\left(\mathrm{f}^1_\lambda, \ldots, \mathrm{f}^Q_\lambda\right)$ by (5.3)

$\Rightarrow x \in \mathsf{M}_{\mathrm{Par}}\left(\mathrm{f}^1_\lambda, \ldots, \mathrm{f}^Q_\lambda\right)$.

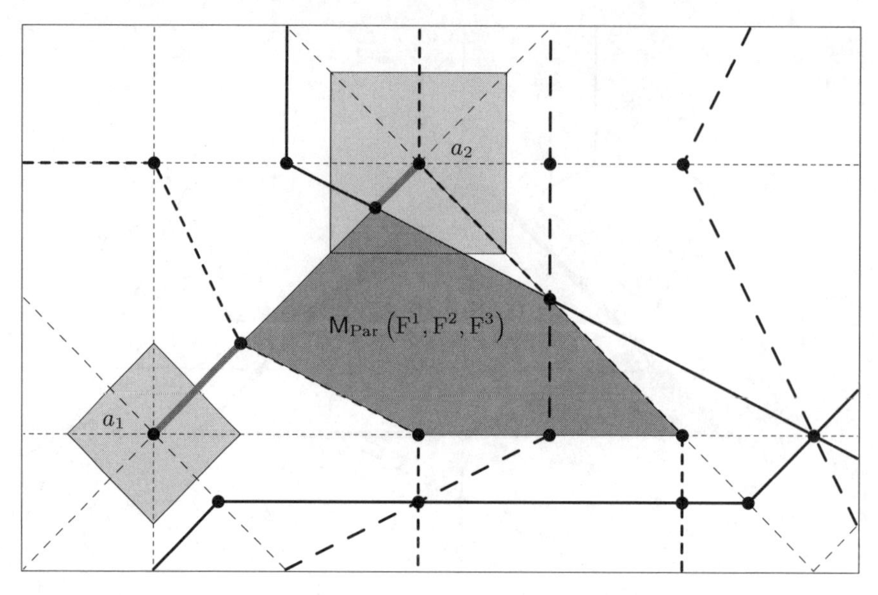

Fig. 5.16. Illustration to Example 5.2

2. $x \in \mathsf{M}_{\text{Par}}\left(\mathrm{f}_\lambda^1, \ldots, \mathrm{f}_\lambda^Q\right)$

$\Rightarrow x \in \mathsf{M}_{\text{w-Par}}\left(\mathrm{f}_\lambda^1, \ldots, \mathrm{f}_\lambda^Q\right)$

$\Leftrightarrow \bigcap_{q=1}^{Q} L_<(f_\lambda^q, f_\lambda^q(x)) = \emptyset$ by (5.1)

$\Leftrightarrow L_<(f_\lambda^p, f_\lambda^p(x)) \cap L_<(f_\lambda^q, f_\lambda^q(x)) \cap L_<(f_\lambda^r, f_\lambda^r(x)) = \emptyset$

for some $p, q, r = 1, \ldots, Q, p < q < r$, by Helly's Theorem

$\Leftrightarrow x \in \mathsf{M}_{\text{w-Par}}(\mathrm{f}_\lambda^\mathrm{p}, \mathrm{f}_\lambda^\mathrm{q}, \mathrm{f}_\lambda^\mathrm{r})$ for some $\mathrm{p, q, r} = 1, \ldots, \mathrm{Q}, \mathrm{p} < \mathrm{q} < \mathrm{r}$, by (5.1)

$\Rightarrow x \in \mathsf{M}_{\text{Par}}^{\text{gen}}(\mathrm{f}_\lambda^\mathrm{p}, \mathrm{f}_\lambda^\mathrm{q}, \mathrm{f}_\lambda^\mathrm{r}) \cup \text{encl}\,(\mathsf{M}_{\text{Par}}^{\text{gen}}(\mathrm{f}_\lambda^\mathrm{p}, \mathrm{f}_\lambda^\mathrm{q}, \mathrm{f}_\lambda^\mathrm{r}))$

for some $p, q, r = 1, \ldots, Q, p < q < r$, by Theorem 3.2 in [178]

$$\Leftrightarrow x \in \bigcup_{\substack{p,q,r=1,\ldots,Q \\ p<q<r}} \mathsf{M}_{\text{Par}}^{\text{gen}}(\mathrm{f}_\lambda^\mathrm{p}, \mathrm{f}_\lambda^\mathrm{q}, \mathrm{f}_\lambda^\mathrm{r}) \cup \bigcup_{\substack{\mathrm{p,q,r}=1,\ldots,\mathrm{Q} \\ \mathrm{p<q<r}}} \text{encl}\,(\mathsf{M}_{\text{Par}}^{\text{gen}}(\mathrm{f}_\lambda^\mathrm{p}, \mathrm{f}_\lambda^\mathrm{q}, \mathrm{f}_\lambda^\mathrm{r}))\,.$$

□

In the Q-criteria case the crucial region is now given by the cells $C \in \mathcal{C}$ with

$$C \subseteq \bigcup_{\substack{p,q,r=1,\ldots,Q \\ p<q<r}} \mathsf{M}_{\text{Par}}^{\text{gen}}(\mathrm{f}_\lambda^\mathrm{p}, \mathrm{f}_\lambda^\mathrm{q}, \mathrm{f}_\lambda^\mathrm{r}) \setminus \bigcup_{\substack{\mathrm{p,q,r}=1,\ldots,\mathrm{Q} \\ \mathrm{p<q<r}}} \text{encl}\,(\mathsf{M}_{\text{Par}}^{\text{gen}}(\mathrm{f}_\lambda^\mathrm{p}, \mathrm{f}_\lambda^\mathrm{q}, \mathrm{f}_\lambda^\mathrm{r}))$$

$$= \bigcup_{\substack{p,q=1,\dots,Q \\ p<q}} \mathsf{M}_{\text{w-Par}}(\mathrm{f}_\lambda^{\mathrm{p}}, \mathrm{f}_\lambda^{\mathrm{q}}) \setminus \bigcup_{\substack{\mathrm{p,q,r}=1,\dots,\mathrm{Q} \\ \mathrm{p<q<r}}} \mathrm{encl}\,(\mathsf{M}_{\text{Par}}^{\text{gen}}(\mathrm{f}_\lambda^{\mathrm{p}}, \mathrm{f}_\lambda^{\mathrm{q}}, \mathrm{f}_\lambda^{\mathrm{r}})) \quad .$$

Similar to Lemma 5.4 it can be tested, by comparing the gradients of the objective functions in $\text{int}(C)$, if the cell $C \in \mathcal{C}$ collapses with respect to $f_\lambda^1, \dots, f_\lambda^Q$ or not. Finally we obtain the following theorem, which can be proven by the same technique as in the 3-criteria case (see the proof of Theorem 5.3).

Theorem 5.5

$$\mathsf{M}_{\text{Par}}\left(\mathrm{f}_\lambda^1, \dots, \mathrm{f}_\lambda^{\mathrm{Q}}\right) =$$

$$\left(\bigcup_{\substack{p,q,r=1,\dots,Q \\ p<q<r}} \mathsf{M}_{\text{Par}}^{\text{gen}}(\mathrm{f}_\lambda^{\mathrm{p}}, \mathrm{f}_\lambda^{\mathrm{q}}, \mathrm{f}_\lambda^{\mathrm{r}}) \cup \bigcup_{\substack{\mathrm{p,q,r}=1,\dots,\mathrm{Q} \\ \mathrm{p<q<r}}} \mathrm{encl}\,(\mathsf{M}_{\text{Par}}^{\text{gen}}(\mathrm{f}_\lambda^{\mathrm{p}}, \mathrm{f}_\lambda^{\mathrm{q}}, \mathrm{f}_\lambda^{\mathrm{r}})) \right)$$

$$\setminus \left\{ x \in \mathbb{R}^2 : \exists C \in \mathcal{C}, C \subseteq \bigcup_{\substack{p,q=1,\dots,Q \\ p<q}} \mathsf{M}_{\text{w-Par}}(\mathrm{f}_\lambda^{\mathrm{p}}, \mathrm{f}_\lambda^{\mathrm{q}}) \setminus \right.$$

$$\left. \bigcup_{\substack{p,q,r=1,\dots,Q \\ p<q<r}} \mathrm{encl}\,(\mathsf{M}_{\text{Par}}^{\text{gen}}(\mathrm{f}_\lambda^{\mathrm{p}}, \mathrm{f}_\lambda^{\mathrm{q}}, \mathrm{f}_\lambda^{\mathrm{r}}))\,, x \in \mathrm{col}_Q(C) \right\} .$$

For the Q-criteria OMP we obtain the following algorithm.

Algorithm 5.3: *Computing* $\mathsf{M}_{\text{Par}}\left(\mathrm{f}_\lambda^1, \dots, \mathrm{f}_\lambda^{\mathrm{Q}}\right)$.

Compute the planar graph generated by the cells $\mathcal{C}$.

Compute $\mathsf{M}_{\text{w-Par}}(\mathrm{f}_\lambda^{\mathrm{p}}, \mathrm{f}_\lambda^{\mathrm{q}}), \mathrm{p}, \mathrm{q} = 1, \dots, \mathrm{Q}, \mathrm{p} < \mathrm{q}$, using Algorithm 5.1.

$$\mathsf{M}_{\text{Par}}^{\text{gen}}(\mathrm{f}_\lambda^{\mathrm{p}}, \mathrm{f}_\lambda^{\mathrm{q}}, \mathrm{f}_\lambda^{\mathrm{r}}) := \mathsf{M}_{\text{w-Par}}(\mathrm{f}_\lambda^{\mathrm{p}}, \mathrm{f}_\lambda^{\mathrm{q}}) \cup \mathsf{M}_{\text{w-Par}}(\mathrm{f}_\lambda^{\mathrm{p}}, \mathrm{f}_\lambda^{\mathrm{r}}) \cup \mathsf{M}_{\text{w-Par}}(\mathrm{f}_\lambda^{\mathrm{q}}, \mathrm{f}_\lambda^{\mathrm{r}}),\ \mathrm{p}, \mathrm{q}, \mathrm{r} = 1, \dots, \mathrm{Q},\ \mathrm{p} < \mathrm{q} < \mathrm{r}.$$

$$\mathsf{M}_{\text{Par}}\left(\mathrm{f}_\lambda^1, \dots, \mathrm{f}_\lambda^{\mathrm{Q}}\right) := \bigcup_{\substack{\mathrm{p,q,r}=1,\dots,\mathrm{Q} \\ \mathrm{p<q<r}}} \mathsf{M}_{\text{Par}}^{\text{gen}}(\mathrm{f}_\lambda^{\mathrm{p}}, \mathrm{f}_\lambda^{\mathrm{q}}, \mathrm{f}_\lambda^{\mathrm{r}}) \cup \bigcup_{\substack{p,q,r=1,\dots,Q \\ p<q<r}} \mathrm{encl}\,(\mathsf{M}_{\text{Par}}^{\text{gen}}(\mathrm{f}_\lambda^{\mathrm{p}}, \mathrm{f}_\lambda^{\mathrm{q}}, \mathrm{f}_\lambda^{\mathrm{r}})).$$

foreach $C \in \mathbf{C}$ *with*

$C \subseteq \bigcup_{\substack{p,q=1,\dots,Q \\ p<q}} \mathsf{M}_{\text{w-Par}}(\mathrm{f}_\lambda^{\mathrm{p}}, \mathrm{f}_\lambda^{\mathrm{q}}) \setminus \bigcup_{\substack{\mathrm{p,q,r}=1,\dots,\mathrm{Q} \\ \mathrm{p<q<r}}} \mathrm{encl}\,(M_{\text{Par}}^{\text{gen}}(\mathrm{f}_\lambda^{\mathrm{p}}, \mathrm{f}_\lambda^{\mathrm{q}}, \mathrm{f}_\lambda^{\mathrm{r}}))$ **do**

  Determine $\mathrm{col}_Q(C)$.

  $\mathsf{M}_{\text{Par}}\left(\mathrm{f}_\lambda^1, \dots, \mathrm{f}_\lambda^{\mathrm{Q}}\right) := \mathsf{M}_{\text{Par}}\left(\mathrm{f}_\lambda^1, \dots, \mathrm{f}_\lambda^{\mathrm{Q}}\right) \setminus \mathrm{col}_{\mathrm{Q}}(\mathrm{C})$.

end

Output: $\mathsf{M}_{\text{Par}}\left(\mathrm{f}_\lambda^1, \dots, \mathrm{f}_\lambda^{\mathrm{Q}}\right)$.

The complexity of Algorithm 5.3 can be obtained by the following analysis. For each cell C, $\mathrm{col}_Q(C)$ takes $O(Q\log(Q))$ since it is equivalent to check whether $0 \in conv\{\nabla f_\lambda^i : i = 1, \dots, Q\}$ at any interior point to

C. Algorithm 5.3 needs to solve $O(Q^3)$ three-criteria problems. Each one of them has the same complexity as the two-criteria problem. Thus, the overall complexity is $O(M^5 G_{\max}^2 Q^3 (\log M + \log G_{\max}) + M^4 G_{\max}^2 Q \log Q) = O(M^5 G_{\max}^2 Q^3 (\log M + \log G_{\max})$.

5.6 Concluding Remarks

This chapter analyzes the multicriteria version of the OMP. We have developed a geometrical description of the Pareto optimal solutions and algorithms to obtain it.

Apart from the results already presented here, there are some others that are worth to be further investigated. The first one is when we allow the weights w_i^q, $i = 1, \dots, M$, $q = 1, \dots, Q$, to be positive or negative and the weights λ_i^q, $i = 1, \dots, M$, $q = 1, \dots, Q$, not to be in non-decreasing sequence. In this situation we cannot apply the procedures presented in the preceding sections. Specifically, we do not have the following properties anymore:

- Convexity of the objective functions f_λ^q, $q = 1, \dots, Q$.
- Connectivity of the set of Pareto optimal points $\mathsf{M}_{\text{Par}}\left(\mathrm{f}_\lambda^1, \dots, \mathrm{f}_\lambda^q\right)$.

As a consequence, a solution algorithm for the multicriteria ordered median problem with attraction and repulsion should have a completely different structure than the algorithm for the convex case. Needless to say, for negative w_i^q we cannot write $w_i^q \gamma_i(x) = \gamma_i(w_i^q x)$ anymore. Instead we have $w_i^q \gamma_i(x) = -\gamma_i(|w_i^q| x)$.

However, the following properties are still fulfilled:

- The cell structure remains the same, since fundamental directions and bisector lines do not depend on λ_i^q.
- Linearity of the objective functions f_λ^q within each cell.

Consequently, we can compute the local Pareto solutions with respect to a single cell as described in the previous sections. Of course, we cannot ensure that the local Pareto solutions remain globally Pareto. Therefore, to obtain the set of global Pareto solutions all local Pareto solutions have to be compared.

A schematic approach for solving the ordered median problem with positive and negative weights would be:

1. Compute the local Pareto solutions for each cell $C \in \mathcal{C}$.
2. Compare all solutions of Step 1 and get $\mathsf{M}_{\text{Par}}\left(\mathrm{f}_\lambda^1, \dots, \mathrm{f}_\lambda^Q\right)$.
3. Output: $\mathsf{M}_{\text{Par}}\left(\mathrm{f}_\lambda^1, \dots, \mathrm{f}_\lambda^Q\right)$.

In general Step 2 might become very time consuming, because we have to compare lots of cells. However, for more special cases efficient algorithms can be developed. If we restrict ourselves to the bicriteria case, we can do a procedure similar to the one used in [89] for network location problems.

In addition, the approach presented in this chapter can be further extended to deal with non-polyhedral gauges using sequences of Pareto solution sets as already done for the multicriteria minisum location problems in [166].

6

Extensions of the Continuous Ordered Median Problem

This chapter contains advanced material that extends the basic model considered so far in previous chapters. We have tried to cover two different goals: 1) extensions of basic location models, and 2) geometrical properties of the OMP in abstract spaces. Most of the material presented here is self-contained and might be skipped without loosing continuity in reading the rest of the book.

The first section considers OMP with forbidden regions and it results as a natural extension of the unconstraint single facility problems considered in Chapter 4. The second section deals with multifacility models where more than one new facility has to be located in order to improve the service of the demand points. The last section analyzes the OMP formulated in abstract spaces. In the analysis we use basic tools of convex analysis and functional analysis. Also a basic knowledge of the geometry of Banach spaces is needed, at some points, to understand completely all the results. Several results obtained in previous chapters are extended or reformulated with great generality giving a panorama view of the geometric insights of location theory.

6.1 Extensions of the Single Facility Ordered Median Problem

6.1.1 Restricted Case

In the last years restricted facility location problems, an extension to classical location models, have attracted considerable attention in the applications and from a methodological point of view. In many applications a forbidden region has to be considered, where it is not allowed to place new facilities. Therefore, when solving restricted location problems this issue has to be taken into account. In the recent literature of location analysis one can find several papers that consider this characteristic in their solution method (see for instance Brady and Rosenthal [27], Drezner [57], Karkazis [120], Aneja and Palar [5]).

Different approaches that are also related to this topic are the contour line approach followed by Francis et al. [77] and the method proposed by Hamacher and Nickel [90] and Nickel [148]. Following the usual trend in location analysis, these papers are model oriented; and thus, they focus on concrete problems and on concrete solution strategies. In this section we want to embed the location problem with forbidden regions within the framework of the OMP. In our analysis we follow the material in Rodríguez-Chía et al. [177].

To simplify the presentation we restrict ourselves to the case of the convex OMP in the plane with only one polyhedral gauge. Needless to say that all the results extend further to the case of different gauges at the different demand points and to the general OMP (not necessarily convex).

We are given a polyhedral gauge γ with unit ball B, a nonnegative lambda vector $\lambda \in S_M^{\leq}$ and a forbidden region $\mathcal{R}$, being an open convex set in $\mathbb{R}^2$. We assume further that $\mathcal{R}$ contains all the optimal solutions of the unrestricted problem. Notice that this hypothesis does not mean loss of generality because otherwise we could solve the problem just considering the unrestricted one.

The problem to be solved is:

$$\min_{x \in \mathbb{R}^2 \setminus \mathcal{R}} f_\lambda(x) . \tag{6.1}$$

We start with a discretization result that is the basis of the solution procedure presented later in Algorithm 6.1.

Theorem 6.1 *The convex OMP with forbidden region $\mathcal{R}$ and polyhedral gauge γ always has an optimal solution on the 0-dimensional intersections between the boundary of $\mathcal{R}$, the fundamental directions of B and the bisector lines.*

Proof.
Since f_λ is convex its lower level sets are convex. Therefore, an optimal solution of the restricted ordered facility location problem is on the boundary of the forbidden region. Moreover, the objective function is linear in each cell, see Theorem 4.1, and thus, there must be a point on the boundary of $\mathcal{R}$ and the boundary of some cell that is an optimal solution of Problem (6.1). □

One of the implications of the above result is that there is no loss of generality in considering only one forbidden region. Otherwise, if they were more than one, we would only need to consider one region which contains some optimal solutions of the unrestricted problem, in order to obtain some optimal solution of the restricted one.

Another immediate consequence of Theorem 6.1 is Algorithm 6.2 for solving the single facility OMP with forbidden region $\mathcal{R}$.

Algorithm 6.1: *Solving the convex ordered median problem with convex forbidden regions*

1. Compute the fundamental directions and bisector lines for all existing facilities.
2. Determine $\{y_1, y_2, \ldots, y_k\}$ the intersection points of fundamental directions or bisector lines and the boundary of the forbidden region, $\mathcal{R}$.
3. Compute $x^*_{\mathcal{R}} \in \arg\min\{f_\lambda(y_1), f_\lambda(y_2), \ldots, f_\lambda(y_k)\}$ ($x^*_{\mathcal{R}}$ is an optimal solution to the restricted location problem).
4. The set of optimal solutions is $\mathsf{M}(\mathrm{f}_\lambda) := \{\mathrm{x} : \mathrm{f}_\lambda(\mathrm{x}) = \mathrm{f}_\lambda(\mathrm{x}^*_{\mathcal{R}})\}$ intersected with the boundary of $\mathcal{R}$.

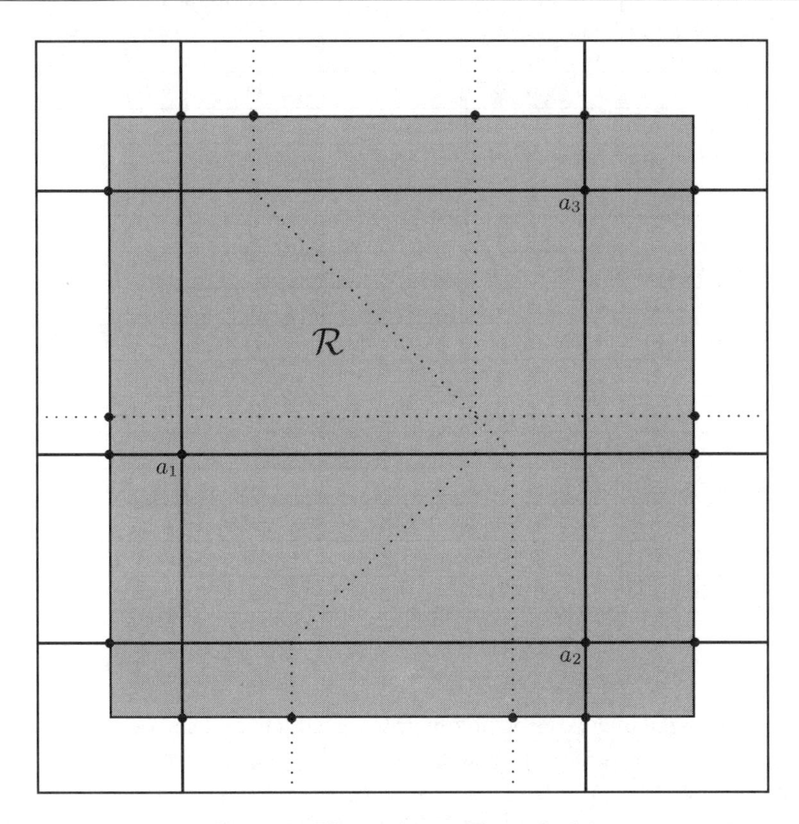

Fig. 6.1. Illustration to Example 6.1

Example 6.1
We apply Algorithm 6.1 to solve the OMP with forbidden regions using the same input data as in Example 4.3: $A = \{a_1 = (0, 2.5), a_2 = (5.5, 0), a_3 = (5.5, 6)\}$ with l_1-norm. In addition, we define the forbidden region $\mathcal{R} = [-1, 7] \times [-1, 7]$.

We get the intersection points $y_1, y_2, \ldots, y_{16}$ as shown by dots in Figure 6.1. In Table 6.1 we list the values of f_λ for different λ-vectors at the intersection points.

For $\lambda = (1, 2, 3)$ the optimal solution is y_7. For $\lambda = (1, 1, 0)$ the set of optimal solutions is $\{y_{12}, y_{16}\}$. Finally, for $\lambda = (1, 0, 0)$ the set of optimal solutions is $\{y_3, y_{12}, y_{16}\}$.

Table 6.1. Data of Example 6.1.

x	(x_1, x_2)	d	$\lambda = (1,2,3)$	$\lambda = (1,1,0)$	$\lambda = (1,0,0)$
y_1	$(-1, 6)$	4.5 12.5 6.5	55	11	4.5
y_2	$(-1, 3)$	1.5 9.5 9.5	49	11	1.5
y_3	$(-1, 2.5)$	1 9 10	39	10	1
y_4	$(-1, 0)$	3.5 6.5 12.5	54	10	3.5
y_5	$(7, 6)$	10.5 7.5 1.5	48	9	1.5
y_6	$(7, 3)$	7.5 4.5 4.5	36	9	4.5
y_7	$(7, 2.5)$	7 4 5	35	9	4
y_8	$(7, 0)$	9.5 1.5 7.5	45	9	1.5
y_9	$(0, 7)$	4.5 12.5 6.5	55	11	4.5
y_{10}	$(1, 7)$	5.5 11.5 5.5	51	11	5.5
y_{11}	$(4, 7)$	8.5 8.5 2.5	45	11	2.5
y_{12}	$(5.5, 7)$	10 7 1	45	8	1
y_{13}	$(0, -1)$	3.5 6.5 12.5	54	10	3.5
y_{14}	$(1.5, -1)$	5 5 11	48	10	5
y_{15}	$(4.5, -1)$	8 2 8	42	10	2
y_{16}	$(5.5, -1)$	9 1 7	42	8	1

For the particular case of polyhedral forbidden regions we get more accurate results. Let $\mathcal{R}$ be a polyhedral forbidden region, defined by its entire collection of facets $\{s_1, s_2, \ldots, s_k\}$; and let $A = \{a_1, \ldots, a_M\}$ be the set of existing facilities. Algorithm 6.2 describes an exact method to obtain an optimal solution of Problem (6.1).

Notice that Algorithm 6.2 can also be used to solve problems with convex forbidden regions not necessarily polyhedral. In order to do so we only have to approximate these regions by polyhedral ones. Since this approximation can be done with arbitrary precision, using for instance the sandwich approximation in [31] and [114], we can get good approximations to the optimal solutions of the original problems.

Algorithm 6.2: *Solving the convex single facility OMP with polyhedral forbidden regions*

Set $p := 1$, $\mathcal{L} := \emptyset$.
Let x^* be an arbitrary feasible solution and set $y^* = x^*$.
1 Let $\mathcal{T}_p$ be the hyperplane defined by the facet s_p of $\mathcal{R}$ and choose x^o belonging to the relative interior of s_p. Let $\mathcal{T}^{\leq}{}_p$ be the halfplane which does not contain $\mathcal{R}$. Set $x^* = x^o$.
2 Determine the ordered region O_{σ^o} where x^o belongs to, and the permutation σ^o which defines this region.
3 Solve the following linear program

$$\begin{array}{ll} \min \sum_{i=1}^{M} \lambda_i z_{\sigma_i^o} & \\ s.t.\ w_i\langle e_g^o, x - a_i\rangle \leq z_i\ \forall e_g^o \in Ext(B^o), i = 1,2,\ldots,M, & \\ \quad z_{\sigma^o(i)} \leq z_{\sigma^o(i+1)} \quad i = 1,2,\ldots,M-1, & \\ \quad x \in \mathcal{T}^{\leq}{}_p. & \end{array} \qquad (P_{\mathcal{T}^{\leq}})$$

Let $u^o = (x^o, z_\sigma^o)$ be an optimal solution of $P_{\mathcal{T}^{\leq}{}_p}$.
if $x^o \notin O_{\sigma^o}$ **then** go to 2
if *x^o belongs to the interior of O_{σ^o}* **then** set $x^* = x^o$ and go to 4
if $f_\lambda(x^o) \neq f_\lambda(x^*)$ **then** $\mathcal{L} := \emptyset$
if *there exist i and j satisfying*

$$\gamma(x^o - a_{\sigma^o(j)}) = \gamma(x^o - a_{\sigma^o(i)}) \quad i < j$$

such that

$$(\sigma^o(1), \ldots, \sigma^o(j), \ldots, \sigma^o(i), \ldots, \sigma^o(M)) \notin \mathcal{L}$$

then
set $x^* := x^o$, $\sigma^o := (\sigma^o(1), \sigma^o(2), \ldots, \sigma^o(j), \ldots, \sigma^o(i), \ldots, \sigma^o(M))$,
$\mathcal{L} := \mathcal{L} \cup \{\sigma^o\}$.
go to 3.
end
4 **if** $f_\lambda(x^*) < f_\lambda(y^*)$ **then** $y^* := x^*$ **else** $p := p + 1$.
if $p < k$ **then** go to 1 **else** the optimal solution is y^*.

6.2 Extension to the Multifacility Case

A natural extension of the single facility model is to consider the location of N new facilities rather than only one. In this formulation the new facilities are chosen to provide service to all the existing facilities minimizing an ordered objective function. These ordered problems are of course harder to handle than the classical ones not considering ordered distances. Therefore, as no detailed complexity results are known for the ordinary multifacility problem, nothing can be said about the complexity of the ordered median problem. Needless to say its resolution is even more difficult than the one for the single facility models. To simplify the presentation we consider that the different demand points use the same polyhedral gauge to measure distances. All the results extend further to the case of different gauges. Moreover, one can also extend the analysis of multifacility models to the case of non-polyhedral gauges using the approach followed in Section 4.3.

The multifacility problem can be analyzed from two different approaches that come from two different interpretations of the new facilities to be located. The first one assumes that the new facilities are not interchangeable, which means that they are of different importance for each one of the existing facilities. The second one assigns the same importance to all new facilities. In the latter case, we are only interested in the size of the distances, which means that we do not consider any order among the new facilities and look for equity in the service, minimizing the largest distances.

6.2.1 The Non-Interchangeable Multifacility Model

Let us consider a set of demand points $A = \{a_1, a_2, \dots, a_M\}$. We want to locate N new facilities $X = \{x_1, x_2, \dots, x_N\}$ which minimize the following expression:

$$f_\lambda^I(x_1, x_2, \dots, x_N) = \sum_{i=1}^{N}\sum_{j=1}^{M} \lambda_{ij} d_{(j)}(x_i) + \sum_{k=1}^{N}\sum_{l=1}^{N} \mu_{kl}\gamma(x_k - x_l) \tag{6.2}$$

where

$$\begin{array}{c} 0 \le \lambda_{11} \le \lambda_{12} \le \dots \le \lambda_{1M} \\ 0 \le \lambda_{21} \le \lambda_{22} \le \dots \le \lambda_{2M} \\ \vdots \\ 0 \le \lambda_{N1} \le \lambda_{N2} \le \dots \le \lambda_{NM}; \end{array}$$

$\mu_{kl} \ge 0$ for any $k = 1, \dots, N$, $l = 1, \dots, N$ and $d_{(j)}(x_i)$ is the expression, which appears at the j-th position in the ordered version of the list

$$L_i^I := (w_1\gamma(x_i - a_1), \dots, w_M\gamma(x_i - a_M)) \quad \text{for } i = 1, 2, \dots, N. \tag{6.3}$$

Note that in this formulation we assign the lambda parameters with respect to each new facility, i.e., x_j is considered to be non-interchangeable with x_i whenever $i \ne j$. For this reason we say that this model has non-interchangeable facilities.

In order to illustrate this approach we show an example which will serve as motivation for the following:
Consider a distribution company with a central department and different divisions to distribute different commodities. A new set of final retailers appears in a new area. Each division wants to locate a new distribution center to supply the demand generated by these retailers with respect to its own strategic objective. The overall operation cost is the sum of the operation cost of the different divisions plus the administrative cost which only depends on the distances between the new distribution centers and which is supported by the

central department. It is evident that for each division the quality of its distribution center only depends on the orders of the distances to the retailers but not on the name given to them, i.e. it depends on the ordered distances from each new facility to the retailers. Therefore, a natural choice to model this situation is to use an ordered multifacility non-interchangeable OMP with a new facility for each commodity (division) of the company.

As in the single facility model we can prove that the objective function (6.2) is convex, which eases the analysis of the problem and the development of an efficient algorithm.

Proposition 6.1 *The objective function f_λ^I is convex.*

Proof.
We know that

$$\sum_{i=1}^{N}\sum_{j=1}^{M}\lambda_{ij}d_{(j)}(x_i) = \sum_{i=1}^{N}\max_{\sigma^i}\sum_{j=1}^{N}\lambda_{ij}w_{\sigma^i(j)}\gamma(x_i - a_{\sigma^i(j)})$$

where σ^i is a permutation of the set $\{1,2,\ldots,M\}$. Therefore, the first part of the objective function is a sum of maxima of convex functions, analogous to Theorem 1.3. Hence, it is a convex function. On the other hand, the second term of the objective function f_λ^I is convex as sum of convex functions. Thus, f_λ^I is a convex function as well, as a sum of convex functions. □

Problem (6.2) can be transformed within the new ordered regions in the same way we did in Section 4.1. It should be noted that in $\mathbb{R}^2$ the subdivisions induced by the ordered regions of this problem are given as intersection of N subdivisions. Each one of these N subdivisions determines the ordered regions of each new facility.

Let $\sigma^k = (\sigma^k(1),\ldots,\sigma^k(M))$ $k = 1,\ldots,N$ be the permutations which give the order of the lists L_i^I, for $i = 1,\ldots,N$ introduced in (6.3). Consider the following linear program (P_σ^I):

$$\min \sum_{k=1}^{N}\sum_{l=1}^{M}\lambda_{kl}z_{k\sigma^k(l)} + \sum_{i=1}^{N}\sum_{j=1}^{N}\mu_{ij}y_{ij}$$

s.t.

$$\begin{aligned}
w_l\langle e_g^o, x_k - a_l\rangle \le z_{kl} \quad &\forall e_g^o \in Ext(B^o), \quad k = 1,2,\ldots,N, \quad l = 1,2,\ldots,M\\
\langle e_g^o, x_i - x_j\rangle \le y_{ij} \quad &\forall e_g^o \in Ext(B^o), \quad i = 1,2,\ldots,N, \quad j = i+1,\ldots,M\\
z_{k\sigma^k(l)} \le z_{k\sigma^k(l+1)} \quad &k = 1,2,\ldots,N, \quad l = 1,2,\ldots,M-1
\end{aligned}$$

P_σ^I plays a similar role as P_σ in Section 4.1.

Algorithm 4.4 can easily be adapted to accommodate the multifacility case. Note that in this case we now look for N points in $\mathbb{R}^2$ or equivalently for one point in $\mathbb{R}^{2N}$ (in the single facility case we look for only one point in $\mathbb{R}^2$). To do that, we only have to choose N starting points instead of one. In addition, we also have to consider that now the ordered regions are defined by N permutations, one for each list L_i^I, for $i = 1, \ldots, N$. Therefore, we have to replace the linear program P_σ by P_σ^I and to adapt its set of optimal solutions.

Since this algorithm is essentially the same than the one proposed for the single facility model, we can conclude that it is also polynomial bounded, hence applicable.

Example 6.2
Consider the two-facility problem.

$$\begin{aligned} \min_{x_1, x_2 \in \mathbb{R}^2} \quad & 2.5d_{(4)}(x_1) + 2d_{(3)}(x_1) + 1.5d_{(2)}(x_1) + d_{(1)}(x_1) + \\ & +0.75d_{(4)}(x_2) + 0.1d_{(3)}(x_2) + 0.1d_{(2)}(x_2) + 0.1d_{(1)}(x_2) + \\ & +0.5\gamma_B(x_1 - x_2) \end{aligned}$$

where $A = \{(3,0), (0,11), (16,8), (-4,-7)\}$, and γ_B is the hexagonal polyhedral norm, already described in the examples of Section 4.4.

Starting with the points $x_1^o = (0,11)$ and $x_2^o = (16,8)$ we obtain in the second iteration the optimal solution: $x_1^* = (2.75, 5.5)$ and $x_2^* = (3.125, 5.875)$. The elementary convex set and the optimal solution can be seen in Figure 6.2.

6.2.2 The Indistinguishable Multifacility Model

The multifacility model that we are considering now differs from the previous one in the sense that the new facilities are similar from the users point of view. Therefore, the new facilities are regarded as equivalent with respect to the existing ones. On the contrary, the weight given to each one of these new facilities depends only on the size of the distances.

Using the same notation as in Section 6.2.1, the objective function of this model is:

$$f_\lambda^{II}(x_1, x_2, \ldots, x_N) = \sum_{j=1}^{NM} \lambda_j d_{(j)}(x) + \sum_{i=1}^{N} \sum_{j=1}^{N} \mu_{ij} \gamma(x_i - x_j)$$

where

$$0 \le \lambda_1 \le \lambda_2 \le \ldots \le \lambda_{NM}$$

and $d_{(j)}(x)$ is the expression which appears at the j-th position in the ordered version of the following ordered list

$$L^{II} = (w_p \gamma(x_k - a_p),\ k = 1, 2, \ldots, N, \quad p = 1, 2, \ldots, M).$$

Also this model is motivated by the following situation:
Consider the same multi-commodity distribution company as in the last subsection. However, assume now that the strategic objective is not fixed by each

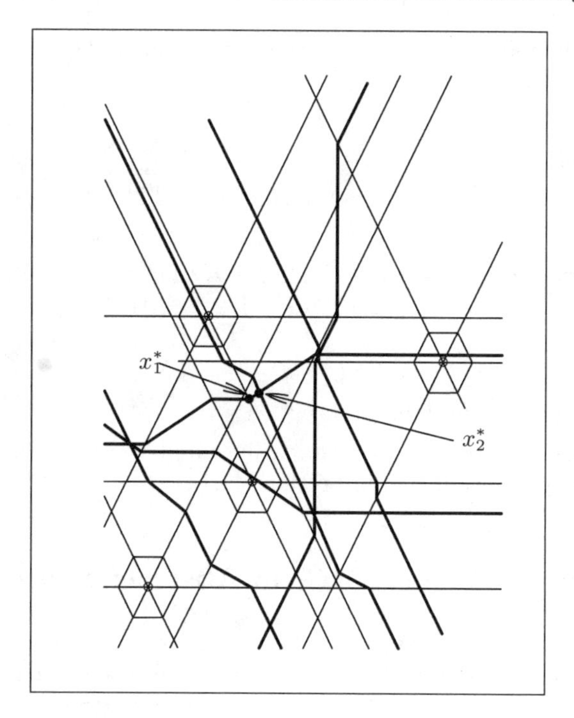

Fig. 6.2. Illustration for Example 6.2

division but it is fixed by the central department. Therefore, the objective is still to locate a distribution center for each commodity but the total cost of the distribution is supported by the central department. Thus, the complete order of the distances to the retailers from the distribution centers is important rather than the distances with respect to each one of the centers for the different commodities. With these hypotheses this situation can be formulated as a multifacility indistinguishable model where we want to locate as many facilities as different commodities in the company depending on the entire list of distances.

Proposition 6.2 *The objective function f_λ^{II} is convex.*

Proof.
The proof is analogous to the one given for Proposition 6.1. □

The same strategy as we have already used for the non-interchangeable multifacility model can be applied here. Therefore this problem can be solved by a modified version of Algorithm 4.4.

Let σ be a permutation of $\{1,\ldots,MN\}$ where $\sigma((k-1)M+j)$ gives the position of $w_j\gamma(x_k - a_j)$ in L^{II}.

Consider the following linear programming problem (P_σ^{II}):

$$(P_\sigma^{II}) \quad \min \sum_{k=1}^{N} \sum_{l=1}^{M} \lambda_{\sigma(kl)} z_{\sigma((k-1)M+l)} + \sum_{p=1}^{N} \sum_{q=1}^{N} \mu_{pq} y_{pq}$$

s.t.

$$\begin{aligned} & w_l \langle e_g^o, x_k - a_l \rangle \le z_{(k-1)M+l}, && \forall e_g^o \in Ext(B^o),\ k = 1, 2, \ldots, N, \\ & && l = 1, 2, \ldots, M \\ & \langle e_g^o, x_p - x_q \rangle \le y_{pq}, && \forall e_g^o \in Ext(B^o),\ p = 1, 2, \ldots, N, \\ & && q = 1, \ldots, N \\ & z_{\sigma((k-1)M+l)} \le z_{\sigma((k-1)M+l+1)}, && k = 1, 2, \ldots, N,\ l = 1, 2, \ldots, M-1. \end{aligned}$$

Once we replace P_σ^I by P_σ^{II}, we can easily adapt the algorithmic approach showed for the previous model in Section 6.2.1.

6.3 The Single Facility OMP in Abstract Spaces

In contrast to the previous sections where we have been dealing with natural extensions of the basic model, we take a more abstract perspective in this section. We study properties of the OMP under very general assumptions in order to link the location theory to the convex analysis and functional analysis point of view.

Let us consider a real Banach space X equipped with a norm $\|.\|$ with unit ball B and a finite set $A = \{a_1, \ldots, a_M\} \subset X$ with at least three different points.

The location problem associated with these data consists of finding a point $x \in X$ that best fits in some sense the points in A. This section studies a single facility location problem assuming that two rationality principles (monotonicity and symmetry) hold. Under these hypotheses a natural way to formulate the location problem consists of minimizing an objective function F of the form

$$F(x) = \gamma(d(x))$$

where γ is a monotone, symmetric norm on $\mathbb{R}^M$ and $d(x) = (\|x-a_1\|, \ldots, \|x-a_M\|)$ is the distance vector. Most of the classical objective functions used in location theory are specific instances of this formulation. It is easy to see that the minimax, the cent-dian or the Weber problem fit into this general model.

Properties of the solution set of location problems have been previously studied by several authors. Durier and Michelot [65] describe the optimal set of Weber problems on a general normed space. Drezner & Goldman [60] as well as Durier & Michelot [66] study dominating sets for the optimal solutions of the Weber problem on $\mathbb{R}^2$ under different norms. Moreover, Baronti & Papini [13] study properties of the size of the solution sets of minisum and minimax problems on general normed spaces, and Durier [62, 63] analyzes a general problem on an arbitrary normed space X considering an objective function which is a monotone but not necessarily a symmetric norm. Puerto and Fernández [169] characterize the complete set of optimal solutions of convex OMP. Recently, Nickel, Puerto and Rodríguez-Chía[151] have characterized the geometrical structure of the set of optimal solutions of location problems with sets as existing facilities.

In this section, we will study the set of optimal solutions of the problem

$$\inf_{x \in X} F(x). \tag{6.4}$$

We provide a geometrical characterization and give existence and uniqueness conditions of these sets using convex analysis. This characterization coincides with the one given in [169]. In addition, we will prove that for any problem of type (6.4) there exists a vector of nonnegative weights $\lambda = (\lambda_1, \ldots, \lambda_M) \in \Lambda_M^{\leq}$, such that Problem (6.4) and

$$(\mathsf{OMP}(\lambda, A)) \quad \inf_{x \in X} f_\lambda(x) := \sum_{i=1}^{M} \lambda_i d_{(i)}(x) \tag{6.5}$$

have the same set of optimal solutions. This result establishes the relationship between the general location problem (6.4) and the convex OMP for particular choices of the vector λ. For the ease of understanding we denote the optimal value of Problem (6.5) as:

$$\mathsf{OM}_X(\lambda, A) := \inf_{x \in X} \sum_{i=1}^{M} \lambda_i d_{(i)}(x).$$

Moreover, let $\mathsf{M}(\gamma, A)$ and $\mathsf{M}(f_\lambda, A)$ be the solution sets of the problems (6.4) and (6.5), respectively, whenever they exist.

$$\mathsf{M}(\gamma, A) = \arg \inf_{x \in X} \gamma(d(x))$$

$$\mathsf{M}(f_\lambda, A) = \arg \inf_{x \in X} \sum_{i=1}^{M} \lambda_i d_{(i)}(x).$$

Finally, for $\varepsilon \geq 0$ and $\lambda \in \Lambda_M^{\leq}$, let:

$$\mathsf{M}(f_\lambda, A, \varepsilon) = \{x \in X : f_\lambda(x) \leq \mathsf{OM}_X(\lambda, A) + \varepsilon\}. \tag{6.6}$$

According to Proposition 6.3, the sets $\mathsf{M}(f_\lambda, A, \varepsilon)$ are always closed, bounded and convex. Therefore, the (possibly empty) set

$$\mathsf{M}(f_\lambda, A) = \bigcap_{\varepsilon > 0} \mathsf{M}(f_\lambda, A, \varepsilon) \tag{6.7}$$

is always closed, bounded and convex.

This section is organized as follows. First, we introduce the notation, the concepts of monotone and symmetric norms and prove some results. Then, we give a geometrical characterization of the set of optimal solutions to Problem (6.4). In addition, we state the equivalence between Problem (6.4) and OMP (6.5). Subsection 6.3.2 deals with a geometrical description of the set of optimal solutions to problems of type (6.4) in terms of their corresponding OMP (6.5). The results are illustrated by several examples.

6.3.1 Preliminary Results

Let us denote by X^* the topological dual of X and $\|.\|^0$ its norm. The unit ball in X^* is denoted B^*. The pairing of X and X^* will be indicated by $\langle \cdot, \cdot \rangle$. For the ease of understanding the reader may replace the space X by $\mathbb{R}^n$, and everything will be more familiar. In this case the topological dual of X can be identified to X, and the pairing is the usual scalar product. Given two points $u, v \in \mathbb{R}^M$, in this section, the natural scalar product is written as (u, v).

Let $C \subset X$ be a convex set. An element $p \in X^*$ is said to be normal to C at $x \in C$ if $\langle p, y - x \rangle \leq 0$ for all $y \in C$. The set of all the vectors normal to C at x is called the normal cone to C at x and is denoted by $N_C(x)$. For the sake of simplicity, we will denote the normal cone to the unit dual ball B^* by N.

We consider $\mathbb{R}^M$ equipped with a norm γ which is assumed to be monotone and symmetric; and we denote γ^0 the dual norm. For the sake of completeness we include the definitions of monotone and symmetric norms.

Definition 6.1

1. *A norm γ on $\mathbb{R}^M$ is monotone (see [15]) if $\gamma(u) \leq \gamma(v)$ for every u, v verifying $|u_i| \leq |v_i|$ for each $i = 1, \ldots, M$.*
2. *A norm γ on $\mathbb{R}^M$ is said to be symmetric (see [169]) if and only if $\gamma(u) = \gamma(u_\sigma)$ for all $\sigma \in \mathcal{P}(1 \ldots M)$.*

[112] gave a characterization of monotone norms as absolute norms, that is $\gamma(u) = \gamma(v)$ if $|u_i| = |v_i|$ for all $i = 1, \ldots, M$.

Notice that there is no general relation of inclusion between the sets of monotone and symmetric norms according to the following example.

Example 6.3
Consider the space $\mathbb{R}^M$.

1. The l_p norms with $p \in [1, \infty]$ are monotone and symmetric.
2. The norm given by $\|u\| = \sum_{i=1}^{M} 1/(i+1)|u_i|$ is monotone but not symmetric.
3. Let $n = 2$ and consider the polyhedral norm defined by the unit ball with extreme points $\{(2,2), (1,0), (-1,0), (-2,-2), (0,-1), (0,1)\}$. It is easy to see that this norm is symmetric but not monotone.

Given an unconstrained minimization problem with convex objective function f, x is an optimal solution if and only if $0 \in \partial f(x)$.

Therefore, studying the subdifferential of the objective function F of (6.4) is an interesting task. First of all, it is straightforward to see that the function $F = \gamma \circ d$ is convex on $\mathbb{R}^M$ provided that γ is monotone (see Prop. 2.1.8 of Chapter IV in [102]). Then, by standard arguments concerning Banach spaces, we obtain that if X is a dual space $\mathsf{M}(\gamma, A) \neq \emptyset$ for any finite set A (see e.g. [129]). In particular, if X has finite dimension, Problem (6.4) always has an optimal solution. In the following we assume that these problems have optimal solutions.

Provided that Problem (6.4) has optimal solutions our goal is to find a geometrical characterization of its solution set. Since the objective function F is convex, one way to achieve this goal is to have a precise description of its subdifferential set. To this end, we study properties of the subdifferential of monotone, symmetric norms.

It is well-known that the subdifferential of the norm $\|.\|$ at x is given by

$$\partial\|x\| = \begin{cases} B^* & \text{if } x = 0 \\ \{p \in B^* : \langle p, x\rangle = \|x\|\} & \text{if } x \neq 0 \end{cases} . \tag{6.8}$$

Let p_σ, x_σ denote the vectors p, x whose entries have been ordered according to the permutation σ.

Lemma 6.1 *Let γ be a symmetric norm on $\mathbb{R}^M$. For any $x \in \mathbb{R}^M$, $p \in \partial\gamma(x)$ iff $p_\sigma \in \partial\gamma(x_\sigma)$ for any permutation $\sigma \in \mathcal{P}(1\ldots M)$.*

Proof.
Let $p \in \partial\gamma(x)$. By the subgradient inequality one has:

$$p \in \partial\gamma(x) \text{ iff } \gamma(y) \geq \gamma(x) + (p, y - x) \quad \forall y \in \mathbb{R}^M.$$

Since γ is a symmetric norm for any $\sigma \in \mathcal{P}(1\ldots M)$ we have $\gamma(y_\sigma) = \gamma(y)$ for any $y \in \mathbb{R}^M$. Besides, $(p, y - x) = (p_\sigma, y_\sigma - x_\sigma)$ for any $\sigma \in \mathcal{P}(1\ldots M)$. Therefore,

$$\gamma(y) \geq \gamma(x) + (p, y - x) \forall y \in \mathbb{R}^M \text{ iff } \gamma(y_\sigma) \geq \gamma(x_\sigma) + (p_\sigma, y_\sigma - x_\sigma)\ \forall y_\sigma \in \mathbb{R}^M.$$

Now, when y covers $\mathbb{R}^M$, y_σ covers $\mathbb{R}^M$ as well. Thus,

$$\gamma(y_\sigma) \geq \gamma(x_\sigma) + (p_\sigma, y_\sigma - x_\sigma) \forall y_\sigma \in \mathbb{R}^M \text{ iff } \gamma(y) \geq \gamma(x_\sigma) + (p_\sigma, y - x_\sigma)\ \forall y \in \mathbb{R}^M.$$

Hence,

$$p \in \partial\gamma(x) \text{ iff } p_\sigma \in \partial\gamma(x_\sigma).$$

□

The objective function of Problem (6.4) is $F(x) = \gamma \circ d(x)$. Since γ is not an increasing componentwise function on the whole space $\mathbb{R}^M$, i.e.

$$y^i \geq z^i \text{ for } i = 1, \ldots, n \Rightarrow \gamma(y) \geq \gamma(z),$$

(see chapter VI in [102] for details about the classical definition of increasing componentwise functions) we cannot use the standard tools of convex analysis to get the subdifferential of F. To avoid this inconvenience we introduce the following auxiliary function defined on $\mathbb{R}^M$:

$$\gamma^+(u) = \gamma(u_1^+, \ldots, u_M^+) \ ,$$

where $u_i^+ = \max(u_i, 0)$. It is straightforward that $F(x) = \gamma^+(d(x))$ because d is a vector of distances. Using composition theorems (see [102]) it is now simple to show that γ^+ is an increasing componentwise, symmetric, convex function on $\mathbb{R}^M$. Moreover, using the properties of γ and the definition of γ^+ we have

$$\begin{aligned} \gamma(v_\sigma^+) = \gamma(v^+) = \gamma^+(v) = \gamma^+(v^+) \leq \gamma^+(|v|) = \gamma(v) = \gamma(v_\sigma) \\ \forall\, v \in \mathbb{R}^M, \ \sigma \in \mathcal{P}(1 \ldots M) \end{aligned} \tag{6.9}$$

where $|v| = (|v_1|, \ldots, |v_M|)$.

Based on that relation, the next lemma gives a precise description of the subdifferential set of γ^+. This characterization will be essential to describe the subdifferential set of the function F.

Lemma 6.2 *Let γ be a symmetric, monotone norm on $\mathbb{R}^M_+$ and assume that $u = (u_1, \overset{r_1}{\ldots}, u_1, \ldots, u_k, \overset{r_k}{\ldots}, u_k) \in \Lambda_M^{\leq}$ for some $k \geq 1$ and $u_1 < u_2 < \ldots < u_k$ if $k > 1$. Then*

$$\partial \gamma^+(u) = \partial \gamma(u) \cap \Lambda_{r_1, \ldots, r_k}.$$

(Recall that the cone $\Lambda_{r_1, \ldots, r_k}$ was introduced in (1.4).)

Proof.

This proof is organized in two different parts. In the first part, we prove that $\partial \gamma^+(u) \subseteq \Lambda_{r_1, \ldots, r_k}$. In the second, we prove $\partial \gamma^+(u) \supseteq \partial \gamma(u) \cap \Lambda_{r_1, \ldots, r_k}$.

Let $u^* \in \partial \gamma^+(u)$ and $v \leq u$. Using that γ^+ is increasing componentwise on $\mathbb{R}^M$, the subgradient inequality and the fact that $v^+ \leq u^+$ we have,

$$0 \leq \gamma^+(u) - \gamma^+(v) \leq (u^*, u - v).$$

Since the first inequality holds for any v such that $v \leq u$, this implies that $u^* \geq 0$. Otherwise, if there exists $u_i^* < 0$ taking v defined by $v_l = u_l$ for $l \neq i$ and $v_i < u_i$ we get a contradiction. Therefore, we already have that $\partial \gamma^+(u) \subseteq \mathbb{R}^M_+$. Thus, the inclusion is proved for $k = 1$ because $\Lambda_M = \mathbb{R}^M_+$. Assume now $k \geq 2$. Using the symmetry property, we have that $\gamma^+(v) = \gamma^+(v_\sigma)$ for all $\sigma \in \mathcal{P}(1 \ldots M)$, and thus

$$0 \leq \gamma^+(u) - \gamma^+(v_\sigma) \leq (u^*, u - v_\sigma) \quad \forall\, v \text{ such that } v \leq u \tag{6.10}$$

which leads us, by looking at the point $v = u$, to $u^* \in \Lambda_{r_1,\dots,r_k}$. Otherwise, if there exists $(u^*)^i_j > (u^*)^{i+1}_l$ for some i and any j, l then let v_σ be the vector u with the j-th entry in block i interchanged with the l-th entry in block $i+1$. Since $u^{i+1}_l > u^i_j$ and we already have that $u^* \geq 0$, we get $(u^*, u - v_\sigma) = (u^{i+1}_l - u^i_j)((u^*)^{i+1}_l - (u^*)^i_j) < 0$. This fact contradicts (6.10). Therefore, $\partial\gamma^+(u) \subset \Lambda_{r_1,\dots,r_k}$.

For the second part of the proof we proceed as follows. Since for any $v \in \mathbb{R}^M$, $\gamma(v) \geq \gamma^+(v)$ (see (6.9)) and $\gamma(u) = \gamma^+(u)$ ($u \in \mathbb{R}^M_+$) we have

$$(u^*, v - u) \leq \gamma^+(v) - \gamma^+(u) \leq \gamma(v) - \gamma(u) \text{ for any } v \in \mathbb{R}^M.$$

Therefore, $u^* \in \partial\gamma(u)$ and hence $\partial\gamma^+(u) \subseteq \partial\gamma(u)$. This inclusion together with the inclusion of the first part of the proof leads us to $\partial\gamma^+(u) \subseteq \partial\gamma(u) \cap \Lambda_{r_1,\dots,r_k}$.

Now, let $u^* \in \partial\gamma(u) \cap \Lambda_{r_1,\dots,r_k}$. For any $v \in \mathbb{R}^M$, let v_{min} be a vector with the same components that v but ordered in such a way that $\langle u^*, u-(v^+)_{min}\rangle \leq \langle u^*, u - (v^+)_\sigma\rangle$ for all $\sigma \in \mathcal{P}(1\dots M)$. Notice that this is always possible since we can order v^+according with the permutation $\pi \in \arg\min_{\sigma\in\mathcal{P}(1\dots M)} \langle u^*, u - (v^+)_\sigma\rangle$. Then, by the subgradient inequality we obtain

$$\gamma(u) - \gamma((v^+)_{min}) \leq (u^*, u - (v^+)_{min}).$$

Therefore, using (6.9) we can conclude that

$$\gamma^+(u)-\gamma^+(v) = \gamma^+(u)-\gamma^+((v^+)_{min}) = \gamma(u)-\gamma((v^+)_{min}) \leq (u^*, u-(v^+)_{min}).$$

But we have chosen $(v^+)_{min}$ such that:

$$(u^*, u - (v^+)_{min}) \leq (u^*, u - v^+).$$

Moreover, $v \leq v^+$ and $u^* \in \Lambda_{r_1,\dots,r_k} \subseteq \mathbb{R}^M_+$, therefore $(u^*, u-v^+) \leq (u^*, u-v)$ and

$$\gamma^+(u) - \gamma^+(v) \leq (u^*, u - v) \text{ for all } v \in \mathbb{R}^M.$$

Hence, u^* belongs to $\partial\gamma^+(u)$ and $\partial\gamma(u) \cap \Lambda_{r_1,\dots,r_k} \subseteq \partial\gamma^+(u)$.

□

A simple consequence of the previous lemma is a description of the subdifferential set of the function γ^+.

Corollary 6.1 *If* $u = (u_1, \overset{r_1}{\dots}, u_1, \dots, u_k, \overset{r_k}{\dots}, u_k) \in \Lambda^{\leq}_M$ *for some* $k \geq 1$ *and* $u_1 < \dots < u_k$ *if* $k > 1$ *then*

$$\partial\gamma^+(u) = \begin{cases} \mathbb{R}^M_+ \cap B^* & u = 0 \\ \{\lambda \in \Lambda_{r_1,\dots,r_k} : \gamma^0(\lambda) = 1, \ (\lambda, u) = \gamma(u)\} & u \neq 0 \end{cases}.$$

The proof follows from (6.8) and Lemma 6.2. As we mentioned before the description of the subdifferential of F is mainly based on these results as we will show in Section 6.3.2. On the other hand, in order to solve (6.5), it is important to know some of its properties. First of all, we start with a simple result about its objective function.

Proposition 6.3 *For any $\lambda \in \Lambda_M^{\leq}$ the function $f_\lambda(x)$ $(x \in X)$ is Lipschitz continuous and convex.*

Proof.
The proof follows directly from Proposition 1.1. □

By standard arguments concerning Banach spaces we thus obtain:

Proposition 6.4 *If X is a dual space (in particular: X is reflexive), then $\mathsf{M}(f_\lambda, A) \neq \emptyset$ for any finite set A and any $\lambda \in \Lambda_M^{\leq}$.*

Remark 6.1 The result above is true, for example, if $X = l^\infty$. Also, the same result holds if X is norm-one complemented in X^{**}. General results of this type have been given in [199].

The next result shows that also other spaces have the same property. This is the case of some spaces as c_0, the space of all sequences converging to 0 equipped with the supremum norm.

Theorem 6.2 *If $X = c_0$, then for every $A = \{a_1, \ldots, a_M\}$ and $\lambda \in S_M^{\leq}$ we have $\mathsf{M}(f_\lambda, A) \neq \emptyset$.*

Proof.
Given A, we may consider it as a subset of l^∞. Since l^∞ is a dual space, there exists $x = (x^{(1)}, x^{(2)}, \ldots, x^{(n)}, \ldots) \in l^\infty$ such that $f_\lambda(x) = \inf\{f_\lambda(y) : y \in l^\infty\}$. Let $\sigma_M(x)$ be the index of the demand point that gives the minimum distance to x, and assume that $\|x - a_{\sigma_M(x)}\| = \varepsilon$. Now find an index h such that $|a_i^{(j)}| \leq \varepsilon$ for all $j > h$ and i between 1 and M. Then set $x_0 = (x^{(1)}, \ldots, x^{(h)}, 0, \ldots, 0, \ldots) \in c_0$. We have:

$$\|x_0 - a_i\| = \max\Big\{ \max\{|a_i^{(m)}| : m > h\};\ \max_{1 \leq m \leq h} \{|x^{(m)} - a_i^{(m)}| : m \leq h\}\Big\} \leq \varepsilon$$

for $i = 1, \ldots, M$. This shows that $f_\lambda(x_0) \leq \|x - a_{\sigma_M(x)}\| \leq f_\lambda(x) = \mathsf{OM}_X(\lambda, A)$, so $x_0 \in \mathsf{M}(f_\lambda, A)$: This concludes the proof. □

Remark 6.2 There are spaces where for some finite sets, centers and/or medians do not always exist; one of these spaces is a hyperplane of c_0 considered in [157] (this does not contradict Theorem 6.2). Examples of four points sets with a center but without median, or with a median but without a center are indicated in [157] and [200]. An example of a three points set without k-centra for any k is shown at the end of this section.

In the following we give a uniqueness result and localization results on the optimal solutions of Problem (6.5) and bounds on its objective values.

Theorem 6.3 *Let X be a strictly convex, reflexive Banach space over the real field and assume that A is not contained in a line, i.e. the points in A are not collinear, then Problem (6.5) has a unique optimal solution.*

Proof.
Let us assume that x_1 and x_2 are two optimal solutions of Problem (6.5). Therefore, $x_1, x_2 \in \mathsf{M}(f_\lambda, A)$. By Theorem 6.8 there exist $\sigma \in \mathcal{P}(1 \dots M)$ and $p_\sigma = (p_{\sigma(1)}, \dots, p_{\sigma(M)})$ with $p_{\sigma(i)} \in B^*$ for all $i = 1, \dots, M$ such that $\mathsf{M}(f_\lambda, A) \subseteq \bigcap_{j=1}^{M} (a_{\sigma(j)} + N(p_{\sigma(j)}))$. Thus, $\langle p_{\sigma(i)}, x_k - a_{\sigma(i)} \rangle = \|x_k - a_{\sigma(i)}\| = d_{(i)}(x_k)$ for k=1,2, $\sigma(i)$ for all $i = 1, \dots, M$.

Since X is strictly convex, if the linear functional $\langle p_{\sigma(i)}, x - a_{\sigma(i)} \rangle$ achieves the value of its norm on two different vectors, i.e., $\langle p_{\sigma(i)}, x_k - a_{\sigma(i)} \rangle = \|x_k - a_{\sigma(i)}\|$ for $k = 1, 2$, these vectors must be proportional (see Theorem 13.3.3 in [213]), i.e. it must exist $t_{\sigma(i)} \in \mathbb{R}$ such that

$$x_1 - a_{\sigma(i)} = t_{\sigma(i)}(x_2 - a_{\sigma(i)}) \quad \forall\, \sigma \in \mathcal{J},\ \forall\, i = 1, \dots, M\ .$$

Hence, $a_{\sigma(i)} = 1/(1 - t_{\sigma(i)})x_1 - t_{\sigma(i)}/(1 - t_{\sigma(i)})x_2$ what implies that the elements of A are collinear. This contradiction concludes the proof. □

We add a simple result on the size of the set $\mathsf{M}(f_\lambda, A)$ of all the optimal solutions for those cases where uniqueness cannot be assured.

Proposition 6.5 *If x_1 and x_2 belong to $\mathsf{M}(f_\lambda, A)$ then*

$$\|x_1 - x_2\| \le 2\mathrm{OM}_X(\lambda, A) / \sum_{i=1}^{M} \lambda_i .$$

Proof.
We have that

$$\|x_1 - x_2\| \le \|x_1 - a_{\sigma(i)}\| + \|x_2 - a_{\sigma(i)}\| \quad \forall\, \sigma \in \mathcal{P}(1 \dots M),\ \forall\, i = 1, \dots, M$$

then we obtain

$$\begin{aligned} \sum_{i=1}^{M} \lambda_i \|x_1 - x_2\| &\le \sum_{i=1}^{M} \lambda_i \|x_1 - a_{\sigma(i)}\| + \sum_{i=1}^{M} \lambda_i \|x_2 - a_{\sigma(i)}\| \\ &\le 2 \max_{\pi \in \mathcal{P}(1 \dots M)} \sum_{i=1}^{M} \lambda_i \|x_1 - a_{\pi(i)}\| = 2\mathrm{OM}_X(\lambda, A)\ . \end{aligned} \tag{6.11}$$

□

It should be noted that this bound is tight. To see this, consider the following example.

Example 6.4
Let us consider the space $\mathbb{R}^2$ equipped with the rectilinear norm (l_1-norm). On this space, let A be the set of demand points defined by $A = \{a_1, a_2\}$ with $a_1 = (1,1)$, $a_2 = (-1,-1)$. It is easy to see that for $\lambda = (0,1) \in S_2^{\leq}$ the set $\mathsf{M}(f_\lambda, A) = [(-1,1),(1,-1)]$ and the optimal objective value $\mathsf{OM}_X(\lambda, A) = 2$.
In this case, if we consider $x_1 = (-1,1)$ and $x_2 = (1,-1)$ then

$$\|x_1 - x_2\|_1 = 2\mathsf{OM}_X(\lambda, A) = 4 \ .$$

In the following we consider a localization property of the ordered medians with respect to $conv(A)$, the convex hull of the set A.

Theorem 6.4 *If X is a two-dimensional space, or if X is a Hilbert space, then for any A and any $\lambda \in \Lambda_M^{\leq}$, we have $\mathsf{M}(f_\lambda, A) \cap conv(A) \neq \emptyset$. Moreover, if X is a Hilbert space, or if dim(X)=2 and X is strictly convex, then $\mathsf{M}(f_\lambda, A) \subset conv(A)$.*

Proof.
The assumptions imply that $\mathsf{M}(f_\lambda, A) \neq \emptyset$. If dim($X$)=2 then (see [208]) for every $x \in X$ there exists $x^* \in conv(A)$ such that $\|x^* - a\| \leq \|x - a\|$ for any $a \in A$; i.e., $\|x^* - a_i\| \leq \|x - a_i\|$ for $i = 1, \ldots, M = |A|$, so $f_\lambda(x^*) \leq f_\lambda(x)$. If we take $x \in \mathsf{M}(f_\lambda, A)$, this shows that there also exists $x^* \in \mathsf{M}(f_\lambda, A) \cap conv(A)$.
Now let X be Hilbert or if dim(X)=2, X strictly convex. If $x \notin conv(A)$, let x^* be the best approximation to x from $conv(A)$. We have $\|x^* - a_i\| < \|x - a_i\|$ for $i = 1, \ldots, M$, so $f_\lambda(x^*) < f_\lambda(x)$. Thus any element of $\mathsf{M}(f_\lambda, A)$ must belong to $conv(A)$. □

Another interesting property of ordered medians of a set A is that they allow to characterize inner product spaces in terms of their intersection with the convex hull of A. Characterizations of inner product spaces were known from the sixties (see [81], [123]). The same property concerning medians was considered in the nineties by Durier [64], where partial answers were given. It has been proved only recently for medians of three-point sets. The next result can be found in [20].

Theorem 6.5 *If dim(X) ≥ 3 and the norm of X is not Hilbertian, then there exists a three-point set A such that $\mathsf{M}(f_{(1,1,1)}, A) \cap conv(A) = \emptyset$.*

Using this theorem, it is not difficult to obtain the following characterization, taken from Papini and Puerto [158], that uses the k-centrum function (different from the center and median functions).

Proposition 6.6 *If $dim(X) \geq 3$ and the norm of X is not Hilbertian, then for every $M \geq 3$ there exists an M-point set D such that $\mathsf{M}(f_\lambda, D) \cap conv(D) = \emptyset$ for $\lambda = (0, \ldots, 0, 1, 1, 1) \in \Lambda_M^{\leq}$.*

Proof.
We prove the result for $M = 4$, the extension to $M \geq 4$ being similar. Also, we can assume without loss of generality that the lambda weights are normalized to 1, i.e. we can consider $\tilde{\lambda} = (0, \ldots, 0, 1/3, 1/3, 1/3)$.

Under the assumptions, according to Proposition 6.4, $\inf\limits_{x \in conv(A)} f_{\tilde{\lambda}}(A, x)$ is always attained. Now take $A = \{a_1, a_2, a_3\}$ and $\hat{\lambda} = (1/3, 1/3, 1/3)$ as given by Theorem 6.5. For some $\sigma > 0$ we have $\inf\limits_{x \in conv(A)} f_{\hat{\lambda}}(A, x) = \mathsf{OM}_X(f_{\hat{\lambda}}, A) + 4\sigma > \mathsf{OM}_X(f_{\hat{\lambda}}, A)$. Take $\bar{x} \in X$ such that $f_{\hat{\lambda}}(A, \bar{x}) < \mathsf{OM}_X(f_{\hat{\lambda}}, A) + \sigma$. It is not a restriction to assume that $\|\bar{x} - a_3\| \leq \min\{\|\bar{x} - a_1\|, \|\bar{x} - a_2\|\}$. Now take $a_4 \notin A$ such that $\|a_3 - a_4\| \leq \sigma$ and let $D = A \cup \{a_4\}$ and $\breve{\lambda} = (0, 1/3, 1/3, 1/3)$. We have $f_{\breve{\lambda}}(D, \bar{x}) \leq f_{\hat{\lambda}}(A, \bar{x}) + \sigma \leq \mathsf{OM}_X(f_{\hat{\lambda}}, A) + 2\sigma$. Now take $y \in conv(F)$. There is $x \in conv(A)$ such that $\|x - y\| \leq \sigma$; therefore $|f_{\tilde{\lambda}}(D, y) - f_{\tilde{\lambda}}(D, x)| \leq \sigma$, so $f_{\tilde{\lambda}}(D, y) \geq f_{\tilde{\lambda}}(D, x) - \sigma \geq f_{\hat{\lambda}}(A, x) - \sigma \geq \mathsf{OM}_X(f_{\hat{\lambda}}, A) + 3\sigma$. Thus $\inf\limits_{y \in conv(F)} f_{\tilde{\lambda}}(D, y) \geq \mathsf{OM}_X(f_{\hat{\lambda}}, A) + 3\sigma \geq f_{\tilde{\lambda}}(D, \bar{x}) + \sigma \geq \mathsf{OM}_X(f_{\tilde{\lambda}}, F) + \sigma$. □

Given X, consider for $\lambda \in S_M^{\leq}$ the parameter

$$J_\lambda(X) = \sup\left\{\frac{2 f_\lambda(A)}{\delta(A)} : A \subset X \text{ finite}; 2 \leq M = |A|\right\}. \qquad (6.12)$$

For $\bar{\lambda} = (0, \ldots, 0, 1)$, the number $J_{\bar{\lambda}}(X) = J(X)$ is called the finite Jung constant and has been studied intensively; in general, $1 \leq J(X) \leq 2$, while the value of $J(X)$ gives information on the structure of X.

We have the following result.

Theorem 6.6 *In every space X, for every $\lambda \in S_M^{\leq}$, we have:*

$$J_\lambda(X) = J(X). \qquad (6.13)$$

Proof.
As shown partially in [12] and later completely in [198], for $\tilde{\lambda} = (1/M, \ldots, 1/M)$ we always have:

$$J(X) = \sup\left\{\frac{2 f_{\tilde{\lambda}}(A)}{\delta(A)} : A \subset X \text{ finite}; 2 \leq M = |A|\right\}.$$

Since $f_{\tilde{\lambda}}(A) \leq f_\lambda(A) \leq f_{\bar{\lambda}}(A)$ always (see Proposition 1.2), we obtain (6.13). □

To conclude our analysis we show an example, adapted from [158], of an OMP without optimal solutions. In order to produce such an example we will use the k-centrum function ($\lambda = (0, \ldots, 0, \overset{(k)}{1/k}, \ldots, 1/k)$). This example is based on properties of k-centra points regarding equilateral sets. Recall that A is called *equilateral* if $\|a_i - a_j\| =$ constant for $i \neq j$, $1 \leq i, j \leq M = |A|$. For equilateral sets there are several nice properties connecting centers, medians and centroids (see [11]). Some of them can be extended further to ordered medians.

Proposition 6.7 *Let A be an equilateral set in a Hilbert space X and let $\lambda \in S_M^{\leq}$; then the centroid of A belongs to $\mathsf{M}(f_\lambda, A)$.*

Proof.
Assume that 0 is the center of A: Then $\langle a_i, a_j \rangle =$ constant for $i \neq j$, $1 \leq i, j \leq M = |A|$. Let $y = \sum_{j=1}^{M} \mu_j a_j$ and the function $g(\mu_1, \ldots, \mu_M) = \sum_{i=1}^{M} \lambda_i d_{(i)}(y)$ is symmetric.

In Hilbert spaces it always exists $x^* \in \mathsf{M}(f_\lambda, A) \cap conv(A)$. Moreover, under the general assumptions of this section $\mathsf{M}(f_\lambda, A)$ is a singleton, then x^* is the unique minimizer of g and $\mu_1 = \mu_2 = \ldots = \mu_M = 1/M$; thus, x^* is the centroid of A. □

Remark 6.3 *Let $A = \{a_1, \ldots, a_M\}$ be an equilateral set with $\|a_i - a_j\| = d$, $\forall\, i \neq j$ and $\lambda^k = (0, \ldots, 0, \overset{(k)}{1/k}, \ldots, 1/k)$ for some $1 \leq k \leq M$; then it is clear that*

$$f_{\lambda^k}(x) \geq \frac{d}{2} \text{ for any } x \in X.$$

Indeed, for any $x \in X$, $kf_{\lambda^k}(x)$ is attained as a sum of distances from x to k points of A. Let us denote by $A_k(x)$ the subset of A containing the k points that define $f_{\lambda^k}(x)$. $A_k(x)$ itself is an equilateral set with $\|a_i - a_j\| = d$, $\forall\, i \neq j$, $a_i, a_j \in A_k(x)$; then

$$kf_{\lambda^k}(x) = \sum_{a \in A_k(x)} \|a - x\| \geq \frac{kd}{2},$$

where inequality comes from Lemma 4.1 in [11] applied to the set $A_k(x)$.

For any equilateral set as the one above, the conditions $\mathsf{val}(\mathrm{P}_{\lambda^1}(\mathrm{A})) = \frac{\mathrm{d}}{2}$ and $\mathsf{val}(\mathrm{P}_{\lambda^k}(\mathrm{A})) = \frac{\mathrm{d}}{2}$ are equivalent, for any $k = 2, 3, \ldots, M$. (Recall that $\mathsf{val}(\mathrm{P})$ denotes the optimal value of Problem P.) This can be proven similarly *as Proposition 5.1 in [11] except for the details of considering partial sums of k-largest distances. Another consequence of this result is that in these cases k-centra for any $k = 1, 2, \ldots, M = |A|$ coincide (see [158]).*

Consider Example 5.2 in [11]. This is an equilateral three-point set without median. Now, we apply the above argument to conclude that this set cannot have k-centra for any $k = 1, 2, 3$.

6.3.2 Analysis of the Optimal Solution Set

In this section we study the solution set of Problem (6.4). Our first result is an exact description of the subdifferential set of the function $F(x)$. This result is mainly based on Lemma 6.2 and on a classical result on the subdifferential of the composition of convex functions.

For any $x \in X$, let $d_{\leq}(x) = (d_{(1)}(x), \ldots, d_{(M)}(x))$ denote the vector $d(x)$ whose components are arranged in non-decreasing sequence. (Recall that $d_{\leq}(x) = sort_M(d(x))$, see 1.1.)Besides, for any $\lambda \in \Lambda_M^{\leq}$, let $I(\lambda) = \{i : \lambda_i \neq 0\}$.

Lemma 6.3 *Let $x \in X$ such that $d_{\leq}(x) = (u_1, \overset{r_1}{\ldots}, u_1, \ldots, u_k, \overset{r_k}{\ldots}, u_k) \in \Lambda_{r_1,\ldots,r_k}$ for some $k \geq 1$ and $u_1 < u_2 \ldots < u_k$ if $k > 1$. An element $x^* \in X^*$ belongs to $\partial F(x)$ if and only if there exists a permutation $\sigma \in \mathcal{P}(1 \ldots M)$, $\lambda \in \Lambda_{r_1,\ldots,r_k} \cap \Lambda_M^{\leq}$ satisfying $\gamma^0(\lambda) = 1$, and elements $\{p_i\}_{i \in I(\lambda)}$ in B^*, such that $d_{(i)}(x) = \|x - a_{\sigma(i)}\|$ for all i; satisfying*

$$x \in \bigcap_{i \in I(\lambda)} (a_{\sigma(i)} + N(p_{\sigma(i)})), \quad \sum_{i \in I(\lambda)} \lambda_i d_{(i)}(x) = F(x) \quad \text{and } x^* = \sum_{i \in I(\lambda)} \lambda_i p_{\sigma(i)}.$$

Proof.
First of all, we can assume without loss of generality that $d(x)$ verifies

$$\|x - a_1\| \leq \|x - a_2\| \leq \ldots \leq \|x - a_M\|.$$

Otherwise, we only need to consider the permutation of A that verifies this relation and to rename the points in A. This fact does not change the proof because γ^+ is symmetric and, by Lemma 6.1 $p \in \partial\gamma^+(x)$ if and only if $p_\sigma \in \partial\gamma^+(x_\sigma)$.

Let $x^* \in \partial F(x)$. Since the distance vector has nonnegative components, we have that $F(x) = \gamma^+(d(x))$. Hence, we can apply Lemma 6.2 together with Theorem 4.3.1 in [102] Part I, chapter VI which gives the subdifferential set of the composition of a increasing componentwise convex function with a vector of convex functions and we have,

$$x^* \in \partial F(x) = \partial(\gamma^+ \circ d)(x) = \Big\{ \sum_{i=1}^{M} \lambda_i p_i : \lambda \in \partial\gamma(d(x)) \cap \Lambda_{r_1,\ldots,r_k}, \quad p_i \in \partial d_i(x), \ \forall\, i = 1, \ldots, M \Big\}. \tag{6.14}$$

Since γ is symmetric $\lambda \in \partial\gamma(d(x))$ if and only if $\lambda_\sigma \in \partial\gamma(d_\sigma(x))$ for any permutation σ. Then for any $\lambda \in \partial\gamma(d(x)) \cap \Lambda_{r_1,\dots,r_k}$, σ exists such that $\lambda_\sigma \in \partial\gamma(d_\sigma(x)) \cap \Lambda_{r_1,\dots,r_k} \cap \Lambda_M^{\leq}$ and $d_\sigma(x) \in \Lambda_{r_1,\dots,r_k} \cap \Lambda_M^{\leq}$. Indeed, to obtain this permutation just arrange the ties within each block of $d(x)$ so that the corresponding entries in λ_σ are ordered in non-decreasing sequence. Therefore, $p_{\sigma(i)} \in d_{\sigma(i)}(x)$ for any i and $\sum_{i=1}^M \lambda_i p_i = \sum_{i=1}^M \lambda_{\sigma(i)} p_{\sigma(i)}$ with $\lambda_\sigma \in \Lambda_M^{\leq}$. This implies that without loss of generality, one can consider in (6.14) $\lambda \in \partial\gamma(x) \cap \Lambda_{r_1,\dots,r_k} \cap \Lambda_M^{\leq}$. Moreover, as γ is a norm and $d(x) \neq 0$ for any $x \in X$ then $\lambda \in \partial\gamma(d(x))$ implies (see (6.8)):

$$1)\, \gamma^0(\lambda) = 1, \text{ and, } 2)\, \gamma(d(x)) = (\lambda, d(x)) = \sum_{i=1}^M \lambda_i \|x - a_i\|.$$

In addition, it is well-known that $p_i \in \partial d_i(x)$ if and only if $p_i \in B^*$ and $x \in a_i + N(p_i)$. Take $I = \{i : \lambda_i \neq 0\}$. Hence, by (6.14) and the above discussion $x^* \in \partial F(x)$ implies that $x^* = \sum_{i \in I} \lambda_i p_i$ with λ and $(p_1, \dots, p_M)$ satisfying the hypotheses.

Conversely, assume that $x^* = \sum_{i \in I(\lambda)} \lambda_i p_i$ for some $\lambda \in \Lambda_{r_1,\dots,r_k} \cap \Lambda_M^{\leq}$ and $p_i \in B^*$ for any $i \in I(\lambda)$ satisfying:

1. $\gamma^0(\lambda) = 1$,
2. $x \in \bigcap_{i \in I(\lambda)} (a_i + N(p_i))$,
3. $F(x) = \sum_{i \in I(\lambda)} \lambda_i \|x - a_i\| = (\lambda, d(x))$.

Since $d(x) \neq 0$, $\lambda \in \Lambda_{r_1,\dots,r_k}$, $\gamma^0(\lambda) = 1$, and $F(x) = (\lambda, d(x))$, we can apply Corollary 6.1 to conclude that $\lambda \in \partial\gamma^+(d(x))$. By 2., $x - a_i \in N(p_i) \quad \forall i \in I(\lambda)$. Thus, $p_i \in \partial d_i(x) \quad \forall i \in I(\lambda)$. Hence, using (6.14) $x^* \in \partial F(x)$. □

The next two results state the relationship between problems (6.4) and (6.5): for any monotone, symmetric norm γ there exists a non-decreasing lambda weight so that both problems have the same set of optimal solutions. Therefore, both families of problems are equivalent.

Theorem 6.7 *Let $x^0 \in X$, $A = \{a_1, \dots, a_M\} \subset X$ and $d_{\leq}(x^0) = (u_1, \overset{r_1}{\dots}, u_1, \dots, u_k, \overset{r_k}{\dots}, u_k) \in \Lambda_{r_1,\dots,r_k}$ for some $k \geq 1$ and $u_1 < u_2 \dots < u_k$ if $k > 1$. Then the following statements are equivalent,*

1. *There exists $\lambda \in \Lambda_{r_1,\dots,r_k} \cap \Lambda_M^{\leq}$ with $\gamma^0(\lambda) = 1$ such that*
$$x^0 \in \mathcal{M}(f_\lambda).$$
2. *There exists a monotone, symmetric norm γ such that*
$$x^0 \in \mathcal{M}(\gamma, A).$$

Proof.
1) $\to$ 2).

Let $f_\lambda(x) = \sum_{i=1}^M \lambda_i d_{(i)}(x)$. Consider $\gamma(x) = \max_{\sigma \in \mathcal{P}(1\ldots M)} \left\{ \sum_{i=1}^M \lambda_i |x_{\sigma(i)}| \right\}$. Then γ is a monotone, symmetric norm. Since by Theorem 386 in [98], we can write

$$f_\lambda(x) = \max_{\sigma \in \mathcal{P}(1\ldots M)} \left\{ \sum_{i=1}^M \lambda_i \|x - a_{\sigma(i)}\| \right\} = \gamma(d(x))$$

the result follows.
2) $\to$ 1).
Let $x^0 \in M_\gamma(A)$. Since F is convex the condition $x^0 \in \mathsf{M}(\gamma, A)$ is equivalent to $0 \in \partial F(x^0)$. Now, by Lemma 6.3, this is equivalent to the existence of $\sigma \in \mathcal{P}(1 \ldots M)$, $\lambda \in \Lambda_{r_1,\ldots,r_k} \cap \Lambda_M^{\leq}$ with $\gamma^0(\lambda) = 1$, $I(\lambda)$ and p_i in B^* for any $i \in I(\lambda)$ such that

$$d_{(i)}(x^0) = \|x^0 - a_{\sigma(i)}\|,\ x^0 \in \bigcap_{i \in I(\lambda)} (a_{\sigma(i)} + N(p_{\sigma(i)})),\ \sum_{i \in I(\lambda)} \lambda_i d_{(i)}(x^0) = F(x^0)$$

$$\text{and } 0 = \sum_{i \in I(\lambda} \lambda_i p_{\sigma(i)}.$$

Consider the function $f_\lambda(x) = \sum_{i \in I(\lambda)} \lambda_i d_{(i)}(x)$. Since, $x^0 \in \bigcap_{i \in I(\lambda)} (a_{\sigma(i)} + N(p_{\sigma(i)}))$ and $d_{(i)}(x^0) = \|x^0 - a_{\sigma(i)}\|$ we get that $p_{\sigma(i)} \in d_{(i)}(x^0)$ for any $i \in I(\lambda)$. Then $\sum_{i \in I(\lambda)} \lambda_i p_{\sigma(i)} \in \partial f_\lambda(x^0)$. Moreover, $0 = \sum_{i \in I(\lambda)} \lambda_i p_{\sigma(i)}$ then $0 \in \partial f_\lambda(x^0)$. Hence, $x^0 \in \mathsf{M}(f_\lambda, A)$ which concludes the proof. □

The above result is important by its own right, and also because it allows to reduce the optimization of monotone, symmetric norms to problems of type (6.5). It should be noted that the direct conclusion of this theorem is only that given a monotone, symmetric norm γ and an optimal solution x^0 to Problem (6.4) there exists a $\lambda \in \Lambda_M^{\leq}$ such that x^0 also belongs to $\mathsf{M}(f_\lambda, A)$. However, this λ may depend on x^0 so that the conclusion of this theorem is that $\mathsf{M}(f_\lambda, A) \subseteq \mathsf{M}(\gamma, A)$. Our next theorem states that this correspondence between λ and γ is one-to-one. Thus, studying the geometrical description of the optimal solution sets of Problem (6.5) is equivalent to study the optimal solution sets of Problem (6.4).

Given a finite set $A \subset X$ and a permutation σ, for each $1 \leq k \leq M$, $p = (p_j)_{j \geq k}$ with $p_j \in B^*$ we let

$$C_{\sigma,k}(p) = \{x \in X : \langle p_{\sigma(j)}, x - a_{\sigma(j)} \rangle = \|x - a_{\sigma(j)}\| \ \forall j \geq k\}$$

or equivalently

$$C_{\sigma,k}(p) = \bigcap_{j \geq k} \left(a_{\sigma(j)} + N(p_{\sigma(j)}) \right)$$

and for $\lambda \in \Lambda_M^{\leq}$

$$D_k(\lambda) = \{x \in X : F(x) = \sum_{j \geq k} \lambda_j d_{(j)}(x)\}.$$

Theorem 6.8 *If $\mathsf{M}(\gamma, A)$ is nonempty then there exist $\sigma \in \mathcal{P}(1 \dots M)$, $p = (p_j)_{j \geq k}$ with $1 \leq k \leq M$ and $p_j \in B^*$ for each j, $\lambda \in \Lambda_M^{\leq}$, $\lambda_j = 0$ if $j < k$, $\lambda_j > 0$ if $j \geq k$ verifying $\gamma^0(\lambda) = 1$ and $0 = \sum_{j \geq k} \lambda_j p_{\sigma(j)}$ such that*

$$\mathsf{M}(\gamma, A) = C_{\sigma,k}(p) \cap D_k(\lambda).$$

Conversely, if for some $\sigma \in \mathcal{P}(1 \dots M)$, $p = (p_j)_{j \geq k}$ with $1 \leq k \leq M$ and $p_j \in B^*$ for each j, and $\lambda \in \Lambda_M^{\leq}$ with $\lambda_j > 0$ if $j \geq k$ and $\gamma^0(\lambda) = 1$; the set $C_{\sigma,k}(p) \cap D_k(\lambda) \neq \emptyset$ then

$$\mathsf{M}(\gamma, A) = C_{\sigma,k}(p) \cap D_k(\lambda).$$

Proof.
The outline of the proof is as follows. First of all, we prove that for any given $x \in \mathsf{M}(\gamma, A)$ there exist σ, k, p and λ in the hypotheses of the theorem such that $x \in C_{\sigma,k}(p) \cap D_k(\lambda)$. Then, we prove that for these σ, k, p, and λ the entire set $C_{\sigma,k}(p) \cap D_k(\lambda)$ is included in $\mathsf{M}(\gamma, A)$. Finally, we prove that the 4-tuple (σ, k, p, λ) does not depend on the chosen $x \in \mathsf{M}(\gamma, A)$ which completes the proof.

Assume without loss of generality that $(d_{(1)}(x), \dots, d_{(M)}(x)) \in \Lambda_{r_1,\dots,r_k}$ for some $1 \leq k \leq M$. The condition $x \in \mathsf{M}(\gamma, A)$ is equivalent to $0 \in \partial F(x)$. Using Lemma 6.3; $0 \in \partial F(x)$ is equivalent to the existence of: 1) $\sigma^* \in \mathcal{P}(1 \dots M)$ such that $\|x - a_{\sigma^*(1)}\| \leq \dots \leq \|x - a_{\sigma^*(M)}\|$; 2)$p = (p_{\sigma^*(j)})_{j \geq k}$ with $1 \leq k \leq M$ and $p_{\sigma^*(j)} \in B^*$ satisfying $x \in \bigcap \left(a_{\sigma^*(j)} + N(p_{\sigma^*(j)}) \right)$; and 3) $\lambda \in \Lambda_{r_1,\dots,r_k} \cap \Lambda_M^{\leq}$ with $\lambda_j > 0$ if $j \geq k$ and $\gamma^0(\lambda) = 1$ such that $0 = \sum_{j \geq k} \lambda_j p_{\sigma^*(j)}$ and $F(x) = \sum_{j \geq k} \lambda_j \|x - a_{\sigma^*(j)}\|$. Using the definitions of $C_{\sigma^*,k}(p)$ and $D_k(\lambda)$ we get $x \in C_{\sigma^*,k}(p) \cap D_k(\lambda)$.

Since the same conditions apply to all the vectors $y \in C_{\sigma^*,k}(p) \cap D_k(\lambda)$ then also $0 \in \partial F(y)$. Therefore, $y \in \mathsf{M}(\gamma, A)$, i.e.

$$C_{\sigma^*,k}(p) \cap D_k(\lambda) \subset \mathsf{M}(\gamma, A). \tag{6.15}$$

However, this does not imply that different σ', k', p' and λ' may determine different optimal solutions. We prove that this is not possible.

Using the first part of this proof, for any given $x \in \mathsf{M}(\gamma, A)$ there exist σ, $p = (p_{\sigma(j)})_{j \geq k}$ with $1 \leq k \leq M$, and $\lambda \in \Lambda_M^{\leq}$ such that $x \in C_{\sigma,k}(p) \cap D_k(\lambda)$. Let $\bar{x} \neq x$ and $\bar{x} \in \mathsf{M}(\gamma, A)$, then we have

$$F(x) = \sum_{j \geq k} \lambda_j \|x - a_{\sigma(j)}\| = \sum_{j \geq k} \lambda_j \langle p_{\sigma(j)}, x - a_{\sigma(j)} \rangle,$$

but using that $\sum_{j \geq k} \lambda_j p_{\sigma(j)} = 0$ we obtain,

$$\sum_{j \geq k} \lambda_j \langle p_{\sigma(j)}, x - a_{\sigma(j)} \rangle = -\sum_{j \geq k} \lambda_j \langle p_{\sigma(j)}, a_{\sigma(j)} \rangle = \sum_{j \geq k} \lambda_j \langle p_{\sigma(j)}, \bar{x} - a_{\sigma(j)} \rangle.$$

Now, since $\langle p_j, \bar{x} - a_j \rangle \leq \|p_j\|^0 \|\bar{x} - a_j\|$ and $\|p_j\|^0 = 1$,

$$\sum_{j \geq k} \lambda_j \langle p_{\sigma(j)}, \bar{x} - a_{\sigma(j)} \rangle \leq \sum_{j \geq k} \lambda_j \|\bar{x} - a_{\sigma(j)}\|.$$

Therefore,

$$F(x) \leq \sum_{j \geq k} \lambda_j \|\bar{x} - a_{\sigma(j)}\| \leq \sum_{j \geq k} \lambda_j d_{(j)}(\bar{x}) \leq F(\bar{x}).$$

Since we assume that $\bar{x} \in \mathsf{M}(\gamma, A)$ then $F(x) = F(\bar{x})$ and we deduce that $\langle p_{\sigma(j)}, \bar{x} - a_{\sigma(j)} \rangle = \|\bar{x} - a_{\sigma(j)}\|$ for all $j \geq k$ what implies that $\bar{x} \in C_{\sigma,k}(p)$. Moreover, as $F(\bar{x}) = \sum_{j \geq k} \lambda_j d_{(j)}(\bar{x})$ then $\bar{x} \in D_k(\lambda)$. This proves that $\mathsf{M}(\gamma, A) \subset C_{\sigma,k}(p) \cap D_k(\lambda)$. This inclusion together with (6.15) completes the proof. □

6.3.3 The Convex OMP and the Single Facility Location Problem in Normed Spaces

In this section we concentrate on obtaining a geometrical description of the set $\mathsf{M}(f_\lambda, A)$, the optimal solution set of the OMP (6.5). This is possible by means of the family of o.e.c.s., in fact we will prove that the solution set of Problem (6.5), as in the planar case, is always a convex union of o.e.c.s..

Let $\lambda \in \Lambda_M^{\leq}$ be a fixed vector. For any $\sigma \in \mathcal{P}(1 \dots M)$ we consider the function

$$W_\sigma(x) = \sum_{i=1}^{M} \lambda_i \|x - a_{\sigma(i)}\|.$$

By Theorem 368 in [98]

$$f_\lambda(x) = \max_{\sigma \in \mathcal{P}(1 \dots M)} \{W_\sigma(x)\}. \tag{6.16}$$

Therefore, f_λ is the composition of a monotone, symmetric norm, the l_∞ norm on $\mathbb{R}^{M!}$, with a vector $(W_\sigma(x))_{\sigma \in \mathcal{P}(M)}$ where each component is a convex function on X. It is also worth noting as a straightforward consequence of Theorem 368 in [98] that

$$W_\sigma(x) = f_\lambda(x) \text{ if and only if } x \in \bigcup_{\pi \in L_\sigma(\lambda)} O_\pi. \tag{6.17}$$

(Recall that the sets $L_\sigma(\lambda)$ were introduced in Section 2.3.2.)

In order to simplify the proof of our next theorem we assume the following convention. Since the set $\mathcal{P}(1\dots M)$ is finite, it is possible to number its elements. Assume that we choose and fix such a numeration in $\mathcal{P}(1\dots M)$. With this process we identify each permutation $\sigma \in \mathcal{P}(1\dots M)$ with a unique number in $\{1,\dots,M!\}$. Let us denote by C such a bijective correspondence, i.e.,

$$\begin{aligned} C : \{1,\dots,M!\} &\mapsto \mathcal{P}(1\dots M) \\ k \quad &\mapsto C(k) = \sigma. \end{aligned}$$

In the same way, let C^{-1} be the inverse correspondence. Therefore, $C^{-1}(\sigma) \in \{1,\dots,M!\}$ for any permutation σ. Then, for any $\sigma \in \mathcal{P}(1\dots M)$ we can write $W_\sigma(x) = W_{C(k)}(x)$ for some $1 \le k \le M!$.

Theorem 6.9 *If $\mathsf{M}(f_\lambda, A) \neq \emptyset$ then there exists*

1. *A set $\mathcal{J}$ of permutations of $\mathcal{P}(1\dots M)$,*
2. *$q = \{q_\sigma\}_{\sigma\in\mathcal{J}}$ with $q_\sigma = \sum_{i=1}^M \lambda_i p_{\sigma(i)}$ for some $p_\sigma = (p_{\sigma(1)},\dots,p_{\sigma(M)})$ such that $p_{\sigma(i)} \in B^*$ for all $\sigma \in \mathcal{J}$ and $i = 1,\dots,M$,*
3. *$w \in \mathbb{R}_+^{M!}$ satisfying $\sum_{\sigma\in\mathcal{J}} w_{C^{-1}(\sigma)} q_\sigma = 0$*

such that

$$\mathsf{M}(f_\lambda, A) = OC_{\mathcal{J},\lambda}(p).$$

Conversely, if there exists $(\mathcal{J}, q, w)$ such that $OC_{\mathcal{J},\lambda}(p) \neq \emptyset$ satisfying:

1. *$\mathcal{J} \subseteq \mathcal{P}(1\dots M)$,*
2. *$q = \{q_\sigma\}_{\sigma\in\mathcal{J}}$ with $q_\sigma = \sum_{i=1}^M \lambda_i p_{\sigma(i)}$ for some $p_\sigma = (p_{\sigma(1)},\dots,p_{\sigma(M)})$ such that $p_{\sigma(i)} \in B^*$ for all $\sigma \in \mathcal{J}$ and $i = 1,\dots,M$,*
3. *$w \in \mathbb{R}_+^{M!}$ satisfying $\sum_{\sigma\in\mathcal{J}} w_{C^{-1}(\sigma)} q_\sigma = 0$*

then

$$OC_{\mathcal{J},\lambda}(p) = \mathsf{M}(f_\lambda, A).$$

Proof.
Let $x \in \mathsf{M}(f_\lambda, A)$. Since f_λ is a convex function then

$$x \in \mathsf{M}(f_\lambda, A) \text{ iff } 0 \in \partial f_\lambda(x).$$

By Theorem 4.3.1 in [102] applied to the expression of f_λ in (6.16) we obtain that $0 \in \partial f_\lambda(x)$ if and only if there exists (q, w) such that:

1. $q = \{q_\sigma\}_{\sigma\in\mathcal{P}(1\dots M)}$ with $q_\sigma \in \partial W_\sigma(x)$ for all $\sigma \in \mathcal{P}(1\dots M)$,

2. $w \in \mathbb{R}_+^{M!}$, $\sum_{1\le j\le M!} w_j = 1$, $\sum_{1\le j\le M!} w_j W_{C(j)}(x) = f_\lambda(x)$ satisfying

$$\sum_{1\le j\le M!} w_j q_{C(j)} = 0.$$

First, note that since $W_{C(j)}(x) \le f_\lambda(x)$ for all j and $\sum_{1\le j\le M!} w_j = 1$ then

$$\sum_{1\le j\le M!} w_j W_{C(j)}(x) = f_\lambda(x) \text{ iff } W_{C(j)}(x) = f_\lambda(x) \text{ for all } j \text{ such that } w_j \neq 0.$$

Additionally, for any $\sigma \in \mathcal{P}(1\dots M)$; $q_\sigma \in \partial W_\sigma(x)$ if and only if there exists $p_\sigma = (p_{\sigma(1)}, \dots, p_{\sigma(M)})$ with $p_{\sigma(i)} \in \partial\|x - a_{\sigma(i)}\|$ for all $i = 1, \dots, M$ satisfying $q_\sigma = \sum_{i=1}^{M} \lambda_i p_{\sigma(i)}$. Now, recall that $p_{\sigma(i)} \in \partial\|x - a_{\sigma(i)}\|$ if and only if $x \in (a_{\sigma(i)} + N(p_{\sigma(i)}))$. Therefore, $0 \in \partial f_\lambda(x)$ if and only if there exists (q, w) such that:

1. $w \in \mathbb{R}_+^{M!}$, $\sum_{1\le j\le M!} w_j = 1$, $W_{C(j)}(x) = f_\lambda(x)$ for all j such that $w_j \neq 0$,
2. $p_\sigma = (p_{\sigma(1)}, \dots, p_{\sigma(M)})$ with $p_{\sigma(i)} \in B^*$ for all $1 \le i \le n$ and for all σ; and $x \in \bigcap_{i=1}^{M} \left(a_{\sigma(i)} + N(p_{\sigma(i)})\right)$ for all $\sigma \in \mathcal{P}(1\dots M)$ if $w_{C^{-1}(\sigma)} \neq 0$; such that
$$q_\sigma = \sum_{i=1}^{M} \lambda_i p_{\sigma(i)}, \text{ and } \sum_{j=1}^{M!} w_j q_{C(j)} = 0.$$

Define the following set of permutations of $\{1\dots, n\}$:

$$\mathcal{J} = \{\sigma \in \mathcal{P}(1\dots M) : w_{C^{-1}(\sigma)} \neq 0\}.$$

Then, using $\mathcal{J}$ and (6.17); $x \in M_\lambda(x)$ if and only if there exists $(\mathcal{J}, q, w)$ such that:

1'. $\mathcal{J} \subseteq \mathcal{P}(1\dots M)$, $\mathcal{J} \neq \emptyset$,
2'. $q_\sigma = \sum_{i=1}^{M} \lambda_i p_{\sigma(i)}$ and $p_{\sigma(i)} \in B^*$ for all $\sigma \in \mathcal{J}$ and $1 \le i \le n$,
3'. $w \in \mathbb{R}_+^{M!}$, such that $\sum_{\sigma\in\mathcal{J}} w_{C^{-1}(\sigma)} q_\sigma = 0$

and

$$x \in \bigcap_{\sigma\in\mathcal{J}} \left(\bigcap_{j=1}^{M} \left(a_{\sigma(j)} + N(p_{\sigma(j)})\right) \cap \bigcup_{\pi \in L_\sigma(\lambda)} O_\pi \right) = OC_{\mathcal{J},\lambda}(p).$$

Since 1', 2' and 3' hold for any $y \in OC_{\mathcal{J},\lambda}(p)$ then it follows that $y \in \mathsf{M}(f_\lambda, A)$. Hence,

$$OC_{\mathcal{J},\lambda}(p) \subset \mathsf{M}(f_\lambda, A). \tag{6.18}$$

Now, we prove that both sets coincide. Indeed, consider $\bar{x} \neq x$ and $\bar{x} \in \mathsf{M}(f_\lambda, A)$, then we have

$$f_\lambda(x) = \sum_{\sigma\in\mathcal{J}} w_{C^{-1}(\sigma)} W_\sigma(x) = \sum_{\sigma\in\mathcal{J}} w_{C^{-1}(\sigma)} \sum_{j=1}^{M} \lambda_j \langle p_{\sigma(j)}, x - a_{\sigma(j)}\rangle,$$

but using that $\sum_{\sigma\in\mathcal{J}} w_{C^{-1}(\sigma)} q_\sigma = \sum_{\sigma\in\mathcal{J}} w_{C^{-1}(\sigma)} \sum_{j=1}^{M} \lambda_j p_{\sigma(j)} = 0$ we obtain,

$$\sum_{\sigma\in\mathcal{J}} w_{C^{-1}(\sigma)} \sum_{j=1}^{M} \lambda_j \langle p_{\sigma(j)}, x - a_{\sigma(j)}\rangle = \sum_{\sigma\in\mathcal{J}} w_{C^{-1}(\sigma)} \sum_{j=1}^{M} \lambda_j \langle p_{\sigma(j)}, \bar{x} - a_{\sigma(j)}\rangle.$$

Now, since $\langle p_j, \bar{x} - a_j\rangle \leq \|p_j\|^0 \|\bar{x} - a_j\|$ and $\|p_j\|^0 = 1$, we obtain

$$f_\lambda(x) \leq \sum_{\sigma\in\mathcal{J}} w_{C^{-1}(\sigma)} \sum_{j=1}^{M} \lambda_j \|\bar{x} - a_{\sigma(j)}\| \leq \sum_{\sigma\in\mathcal{J}} w_{C^{-1}(\sigma)} \sum_{j=1}^{M} \lambda_j d_{(j)}(\bar{x}) \leq f_\lambda(\bar{x}).$$

As we assume that $\bar{x} \in \mathsf{M}(f_\lambda, A)$ then $f_\lambda(x) = f_\lambda(\bar{x})$ and we deduce that: 1) $W_\sigma(\bar{x}) = f_\lambda(\bar{x})$ for all $\sigma \in \mathcal{J}$; 2) $\langle p_{\sigma(j)}, \bar{x} - a_{\sigma(j)}\rangle = \|\bar{x} - a_{\sigma(j)}\|$ for all $j = 1, \ldots, M$ what implies that $\bar{x} \in C_\sigma(p) = \cap_{i=1}^{M}(a_{\sigma(i)} + N(p_{\sigma(i)}))$. Then, using again the same argument that in (6.17) we deduce from 1) above that $\bar{x} \in \bigcup_{\pi\in L_\sigma(\lambda)} O_\pi$ for any $\sigma \in \mathcal{J}$. These conditions together imply that $\bar{x} \in OC_{\mathcal{J},\lambda}(p)$. Therefore, $\mathcal{J}, \lambda, p$ do not depend on the choice of the optimal solution and hence $\mathsf{M}(f_\lambda, A) \subseteq OC_{\mathcal{J},\lambda}(p)$. This inclusion together with (6.18) completes the proof. □

This theorem proves that the solution set of Problem (6.5) is always a convex union of ordered elementary convex sets. In fact, if the weights λ are all distinct then the solution set coincides exactly with an o.e.c.s.. Similar results have been previously obtained in Chapter 4 for the OMP in the plane; and for different single facility location problems (see [63, 65, 101, 151, 169]).

An example illustrates the use of Theorem 6.9.

Example 6.5

Consider the normed space $\mathbb{R}^2$ with the rectilinear (l_1) norm. The set A has three points

$$a_1 = (0,0), \qquad a_2 = (1,2), \qquad a_3 = (3,1).$$

The weights are given by $\lambda_1 = 1/6$, $\lambda_2 = 1/6$ and $\lambda_3 = 2/3$. This formulation leads us to the problem

$$\min_{x\in\mathbb{R}^2} \frac{1}{2}\Big(\sum_{i=1}^{3} \frac{1}{3}\|x - a_i\|_1\Big) + \frac{1}{2}\max\Big(\frac{4}{3}\|x - a_i\|_1\Big)$$

which corresponds to the well-known 1/2-cent-dian problem.

In order to solve the problem, we compute in Table 6.2 the family $q = \{q_\sigma\}_{\sigma\in\mathcal{P}(1\ldots M)}$.

It is easy to see that considering $\mathcal{J}$ as the permutation associated to the ordered regions numbered 1 and 4 in Table 6.2 one has

Table 6.2. O.e.c.s. and q in Example 6.5.

Region n.	O_σ	$p_{\sigma 1}$	$p_{\sigma 2}$	$p_{\sigma 3}$	q_σ
1	$O_{(2,3,1)}$	(1,-1)	(-1,1)	(1,1)	(2/3,2/3)
2	$O_{(2,1,3)}$	(-1,-1)	(1,1)	(-1,1)	(-2/3,2/3)
3	$O_{(1,2,3)}$	(1,1)	(-1,-1)	(-1,1)	(-2/3,2/3)
4	$O_{(2,1,3)}$	(-1,-1)	(1,1)	(-1,-1)	(-2/3,-2/3)
5	$O_{(1,2,3)}$	(1,1)	(-1,-1)	(-1,-1)	(-2/3,-2/3)
6	$O_{(2,1,3)}$	(1,-1)	(1,1)	(-1,-1)	(-1/3,-2/3)
7	$O_{(1,2,3)}$	(1,1)	(1,-1)	(-1,-1)	(-1/3,-2/3)
8	$O_{(1,3,2)}$	(1,1)	(-1,-1)	(1,-1)	(2/3,-2/3)
9	$O_{(3,1,2)}$	(-1,-1)	(1,1)	(1,-1)	(2/3,-2/3)
10	$O_{(2,3,1)}$	(1,-1)	(-1,-1)	(1,1)	(2/3,1/3)
11	$O_{(3,2,1)}$	(-1,-1)	(1,-1)	(1,1)	(2/3,1/3)
12	$O_{(3,2,1)}$	(-1,1)	(1,-1)	(1,1)	(2/3,2/3)

$$OC_{\mathcal{J}}(q) = \{(1,1)\} \; .$$

On the other hand, the weight $w \in S_{3!}^{\leq}$ such that $w_\sigma = 0$ if $\sigma \notin \mathcal{J}$ and $w_1 = w_4 = 1/2$ verifies $w_1 q_1 + w_4 q_4 = 0$. Thus,

$$\mathsf{M}(f_\lambda, A) = OC_{\mathcal{J}}(q) \; .$$

This example also shows that the choice of $\mathcal{J}$ is not unique. For instance, $\mathcal{J} = \{2, 6, 10\}$ with weights $w_2 = 1/5$, $w_6 = 2/5$ and $w_{10} = 2/5$ gives also $w_2 q_2 + w_6 q_6 + w_{10} q_{10} = 0$. Thus, this choice also determines the set of optimal solutions.

6.4 Concluding Remarks

This chapter dealt with three different versions of the OMP: problems with forbidden regions, multifacility and abstract extensions.

Apart from the results already presented here, there are some others that are worth to be further investigated. The first extension consists of considering different norms associated with the different points in the set A. Once we redefine appropriately the elements of this new problem, the same results hold. Another interesting extension concerns the use of gauges of unbounded unit balls following the results in [101]. Moreover, most of the results extend further to the use of star shaped unit balls. Therefore, applications to single facility, positive and negative weights, forbidden regions and multiobjective problems seem to be possible.

The last section dealt with a general location problem where two rationality principles 1) monotony and, 2) symmetry hold. These two principles are justified in terms of the quality of the service provided as well as in terms of the global satisfiability of the demand points. For this problem, we derive an optimality criterion using subdifferential calculus. Moreover, we reduce this kind of problems to OMP. We geometrically characterize the set of optimal solutions of the ordered median problem by means of ordered elementary convex sets as introduced by [169].

Although no algorithms are proposed to solve OMP in general spaces, one can imagine how these tools can be used as part of any algorithmic approach to solve this kind of problems. This section describes the domains where the considered objective function behaves as a fixed weighted sum of distances. Therefore, solving a classical Weber problem within each one of these regions leads us to solve the original problem. Particularly, efficient algorithms for the OMP in finite dimensional spaces with any polyhedral norm were developed in Chapter 4.

On the other hand, since the ordered elementary convex sets describe the linearity domains of these kind of objective functions, another interesting application of the results of this section is the determination of the Pareto-efficient sets of a wide range of location problems in abstract spaces. For instance, the multiobjective minimax, minisum, centdian or any combination of these objectives can be approached using these tools. These topics were already covered in Chapter 5 in the case of finite dimension spaces.

Finally, another interesting open problem is to find alternative characterizations of the set of optimal solutions. One possible way to attach this problem may be using Fenchel duality applied to the OMf.

Part III

Ordered Median Location Problems on Networks

7

The Ordered Median Problem on Networks

7.1 Problem Statement

One of the most important and well-developed branches in Location Theory is the analysis of location problems on networks. Numerous surveys and textbooks (see [59], [142], [165], [127] and the references therein) give evidence for this fact. As well as continuous location the development of network location has been also recognized by the mathematical community reserving the AMS code 90B80 for this area of research. The starting point of this development might be considered the node-dominance result by Hakimi [84] that will be shown to be essential in this part of the book. Nevertheless, in the existing literature of Location Analysis mainly three types of objective functions have been used: the median or sum objective, the center or max objective and the cent-dian objective that is a convex combination of sum and max (see [103] and the references therein for a description of these and many other facility location problems).

In this part of the book (Part III) we will use the OMf to analyze location problems on networks as introduced by Nickel and Puerto [150]. The advantage of using this new objective function will be, apart from the study of new location problems, the possibility of reproving a lot of known results in an easier way and getting more insight into the geometrical structure of the optimal solution sets with respect to different criteria.

Introducing formally the problem, we must identify the elements that constitute the model. In the network case the sets A and X, introduced in Chapter 1, have the following meaning. A is the set of vertices of a graph and X is a set of p points in the graph, usually the location of the new facilities. Moreover, $c(A, X)$, introduced in Section 1.4, will be expressed in terms of weighted distances from A to X.

Specifically, let $G = (V, E)$ be an undirected graph with node set $V = \{v_1, ..., v_M\}$ and edge set $E = \{e_1, \ldots, e_N\}$. Each edge has a nonnegative length and is assumed to be rectifiable. Thus, we will refer to interior points on edges. We let $A(G)$ denote the continuum set of points on the edges of G.

Each edge $e \in E$ has associated a positive length by means of the function $l : E \to \mathbb{R}_+$. The edge lengths induce a distance function on $A(G)$. For any x, y in $A(G)$, $d(x, y)$ will denote the length of a shortest path connecting x and y. Also, if Y is a subset of $A(G)$, we define the distance from x to the set Y by

$$d(x, Y) = d(Y, x) = \inf\{d(x, y) : y \in Y\}.$$

Through $w : V \to \mathbb{R}$, every vertex is assigned a nonnegative weight. We will assume that the weights are nonnegative. Exceptions are stated explicitly.

A point x on an edge $e = [v_i, v_j]$ can be written as a pair $x = (e, t)$, $t \in [0, 1]$, with

$$d(v_k, x) = d(x, v_k) = \min\{d(v_k, v_i) + tl(e), d(v_k, v_j) + (1 - t)l(e)\}.$$

Let $p \geq 1$ be an integer. Then for $X_p := \{x_1, \ldots, x_p\} \subset A(G)$ the distance from a node $v_i \in V$ to the set X_p is

$$d(v_i, X_p) = d(X_p, v_i) = \min_{k=1,\ldots,p} d(v_i, x_k)\,.$$

Now, for $X_p \subset A(G)$, we define

$$d(X_p) := (w_1 d(v_1, X_p), \ldots, w_M d(v_M, X_p)),$$

and

$$d_{\leq}(X_p) := sort_M(d(X_p)) = (w_{(1)} d(v_{(1)}, X_p), \ldots, w_{(M)} d(v_{(M)}, X_p))$$

being $(\cdot)$ a permutation of the the set $\{1, \ldots, M\}$ satisfying

$$w_{(1)} d(v_{(1)}, X_p) \leq w_{(2)} d(v_{(2)}, X_p) \leq \cdots \leq w_{(M)} d(v_{(M)}, X_p)\,. \tag{7.1}$$

To simplify the notation we will denote the entries $w_i d(v_i, X_p)$ and $w_{(i)} d(v_{(i)}, X_p)$ in the above vectors by $d_i(X_p)$ and $d_{(i)}(X_p)$, respectively.

The p–facility OMP on $A(G)$ is defined as

$$\mathsf{OM}_p(\lambda) := \min_{X_p \subset A(G)} f_\lambda(X_p)\,,$$

with

$$f_\lambda(X_p) := \langle \lambda, d_{\leq}(X_p) \rangle = \sum_{i=1}^{M} \lambda_i d_{(i)}(X_p) \text{ and } \lambda = (\lambda_1, \ldots, \lambda_M) \in \mathbb{R}^M_{0+}. \tag{7.2}$$

The function $f_\lambda(X_p)$ is called the **Ordered p-Median Function**. Note that the linear representation of this function is defined point-wise, since it changes when the order of the vector of distances is modified.

The single facility problem is a particular case of the above problem when $p = 1$. Let

$$d(x) := (w_1 d(v_1, x), \ldots, w_M d(v_M, x))$$

and let

$$d_{\leq}(x) := (w_{(1)} d(v_{(1)}, x), \ldots, w_{(M)} d(v_{(M)}, x))$$

a permutation of the elements of $d(x)$, satisfying

$$w_{(1)} d(v_{(1)}, x) \leq w_{(2)} d(v_{(2)}, x) \leq \ldots \leq w_{(M)} d(v_{(M)}, x).$$

Using the above notation, let $d_{(i)}(x) := w_{(i)} d(v_{(i)}, x)$.

The single facility OMP on $A(G)$ is defined as

$$\mathsf{OM}_1(\lambda) := \min_{x \in A(G)} f_\lambda(x) ,$$

with

$$f_\lambda(x) := \sum_{i=1}^{M} \lambda_i d_{(i)}(x) \quad \text{and} \quad \lambda = (\lambda_1, \ldots, \lambda_M) \in \mathbb{R}^M_{0+}. \tag{7.3}$$

For the sake of simplicity, in the single facility case, we denote the objective value by $\mathsf{OM}(\lambda)$ instead of $\mathsf{OM}_1(\lambda)$.

We illustrate the above concepts with a simple example.

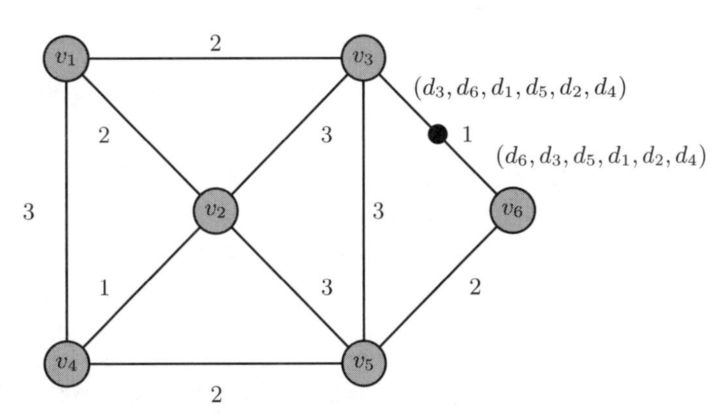

Fig. 7.1. The network used in Example 7.1

Example 7.1

Consider the network $G = (V, E)$ with $V = \{v_1, \ldots, v_6\}$ and $E = \{[v_1, v_2], [v_1, v_3], [v_1, v_4], [v_2, v_3], [v_2, v_4], [v_2, v_5], [v_3, v_5], [v_4, v_5], [v_3, v_6], [v_5, v_6]\}$ given in Figure 7.1. The edge lengths are the numbers depicted on the edges and the node weights are $w_1 = w_2 = w_5 = 1$ and $w_3 = w_4 = w_6 = 2$.

On the edge $[v_3, v_6]$ the black dot marks the point $EQ = ([v_3, v_6], 1/2)$. Notice that $w_1 d(EQ, v_1) = w_5 d(EQ, v_5) = 2.5$ and at the same time $d_3(EQ) = 1 = d_6(EQ)$. For all the points $x \in [v_3, EQ]$ the ordering of the weighted distances from the nodes to x is $d_3(x) \leq d_6(x) \leq d_5(x) \leq d_1(x) \leq d_2(x) \leq d_4(x)$. Similarly, the points $y \in [EQ, v_6]$ satisfy $d_6(y) \leq d_3(y) \leq d_5(y) \leq d_1(y) \leq d_2(y) \leq d_4(y)$. We take two different λ-parameters $\lambda_1 = (0,0,0,2,2,1)$ and $\lambda_2 = (3,0,0,0,0,1)$ in order to evaluate the objective function at EQ. Since $d_\leq(EQ) = (1,1,2.5,2.5,3.5,10)$ the result is:

$$f_{\lambda_1}(EQ) = 2 \times 2.5 + 2 \times 3.5 + 10 = 21,$$
$$f_{\lambda_2}(EQ) = 3 \times 1 + 10 = 13.$$

7.2 Preliminary Results

In this section we proceed to introduce some concepts related to the OMP that will be important in the development of our analysis for this family of problems on networks. The general idea is to identify some sets of points, defined by their geometrical properties, that will be useful to find the optimal solutions of different instances of OMP.

We start recalling the concept of a bottleneck point. A point x on an edge $e = [v_i, v_j] \in E$ is called a bottleneck point of node v_k, if $w_k \neq 0$, and

$$d(x, v_k) = d(x, v_i) + d(v_i, v_k) = d(x, v_j) + d(v_j, v_k).$$

Let BN_i denote the set of all bottleneck points of a node $v_i \in V$ and let $\mathsf{BN} := \bigcup_{i=1}^{M} BN_i$ be the set of all bottleneck points of the graph.

Define $\mathsf{NBN} := \bigcup_{\substack{i=1 \\ w_i<0}}^{M} BN_i$. A point in NBN is called a negative bottleneck point.

For all $v_i,\ v_j \in V$, $i \neq j$ define

$$EQ'_{ij} := \{x \in A(G) : w_i d(v_i, x) = w_j d(v_j, x)\}. \tag{7.4}$$

Let EQ_{ij} be the relative boundary of EQ'_{ij}, i.e. the set of endpoints of the closed subedges forming the elements in EQ'_{ij} and let $\mathsf{EQ} := \bigcup_{\substack{i,j \\ i\neq j}} EQ_{ij}$. The points in EQ are called equilibrium points of G (see [150]). The reader may note that we take the relative boundary of these subsets to avoid the fact that EQ includes a continuum of points (see Figure 7.2). In order to simplify the presentation, we denote by $EQ_{ij}^{kl} \subseteq EQ_{ij}$ the equilibrium points of nodes v_i, v_j on the edge $[v_k, v_l]$, for any $i, j \in \{1, \ldots M\}$ and k, l so that $[v_k, v_l] \in E$. Note that $|EQ_{ij}^{kl}| \leq 2$. In the case that $|EQ_{ij}^{kl}| = 1$ we denote by EQ_{ij}^{kl} the unique element of the set EQ_{ij}^{kl}.

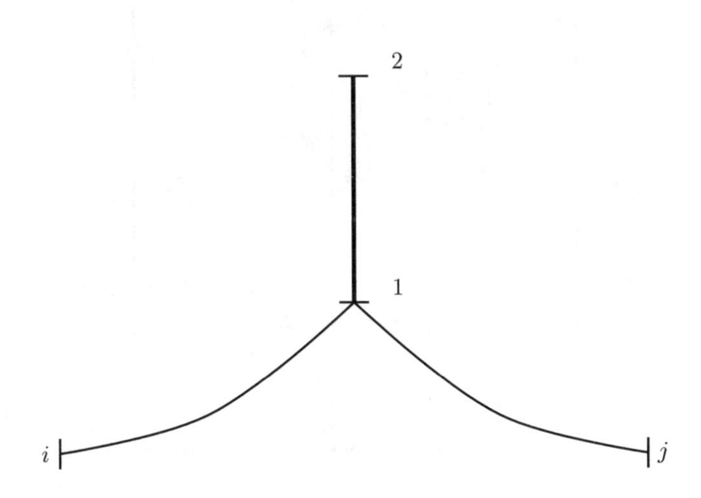

Fig. 7.2. Infinite points in EQ'_{ij}

Two points $a, b \in \mathsf{EQ}$ are called consecutive, if there is no other $c \in \mathsf{EQ}$ on a shortest path between a and b. The points in EQ establish a partition on G with the property, that for two consecutive elements $a, b \in \mathsf{EQ}$ the permutation, which gives the order of the vector $d_{\leq}(x)$, is the same for all x on a shortest path from a to b. Note that the equilibrium points include the well-known center-bottleneck points [103]. It can be easily seen by examples that this inclusion is strict. In Figure 7.3 the weighted distance functions from vertices v_i and v_j to another point z within the edge $[u, v]$ are shown. We get two equilibrium points x and y, where only y is a center-bottleneck.

We also define the set of ranges (canonical set of distances) $\mathsf{R} \subseteq \mathbb{R}_+$ by

$$\begin{aligned} \mathsf{R} := \{r \in \mathbb{R}_+ \mid \exists EQ \in EQ_{ij} \, : \, d_i(EQ) = r = d_j(EQ) \quad \textit{or} \\ \exists v_i, v_j \in V, \, v_i \neq v_j \, : \, r = w_i d(v_j, v_i)\}. \end{aligned} \tag{7.5}$$

Ranges correspond to weighted distance values between equilibrium points and nodes.

A point x is called an r-extreme point or pseudo-equilibrium with range $r \in \mathsf{R}$, if there exists a node $v_i \in V$ with $r = w_i d(x, v_i)$.

Finally, let us denote by PEQ the set of all pseudo-equilibria with respect to all ranges $r \in \mathsf{R}$.

$$\begin{aligned} \mathsf{PEQ}(r) &= \{y \in A(G) \mid w_i d(v_i, y) = r, v_i \in V\} \text{ with } r \in \mathsf{R}, \\ \mathsf{PEQ} &= \bigcup_{r \in \mathsf{R}} \mathsf{PEQ}(r). \end{aligned} \tag{7.6}$$

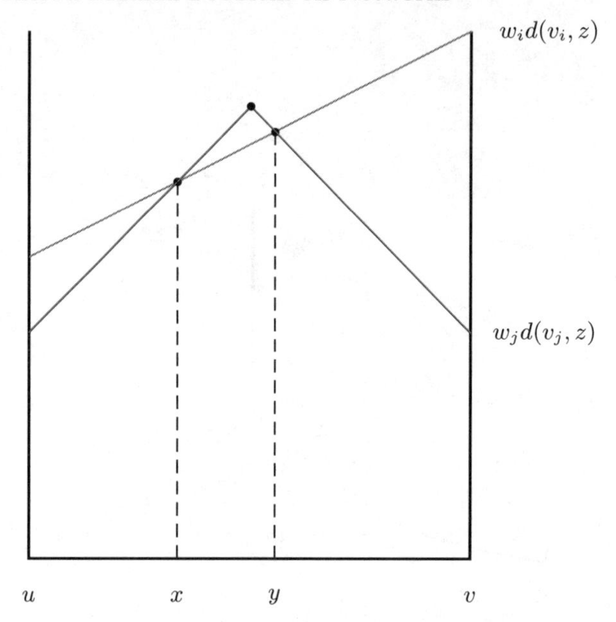

Fig. 7.3. x, y are equilibria but only y is a center-bottleneck

Note that $V \subseteq \mathsf{PEQ}$, which follows directly from the above definition, and also $\mathsf{EQ} \subseteq \mathsf{PEQ}$, since every equilibrium $EQ \in EQ_{ij}$ of two nodes v_i and v_j is an r-extreme point with $r = d_i(EQ) = d_j(EQ)$. The above definition extends the concept of $r-$extreme in Pérez-Brito et al. [162].

All the above introduced sets will be necessary to characterize optimal solution sets for different instances of OMP. We identify finite sets of points where some optimal solution of a problem can always be found. These sets are called *finite dominating sets* (FDS) (see [103]), and they allow us to design algorithms to find exact solutions. Moreover, we can calculate the complexity of these algorithms, which depends strongly on the cardinality of the FDS.

Another interesting peculiarity of the ordered median problem is, that the objective function does not have a unified linear representation on the whole space we are working in. The representation may change in the neighborhood of x when $w_i d(x, v_i) = w_j d(x, v_j)$ for some $i, j \in \{1, \ldots, M\}$. It is well-known that the set of points at the same distance of two given points v_i, v_j belongs to the bisector between them.

Bisectors play a crucial role to develop the theory of the ordered median problems in networks, because they will be always related to the solution set.

Following we identify these bisectors, and we describe how to compute them efficiently.

Definition 7.1 *The set* $\mathsf{Bisec}(v_i, v_j) = EQ'_{ij}$ *consisting of points* $\{x : w_i d(x, v_i) = w_j d(x, v_j), i \neq j\}$ *is called bisector of* v_i *and* v_j *with respect to* d.

It is obvious from the definition, that the set EQ_{ij} (the equilibrium points of v_i and v_j) is included in the bisector of v_i and v_j with respect to d. Note as well that EQ is different from the union of the sets BN_{ij}, of the center-bottleneck points on the edge $[v_i, v_j]$, introduced by Kariv and Hakimi [118].

Let us recall, that in the planar case bisectors are curves (or lines). On the contrary, in our case, bisectors can take many different shapes. They can be connected subedges, as well as unions of a disconnected group of subedges or even isolated points.

For algorithmic purposes one should note that the set EQ can be computed by intersecting all distance functions on all edges. Since a distance function has maximally one breakpoint on every edge, we can use a sweep-line technique to determine EQ on one edge in $O((M + k)\log M)$, where $k \leq M^2$ is the number of intersection points. The sweep-line methods is one of the fastest ways to compute intersections of line segments (see [21]). Therefore, we can compute EQ for the whole network in $O(N(M + k)\log M)$ time. Of course, this is a worst-case bound and the set of candidates can be further reduced by some domination arguments: Take for two candidates x, y the corresponding weighted (and sorted) distance vectors $d_{\leq}(x)$, $d_{\leq}(y)$. If $d_{\leq}(x)$ is in every component strictly smaller than $d_{\leq}(y)$, then there is no positive λ for which $f_\lambda(y) \leq f_\lambda(x)$. This domination argument can be integrated in any sweep-line technique reducing, in most cases, the number of candidates.

Apart from the standard sweep-line technique we can also use a different machinery to generate the equilibrium points. Consider a particular edge $e_i = (v_s, v_t)$. Let x denote a point on e_i (for convenience x will also denote the distance, along e_i, of the point from v_s). For each $v_j \in V$, $d(x, v_j)$ is a piecewise linear concave function with at most one breakpoint (if the maximum of $d(v_j, x)$ is attained at an interior point, this is a bottleneck point). Assuming that all internodal distances have already been computed, it clearly takes constant time to construct $d_j(x)$ and the respective bottleneck point. If $w_j > 0$, $d_j(x)$ is concave and otherwise $d_j(x)$ is convex. To compute all the equilibrium points on e_i, we calculate in $O(M^2)$ total time the solutions to the equations $d_j(x) = d_k(x)$, where $v_j, v_k \in V$, $v_j \neq v_k$. To conclude, in $O(M^2)$ total time we identify the set of $O(M)$ bottleneck points and the set of $O(M^2)$ equilibrium points on the edge e_i.

7.3 General Properties

In this section we collect some geometrical properties of the OMf and OMP that will be used later when analyzing the OMP in networks.

We start by proving the NP-hardness of the multifacility version of the problem on general networks.

Theorem 7.1 *The p-facility ordered median problem is NP-hard.*

Proof.
Taking the λ weights all equal to one, the p-facility ordered median problem reduces to the p-facility median problem for which Kariv and Hakimi [119] proved NP-hardness. Hence, the result follows. □

In the following we will address the interesting question of how the optimal value of the ordered median problem behaves as a function of the parameters λ. To do this, we consider a closed, convex set Ω and allow any λ to vary in Ω.

Lemma 7.1 *The function* $\mathsf{OM}(\lambda)$ *is a continuous, concave function of* $\lambda \in \Omega$. *Moreover, if the set Cand of candidates to be optimal solutions is finite then* $\mathsf{OM}(\lambda)$ *is piecewise linear.*

Proof.
One has

$$\mathsf{OM}(\lambda) = \min_{x \in Cand} \sum_{i=1}^{M} \lambda_i d_{(i)}(x).$$

Let us denote by σ_x the permutation that gives the order in the weighted distance vector for the point x. Then, we have

$$\mathsf{OM}(\lambda) = \min_{x \in Cand} \sum_{i=1}^{M} \lambda_i w_{\sigma_x(i)} d(v_{\sigma_x(i)}, x).$$

Therefore, $\mathsf{OM}(\lambda)$ is the pointwise minimum of linear functions. Hence, it is concave. Moreover, if *Cand* is finite, $\mathsf{OM}(\lambda)$ is the minimum of a finite number of linear functions and then the function is piecewise linear and concave. □

A straightforward consequence of this result is, that $\mathsf{OM}(\lambda)$ always achieves its minimum value at some extreme points of Ω.

In Chapter 4, we proved that when $w_i \geq 0$, for $i = 1, ..., M$, and the λ-vector satisfies $0 \leq \lambda_1 \leq \lambda_2 \leq \ldots \leq \lambda_M$, the ordered median function is convex on the plane with respect to any metric generated by norms. Next we present an analogous result for trees using the convexity concept by Dearing et al. [52].

Lemma 7.2 *Let* $T = (V, E)$ *be a tree network and* $w_i \geq 0$, $i = 1, \ldots, M$. *If* $0 \leq \lambda_1 \leq \lambda_2 \leq \ldots \leq \lambda_M$ *then the function* $f_\lambda(x)$ *is convex on* T.

Proof.
Let λ satisfy $0 \leq \lambda_1 \leq \lambda_2 \leq \ldots \leq \lambda_M$. By [52], $w_i d(x, v_i)$ is convex for $x \in A(T)$ and $i = 1, \ldots, M$. Let $\sigma \in \mathcal{P}(1 \ldots M)$ be a fixed permutation of the

set $\{1,2,\ldots,M\}$ and $x \in A(T)$. The function $\sum_{i=1}^{n} \lambda_i w_{\sigma(i)} d(x, v_{\sigma(i)})$ is convex. Therefore,

$$g(x) = \max_{\tau \in \mathcal{P}(1\ldots M)} \left\{ \sum_{i=1}^{M} \lambda_i w_{\tau(i)} d(x, v_{\tau(i)}) \right\},$$

is also convex as maximum of convex functions.

Since the λ_i's are non-decreasing, the permutation $\sigma \in \mathcal{P}(1\ldots M)$, which sorts the weighted distance functions $w_i d(x, v_i)$ in the vector $d_{\leq}(x)$ for a given $x \in A(T)$, is identical to the permutation τ^*, which defines $g(x)$ for this x (see e.g. Theorem 368 in Hardy et al. [98]). Therefore, we obtain

$$f_\lambda(x) = \max_{\tau \in \mathcal{P}(1\ldots M)} \left\{ \sum_{i=1}^{M} \lambda_i w_{\tau(i)} d(x, v_{\tau(i)}) \right\},$$

and hence the desired result follows. □

The above result can be partially extended to the p-facility OMP. Let $\mathsf{OM}_p(\lambda)$ denote the optimal objective value of the p-facility ordered median problem. Clearly, for $p = 1,\ldots,M$, $\mathsf{OM}_p(\lambda)$ is a monotone function of p with $\mathsf{OM}_M(\lambda) = 0$. We know that $\mathsf{OM}_p(\lambda)$ may not be convex even for metric spaces induced by tree networks (examples can be found in [79]). We will try now to derive some partial convexity results by relating the objective value $\mathsf{OM}_p(\lambda)$ for two distinct values of p. The following lemma is needed.

Lemma 7.3 *Let $c_1 \geq \cdots \geq c_m$ be a sequence of real numbers, and let $b_1 \geq \cdots \geq b_m$ be a sequence of m nonnegative real numbers. For $j = 1,\ldots,m-1$,*

$$(b_1 + \cdots + b_m)(b_1 c_{j+1} + \cdots b_{m-j} c_m) \leq (b_1 + \cdots b_{m-j})(b_1 c_1 + b_2 c_2 + \cdots + b_m c_m).$$

The proof of this lemma can be found in Francis et al. [79]. Based on this result we can now state some interesting consequences also taken from [79].

Theorem 7.2 *Let $\lambda = (\lambda_1,\ldots,\lambda_M)$ be a nonnegative vector satisfying $\lambda_1 \geq \cdots \geq \lambda_M$. For $p = 2,\ldots,M$, suppose that the solution to the p-facility ordered median problem is attained by setting p new facilities at $X^* = \{x_1, x_2, \ldots, x_p\}$, where exactly t new facilities $0 \leq t \leq p$, are in $A(G) \setminus V$. Then, for any q, $1 \leq p < q < M$,*

$$\mathsf{OM}_q(\lambda) \left(\sum_{i=1}^{M-p+t} \lambda_i \right) \leq \mathsf{OM}_p(\lambda) \left(\sum_{i=1}^{M-q+t} \lambda_i \right).$$

Proof.
Without loss of generality, assume that $\{x_1,\ldots,x_t\} \subset A(G) \setminus V$, $(x_{t+1},\ldots,x_p) = (v_{M-(p-t)+1},\ldots,v_M)$ and

$$w_1 d(X^*, v_1) \geq w_2 d(X^*, v_2) \geq \cdots \geq w_M d(X^*, v_M)$$

note that $w_{M-(p-t)+1} d(X^*, v_{M-(p-t)+1}) = \cdots = w_M d(X^*, v_M) = 0$. Consider now a feasible solution to the ordered q-median problem where new facilities are located at each point of the subset $X' = \{v_1, \ldots, v_{q-p}, x_1, \ldots, x_p\}$. (We augment $q - p$ new facilities to X^* and set them at $\{v_1, \ldots, v_{q-p}\}$.) Using Lemma 7.3 with $m = M - (p - t)$, $j = q - p$, $(c_1, \ldots, c_m) = (w_1 d(X^*, v_1), \ldots, w_m d(X^*, v_m))$, and $\lambda_i = b_i$ for $i = 1, \ldots, m$, we have

$$\begin{aligned}
&(\lambda_1 + \cdots \lambda_m)(w_{q-p+1} d(X^*, v_{q-p+1})\lambda_1 + \cdots + w_m d(X^*, v_m)\lambda_{m-j}) \leq \\
\leq \quad &(\lambda_1 + \cdots + \lambda_{m-j})(w_1 d(X^*, v_1)\lambda_1 + \cdots + w_m d(X^*, v_m)\lambda_m) \quad = \\
= \quad &\mathsf{OM}_p(\lambda)(\lambda_1 + \cdots \lambda_{m-j}).
\end{aligned}$$

By definition, the last inequality is now written as

$$\begin{aligned}
(\lambda_1 + \cdots + \lambda_m) g_\lambda(0, \ldots, 0, w_{q-p+1} d(X^*, v_{q-p+1}), \ldots, w_m d(X^*, v_m), 0, \ldots, 0)& \\
\leq \mathsf{OM}_p(\lambda)(\lambda_1 + \cdots \lambda_{m-j})&\ ,
\end{aligned}$$

where g_λ is such that

$$f_\lambda(X) = g_\lambda(w_1 d(v_1, X), \ldots, w_M d(v_M, X))$$

is the ordered median function. Since $X^* \subset X'$, we obtain

$$w_j d(X', v_j) \leq w_j d(X^*, v_j) \quad j = 1, \ldots, M.$$

Using Theorem 368 in Hardy et al. [98],

$$\begin{aligned}
f_\lambda(X') = g_\lambda(0, \ldots, 0, w_{q-p+1} d(X', v_{q-p+1}), \ldots, w_m d(X', v_m), 0, \ldots, 0) \leq \\
\leq g_\lambda(0, \ldots, 0, w_{q-p+1} d(X^*, v_{q-p+1}), \ldots, w_m d(X^*, v_m), 0, \ldots, 0).
\end{aligned}$$

Therefore, $(\lambda_1 + \cdots + \lambda_m) f_\lambda(X') \leq \mathsf{OM}_p(\lambda)(\lambda_1 + \cdots + \lambda_{m-j})$.
Finally, note that from optimality $\mathsf{OM}_q(\lambda) \leq f_\lambda(X')$. Therefore,

$$(\lambda_1 + \cdots + \lambda_m)\mathsf{OM}_q(\lambda) \leq \mathsf{OM}_p(\lambda)(\lambda_1 + \cdots + \lambda_{m-j}).$$

Substituting $m = M - (p - t)$, $j = (q - p)$, we have

$$\mathsf{OM}_q(\lambda)\left(\sum_{i=1}^{M-p+t} \lambda_i\right) \leq \mathsf{OM}_p(\lambda)\left(\sum_{i=1}^{M-q+t} \lambda_i\right).$$

□

Recall that in the node restricted (discrete) p-facility ordered median problem all new facilities must be in V. Applying the above theorem to this discrete case with $q = p + 1$, we obtain:

$$\frac{\mathsf{OM}_{M-1}(\lambda)}{\left(\sum_{i=1}^{1}\lambda_i\right)} \leq \frac{\mathsf{OM}_{M-2}(\lambda)}{\left(\sum_{i=1}^{2}\lambda_i\right)} \leq \cdots \leq \frac{\mathsf{OM}_1(\lambda)}{\left(\sum_{i=1}^{M-1}\lambda_i\right)}.$$

In particular, for the p-median problem, where $\lambda = \mathbf{1} = (1,\ldots,1)$, for $p = 2,3,\ldots,M-1$, we have

$$\frac{\mathsf{OM}_p(\mathbf{1})}{(M-p)} \leq \frac{\mathsf{OM}_{p-1}(\mathbf{1})}{(M-p+1)}.$$

The following "hub-and-spoke" tree example demonstrates, that the above inequalities are the tightest possible for the discrete ordered median problem.

Example 7.2
Consider a star tree network ("hub-and-spoke tree"), $T = (V,E)$, where $V = (v_1,\ldots,v_M)$ and $E = \{[v_1,v_2],\ldots,[v_1,v_M]\}$. Each edge has a unit length, and $w_i = 1$, $i = 1,\ldots,M$. For any vector λ satisfying the hypothesis of Theorem 7.2, it is easy to verify that $\mathsf{OM}_p(\lambda) = \sum_{i=1}^{M-p}\lambda_i$, for $p = 1,\ldots,M-1$, and $\mathsf{OM}_M(\lambda) = 0$ (another equivalent example is a complete graph with unit lengths and node weights).

More generally, we have the following property for the discrete p-facility ordered median problem:

Proposition 7.1 *Let $\lambda = (\lambda_1,\ldots,\lambda_M)$ be a nonzero, nonnegative vector, satisfying $\lambda_1 \geq \cdots \geq \lambda_M$. Consider the discrete p-facility ordered median problem. For each i, define $c_i = w_i \min_{j\neq i} d(v_i,v_j)$ and let $c = \min_{i=1,\ldots,M} c_i$. Suppose that $M \geq 3$. The following are equivalent:*

1.
$$\mathsf{OM}_{M-1}(\lambda)\left(\sum_{i=1}^{M-1}\lambda_i\right) = \mathsf{OM}_1(\lambda)\lambda_1.$$

2. *For all $p = 1,\ldots,M-1$,*
$$\mathsf{OM}_p(\lambda)\left(\sum_{i=1}^{M-1}\lambda_i\right) = \mathsf{OM}_1(\lambda)\left(\sum_{i=1}^{M-p}\lambda_i\right).$$

3. *There exists a node $v_k \in V$, such that $w_i d(v_k,v_i) = c$, for all $v_i \in V$, $i \neq k$.*

Proof.
We start by showing that (3) implies (2). For all j, $j = 1,\ldots,M-1$, let $\lambda_j^* = \sum_{i=1}^{M-j}\lambda_i$. Indeed, if there is a node v_k with $w_i d(v_k,v_i) = c$ for $i \neq k$, then by establishing new facilities at any subset $X \in V$ which contains v_k, $|X| = p$, the objective value of the corresponding solution will be $c\lambda_p^*$. But since $c\lambda_p^*$ is a lower bound on $\mathsf{OM}_p(\lambda)$, we get $\mathsf{OM}_p(\lambda) = c\lambda_p^*$. Clearly, for

$p = 1$, the new facility is optimally located at v_k, and so $\mathsf{OM}_1(\lambda) = c\lambda_1^*$. Thus, $\mathsf{OM}_p(\lambda)\lambda_1^* = \mathsf{OM}_1(\lambda)\lambda_p^*$.

Clearly, (2) implies (1). We conclude by showing that (1) implies (3).

Note that $\mathsf{OM}_{M-1}(\lambda)$ is attained by establishing new facilities at all nodes but one, say v_t. Moreover, $\mathsf{OM}_{M-1}(\lambda) = \lambda_1 w_t \min_{j\neq t} d(v_j, v_t)$. The optimality of $\mathsf{OM}_{M-1}(\lambda)$ implies that v_t must satisfy $w_t \min_{j\neq t} d(v_j, v_t) = c$, where c is defined above. Thus, $\mathsf{OM}_{M-1}(\lambda) = c\lambda_1$. Using (1) we conclude that

$$\mathsf{OM}_1(\lambda) = c\lambda_1^*.$$

Suppose that $\mathsf{OM}_1(\lambda)$ is attained by establishing a new facility at a node v_k. From the definition $w_i d(v_i, v_k) \geq c$ for each $i = 1, \ldots, M$, $i \neq k$. Therefore, if $\max_{i\neq k} w_i d(v_i, v_k) > c$, we would get that $\mathsf{OM}_1(\lambda) > c\lambda_1^*$, contradicting $\mathsf{OM}_1(\lambda) = c\lambda_1^*$. We conclude that $\max_{i\neq k} w_i d(v_i, v_k) = c$, and, therefore $w_i d(v_i, v_k) = c$ for each $i = 1, \ldots, M$, $i \neq k$. This proves (3). □

To conclude this analysis we remark that for the special case of the discrete p-median problem, the following stronger property holds.

Proposition 7.2 *Consider the discrete p-median problem. For each* $i = 1, \ldots, M$, *define* $c_i = w_i \min_{j\neq i} d(v_i, v_j)$, *and let* $c = \min_{i=1,\ldots M} c_i$. *Suppose that* $M \geq 3$. *The following are equivalent:*

1. *There exists an integer* p', $1 < p' < M$, *such that*

$$OM_{p'}(\mathbf{1}) = OM_1(\mathbf{1})\frac{(M-p')}{(M-1)}.$$

2. *For all* $p = 1, \ldots, M$

$$OM_p(\mathbf{1}) = OM_1(\mathbf{1})\frac{(M-p)}{(M-1)}.$$

3. *There exists a node* $v_k \in V$, *such that* $w_i d(v_k, v_i) = c$, *for all* $v_i \in V$, $i \neq k$.

8

On Finite Dominating Sets for the Ordered Median Problem

8.1 Introduction

Since the seminal paper by Hakimi [84], much of the work related to location problems on networks has been devoted to identify a finite set of points where some optimal solution of a problem must belong to. This set, called *finite dominating set* (FDS), is very useful in a wide range of optimization problems, in order to restrict the number of possible candidates to be an optimal solution.

The goal of this chapter is to identify the set of candidates for optimal solutions to the OMP on networks and to study its structure as well as to show that the properties of this set depend on the number of facilities to be located.

Previous to the paper by Puerto and Rodríguez-Chía [170], it was an open problem whether polynomial size FDS would exist for the general p-facility version of the ordered median problem. In that paper, the authors prove that for the general ordered p-median problem on networks it is not possible to obtain an FDS with polynomial cardinality even on path graphs. In this chapter we describe known FDS for different instances of OMP. Moreover, we include examples that show how those FDS obtained for particular cases are no longer valid even for slightly modified models. An overview of the literature involving characterizations of FDS shows a lot of papers that succeed to find such kind of sets. An excellent paper on this subject, is the one by Hooker et al. [103] that contains characterizations of FDS for a large number of problems of Location Theory. This chapter, following the line of [170], also provides a negative result regarding the polynomiality of the FDS for the general OMP. The reader may notice that the main difference with respect to existence proofs is that there is not a standard line to follow in proving negative results, which essentially rely on counterexamples.

Recently, much effort have been spent to obtain FDS for some instances of the OMP (see [150, 117, 172, 116]). Obtaining an FDS allows the development of different types of algorithms to solve problems by enumerating a (finite) candidate set. The main results in this chapter can be summarized in the following.

- The node set V constitutes an FDS for the ordered p-median problem when $\lambda_1 \geq \ldots \geq \lambda_M \geq 0$. For arbitrary nonnegative λ-weights, we also obtain that $V \cup \mathsf{EQ}$ is an FDS for the single-facility ordered median problem. ([150])
- The set PEQ is an FDS for the ordered p-median problem where the λ-weights are defined as:

$$a = \lambda_1 = \ldots = \lambda_k \neq \lambda_{k+1} = \ldots = \lambda_M = b, \tag{8.1}$$

  for a fixed k, such that, $1 \leq k < n$ [116] (see thc definition of PEQ in (7.6)).
- The set $F = ((V \cup \mathsf{EQ}) \times \mathsf{PEQ}) \cup T \subset A(G) \times A(G)$ contains an optimal solution of the ordered 2-median problem in any network for any choice of nonnegative λ-weights [172] (see definition of T in Theorem 8.6).
- The set $V \cup \mathsf{EQ} \cup \mathsf{NBN}$ is an FDS for the single facility ordered median problem with general node weights (the w-weights can be negative). Moreover, for the case of a directed network with nonnegative w-weights, there is always an optimal solution in the node-set V [117].
- For any integer p there exists a path graph with $2p$ nodes and a nonnegative vector $\lambda \in \mathbb{R}^{2p}$ with the following properties: There is a family of 2^p ordered median problems on the above path, such that the solution to each one of these problems contains a point (facility), which is not in the solution of any other problem in this family. In particular, an FDS for this family is of exponential cardinality [170].

8.2 FDS for the Single Facility Ordered Median Problem

As we have mentioned before, the paper by Nickel and Puerto [150] contains the first results concerning FDS for some instances of the ordered median problem. The first one is an FDS for the single facility ordered median problem, a set that we denote as *Cand*.

Theorem 8.1 *An optimal solution for the single facility ordered median problem with nonnegative λ-weights can always be found in the set Cand := $\mathsf{EQ} \cup V$.*

Proof.
Starting from the original graph G, build a set of new subgraphs $G_1, \ldots, G_K$ by inserting all points of EQ as new nodes. Now every subgraph G_i is defined by either

1. Two consecutive elements of EQ on an edge or
2. An element $v_i \in V \backslash \mathsf{EQ}$ and the adjacent elements of EQ

and the corresponding edges. In this situation for every subgraph G_i the permutation of $d_{\leq}(x)$ is constant (by definition of EQ). Therefore, for all $x \in A(G_i)$ we have

$$\sum_{i=1}^{M} \lambda_i d_{(i)}(x) = \sum_{i=1}^{M} \lambda_i w_{\pi(i)} d(v_{\pi(i)}, x) ,$$

where $\pi \in \mathcal{P}(1 \ldots M)$. Therefore, we can replace the objective function by a classical median-objective. Now we can apply Hakimi's node dominance result in every G_i and the result follows. □

In the rest of this section, we present two extensions of this FDS for the OMP on networks. The first one describes the structure of the FDS for the single facility ordered median problem with positive and negative node weights. The second result refers to the case of directed networks and nonnegative node weights. The interested reader is referred to [117] for further details.

For the first case we have the following result:

Theorem 8.2 *The set $V \cup \mathsf{EQ} \cup \mathsf{NBN}$ is a finite dominating set for the single facility ordered median problem with general node weights.*

Proof.
Let G be an undirected graph. Augment G by inserting the equilibria in EQ and negative bottleneck points from NBN as new nodes. $A(G)$ is now decomposed into sub-edges, where each sub-edge connects two adjacent elements of $V \cup \mathsf{EQ} \cup \mathsf{NBN}$.

From the definition of equilibrium points, it follows that there exists a permutation of the weighted distance functions in $\{d_j(x)\}_{j=1}^{M}$, which is fixed for all the points x on every sub-edge. Therefore, the ordered median function reduces to the classical median function on every sub-edge. Since we included the negative bottleneck points NBN in the decomposition of the network, the distance functions are now piecewise linear and concave on every sub-edge. Therefore, the desired result follows. □

In order to analyze the second case we need to introduce some additional concepts related to directed networks.

Let $G_D = (V, E)$ be a directed network, and let x be a point on the edge $e = [v_i, v_j] \in E$. Following Mirchandani and Francis [142], the distance between a point $x \in A(G_D)$ and $v_i \in V$ is given by $\bar{d}_i(x) = w_i(d(x, v_i) + d(v_i, x))$. Denote by

$$\bar{d}(x) := (\bar{d}_1(x), \ldots, \bar{d}_M(x))$$

the vector of weighted distances from x to the nodes $v_i \in V$ and back. Again, we sort the entries in $\bar{d}(x)$ in non-decreasing order. The resulting vector is denoted by

$$\bar{d}_{\leq}(x) := (\bar{d}_{(1)}(x), \ldots, \bar{d}_{(M)}(x)), \tag{8.2}$$

where $(\cdot)$ is a permutation of $\{1, \ldots, M\}$ such that $\bar{d}_{(1)}(x) \leq \ldots \leq \bar{d}_{(M)}(x)$ holds.

For any given M-dimensional, nonnegative vector $\lambda = (\lambda_1, \ldots, \lambda_M) \in \mathbb{R}^M_{0+}$ the ordered median problem on the directed graph G_D is now defined as:

$$\min_{x \in A(G_D)} f_\lambda(x) := \sum_{i=1}^{M} \lambda_i \bar{d}_{(i)}(x). \tag{8.3}$$

In order to derive an FDS for Problem (8.3), we have to take a closer look at the distance functions. Based on their definitions we get the following lemma.

Lemma 8.1 *Let $G_D = (V, E)$ be a directed network, $e = [v_i, v_j] \in E$ an edge of the network and $x \in A(G_D)$ a point in the interior of this edge. Then for a node $v_k \in V$, the distance function $\bar{d}_k$ is constant on the interior of the edge $[v_i, v_j]$, i.e. $\bar{d}_k(x) = c \in \mathbb{R}$, $\forall x \in [v_i, v_j] \backslash \{v_i, v_j\}$. Moreover, if $w_k \geq 0$ (respectively $w_k < 0$) then $\bar{d}_k(v_i)$, $\bar{d}_k(v_j) \leq \bar{d}_k(x)$ (respectively $\bar{d}_k(v_i)$, $\bar{d}_k(v_j) \geq \bar{d}_k(x)$) for all $x \in [v_i, v_j]$.*

Proof.
Let x be a point in the interior of the edge $e = [v_i, v_j]$ and $v_k \in V$. For $x \in [v_i, v_j]$ it follows that

$$\begin{aligned} \bar{d}_k(x) &= w_k(d(x, v_k) + d(v_k, x)) \\ &= w_k(d(x, v_j) + d(v_j, v_k) + d(v_k, v_i) + d(v_i, x)) \\ &= w_k(d(v_j, v_k) + d(v_k, v_i) + d(v_i, v_j)) = c. \end{aligned}$$

If $w_k \geq 0$ then for v_i we have

$$\begin{aligned} \bar{d}_k(v_i) &= w_k(d(v_i, v_k) + d(v_k, v_i)) \\ &\leq w_k(d(v_i, v_j) + d(v_j, v_k) + d(v_k, v_i)) = \bar{d}_k(x), \end{aligned}$$

where $x \in [v_i, v_j]$, $x \neq v_i, v_j$. Analogously for v_j we obtain that $\bar{d}_k(v_j) \leq \bar{d}_k(x)$. If $w_k < 0$, the reverse inequality holds and the desired result follows. □

From the above lemma it is clear, that there are 4 possibilities for the shape of a distance function $\bar{d}_k$ with $w_k > 0$ on a directed edge $e = [v_i, v_j]$, see also Figure 8.1:

1. $\bar{d}_k$ is constant on e, i.e. $\bar{d}_k(x) = c$, $\forall x \in [v_i, v_j]$.
2. $\bar{d}_k(v_i) < \bar{d}_k(x) = c$, $\forall x \in [v_i, v_j]$ and $x \neq v_i$.
3. $\bar{d}_k(v_j) < \bar{d}_k(x) = c$, $\forall x \in [v_i, v_j]$ and $x \neq v_j$.

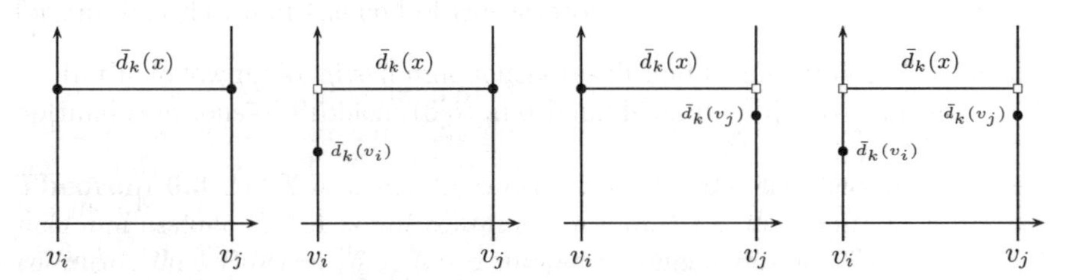

Fig. 8.1. The 4 possible positive distance functions on a directed edge with $w_k > 0$

4. $\bar{d}_k(v_i), \bar{d}_k(v_j) < \bar{d}_k(x) = c$, $\forall x \in [v_i, v_j]$ and $x \neq v_i, v_j$.

In case $w_k < 0$, $d_k(\cdot)$ may have one of the above 4 possible shapes with $<$ replaced by $>$. Based on the previous observations we get, that the objective function $f_\lambda(\cdot)$ is constant on the interior of an edge, since it is a sum of constant functions. Furthermore, if all node weights are nonnegative, we get a possibly smaller objective value only at the endpoints as shown by the following theorem.

Theorem 8.3 *The ordered median problem on directed networks with nonnegative node weights always has an optimal solution in the node-set V. If in addition $\lambda_1 > 0$ and $w_i > 0$, $\forall i = 1, \ldots, M$, then any optimal solution is in V.*

Proof.
Let $e = [v_i, v_j] \in E$ and let x be an interior point of the edge $[v_i, v_j]$. It is sufficient to show that

$$\max\{f_\lambda(v_i),\, f_\lambda(v_j)\} \leq f_\lambda(x) = \sum_{k=1}^{M} \lambda_k \bar{d}_{(k)}(x)\,. \tag{8.4}$$

Without loss of generality we prove $f_\lambda(v_i) \leq f_\lambda(x)$. Let us consider v_i and let $v_k \in V$ be an arbitrary node. From the observation above we know that $\bar{d}_k(v_i) \leq \bar{d}_k(x)$, $k = 1, \ldots, M$. Therefore, $\bar{d}(v_i) \leq \bar{d}(x)$, and then also $\bar{d}_{\leq}(v_i) \leq \bar{d}_{\leq}(x)$.

Equation (8.4) follows by taking the scalar product with λ.

Next, suppose that $\lambda_1 > 0$, and $w_j > 0$, for $j = 1, ..., M$. Therefore, we obtain $0 \leq \bar{d}_{(j)}(v_i) \leq \bar{d}_{(j)}(x)$, for $j = 1, ..., M$, and $0 = \bar{d}_{(1)}(v_i) < \bar{d}_{(1)}(x)$. Hence, since $\lambda_1 > 0$, we get $f_\lambda(v_i) < f_\lambda(x)$, and the result follows. □

8.3 Polynomial Size FDS for the Multifacility Ordered Median Problem

This section is devoted to characterize finite dominating sets for some particular cases of the multifacility OMP. We obtain three different types of results. First, we establish a generalization of the well-known Theorem of Hakimi for median problems, which states that there always exists an optimal solution for the OMP with $\lambda_1 \geq \lambda_2 \geq \ldots \geq \lambda_M \geq 0$ in V. Then, we consider the case where the set of λ parameters only takes two different values. Finally, we also characterize an FDS for the two-facility case without any hypothesis on the structure of the λ weights (see [172]).

Let us start considering the problem of finding an FDS for the $p-$facility case when $\lambda_1 \geq \lambda_2 \geq \ldots \geq \lambda_M \geq 0$. The result is the following.

Theorem 8.4 *The ordered p-median problem with $\lambda_1 \geq \lambda_2 \geq \ldots \geq \lambda_M \geq 0$ has always an optimal solution X_p^* contained in V.*

Proof.
Since by hypothesis $\lambda_1 \geq \lambda_2 \geq \ldots \geq \lambda_M \geq 0$ we have (see [98]) that

$$f_\lambda(X_p) = \sum_{i=1}^{M} \lambda_i d_{(i)}(X_p) = \min_{\pi \in \mathcal{P}(1\ldots M)} \left\{ \sum_{i=1}^{M} \lambda_i d_{\pi(i)}(X_p) \right\}.$$

Assume that $X_p \not\subset V$.

Then, there must exist $x_i \in X_p$ with $x_i \notin V$. Let $e = [v, w]$ be the edge containing x_i and $l(e)$ its length. Denote by $X_p(s) = X_p \setminus \{x_i\} \cup \{x(s)\}$ where $x(s) := (e, s)$ for $s \in [0, 1]$. Therefore, $d(v, x(s)) = sl(e)$.

The function g defined as $g(s) = \sum_{i=1}^{M} \lambda_i d_{(i)}(X_p(s))$ is concave for all $s \in [0, 1]$ because it is the composition of a concave and a linear function, i.e.

$$g(s) = \min_{\pi \in \mathcal{P}(1\ldots M)} \left\{ \sum \lambda_i d_{\pi(i)}(X_p(s)) \right\}$$

and each

$$d_{\pi(j)}(X_p(s)) = \min \left\{ d(v_{\pi(j)}, x_1), \ldots, d(v_{\pi(j)}, x(s)), \ldots, d(v_{\pi(j)}, x_p) \right\}$$

with

$$d(v_{\pi(j)}, x(s)) = \min \left\{ d(v_{\pi(j)}, v) + sl(e), d(v_{\pi(j)}, w) + (1 - s)l(e) \right\}$$

is concave.

Hence, $g(s) = f_\lambda(X_p(s)) \geq \min\{f_\lambda(X_p(0)), f_\lambda(X_p(1))\}$ and the new solution set $X_p(s)$ contains instead of x_i a vertex of G.

Repeating this scheme a finite number of times the result follows. □

Although different, this result resembles the correction given by Hooker et al. [103] to the dominance result (see Result 7 in [103]) of Weaver and Church [205] for the so called vector assignment N-median model. In that model, V is found to be a FDS for the vector assignment p-median model assuming, that the fractions of time a node v_i is served by the l-th closest facility verify, that the nearer servers are used at least as frequently as the more distant ones. In other words, the fractions are ranked in non-decreasing sequence of the distance to the servers. This is exactly the hypothesis of Theorem 8.4.

8.3.1 An FDS for the Multifacility Ordered Median Problem when $a = \lambda_1 = \ldots = \lambda_k \neq \lambda_{k+1} = \ldots = \lambda_M = b$

We have just seen that for $\lambda_1 \geq \cdots \geq \lambda_M \geq 0$ the node set V constitutes an FDS for the ordered p-median problem and that for arbitrary $\lambda \geq 0$, $V \cup \mathsf{EQ}$ is an FDS for the single facility ordered median problem. We demonstrate by a simple counter-example that this latter dominance result for the single facility case does not hold for the $p-$facility case.

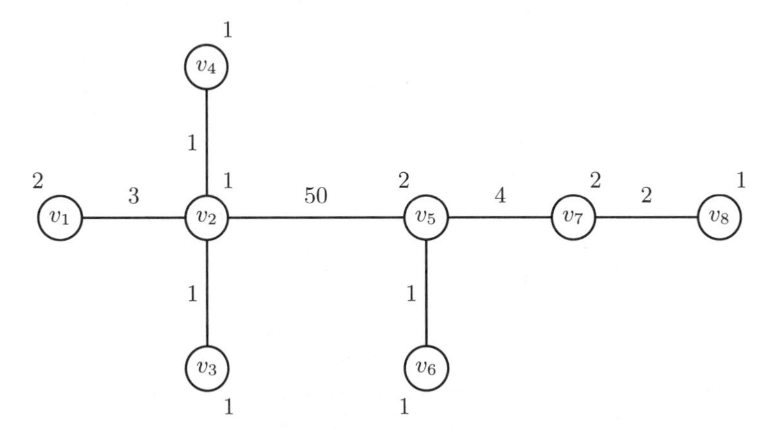

Fig. 8.2. Tree network of Example 8.1

Example 8.1
Consider the tree network of Figure 8.2. Observe, that $V \cup \mathsf{EQ}$ is not an FDS for the ordered 2-median problem with $\lambda = (0.2, 0.2, \ldots, 0.2, 1)$. If we restrict X_2 to be in $V \cup \mathsf{EQ}$, the optimal solution is given by

$$X_2 = \left\{EQ_{13}^{12} = \left([v_1, v_2], \tfrac{4}{9}\right), EQ_{57}^{57} = \left([v_5, v_7], \tfrac{1}{2}\right)\right\} ,$$

with objective value $f_\lambda(X_2) = 8\frac{2}{15} = 8.1\bar{3}$. If we drop this restriction we obtain a better solution, namely

$$X_2^* = \{x^* = ([v_1, v_2], \tfrac{2}{3}),\ EQ_{57}^{57} = ([v_5, v_7], \tfrac{1}{2})\},$$

with an optimal objective function value of 8.0 (see also Figure 8.3). Note that x^* is neither an equilibrium point nor a vertex.

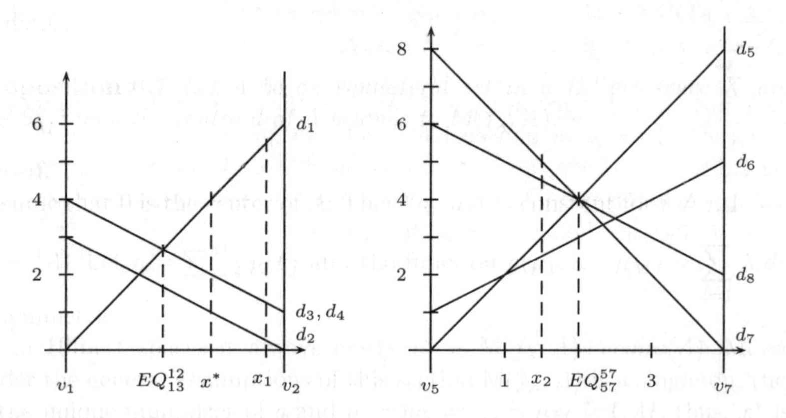

Fig. 8.3. The distance functions along $[v_1, v_2]$ and $[v_5, v_7]$ of Example 8.1

Despite this negative result, we are able to characterize a polynomial size FDS for an important class of ordered p-median problems (see Kalcsics et al. [116]). Let $1 \le k < M$, $\lambda^k = (a, \dots, a, b \dots, b) \in \mathbb{R}_{0+}^M$, where

$$a = \lambda_1 = \dots = \lambda_k \neq \lambda_{k+1} = \dots = \lambda_M = b\,.$$

Note that the $\lambda-$vectors corresponding to the center, centdian or $k-$centrum problem are of this type.

For $a \ge b$ Theorem 8.4 ensures that V is an FDS for the problem. Thus, only the case with $a < b$ has to be further investigated.

Example 8.1 already points out the two main characteristics of a potential FDS for the case $a < b$. First, one of the solution points belongs to the set $V \cup \mathsf{EQ}$: EQ_{57}^{57}. Second, the other solution point, x^*, is related to the one in $V \cup \mathsf{EQ}$. Namely, there exist two nodes v_1 and v_5 allocated to x^* and EQ_{57}^{57}, respectively, such that $d_1(x^*) = d_5(EQ_{57}^{57})$. In general, we will show in the following that there always exists an optimal solution X_p^* such that

- one or more solution points belong to the set $V \cup \mathsf{EQ}$, i.e. $X_p^* \cap (V \cup \mathsf{EQ}) \neq \emptyset$, and

- for every point in $X_p^* \backslash (V \cup \mathsf{EQ})$ there exists another solution point in $X_p^* \cap (V \cup \mathsf{EQ})$ and two nodes allocated to each of the two points such that the weighted distances of these two nodes to their respective solution points are equal.

The important point is not just to prove existence of a finite size FDS for the problem but to identify it. This dominating set can then be used for developing algorithms to solve the problem while the existence itself is of limited use. For the sake of readability the following proof of the FDS is split into a sequence of four results. Moreover, we first give an informal description of the main arguments of the proofs.

Let $X_p = \{x_1, \dots, x_p\} \subseteq A(G)$. We define the following sets:

$$I_l := \{i \in \{1, \dots, M\} \mid d(X_p, v_i) = d(x_l, v_i)\} \setminus \bigcup_{j=0}^{l-1} I_j \,, \quad l = 1, \dots, p,$$

of the indices of nodes which are allocated to x_l, $l = 1, \dots, p$, where $I_0 := \emptyset$. Note that ties are resolved by allocating nodes to solution points with smallest indices. The objective function $f_\lambda(X_p)$ can now be rewritten with respect to I_l as

$$f_\lambda(X_p) = \sum_{i=1}^{M} \lambda_i w_{(i)} d(X_p, v_{(i)}) = \underbrace{\sum_{i \in I_1} \lambda_i w_{(i)} d(x_1, v_{(i)})}_{=: f_1(x_1)} + \dots + \underbrace{\sum_{i \in I_p} \lambda_i w_{(i)} d(x_p, v_{(i)})}_{=: f_p(x_p)} .$$

For a fixed permutation $(\cdot)$ and fixed allocations I_l, the functions f_l, $l \in \{1, \dots, p\}$, are, as a sum of concave functions, also concave in x_l.

Now consider the tree network given in Figure 8.2. Let $X_2 = \{x_1, x_2\}$ with $x_1 = \left([v_1, v_2], \frac{11}{12}\right)$ and $x_2 = \left([v_5, v_7], \frac{3}{8}\right)$, see Figure 8.3. The vector of ordered distance functions for this set of points is $d_\leq(X_2) = (0.25, 1.25, 1.25, 2.5, 3, 4.5, 5, 5.5)$. Using $\lambda = (0.2, 0.2, \dots, 0.2, 1)$ we get $f_\lambda(X_2) = 9.05$.

Starting from X_2 we try to obtain a better solution. From Figure 8.3 observe that fixing x_2 and moving x_1 on its edge a little to the left or right will neither change the order of the distance functions in the vector $d_\leq(X_2)$ nor the allocation of nodes to the solution points. Hence, the permutation $(\cdot)$ and the index sets I_1 and I_2 remain the same and $f_\lambda(\cdot)$ is concave with respect to x_1. As a result we can find a descent direction and obtain a better solution.

The formal argument is as follows. Define for $t \in \mathbb{R}$:

$$x_1(t) := \left([v_1, v_2], \tfrac{11}{12} + t\right) \text{ and}$$

$$X_2(t) := \{x_1(t), x_2\}.$$

Let $d_i(t) := w_i d(X_2(t), v_i)$, $i \in \{1, \ldots, M\}$. The vector of distance functions with respect to t is $d(X_2(t)) = (d_i(t))_{i=1,\ldots,M} = (5.5 + 6t, 0.25 - 3t, 1.25 - 3t, 1.25 - 3t, 3, 2.5, 5, 4.5)$. For $t = 0$ we obtain $d_{\leq}(X_2(t)) = (0.25 - 3t, 1.25 - 3t, 1.25 - 3t, 2.5, 3, 4.5, 5, 5.5 + 6t)$. Now observe, that the order of the distance functions does not change for $-\frac{1}{12} \leq t \leq \frac{1}{12}$. This means that we can move the point $x_1 \to x_1(t)$ by a small amount on its edge (to the left) without disturbing the permutation. Therefore, we can write the objective function as $f_\lambda(X_2) := 9.05 + 4.2t$ which is a concave function for $t \in [-\frac{1}{12}, \frac{1}{12}]$. Moreover, any value of t, $-\frac{1}{12} \leq t < 0$, will yield a lower objective value.

In the above example the order of the distance functions did not change at all for $t \in [-\frac{1}{12}, \frac{1}{12}]$. But obviously, even a change in the ordering of only the first $k - 1$ or last $k + 1$ vertices is not going to be a problem, and we can still argue that the objective function value will not increase. The following lemma addresses the circumstances, under which we can move a point while not increasing the objective function value, and how far we can move it.

Lemma 8.2 *Let $G = (V, E)$ be an undirected network with nonnegative node weights, $X_p = \{x_1, \ldots, x_p\} \subseteq A(G)$, $\tilde{x} = (e, t) \in X_p$ with $e \in E$ and $t \in [0, 1]$ an arbitrary solution point and $\lambda^k = (a, \ldots, a, b, \ldots, b) \in \mathbb{R}^M_{0+}$. Then there exists a point $x' = (e, t')$, $t' \in [0, 1]$, such that $f_\lambda(X'_p) \leq f_\lambda(X_p)$, where $X'_p := X_p \backslash \{\tilde{x}\} \cup \{x'\}$, and either*

$$x' \in V \quad \textbf{or} \quad d_{(k)}(X'_p) = d_{(k+1)}(X'_p) \tag{8.5}$$

holds.

Proof.
Let $X_p = \{x_1, \ldots, x_p\} \subseteq A(G)$ with $x_l = (e_l, s_l)$, $l = 1, \ldots, p$, such that X_p does not satisfy one of the relations in (8.5). W.l.o.g. let $x_1 = \tilde{x} \in A(G)$. Furthermore, we assume that
i) $d_{(n)}(X_p) \neq d_{(n+1)}(X_p)$ for all $n \in \{1, \ldots, M - 1\}$ and
ii) $\nexists v_i \in V : w_i d(v_i, x_1) = w_i d(v_i, x_l)$, i.e. none of the nodes is at the same distance from x_1 as to another solution point $x_l \neq x_1$.

Define for $t : X_p(t) := \{x_1(t), x_2, \ldots, x_p\}$, where $x_1(t) := (e_1, s_1 + t)$. Let $T := [\underline{t}, \overline{t}]$ be an interval with $-s_1 \leq \underline{t} < 0 < \overline{t} \leq 1 - s_1$, such that *i)* and *ii)* hold for $X_p(t)$ for all $t \in T$. This interval exists since *i)* and *ii)* are satisfied for $t = 0$ and all distance functions $d_i(\cdot)$ are continuous on an edge.

Let $v_i := v_{(n)}$, for some $n = 1, \ldots, M$, be allocated to x_1, i.e. $d_{(n)}(X_p) = d_i(x_1)$. Then by the above assumptions on X_p and the definition of T we have $d_{(n)}(X_p(t)) = d_i(x_1(t))$, $\forall t \in T$. For all nodes v_j allocated to $x_l \in X_p$, $x_l \neq x_1$, $d_{(r)}(X_p(t)) = d_j(x_l)$, for some $r \neq n$, is constant. Therefore, $d_{(\cdot)}(X_p(t))$ is either concave or constant with respect to $t \in T$, since $d_i(x_1(t))$ is concave on e_1.

In summary, we have that $d_i(X_p(t)) = d_{(n)}(X_p(t))$ is a concave function for $t \in T$, $n = 1, \ldots, M$. Moreover, since the inequality $d_{(n)}(X_p(t)) < d_{(n+1)}(X_p(t))$, $n = 1, \ldots, M-1$, holds for all $t \in T$, it follows that the order of the distance functions does not change. As a result, $f_\lambda(X_p(t))$ is also concave in the interval T. Assume w.l.o.g. that the objective function is non-increasing for $t < 0$. Hence, we may decrease $\underline{t}$ until either $x_1(t) \in V$ or $d_{(k)}(X_p(t)) = d_{(k+1)}(X_p(t))$.
Now, we prove that the two assumptions made on X_p do not imply any loss of generality:

i) Let $n \in \{1, 2, \ldots, M-1\}$ such that $d_i(X_p) = d_{(n)}(X_p) = d_{(n+1)}(X_p) = d_j(X_p)$, where $i := (n)$ and $j := (n+1)$. Note that $n \neq k$, since otherwise X_p would satisfy (8.5). Hence, the elements $d_i(X_p)$ and $d_j(X_p)$ possibly swap positions in the vector $d_\leq(X_p(t))$, i.e. $d_i(X_p(t)) \leq d_j(X_p(t))$ for $t \leq 0$ and $d_i(X_p(t)) > d_j(X_p(t))$ for $t > 0$. However, both functions $d_i(X_p)$ and $d_j(X_p)$ are still concave with respect to $t \in T$ and both are still multiplied by the same λ-value, a or b (since $n \neq k$). Therefore, this change has no influence on the concavity and the slope of the objective function $f_\lambda(X_p(t))$.

ii) Concerning the re-allocation of nodes, let $v_i \in V$ be a node such that $w_i d(v_i, x_1) = w_i d(v_i, x_l)$ holds for another solution point $x_l \neq x_1$ and x_1 is not the bottleneck point of this node on edge e_1 (otherwise the allocation will not change with respect to $x_1(t)$, $t \in T$). One of the following two cases can occur:

1. v_i is allocated to $x_1(t)$ for $t \leq 0$ and to x_l, $l = 2, \ldots, p$, for $t > 0$. Thus, we have $d_i(X_p(t)) = d_i(x_1(t))$ for $t \leq 0$ on edge e_1 and $d_i(X_p(t)) = d_i(x_l)$, $t > 0$, on edge e_l. In order to be re-allocated, the distance function of v_i on edge e_1, $d_i(x_1(t))$ has to be increasing for $t \leq 0$. After the change of allocations we obtain $d_i(X_p(t)) = d_i(x_l)$ on e_l, which is constant with respect to t (see the two left most edges of Figure 8.4). Thus, $d_i(X_p(t))$ is concave for $t \in T$.

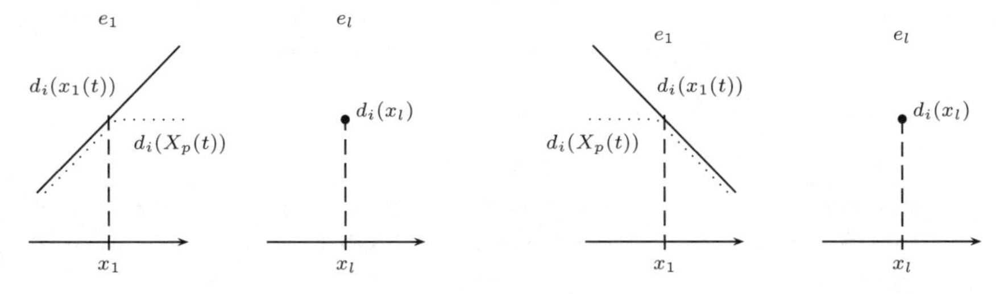

Fig. 8.4. v_i changes its allocation from $x_1(t)$ to x_l respectively from x_l to $x_1(t)$

2. v_i is allocated to x_l for $t \leq 0$ and to $x_1(t)$ for $t > 0$. As above, $d_i(X_p(t))$ is concave for $t \in T$ since it is the minimum of a linear and a constant function (see the two edges on the right-hand side of Figure 8.4).

□

Note that the above result does not hold if one or more node weights are negative. In this case $d_i(X_p(t))$ may be convex with respect to t and hence the ordered median function may no longer be concave in the interval T.

From Lemma 8.2 it follows that we can move an arbitrary solution point on its edge either to the left or to the right without increasing the objective function value until a point is attained for which (8.5) holds. This is illustrated in the following example continuing the discussion preceding Lemma 8.2.

Example 8.2
Consider the network of Example 8.1 and the initial solution $X_2 = \{x_1 = \left([v_1, v_2], \frac{11}{12}\right), x_2 = \left([v_5, v_7], \frac{3}{8}\right)\}$. For $\lambda^7 = (0.2, 0.2, \ldots, 0.2, 1)$ it is possible to improve the objective function value by moving the point x_1 to the left up to the point $\left([v_1, v_2], \frac{10}{12}\right)$. This yields $d_{(7)}(X_2(-\frac{1}{12})) = d_7(x_2) = 5 = d_1(x_1(-\frac{1}{12})) = d_{(8)}(X_2(-\frac{1}{12}))$. Hence, (8.5) holds.

In the previous lemma we could choose a single solution point and move it along its edge until condition (8.5) was fulfilled. Obviously, we can repeat this procedure for all points in X_p. This leads to the following corollary:

Corollary 8.1 *Let $X_p = \{x_1, \ldots, x_p\} \subseteq A(G)$ be a solution to the ordered p-median problem with nonnegative node weights, $p \geq 2$ and $\lambda^k \in \mathbb{R}^M_{0+}$, $1 \leq k \leq M-1$. Then there exists a solution $X_p^{'}$ with $f_\lambda(X_p^{'}) \leq f_\lambda(X_p)$ such that either $X_p^{'} \subseteq V$ or $d_{(k)}(X_p^{'}) = d_{(k+1)}(X_p^{'})$ holds.*

Proof.
Assume that $X_p \not\subseteq V$ and $d_{(k)}(X_p) < d_{(k+1)}(X_p)$. Then, according to Lemma 8.2, we start by moving one solution point after the other until we obtain a new solution $X_p^{'}$, where either all solution points are nodes or finally $d_{(k)}(X_p^{'}) = d_{(k+1)}(X_p^{'})$ holds. □

In the following we will deal with the difficulties which occur when the kth and $(k+1)^{st}$ vertices in the ordered vector of the distance functions have the same value. Resolving these difficulties will lead us to the proof of the FDS.

We first give a formalized description of the set of solution points addressed in the second part of the characterization of our FDS introduced at the beginning of this section. For every point in $X_p^* \backslash (V \cup \mathsf{EQ})$, there exists another solution point in $X_p^* \cap (V \cup \mathsf{EQ})$ and two nodes allocated to each of the two points such that the weighted distances of these nodes to their respective solution points are equal. Let us recall now the definition of the set of ranges R previously defined in (7.5). Ranges correspond to function values of equilibria or node to node distances. In terms of the general characterization of the FDS, let R' be the set of ranges of the points in $X_p^* \cap (V \cup \mathsf{EQ})$. Then, for every other solution point x not in this set there exists a node v_i allocated to x and a range $r \in R'$ such that $w_i d(x, v_i) = r$.

Recall that by PEQ we denote the set of all pseudo-equilibria with respect to all ranges $r \in \mathsf{R}$. Note that $V \subseteq \mathsf{PEQ}$, which follows directly from the above definition, and also $\mathsf{EQ} \subseteq \mathsf{PEQ}$, since every equilibrium $EQ \in EQ_{ij}$ of two nodes v_i and v_j is an r-extreme point with $r = d_i(EQ) = d_j(EQ)$.

Example 8.3

The set of ranges on the edge $[v_1, v_2]$ of the network given in Example 8.1 is $\{0, 1, 2, 2.\bar{6}, 3, 4, 6\}$ (see Figure 8.3). The point $x^* = ([v_1, v_2], \frac{2}{3})$ is a pseudo-equilibrium of range 4.

Our goal is to prove that PEQ is an FDS for the ordered p-median problem with $p \geq 2$ and $\lambda^k \in \mathbb{R}^M_{0+}$, $1 \leq k \leq M-1$. In addition, any optimal solution X_p^* must satisfy $X_p^* \cap (V \cup \mathsf{EQ}) \neq \emptyset$. The first result proves the existence while the latter allows us to identify an FDS for any given problem.

Using Lemma 8.2 we could move an arbitrary solution point along its edge to the left or to the right until we have $d_{(k)}(X_p) = d_{(k+1)}(X_p)$. The goal is to find a method to continue this process without increasing the objective function value. If we are in this situation of equality, the idea is to move two or more solution points simultaneously preserving the relationship $d_{(k)}(X_p) = d_{(k+1)}(X_p)$ and the permutation of the distance functions at the positions k and $k+1$. In Example 8.2 we have $d_{(7)}(X_2(-\frac{1}{12})) = 5 = d_{(8)}(X_2(-\frac{1}{12}))$. Here we can continue moving x_1 to the left if we simultaneously move x_2 to the right in such a way that $d_{(7)}(X_2(\cdot)) = d_{(8)}(X_2(\cdot))$ is preserved.

Before formalizing this approach, we introduce additional notation. Observe that we stop moving a point if for two vertices $v_{(k)}$ and $v_{(k+1)}$ we had $r_k := d_{(k)}(X_p) = d_{(k+1)}(X_p)$. But obviously there may be more than two nodes allocated to solution points whose weighted distance to their respective points is r_k. Therefore, let $X_p = \{x_1, \ldots, x_p\} \subseteq A(G)$ be a solution to the ordered p-median problem with $X_p \not\subseteq V$ and $d_{(k)}(X_p) = d_{(k+1)}(X_p) = r_k$, $r_k \in \mathsf{R}$. Define $\underline{n}, \overline{n}$ as the two indices with $1 \leq \underline{n} \leq k < k+1 \leq \overline{n} \leq M$ such that

$$d_{(\underline{n}-1)}(X_p) < d_{(\underline{n})}(X_p) = \ldots = r_k = \ldots = d_{(\overline{n})}(X_p) < d_{(\overline{n}+1)}(X_p),$$

where $d_{(0)}(X_p) := -\infty$ and $d_{(M+1)}(X_p) := +\infty$. Note that $\overline{n} - \underline{n} \geq 1$, i.e. $\underline{n} < \overline{n}$, by the assumption on X_p.

Furthermore, define $X_L \subseteq X_p$ as the (sub)set of points of X_p such that for every $x_l \in X_L$ there exists a node $v_i \in V$ allocated to x_l with $r_k = d_i(x_l) = d_{(n)}(X_p)$ and $(n) = i$. W.l.o.g. we assume that $X_L := \{x_1, \ldots, x_L\}$ consists of the first L solution points of X_p, $1 \leq L \leq p$. Note that by assumption $L \geq 2$.

We first state a lemma for the special case $\overline{n} - \underline{n} + 1 > L$, which means that there exists a solution point x which has at least two nodes, v_i and v_j, allocated to it with weighted distance r_k. In this case $x \in X_p$ is an equilibrium of these two nodes, yielding $x \in EQ_{ij}$. Therefore, it is possible to prove the FDS using the results of Lemma 8.2.

Lemma 8.3 *Let $X_p = \{x_1, \ldots, x_p\} \subseteq A(G)$ be a solution to the ordered p-median problem with nonnegative node weights, $p \geq 2$ and $\lambda^k \in \mathbb{R}^M_{0+}$, $1 \leq k \leq M-1$. Then there exists a solution $X_p^{'}$ with $f_\lambda(X_p{'}) \leq f_\lambda(X_p)$ such that either $\overline{n} - \underline{n} + 1 = L$ holds for the new solution or $X_p^{'} \subseteq \mathsf{PEQ}$ with $X_p^{'} \cap (V \cup \mathsf{EQ}) \neq \emptyset$.*

Proof.
Let $\overline{X}_p$, $p \geq 2$, be a solution. Given $\overline{X}_p$ we know from Corollary 8.1 that there exists another solution $X_p = \{x_1, \ldots, x_p\} \subseteq A(G)$, $x_l = (e_l, s_l)$, $l = 1, \ldots, p$, with $f_\lambda(X_p) \leq f_\lambda(\overline{X}_p)$ such that $X_p \subseteq V$ or $d_{(k)}(X_p) = d_{(k+1)}(X_p) =: r_k$. Note that if the former case holds the desired result follows.
Let $V_X := X_p \cap V$ and define R_X as the set of ranges of the nodes in V_X, i.e.

$$\mathsf{R}_X := \begin{cases} \{r \in \mathbb{R} \mid \exists v_i \in V, \exists v_j \in V_X, v_i \neq v_j : r = w_i d(v_i, v_j)\} & \text{if } V_X \neq \emptyset \\ \emptyset & \text{otherwise} . \end{cases}$$

Now for X_p let $L, X_L, \underline{n}$ and $\overline{n}$ be defined as above and assume that $\overline{n} - \underline{n} + 1 > L$.

Then, there exist $x_l \in X_L$ and $v_i, v_j \in V$, $v_i \neq v_j$, such that $d_{(n_1)}(X_p) = d_i(x_l) = r_k = d_j(x_l) = d_{(n_2)}(X_p)$, where $(n_1) = i, (n_2) = j$, and v_i and v_j are both allocated to x_l (with respect to X_p). Thus, $x_l \in EQ_{ij}$ is an equilibrium of the two nodes v_i and v_j with range r_k. As a result, $r_k \in \mathsf{R}$ and all the points of the set X_L are r_k-extreme points. Let $x_m \in X_p \backslash (X_L \cup V_X)$. Using the same arguments as in Lemma 8.2, we can fix all other solution points and just move $x_m \to x_m(t)$ on its edge until $x_m(t)$ is a node or $d_i(x_m(t)) =: r \in (\{r_k\} \cup \mathsf{R}_X) \subseteq \mathsf{R}$ for some node v_i allocated to $x_m(t)$. In this case, $x_m(t)$ is a pseudo-equilibrium with range r. This procedure can be applied to all solution points not belonging to $X_L \cup V$.

It is also obvious that for those solution points $x_l \in V_X \cap X_L$, the above procedure can be applied. Therefore, if $\overline{n} - \underline{n} + 1 > L$ we have $X_p \subseteq \mathsf{PEQ}$ and $X_p \cap (V \cup \mathsf{EQ}) \neq \emptyset$ and the desired result follows. □

Lemma 8.3 characterizes an FDS for the ordered p-median problem with $\lambda^k = (a, \ldots, a, b \ldots, b)$ except for $\overline{n} - \underline{n} + 1 = L$. Dealing with this case will finally complete the identification of the FDS. Here we really have to move solution points simultaneously in order to find a non-ascent direction for the objective function.

Theorem 8.5 *The ordered p-median problem with nonnegative node weights,* $p \geq 2$ *and* $\lambda^k \in \mathbb{R}^M_{0+}$, $1 \leq k \leq M-1$, *always has an optimal solution* $X_p^* \subseteq A(G)$ *in the set* PEQ. *Moreover,* $X_p^* \cap (V \cup \mathsf{EQ}) \neq \emptyset$.

Proof.
Let $\overline{X}_p$, $p \geq 2$, be an optimal solution. We know from Lemma 8.3 that there exists another optimal solution $X_p = \{x_1, \ldots, x_p\} \subseteq A(G)$, $x_l = (e_l, s_l)$, $l = 1, \ldots, p$, with $f_\lambda(X_p) = f_\lambda(\overline{X}_p)$ such that either $\overline{n} - \underline{n} + 1 = L$ holds for the new solution or $X_p \subseteq \mathsf{PEQ}$ with $X_p \cap (V \cup \mathsf{EQ}) \neq \emptyset$. Note that if the latter case holds, the desired result follows.
Consider for X_p the elements V_X, R_X, $L, X_L, \underline{n}$ and $\overline{n}$ as defined above. Now we analyze the case $\overline{n} - \underline{n} + 1 = L$. Observe that $V_X \cap X_L = \emptyset$ (see the proof of Lemma 8.3). Thus, for every $x_l \in X_L$ there exists a unique $v_{i_l} \in V$ allocated to x_l with $d_{i_l}(x_l) = r_k$. First, we assume that:
i) $d_{(n)}(X_p) \neq d_{(n+1)}(X_p)$ for all $n \in \{1, \ldots, M\} \setminus (\{M\} \cup \{\underline{n}, \ldots, \overline{n}\})$,
ii) none of the solution points $x_l \in X_L$ is a bottleneck point of some node $v_i \in V$ and
iii) $\nexists v_i \in V : w_i d(v_i, x_{l_1}) = w_i d(v_i, x_{l_2})$, i.e. no node is at the same distance from two solution points $x_{l_1}, x_{l_2} \in X_p$.

Define for $t \in \mathbb{R}$: $X_p(t) := \{x_1(\frac{t}{\nabla_1}), \ldots, x_L(\frac{t}{\nabla_L}), x_{L+1}, \ldots, x_p\}$, where $\nabla_l := \pm w_{i_l} l(e_l)$ is the slope of the distance function $d_{i_l}(\cdot)$ of node v_{i_l} at the point x_l on edge e_l, and $x_l(\frac{t}{\nabla_l}) := (e_l, s_l + \frac{t}{\nabla_l})$, $l = 1, \ldots, L$. Note that the distance functions $d_{i_l}(\cdot)$ are all linear in a sufficiently small interval around the points x_l. Otherwise x_l would be a bottleneck point of node v_{i_l}.

Let $T := [\underline{t}, \overline{t}] \in \mathbb{R}$ be an interval with $\underline{t} < 0 < \overline{t}$ and $s_l + \frac{t}{\nabla_l} \in (0,1)$, $\forall l \in \{1, \ldots, L\}$, $t \in T$, such that *i)* and *iii)* hold for $X_p(t)$ and *ii)* for $x_l(\frac{t}{\nabla_t})$, $1 \leq l \leq L$, for all $t \in T$. This interval exists since *i)*, *ii)* and *iii)* hold for $t = 0$ and all distance functions $d_i(\cdot) = w_i d(\cdot, v_i)$ are continuous on any edge.

Let $n \notin \{\underline{n}, \ldots, \overline{n}\}$ and $v_i = v_{(n)}$. By the above assumptions on X_p and the definition of T, we have that $d_{(n)}(X_p(t)) = d_i(x_l(\frac{t}{\nabla_l}))$, for all $t \in T$, if v_i is allocated to $x_l \in X_L$ and $d_{(n)}(X_p(t)) = d_i(x_l)$ if v_i is allocated to a solution point in $X_p \backslash X_L$. In both cases $d_{(n)}(X_p(t))$ is linear with respect to $t \in T$.

On the other hand, let $n_1, n_2 \in \{\underline{n}, \ldots, \overline{n}\}$, $n_1 \neq n_2$, and let node $v_{(n_1)} =: v_i$ and node $v_{(n_2)} =: v_j$ be allocated to x_{l_1} and x_{l_2}, respectively. By definition of T we have that $x_{l_1}, x_{l_2} \in X_L$. Note that $x_{l_1} \neq x_{l_2}$, since $\overline{n} - \underline{n} + 1 = L$. Thus,

$$\begin{aligned} d_i(x_{l_1}(\frac{t}{\nabla_{l_1}})) &= d_i(x_{l_1}) + sgn(\nabla_{l_1}) w_i \frac{tl(e_{l_1})}{\nabla_{l_1}} = d_{(n_1)}(X_p) + t = r_k + t \\ &= d_{(n_2)}(X_p) + t = d_j(x_{l_2}) + sgn(\nabla_{l_2}) w_j \frac{tl(e_{l_2})}{\nabla_{l_2}} \\ &= d_j(x_{l_2}(\frac{t}{\nabla_{l_2}})) \quad \text{for all } t \in T, \end{aligned}$$

where $sgn(x)$ is the sign function of x. Hence, $d_i(x_{l_1}(\frac{t}{\nabla_{l_1}})) = d_{(n_1)}(X_p(t)) = d_{(n_2)}(X_p(t)) = d_j(x_{l_2}(\frac{t}{\nabla_{l_2}}))$, $\forall t \in T$, and we are increasing or decreasing r_k by $|t|$. This means that we are simultaneously moving each solution point $x_l \in X_L$ on e_l by $x_l \to x_l(\frac{t}{\nabla_t})$, while preserving the relationship $d_{(\underline{n})}(X_p(t)) = \ldots = d_{(\overline{n})}(X_p(t))$. See Figure 8.5 for an example with $\underline{n} = k$ and $\overline{n} = k+1$.

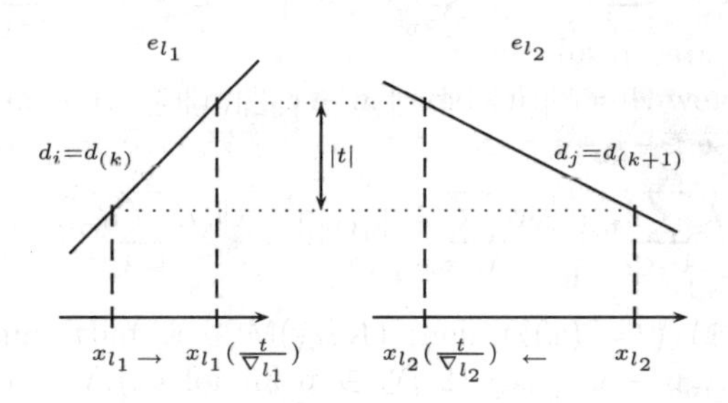

Fig. 8.5. Simultaneous movement of x_{l_1} and x_{l_2} on e_{l_1} and e_{l_2}

As a result, all entries $d_{(n)}(X_p(t))$ are linear with respect to $t \in T$. This fact together with the assumption that $d_{(n)}(X_p(t)) \neq d_{(n+1)}(X_p(t))$, $\forall t \in T$, $n \in \{1, \ldots, M\} \backslash (\{M\} \cup \{\underline{n}, \ldots, \overline{n}\})$, implies that the objective function $f_\lambda(X_p(t))$ is also linear with respect to $t \in T$ and hence, constant over T, since $X_p = X_p(0)$ is already optimal. Consequently, any $X_p(t)$ with $t \in T$ is also optimal.

In summary, the objective function $f_\lambda(X_p(t))$ is constant over the interval T, and we can either decrease $\underline{t}$ or increase $\overline{t}$ by an arbitrarily small value without changing the objective function value $f_\lambda(X_p(t))$. Assume w.l.o.g. that we increase $\overline{t}$. Then, one of the following two cases can occur:

$\mathbf{r_k} + \overline{\mathbf{t}} \in \mathsf{R}_\mathbf{X}$. Then, either $x_l(\frac{\overline{t}}{\nabla_l}) = v_i \in V$ is a node for some $l \in \{1, \ldots, L\}$ or $x_l(\frac{\overline{t}}{\nabla_l})$ is an equilibrium EQ_{ij} of two nodes v_i and v_j which are both allocated to $x_l(\frac{\overline{t}}{\nabla_l})$ such that $d_{(n_1)}(X_p(\overline{t})) = d_i(x_l(\frac{\overline{t}}{\nabla_l})) = r_k + \overline{t} = d_j(x_l(\frac{\overline{t}}{\nabla_l})) = d_{(n_2)}(X_p(\overline{t}))$, where $v_i = v_{(n_1)}$ and $v_j = v_{(n_2)}$. Then all the remaining points $x_l(\frac{\overline{t}}{\nabla_l})$, $l \in \{1, \ldots, L\}$, are pseudo-equilibria with range $r_k + \overline{t}$. In the latter case we extend R_X by the ranges of v_i. Now we can again, as already described above, move the remaining solution points $x_m \in X_p \backslash X_L$ independently from each other until we obtain a new optimal solution $X_p^* \in \mathsf{PEQ}$ such that all solution points are pseudo-equilibria with respect to a range of one of the points in $X_p^* \cap (V \cup \mathsf{EQ})$.

$\mathbf{r_k} + \overline{\mathbf{t}} \notin \mathsf{R}_\mathbf{X}$. In this case it must exist a solution point $x_l \in X_p \backslash X_L$ together with a node v_{i_l} allocated to x_l (with respect to $X_p(\overline{t})$) such that $d_{(n)}(X_p(\bar{t})) = d_{i_l}(x_l) = r_k + \bar{t}$, where $v_{i_l} = v_{(n)}$. We redefine $X_L := X_L \cup \{x_l\}$ and also $X_p(\bar{t}) := \{x_1(\frac{t}{\nabla_1}), \dots, x_{L+1}(\frac{t}{\nabla_{L+1}}), x_{L+2}, \dots, x_p\}$, where w.l.o.g. $l = L+1$. Then, we can apply the same argument as above in order to move the $L+1$ solution points simultaneously.

Now we show that the assumptions *i*) - *iii*) previously made for X_p do not imply any loss of generality.

i) As in Lemma 8.2, a possible swap of the elements in the vector $d_{\leq}(X_p)$ has no influence on the slope of the objective function $f_\lambda(X_p(t))$.

ii) If $x_l \in X_L$ would be a bottleneck point of some node v_i allocated to x_l, then the distance function of this node $d_i(x_l(\frac{t}{\nabla_l})) = d_{(n)}(X_p(t)), v_i = v_{(n)}$, would be concave with respect to t and therefore also $f_\lambda(X_p(t))$, i.e. we could find a descent direction for t, which contradicts the assumption that X_p is optimal.

iii) Let $v_i \in V$ be a node such that $w_i d(v_i, x_{l_1}) = w_i d(v_i, x_{l_2})$, i.e. v_i possibly changes its allocation between the solution points x_{l_1} and x_{l_2} for $t < 0$ or $t > 0$, where one or both points are in X_L, w.l.o.g. $x_{l_1} \in X_L$ (otherwise, x_{l_1} and x_{l_2} are fixed with respect to $t \in T$). But, similar to Lemma 8.2, a re-allocation of $v_i \in V$ from, w.l.o.g., the solution point on e_{l_1} to the solution point on edge e_{l_2} can only occur if the distance function of node v_i on e_{l_1}, that is $d_i(x_{l_1}(\frac{t}{\nabla_1}))$, has, with respect to t, a greater slope than $d_i(x_2(\frac{t}{\nabla_2}))$ on edge e_2. Hence, the function $d_{(n)}(X_p(t)) = d_i(X_p(t))$, $v_{(n)} = v_i$, is concave over T for some $n = 1, \dots, M$, which leads again to a contradiction to the optimality of X_p. □

Since PEQ is an FDS for the p-facility OMP with $\lambda = (a, \dots, a, b, \dots, b)$, $a < b$, a natural question that arises now refers to the number of elements contained in the set PEQ. Taking $K = |\mathsf{EQ}|$ and $M = |V|$, we obtain a range r for every equilibrium and every pair of nodes $u, v \in V$, $u \neq v$, yielding in total $|\mathsf{R}| = O(K + M^2)$ ranges. Since every distance function $d_i(\cdot)$ can assume a value $r \in \mathsf{R}$ in at most two points on an edge $e \in E$, we have $O(NM)$ r-extreme points and as a result $|\mathsf{PEQ}| = O(NM(K + M^2))$.

8.3.2 An FDS for the Ordered 2-Median Problem with General Nonnegative λ-Weights

The previous section gives an FDS for the p-facility ordered median problem with a special structure on the λ-weights. Nevertheless, determining an FDS without additional hypothesis on the vector λ was not addressed. In this section, following the presentation in Puerto et al. [172], we study partially this problem identifying a finite set of candidates to be optimal solutions of the 2-facility ordered median problem without additional hypothesis on the structure of λ.

Before starting to prove the main result in this section, we define several sets which will be used later.

Let $X_p = (x_1, \ldots, x_p) \in A(G) \times \ldots \times A(G)$ and $x_k \in [v_{i_k}, v_{j_k}]$, for any $k = 1, \ldots, M$, we define the following sets:

$$\begin{aligned}
U_k(X_p) &= \{v \in V \, : \, d(v, X_p) = d(v, x_k)\}, \\
U_k^{=}(X_p) &= \{v \in U_k(X_p) : d(v, x_k) = d(v, x_m) \\
&= \min_{i=1,\ldots,p} d(v, x_i), \text{ for some } m \neq k\}, \\
U_k^{<}(X_p) &= U_k(X_p) \setminus U_k^{=}(X_p), \\
\overline{U}_{i_k} &= \{v \in U_k(X_p) : d(v, x_k) \\
&= l(x_k, v_{i_k}) + d(v_{i_k}, v) \leq l(x_k, v_{j_k}) + d(v_{j_k}, v)\}, \\
U_{i_k} &= \overline{U}_{i_k} \setminus (U_k^{=}(X_p) \cup \overline{U}_{j_k}).
\end{aligned}$$

Remark 8.1

- $U_k(X_p)$ is the set of nodes whose demand can be covered optimally by x_k, that is, the set of nodes that can be allocated to x_k.
- $U_k^{=}(X_p)$ is the set of nodes that can be allocated either to x_k or to x_m for some $m \neq k$.
- $U_k^{<}(X_p)$ is the set of nodes allocated to x_k that cannot be allocated to a different service facility.
- $\overline{U}_{i_k}$ is the set of nodes which can be served optimally by x_k through v_{i_k}.
- U_{i_k} is the set of nodes included in $U_k^{<}(X_p)$, such that, their corresponding distances to their service, x_k, increase when x_k is displaced towards v_{j_k}.
- Notice that based on their definitions, then $U_{i_k} \subseteq \overline{U}_{i_k}$ and $U_{i_k} \cap \overline{U}_{j_k} = \emptyset$. Moreover, U_{i_k}, $\overline{U}_{j_k}$ and $(U_k^{=}(X_p) \cap \overline{U}_{i_k}) \setminus \overline{U}_{j_k}$ constitutes a partition of the set $U_k(X_p)$, i.e., these sets are pairwise disjoint and their union is $U_k(X_p)$.

We say, that there exist ties in the vector of weighted distances between the services $x_1, \ldots, x_p$ and their demand nodes if there exist $v_k, v_l \in V$ such that $w_k d(v_k, X_p) = w_l d(v_l, X_p)$.

The nodes of $U_k^{=}(X_p)$ can be allocated to x_k and to some other element of the vector X_p. However, if x_k is moved a small enough amount ξ either towards v_{j_k} or towards v_{i_k}, then some of these nodes will not be still assigned to x_k and some of them will be assigned only to x_k (since the existing tie is destroyed). Let $x_k(\xi)$ denote the position of x_k after it has been moved an amount ξ towards v_{j_k}; i.e., if $x_k = ([v_{i_k}, v_{j_k}], t)$ then $x_k(\xi) = ([v_{i_k}, v_{j_k}], t + \xi)$.

Lemma 8.4 *If a point $x_k \in (v_{i_k}, v_{j_k})$ is moved an amount ξ towards v_{j_k} the contribution to the slope of $f_\lambda(X_p)$ is $\left(\sum_{v_l \in U_{i_k}} w_l \lambda_{\sigma(l)} - \sum_{v_l \in \overline{U}_{j_k}} w_l \lambda_{\sigma(l)}\right)$ provided that no ties exist in the vector of weighted distances and being $\sigma(\cdot)$ the permutation of the weighted distances defined in (7.1).*

Proof.
Since we have assumed that there are no ties in the vector of distances, the weights $\lambda_1, \ldots, \lambda_M$ are not reallocated after moving x_k a small enough amount. This is, because the order of the sequence of weighted distances (7.1) does not change.

The nodes of $U_k^=(X_p)$ can be allocated to x_k or to x_m, for some $m \in \{1, 2, ..., p\} \setminus \{k\}$, but if x_k is moved an amount ξ towards j_k, a new point $x_k(\xi)$ will be generated.

1. The nodes of U_{i_k} are still assigned to $x_k(\xi)$, but the distances to $x_k(\xi)$ increase and their contribution to $f_\lambda(X_p)$ has slope $\sum_{v_l \in U_{i_k}} w_l \lambda_{\sigma(l)}$.
2. The nodes of $\overline{U}_{j_k}$ are allocated to $x_k(\xi)$, because the distance from $x_k(\xi)$ to v_{j_k} decreases. Thus, $-\sum_{v_l \in \overline{U}_{j_k}} w_l \lambda_{\sigma(l)}$ is the contribution of $\overline{U}_{j_k}$ to the slope of $f_\lambda(X_p)$.
3. The nodes of $(U_k^=(X_p) \cap \overline{U}_{i_k}) \setminus \overline{U}_{j_k}$ are allocated to x_m for some $m \in \{1, \ldots, p\} \setminus \{k\}$ and their contribution to $f_\lambda(X_p)$ is null.

Hence, the result follows. □

Let $m^{v_{i_k}}$, $m^{v_{j_k}}$ be the slopes of f_λ when x_k is displaced a small enough amount towards the node v_{i_k} and the node v_{j_k}, respectively. Lemma 8.5 shows that there is always a direction (either towards v_{ik} or towards v_{jk}) where the value of the objective function of the problem does not get worse when x_k is moved an amount ξ; provided that no ties are allowed in the vector of ordered distances.

Lemma 8.5 *If $x_k \in (v_{i_k}, v_{j_k})$, then either $m^{v_{i_k}}$ or $m^{v_{j_k}}$ is non positive provided that no ties exist in the vector of distances.*

Proof.
Using the sets of Remark 8.1, we can write down the slopes $m^{v_{i_k}}$ and $m^{v_{j_k}}$ as:

$$m^{v_{i_k}} = \sum_{v_l \in U_{j_k}} w_l \lambda_{\sigma(l)} - \sum_{v_l \in \overline{U}_{i_k}} w_l \lambda_{\sigma(l)}$$

$$m^{v_{j_k}} = \sum_{v_l \in U_{i_k}} w_l \lambda_{\sigma(l)} - \sum_{v_l \in \overline{U}_{j_k}} w_l \lambda_{\sigma(l)}$$

Then:

$$m^{v_{i_k}} + m^{v_{j_k}} = \left(\sum_{v_l \in U_{i_k}} w_l \lambda_{\sigma(l)} + \sum_{v_l \in U_{j_k}} w_l \lambda_{\sigma(l)} \right) - \left(\sum_{v_l \in \overline{U}_{i_k}} w_l \lambda_{\sigma(l)} + \sum_{v_l \in \overline{U}_{j_k}} w_l \lambda_{\sigma(l)} \right).$$

Since $U_{i_k} \subseteq \overline{U}_{i_k}$ and $U_{j_k} \subseteq \overline{U}_{j_k}$ then $m^{v_{i_k}} + m^{v_{j_k}} \leq 0$ and the result follows. □

Remark 8.2 From the above lemma one concludes that there always exists a movement that strictly decreases the objective function except when $m^{v_{i_k}} = m^{v_{j_k}} = 0$. Nevertheless, in the latter case we can move x_k towards v_{i_k} as well as towards v_{j_k} without increasing the objective function.

Now we are in position to state the result that describes an FDS for the ordered 2-median problem. (See [172].)

Theorem 8.6 *Consider the following sets:*

> $T = \{X_2 = (x_1, x_2) \in A(G) \times A(G)$: $\exists v_r, v_s \in U_1^<(X_2)$ *and* $v_{r'}, v_{s'} \in U_2^<(X_2)$ *such that* $w_r d(v_r, x_1) = w_{r'} d(v_{r'}, x_2)$ *and* $w_s d(v_s, x_1) = w_{s'} d(v_{s'}, x_2)$. *Moreover, if* $w_r = w_{r'}$ *and* $w_s = w_{s'}$, *then the slopes of the functions* $d(v_r, \cdot)$ *and* $d(v_s, \cdot)$, *in the edge that* x_1 *belongs to, must have the same signs at* x_1 *and the slopes of the functions* $d(v_{r'}, \cdot)$ *and* $d(v_{s'}, \cdot)$, *in the edge that* x_2 *belongs to, must have different signs at* $x_2\}$, *and*

$$F = ((V \cup \mathsf{EQ}) \times \mathsf{PEQ}) \cup T \subset A(G) \times A(G). \tag{8.6}$$

The set F is a finite set of candidates to be optimal solutions of the 2-facility ordered median problem in the graph $G = (V, E)$.

Remark 8.3 The structure of the set F is different from previously discussed FDS. Indeed, the set F is itself a set of candidates for optimal solutions because it is a set of pairs of points. That means that we do not have to choose the elements of this set by pairs to enumerate the whole set of candidates. The candidate solutions may be either a pair of points belonging to $(V \cup \mathsf{EQ}) \times \mathsf{PEQ}$ or a pair belonging to T, but they never can be one point of PEQ and another point of any pair in T. For this reason, we consider pairs of points in $A(G) \times A(G)$.

Proof.
The proof consists of the following. For any $X_2 = (x_1, x_2) \neq F$ there exist movements of its elements that transform the pair X_2 into a new pair $X_2^* = (x_1^*, x_2^*)$ without increasing the objective value of the problem.

Let $X_2 = (x_1, x_2)$ be a candidate to be optimal solution for the 2-facility ordered median problem. First, we assume that $x_1 \in V \cup \mathsf{EQ}$, and $x_2 \notin Y$ and $(x_1, x_2) \notin T$. Then, x_2 belongs to the subedge (y, y'), such that, $y, y' \in \mathsf{PEQ}$ are two different and consecutive points of PEQ on the edge where x_2 belongs to.

Since $x_1 \in (V \cup \mathsf{EQ})$ and $x_2 \notin \mathsf{PEQ}$ then there is no $v_i \in V$, such that,

$$w_i d(v_i, x_1) = w_j d(v_i, x_2).$$

Moreover, the above inequality does not hold for any pair (x_1, x) for any $x \in (y, y')$. This means that the set $U_2^<(X_2)$ does not change when we move

x_2 in the subedge (y, y'). In addition, since $x_1 \in (V \cup \mathsf{EQ})$ and $x_2 \notin \mathsf{PEQ}$, the vector of ordered weighted distances can only have ties between the elements of the set $U_1^<(X_2)$. Thus, there are no reassignments of the λ-weights after any movement of x_2 in the subedge (y, y'). Hence, the problem of finding the best location of the second facility in (y, y') is a 1-facility median problem in $U_2^<(X_2)$, and an optimal solution always exists on the extreme points of the interval (y, y'). This implies that the new pair $X_2' = (x_1, x_2') \in (V \cup \mathsf{EQ}) \times \mathsf{PEQ}$ is not worse than (x_1, x_2).

In the following we analyze the situation in which neither x_1 nor x_2 belong to $(V \cup \mathsf{EQ})$. We distinguish four possible cases:

Case 1: There exist no ties in the vector of weighted distances between the nodes and their service facilities.
Case 2: There exists one tie in the vector of weighted distances between the nodes and their service facilities.
Case 3: There exist two ties in the vector of weighted distances between the nodes and their service facilities.
Case 4: There exist more than two ties in the vector of weighted distances between the nodes and their service facilities.

It is worth noting that in these cases, since neither x_1 nor x_2 belong to PEQ, the ties in the vector of weighted distances (see (7.1)) have to occur between weighted distances from two nodes: one associated to x_1 and the other to x_2. Indeed, if there would exist two equal weighted distances between two nodes associated with the same service, this service would be an equilibrium point.

Case 1: Using Lemma 8.5, we can move x_1 or x_2 without increasing the objective value while a tie does not occur.

Case 2: There exist $v_r \in U_1(X_2)$ and $v_{r'} \in U_2(X_2)$ such that $w_r d(v_r, x_1) = w_{r'} d(v_{r'}, x_2)$. Assume that x_1 belongs to the edge $[v_{i_1}, v_{j_1}]$ and that x_2 belongs to the edge $[v_{i_2}, v_{j_2}]$. Moreover, denote by $\lambda_{\sigma(r)}$ and $\lambda_{\sigma(r')}$ the λ-weights assigned to v_r and $v_{r'}$, respectively. We can assume without loss of generality that $\sigma(r') = \sigma(r)+1$, $v_r \in U_{j_1}$ and $v_{r'} \in U_{j_2}$. For sake of simplicity, we denote by $V_T = \{v_r, v_{r'}\} \cup U_1^=(X_2)$.

In this case, if we move x_1, a small enough amount ξ_1, towards v_{j_1} and x_2, a small enough amount ξ_2, towards v_{j_2}, such that $\xi_1 w_r = \xi_2 w_{r'}$, we have that the change in the objective value of the problem is:

$$m^{v_{j_1}}(\xi_1) + m^{v_{j_2}}(\xi_2) = \xi_1\Big(\sum_{v_t \in U_{i_1} \setminus V_T} w_t\lambda_{\sigma(t)} - \sum_{v_t \in \overline{U}_{j_1} \setminus V_T} w_t\lambda_{\sigma(t)}$$

$$- \sum_{v_t \in U_1^{=}(X_2) \cap (\overline{U}_{j_1} \setminus \overline{U}_{j_2})} w_t\lambda_{\sigma(t)} - w_r\lambda_{\sigma(r)}\Big)$$

$$+\xi_2\Big(\sum_{v_t \in U_{i_2} \setminus V_T} w_t\lambda_{\sigma(t)} - \sum_{v_t \in \overline{U}_{j_2} \setminus V_T} w_t\lambda_{\sigma(t)}$$

$$- \sum_{v_t \in U_1^{=}(X_2) \cap (\overline{U}_{j_2} \setminus \overline{U}_{j_1})} w_t\lambda_{\sigma(t)} - w_{r'}\lambda_{\sigma(r)+1}\Big)$$

$$+ \sum_{v_t \in U_1^{=}(X_2) \setminus (\overline{U}_{j_1} \cup \overline{U}_{j_2})} \min\{\xi_1, \xi_2\} w_t\lambda_{\sigma(t)}$$

$$- \sum_{v_t \in U_1^{=}(X_2) \cap (\overline{U}_{j_1} \cap \overline{U}_{j_2})} \max\{\xi_1, \xi_2\} w_t\lambda_{\sigma(t)} ,$$

and if we move the same amounts as before, x_1 and x_2 towards v_{i_1} and v_{i_2} respectively, we have that the change in the objective value of the problem is:

$$m^{v_{i_1}}(\xi_1) + m^{v_{i_2}}(\xi_2) = \xi_1\Big(\sum_{v_t \in U_{j_1} \setminus V_T} w_t\lambda_{\sigma(t)} - \sum_{v_t \in \overline{U}_{i_1} \setminus V_T} w_t\lambda_{\sigma(t)}$$

$$- \sum_{v_t \in U_1^{=}(X_2) \cap (\overline{U}_{i_1} \setminus \overline{U}_{i_2})} w_t\lambda_{\sigma(t)} + w_r\lambda_{\sigma(r)}\Big)$$

$$+\xi_2\Big(\sum_{v_t \in U_{j_2} \setminus V_T} w_t\lambda_{\sigma(t)} - \sum_{v_t \in \overline{U}_{i_2} \setminus V_T} w_t\lambda_{\sigma(t)}$$

$$- \sum_{v_t \in U_1^{=}(X_2) \cap (\overline{U}_{i_2} \setminus \overline{U}_{i_1})} w_t\lambda_{\sigma(t)} w_t\lambda_{\sigma(t)} + w_{r'}\lambda_{\sigma(r)+1}\Big)$$

$$+ \sum_{v_t \in U_1^{=}(X_2) \setminus (\overline{U}_{i_1} \cup \overline{U}_{i_2})} \min\{\xi_1, \xi_2\} w_t\lambda_{\sigma(t)}$$

$$- \sum_{v_t \in U_1^{=}(X_2) \cap (\overline{U}_{i_1} \cap \overline{U}_{i_2})} \max\{\xi_1, \xi_2\} w_t\lambda_{\sigma(t)} .$$

Since $U_{j_q} \subseteq \overline{U}_{j_q}$, $U_{i_q} \subseteq \overline{U}_{i_q}$ for $q = 1, 2$, $U_1^{=}(X_2) \setminus (\overline{U}_{j_1} \cup \overline{U}_{j_2}) \subseteq U_1^{=}(X_2) \cap (\overline{U}_{i_1} \cap \overline{U}_{i_2})$, $U_1^{=}(X_2) \setminus (\overline{U}_{i_1} \cup \overline{U}_{i_2}) \subseteq U_1^{=}(X_2) \cap (\overline{U}_{j_1} \cap \overline{U}_{j_2})$ then $m^{v_{j_1}}(\xi_1) + m^{v_{j_2}}(\xi_2) + m^{v_{i_1}}(\xi_1) + m^{v_{i_2}}(\xi_2)$ is a non positive. Therefore, there exists a movement of x_1 and x_2 that does not increase the objective value.

Notice that the initial assumption $v_r \in U_{j_1}$ and $v_{r'} \in U_{j_2}$ is not restrictive because if $v_r \in \overline{U}_{j_1} \setminus U_{j_1}$ ($v_{r'} \in \overline{U}_{j_2} \setminus U_{j_2}$) then the term $w_r\lambda_{\sigma(r)}$ ($w_{r'}\lambda_{\sigma(r)+1}$) in the above two expressions would appear with negative sign. Moreover, if $v_r \in U_1^{=}(X_2)$, we would have that $w_r d(v_r, v_1) = w_r(v_r, x_2)$. However, since we have assumed that $w_r d(v_r, x_1) = w_{r'} d(v_{r'}, x_2)$, it follows that $w_r d(v_r, x_2) = w_{r'} d(v_{r'}, x_2)$, that is, $x_2 \in (V \cup \mathsf{EQ})$, what contradicts the hypothesis that $x_2 \notin (V \cup \mathsf{EQ})$. A similar argument can be used if $v_{r'} \in U_1^{=}(X_2)$.

Case 3: There exist $v_r, v_s \in U_1(X_2)$ and $v_{r'}, v_{s'} \in U_2(X_2)$ such that $w_r = w_{r'}$, $w_s = w_{s'}$, $d(v_r, x_1) = d(v_{r'}, x_2)$ and $d(v_s, x_1) = d(v_{s'}, x_2)$, we will distinguish two different subcases.
Case 3.1: If $w_r \neq w_{r'}$ or $w_s \neq w_{s'}$ the pairs in this subcase are included in T and therefore belong to the set of candidates to be optimal solutions.
Case 3.2: If $w_r = w_{r'}$ and $w_s = w_{s'}$, what in turns implies $d(v_r, x_1) = d(v_{r'}, x_2)$ and $d(v_s, x_1) = d(v_{s'}, x_2)$. In order to obtain an easy understanding, let $V_T = \{v_r, v_{r'}, v_s, v_{s'}\} \cup U_1^{=}(X_2)$ and assume without loss of generality that:

i) $x_1 \in [v_{i_1}, v_{j_1}]$, $\sigma(r') = \sigma(r) + 1$.
ii) $x_2 \in [v_{i_2}, v_{j_2}]$, $\sigma(s') = \sigma(s) + 1$.

We will distinguish four additional subcases. It should be noted that any other configuration reduces to one of them interchanging the name of the points x_1 and x_2.

3.2.1 The slopes of the functions $d(v_r, \cdot)$ and $d(v_s, \cdot)$ on the edge $[v_{i_1}, v_{j_1}]$ have the same sign at x_1 and the slope of the functions $d(v_{r'}, \cdot)$ and $d(v_{s'}, \cdot)$ on the edge $[v_{i_2}, v_{j_2}]$ have the same sign at x_2.
3.2.2 The slopes of the functions $d(v_r, \cdot)$ and $d(v_s, \cdot)$ on the edge $[v_{i_1}, v_{j_1}]$ have different sign at x_1 and the slopes of the functions $d(v_{r'}, \cdot)$ and $d(v_{s'}, \cdot)$ on the edge $[v_{i_2}, v_{j_2}]$ have different sign at x_2.
3.2.3 The slopes of the functions $d(v_r, \cdot)$ and $d(v_s, \cdot)$ on the edge $[v_{i_1}, v_{j_1}]$ have different sign at x_1 and the slopes of the functions $d(v_{r'}, \cdot)$ and $d(v_{s'}, \cdot)$ on the edge $[v_{i_2}, v_{j_2}]$ have the same sign at x_2.
3.2.4 The slope of the function $d(v, \cdot)$ for some $v \in \{v_r, v_{r'}, v_s, v_{s'}\}$ is not defined at the service facility that covers v.

Now, we prove that there exists a movement for the first two cases that does not increase the objective value.

3.2.1 Since the sign of the slopes of the functions $d(v_r, \cdot)$ and $d(v_s, \cdot)$ at x_1 are the same, we can assume without loss of generality that $v_r, v_s \in U_{j_1}$. In the same way, we assume that $v_{r'}, v_{s'} \in U_{j_2}$.
If we move, the same small enough amount, x_1 and x_2 towards v_{j_1} and v_{j_2} respectively, we have that the slope of these movements is

$$\begin{aligned} m^{v_{j_1}} + m^{v_{j_2}} = & \sum_{v_t \in U_{i_1} \setminus V_T} w_t \lambda_{\sigma(t)} - \sum_{v_t \in \overline{U}_{j_1} \setminus V_T} w_t \lambda_{\sigma(t)} + \sum_{v_t \in U_{i_2} \setminus V_T} w_t \lambda_{\sigma(t)} - \sum_{v_t \in \overline{U}_{j_2} \setminus V_T} w_t \lambda_{\sigma(t)} \\ & - \sum_{v_t \in U_1^{=}(X_2) \cap (\overline{U}_{j_1} \cup \overline{U}_{j_2})} w_t \lambda_{\sigma(t)} + \sum_{v_t \in U_1^{=}(X_2) \setminus (\overline{U}_{j_1} \cup \overline{U}_{j_2})} w_t \lambda_{\sigma(t)} \\ & - w_r \lambda_{\sigma(r)} - w_{r'} \lambda_{\sigma(r)+1} - w_s \lambda_{\sigma(s)} - w_{s'} \lambda_{\sigma(s)+1}, \end{aligned}$$

and if we move, by the same amount, x_1 and x_2 towards v_{i_1} and v_{i_2} respectively, we have that the slope of these movements is

$$m^{v_{i_1}} + m^{v_{i_2}} = \sum_{v_t \in U_{j_1} \setminus V_T} w_t \lambda_{\sigma(t)} - \sum_{v_t \in \overline{U}_{i_1} \setminus V_T} w_t \lambda_{\sigma(t)} + \sum_{v_t \in U_{j_2} \setminus V_T} w_t \lambda_{\sigma(t)} - \sum_{v_t \in \overline{U}_{i_2} \setminus V_T} w_t \lambda_{\sigma(t)}$$
$$- \sum_{v_t \in U_1^{=}(X_2) \cap (\overline{U}_{i_1} \cup \overline{U}_{i_2})} w_t \lambda_{\sigma(t)} + \sum_{v_t \in U_1^{=}(X_2) \setminus (\overline{U}_{i_1} \cup \overline{U}_{i_2})} w_t \lambda_{\sigma(t)}$$
$$+ w_r \lambda_{\sigma(r)} + w_{r'} \lambda_{\sigma(r)+1} + w_s \lambda_{\sigma(s)} + w_{s'} \lambda_{\sigma(s)+1}.$$

Hence, since $U_{j_q} \subseteq \overline{U}_{j_q}$, $U_{i_q} \subseteq \overline{U}_{i_q}$ for $q = 1, 2$; $U_1^{=}(X_2) \setminus (\overline{U}_{j_1} \cup \overline{U}_{j_2}) \subseteq U_1^{=}(X_2) \cap (\overline{U}_{i_1} \cup \overline{U}_{i_2})$ and $U_1^{=}(X_2) \setminus (\overline{U}_{i_1} \cup \overline{U}_{i_2}) \subseteq U_1^{=}(X_2) \cap (\overline{U}_{j_1} \cup \overline{U}_{j_2})$, we have that $m^{v_{j_1}} + m^{v_{j_2}} + m^{v_{i_1}} + m^{v_{i_2}}$ is non positive. Therefore, at least one of these two movements cannot increase the value of the objective function.

Notice that using the arguments of Case 2, the initial assumption $v_r, v_s \in U_{j_1}$ and $v_{r'}, v_{s'} \in U_{j_2}$ is not restrictive.

3.2.2 Since the sign of the slopes of the functions $d(v_r, \cdot)$ and $d(v_s, \cdot)$ at x_1 are different, we can assume without loss of generality that $v_r \in U_{j_1}$ and $v_s \in U_{i_1}$. In the same way, we assume that $v_{r'} \in U_{j_2}$ and $v_{s'} \in U_{i_2}$.
If we move the same small enough amount, x_1 and x_2 towards v_{j_1} and v_{j_2} respectively, we have that the slope of these movements is

$$m^{v_{j_1}} + m^{v_{j_2}} = \sum_{v_t \in U_{i_1} \setminus V_T} w_t \lambda_{\sigma(t)} - \sum_{v_t \in \overline{U}_{j_1} \setminus V_T} w_t \lambda_{\sigma(t)} + \sum_{v_t \in U_{i_2} \setminus V_T} w_t \lambda_{\sigma(t)} - \sum_{v_t \in \overline{U}_{j_2} \setminus V_T} w_t \lambda_{\sigma(t)}$$
$$- \sum_{v_t \in U_1^{=}(X_2) \cap (\overline{U}_{j_1} \cup \overline{U}_{j_2})} w_t \lambda_{\sigma(t)} + \sum_{v_t \in U_1^{=}(X_2) \setminus (\overline{U}_{j_1} \cup \overline{U}_{j_2})} w_t \lambda_{\sigma(t)}$$
$$- w_r \lambda_{\sigma(r)} - w_{r'} \lambda_{\sigma(r)+1} + w_s \lambda_{\sigma(s)} + w_{s'} \lambda_{\sigma(s)+1}.$$

Besides, if we move the same amount, x_1 and x_2 towards v_{i_1} and v_{i_2} respectively, we have that the slope of these movements is

$$m^{v_{i_1}} + m^{v_{i_2}} = \sum_{v_t \in U_{j_1} \setminus V_T} w_t \lambda_{\sigma(t)} - \sum_{v_t \in \overline{U}_{i_1} \setminus V_T} w_t \lambda_{\sigma(t)} + \sum_{v_t \in U_{j_2} \setminus V_T} w_t \lambda_{\sigma(t)} - \sum_{v_t \in \overline{U}_{i_2} \setminus V_T} w_t \lambda_{\sigma(t)}$$
$$- \sum_{v_t \in U_1^{=}(X_2) \cap (\overline{U}_{i_1} \cup \overline{U}_{i_2})} w_t \lambda_{\sigma(t)} + \sum_{v_t \in U_1^{=}(X_2) \setminus (\overline{U}_{i_1} \cup \overline{U}_{i_2})} w_t \lambda_{\sigma(t)}$$
$$+ w_r \lambda_{\sigma(r)} + w_{r'} \lambda_{\sigma(r)+1} - w_s \lambda_{\sigma(s)} - w_{s'} \lambda_{\sigma(s)+1}.$$

Hence, since $U_{j_q} \subseteq \overline{U}_{j_q}$, $U_{i_q} \subseteq \overline{U}_{i_q}$ for $q = 1, 2$; $U_1^{=}(X_2) \setminus (\overline{U}_{j_1} \cup \overline{U}_{j_2}) \subseteq U_1^{=}(X_2) \cap (\overline{U}_{i_1} \cup \overline{U}_{i_2})$ and $U_1^{=}(X_2) \setminus (\overline{U}_{i_1} \cup \overline{U}_{i_2}) \subseteq U_1^{=}(X_2) \cap (\overline{U}_{j_1} \cup \overline{U}_{j_2})$, we have that $m^{v_{j_1}} + m^{v_{j_2}} + m^{v_{i_1}} + m^{v_{i_2}}$ is a non positive amount. Therefore, at least one of these two movements cannot increase the objective function. Notice that using the arguments of Case 2, the initial assumption is not restrictive.

3.2.3 The pairs in this subcase are included in T and therefore, belong to the set of candidates to be an optimal solution (see Example 8.5).

3.2.4 Assume without loss of generality that the slope of the function $d(v_r, \cdot)$ is not defined at x_1 when $v_r \in \overline{U}_{j_1} \cap \overline{U}_{i_1}$ (the distance $d(v_r, \cdot)$ has a breakpoint at x_1). In this case, if we move x_1 and x_2 as in 3.1 or 3.2, we have that the expressions of $m^{v_{j_1}} + m^{v_{j_2}}$ and $m^{v_{i_1}} + m^{v_{i_2}}$ are equal to the ones obtained in cases 3.1 or 3.2, respectively (depending on the relative position of the nodes $\{v_{r'}, v_s, v_{s'}\}$ and their corresponding service facility), except the term $w_r \lambda_{\sigma(r)}$ that appears in these two expressions with negative sign. Therefore, $m^{v_{j_1}} + m^{v_{j_2}} + m^{v_{i_1}} + m^{v_{i_2}}$ is again non positive. A similar argument can be used when more than one of the slopes of the distance functions are not defined.

Notice that in 3.2.1, 3.2.2 and 3.2.4, we have assumed that $\{v_r, v_{r'}, v_s, v_{s'}\} \cap U_1^=(X_2) = \emptyset$. However, this hypothesis is not restrictive, because using the arguments of Case 2, we obtain that this intersection is always empty.

Case 4: There exist $v_{r_1}, \ldots, v_{r_Q} \in U_1(X_2)$ and $v_{r'_1}, \ldots, v_{s'_Q} \in U_2(X_2)$, with $q > 2$, such that $w_{r_l} d(v_{r_l}, x_1) = w_{r'_l} d(v_{r'_l}, x_2)$ and $w_{r_l} = w_{r'_l}$ for $l = 1, \ldots, Q$ (notice that, if $w_{r_l} \neq w_{r'_l}$ for some $l = 1, \ldots, Q$, we are in an instance of Case 3.1 and the pair (x_1, x_2) belongs to T). We assume, that x_1 belongs to the edge $[v_{i_1}, v_{j_1}]$ and that x_2 belongs to the edge $[v_{i_2}, v_{j_2}]$. We distinguish two subcases:

4.1 There exist no $v_{r_{l_c}}, v_{r_{l_d}} \in U_1(X_2)$ with l_c, $l_d \in \{1, \ldots, Q\}$ and $v_{r'_{l_c}}, v_{r'_{l_d}} \in U_2(X_2)$ such that the slopes of the functions $d(v_{r_{l_c}}, \cdot)$ and $d(v_{r_{l_d}}, \cdot)$ on the edge $[v_{i_1}, v_{j_1}]$ have different signs at x_1 and the slopes of the functions $d(v_{r'_{l_c}}, \cdot)$ and $d(v'_{r_{l_d}}, \cdot)$ on the edge $[v_{i_2}, v_{j_2}]$ have different signs at x_2.

4.2 There exist $v_{r_{l_c}}, v_{r_{l_d}} \in U_1(X_2)$ with l_c, $l_d \in \{1, \ldots, Q\}$ and $v_{r'_{l_c}}, v_{r'_{l_d}} \in U_2(X_2)$ such that the slopes of the functions $d(v_{r_{l_c}}, \cdot)$ and $d(v_{r_{l_d}}, \cdot)$ on the edge $[v_{i_1}, v_{j_1}]$ have different sign at x_1 and the slopes of the functions $d(v_{r'_{l_c}}, \cdot)$ and $d(v_{r'_{l_d}}, \cdot)$ on the edge $[v_{i_2}, v_{j_2}]$ have the same sign at x_2.

In the first case, the four nodes that define each two ties in the sequence of ordered weighted distances are either in Case 3.2.1 or 3.2.2. Therefore, using the same arguments as in cases 3.2.1 and 3.2.2, there exists a movement of x_1 and x_2 that does not get a worse objective value. The second case is a particular instance of 3.2.3 and X_2 is also included in T.

We have proved that in all the cases $m^{v_{i_1}} + m^{v_{i_2}} + m^{v_{j_1}} + m^{v_{j_2}} \leq 0$ when $X_2 = (x_1, x_2) \notin F$. Thus, if $m^{v_{i_1}} + m^{v_{i_2}}$ or $m^{v_{j_1}} + m^{v_{j_2}}$ are different from zero there exists a movement of $X_2 = (x_1, x_2)$ to a new pair X'_2 which strictly decreases the objective value. Otherwise, if $m^{v_{i_1}} + m^{v_{i_2}} = m^{v_{j_1}} + m^{v_{j_2}} = 0$, then the movements of x_1 and x_2, respectively, towards v_{i_1} and v_{i_2} as well as towards v_{j_1} and v_{j_2} do not increase the objective value. One of these two displacements avoid cycling since one of them has not been used in the opposite direction in the previous step (see Remark 8.2).

The movement from X_2 to X_2' is valid whenever the sets $U_k(X_2')$, $U_k^=(X_2')$, $U_k^<(X_2')$, $\overline{U}_{i_k}$ and $\overline{U}_{j_k}$ (associated to X_2') for $k = 1, 2$ do not change. Hence, if the maximal displacement without increasing the objective value transforms X_2 into X_2'' and $X_2'' \notin F$, we repeat the same process a finite number of times until a pair $X_2^* \in F$ is reached.

□

The following examples show that the set F cannot be reduced, because even in easy cases on the real line all the points are necessary. The first example shows a graph where the optimal solution $X_2 = (x_1, x_2)$ verifies that x_1 is an equilibrium point and x_2 is not a equilibrium point, which belongs to $\mathsf{PEQ} \setminus (V \cup \mathsf{EQ})$ for a given r. In the second example the optimal solution $X_2 = (x_1, x_2)$ belongs to the set T.

Example 8.4
Consider the graph $G = (V, E)$ where $V = \{v_1, v_2, v_3, v_4\}$ and $E = \{[v_1, v_2], [v_2, v_3], [v_3, v_4]\}$. The length function is given by $l([v_1, v_2]) = 3, l([v_2, v_3]) = 20, l([v_3, v_4]) = 6$. The $w-$weights are all equal to one and the λ-modeling weights are $\lambda_1 = 0.1, \lambda_2 = 0.2, \lambda_3 = 0.4, \lambda_4 = 0.3$, see Figure 8.6.

It should be noted that this example does not have an optimal solution on the edge $[v_2, v_3]$, because any point of this edge is dominated by v_2 or v_3. In addition, by symmetry many of the elements of PEQ have been eliminated from consideration.

In Figure 8.6 we represent the nodes (dots), the equilibrium points (ticks) and element of PEQ (small ticks). Notice that in this example there are no pairs in T.

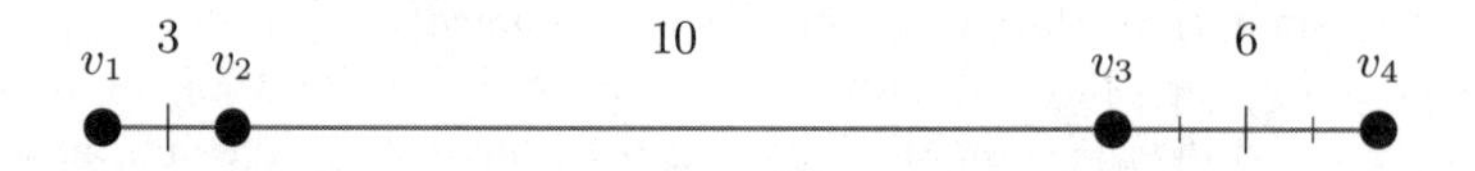

Fig. 8.6. Illustration of Example 8.4

The optimal solution is given by $x_1 = ([v_1, v_2], 1.5)$ and $x_2 = ([v_3, v_4], 1.5)$ (see Table 8.1). It is easy to check that x_1 is an equilibrium point between v_1 and v_2, and $x_2 \in \{y \in A(G) \; : \; w_i d(v_i, y) = 1.5, \, v_i \in V\}$. It is worth noting that the radius 1.5 is given by the distance from the equilibrium point generated by v_1 and v_2 to any of these nodes.

Example 8.5
Let $G = (V, E)$ be a graph with

$$V = \{v_1, v_2, v_3, v_4, v_5\} \text{ and}$$

Table 8.1. Evaluation of the candidate pairs of Example 8.4.

Candidate pair X_2	Objective value
$([v_1, v_2], 0), ([v_3, v_4], 0)$	3
$([v_1, v_2], 0), ([v_3, v_4], 1.5)$	2.85
$([v_1, v_2], 0), ([v_3, v_4], 3)$	2.7
$([v_1, v_2], 1.5), ([v_3, v_4], 0)$	2.7
$([v_1, v_2], 1.5), ([v_3, v_4], 1.5)$	2.4
$([v_1, v_2], 1.5), ([v_3, v_4], 3)$	2.55

$$E = \{[v_1, v_2], [v_2, v_3], [v_3, v_4], [v_4, v_5]\}.$$

The length function is given by $l([v_1, v_2]) = 5$, $l([v_2, v_3]) = 20$, $l([v_3, v_4]) = 5.1$, $l([v_4, v_5]) = 1$. The w-weights are all equal to one and the λ-modeling weights are $\lambda_1 = 0, \lambda_2 = 1, \lambda_3 = 0, \lambda_4 = 1, \lambda_5 = 1.1$, see Figure 8.7.

In Figure 8.7, we use the same notation as in Figure 8.6 and in addition, pairs in T are represented by $(\star)$. By domination and symmetry arguments not all the candidates are necessary and therefore, they are not depicted.

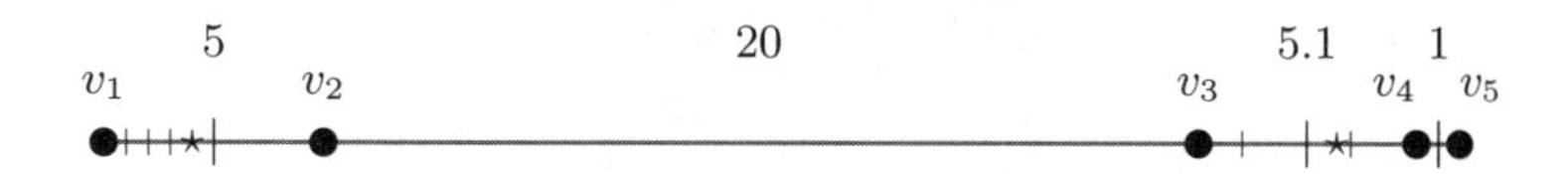

Fig. 8.7. Illustration of Example 8.5

In this example the optimal solution is given by $x_1 = ([v_1, v_2], 2)$ and $x_2 = ([v_3, v_4], 3.1)$ (see Table 8.2). Therefore, the optimal pair (x_1, x_2) belongs to the set T. Indeed, $d(x_1, v_1) = d(x_2, v_4)$ and $d(x_1, v_2) = d(x_2, v_5)$ and the slopes of $d(\cdot, v_1), d(\cdot, v_2)$ in the edge $[v_1, v_2]$ at x_1 are $1, -1$, respectively; and the slopes of $d(\cdot, v_4), d(\cdot, v_5)$ in the edge $[v_3, v_4]$ at x_2 are $-1, -1$, respectively.

We conclude this section with a remark on the structure of the set F. The reader may notice that F itself is a set of pairs of candidates to optimal solutions for the 2-facility OMP (it is straightforward to check that $|F|$ is $O(N^3M^6)$). The difference with previous approaches is that this set is not a set of candidates for each individual facility; on the contrary, it is the set of candidates for any combination of facilities in the optimal solution.

Table 8.2. Evaluation of the candidate pairs of Example 8.5.

Candidate pair X_2	Objective value
$([v_1, v_2], 0), ([v_3, v_4], 0)$	11.81
$([v_1, v_2], 0), ([v_3, v_4], 2.55)$	11.6
$([v_1, v_2], 0), ([v_3, v_4], 3.05)$	10.6
$([v_1, v_2], 0), ([v_4, v_5], 0)$	10.61
$([v_1, v_2], 0), ([v_4, v_5], 0.5)$	11.66
$([v_1, v_2], 0), ([v_4, v_5], 1)$	11.71
$([v_1, v_2], 0.5), ([v_4, v_5], 0.5)$	11.16
$([v_1, v_2], 1), ([v_4, v_5], 0)$	10.61
$([v_1, v_2], 1), ([v_4, v_5], 1)$	11.71
$([v_1, v_2], 1.45), ([v_3, v_4], 2.55)$	10.005
$([v_1, v_2], 1.95), ([v_3, v_4], 3.05)$	8.455
$([v_1, v_2], 2), ([v_3, v_4], 3.1)$	8.41
$([v_1, v_2], 2.05), ([v_3, v_4], 3.05)$	8.455
$([v_1, v_2], 2.45), ([v_3, v_4], 2.55)$	9.005
$([v_1, v_2], 2.5), ([v_3, v_4], 0)$	14.31
$([v_1, v_2], 2.5), ([v_3, v_4], 2.5)$	9.06
$([v_1, v_2], 2.5), ([v_3, v_4], 2.55)$	8.955
$([v_1, v_2], 2.5), ([v_3, v_4], 2.6)$	8.95
$([v_1, v_2], 2.5), ([v_3, v_4], 3.05)$	8.905
$([v_1, v_2], 2.5), ([v_3, v_4], 3.6)$	8.96
$([v_1, v_2], 2.5), ([v_4, v_5], 0)$	9.11
$([v_1, v_2], 2.5), ([v_4, v_5], 0.5)$	9.16
$([v_1, v_2], 2.5), ([v_4, v_5], 1)$	10.21

8.4 On the Exponential Cardinality of FDS for the Multifacility Facility Ordered Median Problem

In this section, based on the results in [170], we prove that there is no FDS of polynomial cardinality for the general ordered p-median problem on networks. In order to do that, we consider a network N with underlying graph $G = (V, E)$ where $V = \{v_1, \ldots, v_{2p}\}$, $E = \{[v_1, v_2], [v_2, v_3], \ldots, [v_{2p-1}, v_{2p}]\}$ and p is a fixed natural number. In this graph we impose that $d(v_{2i-1}, v_{2i}) = 2^i$ for $i = 1, \ldots, p$; and $d(v_{2i}, v_{2i+1}) = K$ for $i = 1, \ldots, p-1$ where $K = 4\sum_{i=1}^{p} 2^i + 1$ (a large enough amount) (see Figure 8.8).

The w-weights associated to each node are assumed to be equal to one and the weights defining the objective function, the λ-weights, are given as

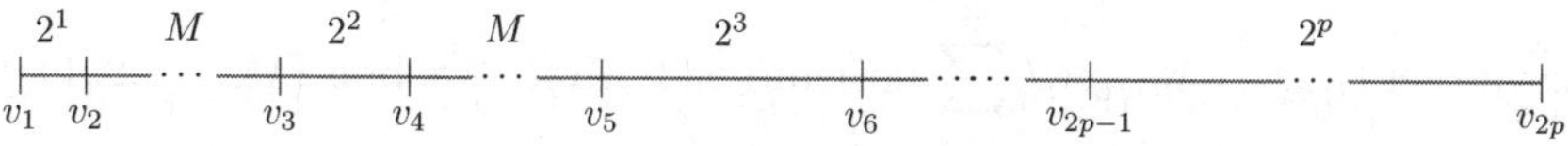

Fig. 8.8. Illustration of the graph

follows:

$$\lambda_1 = 0,\ \lambda_2 = \lambda_3 = 2p,\ \lambda_4 = p \text{ and } \lambda_i = \frac{2^{2p}+1}{2^{2p+1}}(\lambda_{i-2}+\lambda_{i-1}) \text{ for } i = 5,\dots,2p. \tag{8.7}$$

Under these conditions our goal is to find p points on $A(G)$, $X_p = \{x_1, \dots, x_p\}$, solving the following problem:

$$\min_{X_p \subseteq A(G)} f_\lambda(X_p) := \sum_{i=1}^{2p} \lambda_{\sigma_i} d(v_i, X_p), \tag{8.8}$$

where σ is a permutation of $\{1, \dots, 2p\}$, such that, $\sigma_k < \sigma_l$ if $d(v_k, X_p) \le d(v_l, X_p)$ for each $k, l \in \{1, \dots, 2p\}$ (in this case, we say that the λ-weight λ_{σ_i} is assigned to the node v_i).

Remark 8.4 Notice that, the λ-weights defined in (8.7) satisfy the relationships:

$$2\max\{\lambda_{i-2}, \lambda_{i-1}\} > 2\lambda_i > \lambda_{i-2} + \lambda_{i-1} \quad \text{for all } i = 5, \dots, 2p, \tag{8.9}$$

$$2\lambda_4 = \lambda_2 = \lambda_3 > \lambda_1 = 0, \tag{8.10}$$

$$2\lambda_2 > \lambda_4 + \lambda_5 + \lambda_8. \tag{8.11}$$

Moreover, the components of the vector $\lambda = (\lambda_1, \dots, \lambda_{2p})$ satisfy the following chain of inequalities:

$$\lambda_2 = \lambda_3 > \dots > \lambda_{2p-3} > \lambda_{2p-1} > \lambda_{2p} > \lambda_{2p-2} > \lambda_{2p-4} > \dots > \lambda_4 > \lambda_1 = 0.$$

Therefore:

i) $\lambda_{2j+1} > \lambda_{2(j+1)+1}$, for $j = 1, \dots, p-2$ (the sequence of λ-weights with odd indexes (> 1) is decreasing).
ii) $\lambda_{2(j+1)} > \lambda_{2j}$, for $j = 2, \dots, p-1$ (the sequence of λ-weights with even indexes (> 2) is increasing).
iii) $\lambda_{2j-1} > \lambda_{2i}$, for any $i, j \in \{2, \dots, p\}$ (a λ-weight with odd index (> 1) is always greater than any other with an even index (> 2)).
iv) $\lambda_{2j} < \lambda_k$ if $k > 2j$, $j > 1$ and $\lambda_{2j+1} > \lambda_k$ if $k > 2j+1$, $j \ge 1$.

We will prove that the optimal policy to solve Problem (8.8) is to locate a service facility on each edge $[v_{2i-1}, v_{2i}]$ for $i = 1, \dots, p$ and to assign λ_i to v_i for $i = 1, \dots, 2p$.

Lemma 8.6 *If $X_p = \{x_1, \dots, x_p\}$ is an optimal solution of Problem (8.8) then $x_i \in [v_{2i-1}, v_{2i}]$ for $i = 1, \dots, p$.*

Proof.
First, we prove that the nodes v_{2i} and v_{2i+1} for any $i = 1, \ldots, p$, are not covered by the same service facility. Suppose, on the contrary, that there exists $j \in \{1, \ldots, p\}$, such that, v_{2j} and v_{2j+1} are served by the same service facility $x \in X_p$. This implies that the following terms would appear in the objective function:

$$\lambda_{\sigma_{2j}} d(v_{2j}, x) + \lambda_{\sigma_{2j+1}} d(v_{2j+1}, x).$$

Notice that $d(v_{2j}, x) + d(v_{2j+1}, x) \geq K$. Moreover:

1. If both σ_{2j} and σ_{2j+1} are different from 1, we have that $\lambda_{\sigma_{2j}} \geq \lambda_4$, $\lambda_{\sigma_{2j+1}} \geq \lambda_4$, and at least one of these inequalities is strict.
2. If $\sigma_{2j} = 1$, then $d(v_{2j+1}, x) \geq \frac{K}{2}$. In a similar way, the case $\sigma_{2j+1} = 1$ implies that $d(v_{2j}, x) \geq \frac{K}{2}$.

In all cases

$$\lambda_{\sigma_{2j}} d(v_{2j}, x) + \lambda_{\sigma_{2j+1}} d(v_{2j+1}, x) > \frac{K-1}{2} \lambda_4 = 2\lambda_4 \sum_{i=1}^{p} 2^i = \lambda_2 \sum_{i=1}^{p} 2^i.$$

The inequality above contradicts the optimality of X_p. Indeed, consider $X'_p = \{x'_1, \ldots, x'_p\}$ such that x'_i is located at the midpoint of the edge $[v_{2i-1}, v_{2i}]$ for $i = 1, \ldots, p$. Then, since $\lambda_2 \geq \lambda_i$ for $i = 1, \ldots, 2p$, we have that $f_\lambda(X'_p) \leq \lambda_2 \sum_{i=1}^{p} 2^i$. Therefore, v_{2i} and v_{2i+1} for $i = 1, \ldots, p$, can not be covered by the same service facility.

Hence, in what follows, we can assume without loss of generality that each service facility x_i covers the demand of v_{2i-1} and v_{2i} for $i = 1, \ldots, p$.

The fact that $x_i \in [v_{2i-1}, v_{2i}]$ follows directly from the isotonicity property of the ordered median objective with nonnegative λ-weights (see Theorem 1 in [78]). Indeed, if x_i is moved to its closest node in the interval $[v_{2i-1}, v_{2i}]$ for $i = 1, \ldots, p$, the new vector of distances of $\{v_1, \ldots, v_{2p}\}$ from the servers is smaller than the old vector. □

Remark 8.5 Since all the $w-$weights are equal to one, by symmetry arguments and without loss of generality, we only consider solutions of this problem satisfying that $d(v_{2i-1}, x_i) \leq d(v_{2i}, x_i)$ for $i = 1, \ldots, p$, and consequently, by the structure of the graph, $d(v_{2i}, x_i) \leq d(v_{2i+2}, x_{i+1})$ for $i = 1, \ldots, p-1$. Hence,

i) $\sigma_{2i-1} < \sigma_{2i}$ for $i = 1, \ldots, p$,
ii) $\sigma_{2i} < \sigma_{2i+2}$ for $i = 1, \ldots, p-1$.

The above assertions imply that $\sigma_{2p} = 2p$. Moreover, by Lemma 8.6 and for the sake of the readability, we can represent the graph of Figure 8.8 as a graph with only p edges where the edges with length K are omitted (see Figure 8.9).

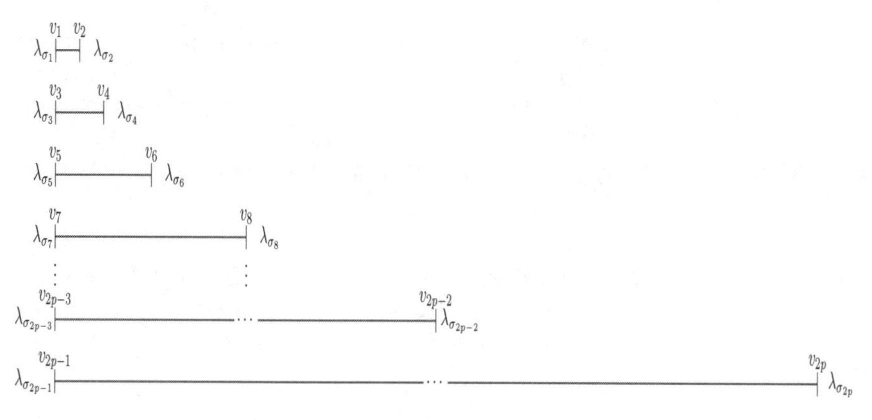

Fig. 8.9. The new representation of the graph where $\sigma_{2i} > \sigma_{2i-1}$ for $i = 1, \ldots, p$ and $\sigma_{2i+2} > \sigma_{2i}$ for $i = 1, \ldots, p-1$

8.4.1 Some Technical Results

To get to the final result about the exponential cardinality of the FDS for the $p-$facility ordered median problems, we need some technical results. These results are not necessary for the understanding of the further statements but we include them here for the sake of the completeness of the chapter.

Lemma 8.7 *If X_p is an optimal solution of Problem (8.8) then for $k = 2, \ldots, p$, λ_{2k} is assigned to v_{2i} for some i, $i = 1, \ldots, p$.*

Proof.
Suppose on the contrary that λ_{2k} is assigned to v_{2j-1} for some j, $j = 1, \ldots, p$. We can assume without loss of generality that $2k$ is the maximum possible even index of a λ-weight assigned to v_{2j-1} with $j = 1, \ldots, p$. Recall that k must be less than p since $\sigma_{2p} = 2p$ (see Remark 8.5). In what follows we distinguish two cases depending on the type of node where λ_{2k+1} has been assigned to.

Case 1: λ_{2k+1} is assigned to $v_{2j'}$ for some j', $j' = 1, \ldots, p$.
By Remark 8.5 i) we have that $\sigma_{2j} > \sigma_{2j-1} = 2k$. Thus, by Remark 8.4 iv), we must have that $\lambda_{\sigma_{2j}} > \lambda_{\sigma_{2j-1}} = \lambda_{2k}$. Hence, since X_p is optimal, x_j must be located as far as possible from v_{2j-1}. Besides, since $\sigma_{2j-1} = 2k < 2k+1 = \sigma_{2j'}$ then $d(v_{2j-1}, x_j) \leq d(v_{2j'}, x_{j'})$. Therefore, we have that $d(v_{2j-1}, x_j) = d(v_{2j'}, x_{j'})$ and we can reassign the λ-weights, so that λ_{2k} is assigned to $v_{2j'}$ and λ_{2k+1} to v_{2j-1}.

Case 2: λ_{2k+1} is assigned to $v_{2j'-1}$ for some j', $j' = 1, \ldots, p$.
Assume that $d(v_{2j-1}, x_j) \neq d(v_{2j}, x_j)$ and $d(v_{2j'-1}, x_{j'}) \neq d(v_{2j'}, x_{j'})$. Under this assumption, we can move x_j and $x_{j'}$ towards v_{2j} and $v_{2j'}$, respectively,

by the same small enough amount, ξ, without any reassignment of the λ-weights (see Figure 8.10).This is possible because $2k + 2 = \sigma_{2j''}$ for some j'', $j'' = 1, \ldots, p$ (recall that λ_{2k} is the maximum index of a λ-weight assigned to a node with odd index) and $d(v_{2j'-1}, x_{j'})$ as well as $d(v_{2j-1}, x_j)$ are strictly smaller than $d(v_{2j''}, x_{j''})$. These movements imply the following change in the objective function:

$$\xi(\lambda_{2k} + \lambda_{2k+1} - \lambda_{\sigma_{2j}} - \lambda_{\sigma_{2j'}}).$$

This amount is negative. Indeed, since $\sigma_{2j} > 2k + 1$ and $\sigma_{2j'} > 2k + 1$, by Remark 8.4 iv), we get $\lambda_{\sigma_{2j}} + \lambda_{\sigma_{2j'}} > 2\lambda_{2k+2}$ and, by (8.9), we have that $2\lambda_{2k+2} > \lambda_{2k} + \lambda_{2k+1}$. This is a contradiction because X_p was an optimal solution.

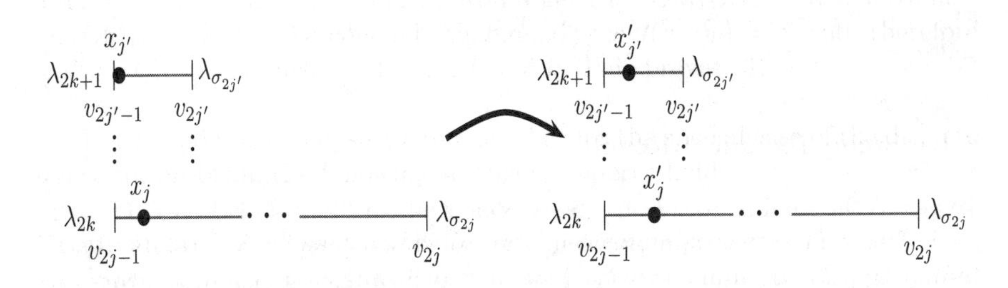

Fig. 8.10. Illustration to Case 2 of Lemma 8.7

In what follows, we study the cases $d(v_{2j-1}, x_j) = d(v_{2j}, x_j)$ and $d(v_{2j'-1}, x_{j'}) = d(v_{2j'}, x_{j'})$.

Case 2.1: $d(v_{2j'-1}, x_{j'}) = d(v_{2j'}, x_{j'})$.
Since $\sigma_{2j'-1} = 2k+1$ we can assume without loss of generality that $\sigma_{2j'} = 2k+2$. Now, since λ_{2k+1} and λ_{2k+2} have been already assigned and, by Remark 8.5 i), $\sigma_{2j} > 2k$ we get that $\sigma_{2j} > 2k + 2 = \sigma_{2j'}$. This means, by Remark 8.5 ii), that $j > j'$. Moreover, since $\sigma_{2j} > 2k$ then, by Remark 8.4 iv), $\lambda_{\sigma_{2j}} > \lambda_{\sigma_{2j-1}} = \lambda_{2k}$. Hence, x_j must be located as far as possible from v_{2j-1} because X_p is optimal. Besides, the relationship $\sigma_{2j-1} = 2k < 2k+1 = \sigma_{2j'-1}$ implies that $d(v_{2j-1}, x_j) \leq d(v_{2j'-1}, x_{j'})$. Therefore, we obtain that $d(v_{2j-1}, x_j) = d(v_{2j'-1}, x_{j'})$. This permits reassigning the λ-weights so that λ_{2k} is assigned to $v_{2j'-1}$, λ_{2k+1} to $v_{2j'}$ and λ_{2k+2} to v_{2j-1} (see Figure 8.11). However, this allocation induces a contradiction because $2k$ is the maximum even index of a λ-weight assigned to a node with odd index.

Case 2.2: $d(v_{2j-1}, x_j) = d(v_{2j}, x_j)$. The analysis of this case is analogous to the Case 2.1 and also induces a contradiction.

After this case analysis, we conclude that the optimal assignment of the λ-weights satisfies that each λ_{2k} for any $k = 2, \ldots, p$, is allocated to v_{2i} for some i, $i = 1, \ldots, p$. □

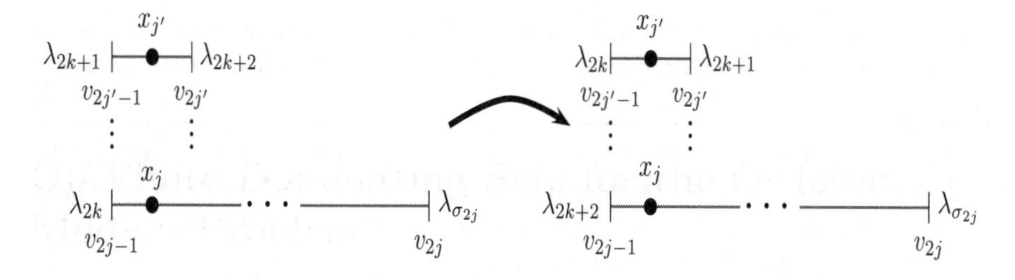

Fig. 8.11. Illustration to Case 2.1 of Lemma 8.7

The result above describes the optimal assignment of the λ-weights with even index, $k > 2$. However, it is still missing the case λ_2. The following result analyzes this case.

Lemma 8.8 *If X_p is an optimal solution of Problem (8.8) then λ_2 must be assigned to v_2.*

Proof.
First, notice that if λ_2 were assigned to v_{2i} for some i, $i = 1, \ldots, p$ then, by Remark 8.5 ii) and since λ_1 is already assigned to a v_{2j-1} for some j, $j = 1, \ldots, p$, we would have that $i = 1$.

In order to prove the result, we proceed by contradiction assuming that λ_2 is assigned to v_{2j-1} for some j, $j = 1, \ldots, p$. Therefore, since by Lemma 8.7, for $k = 2, \ldots, p$, λ_{2k} is assigned to v_{2i} for some i with $i = 1, \ldots, p$, and λ_2 is assigned to v_{2j-1}, then there exists only one $j_o \in \{1, \ldots, p\}$ such that λ_{2j_o-1} is assigned to a node v_{2i} for some i, $i = 1, \ldots, p$. Depending on the value of j_o, we distinguish the following cases:

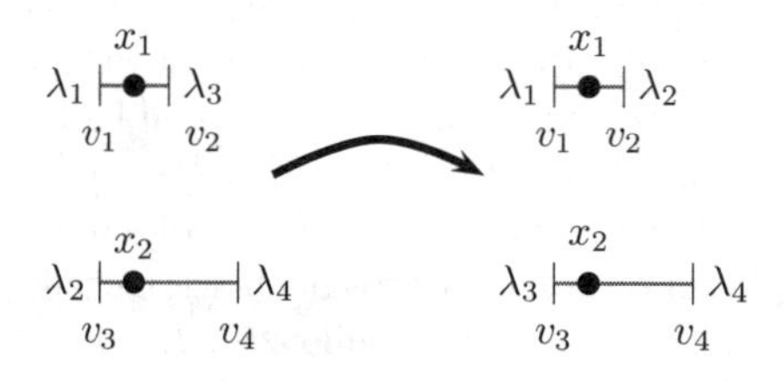

Fig. 8.12. Illustration to Case 1 of Lemma 8.8

Case 1: $j_o = 2$ (see Figure 8.12).
If λ_3 is assigned to v_{2i} for some i, $i = 1, \ldots, p$, then, by Remark 8.5 ii) and Lemma 8.7, λ_3 must be assigned to v_2 $(i = 1)$ and λ_4 to v_4. By Remark 8.5 i), $\sigma_1 < 3$ and $\sigma_3 < 4$ then $\sigma_1 \leq 2$ and $\sigma_3 \leq 2$. Therefore, λ_1 is assigned either to v_1 or v_3 and the same occurs with λ_2. In any case, to minimize the objective function, we must have that $d(v_1, x_1) = d(v_2, x_2) = d(v_3, x_3)$, and it implies

that we can reassign the λ-weights such that λ_1 goes to v_1, λ_2 to v_2 and λ_3 to v_3. Since the objective value does not change, we get the thesis of the Lemma.

Case 2: $j_o = 3$ (see Figure 8.13).
If λ_5 is assigned to v_{2i} for some i, $i = 1, \ldots, p$, then, by Remark 8.5 ii) and Lemma 8.7, λ_4 must be assigned to v_2, λ_5 to v_4 and λ_{2i} to v_{2i} for any $i = 3, \ldots, p$. Since λ_4, λ_5 and λ_6 have been already allocated and Remark 8.5 i) ensures that $\sigma_1 < 4$, $\sigma_2 < 5$ and $\sigma_3 < 6$ then $\sigma_1 \leq 3$, $\sigma_3 \leq 3$ and $\sigma_5 \leq 3$. Moreover:

i) Since $\lambda_4 < \lambda_5$ then $d(v_1, x_1) \leq d(v_3, x_2)$. (Otherwise the objective function may decrease). Indeed, if $d(v_1, x_1) > d(v_3, x_2)$ we move x_1 and x_2 towards v_1 and v_4, respectively, such that, x_1' and x_2', the new locations of x_1 and x_2, satisfy that $d(v_1, x_1') = d(v_3, x_2)$ and $d(v_3, x_2') = d(v_1, x_1)$. This movement induces the following change in the objective function:
$$\Big(d(v_1, x_1) - d(v_3, x_2)\Big)(\lambda_4 - \lambda_5) < 0,$$
what contradicts the optimality of X_p.
ii) Since $\sigma_3 \leq 3$, $\sigma_5 \leq 3$ and $\sigma_2 = 4$, X_p must satisfy that $d(v_2, x_1) \geq d(v_3, x_2)$ and $d(v_2, x_1) \geq d(v_5, x_3)$. In addition, we have by construction that $d(v_2, x_1) \leq 2$ then $d(v_3, x_2) \leq 2$ and $d(v_5, x_3) \leq 2$. This allows us to use the same arguments of Case 2.i to prove that $d(v_3, x_2) \geq d(v_5, x_3)$ because $\lambda_5 > \lambda_6$.

Therefore, λ_1 must be assigned to v_1, λ_2 to v_5 and λ_3 to v_3. In addition, since λ_1, λ_2, λ_3, λ_4, λ_5, λ_6 have been already assigned and λ_8 is assigned to v_8; Remark 8.5 i) implies that λ_7 is assigned to v_7. Repeating this argument for any $i > 4$ we have that λ_{2i-1} is assigned to v_{2i-1}. Thus, $\sigma_{2i-1} = 2i - 1$ and $\sigma_{2i} = 2i$ for any $i = 4, \ldots, p$.

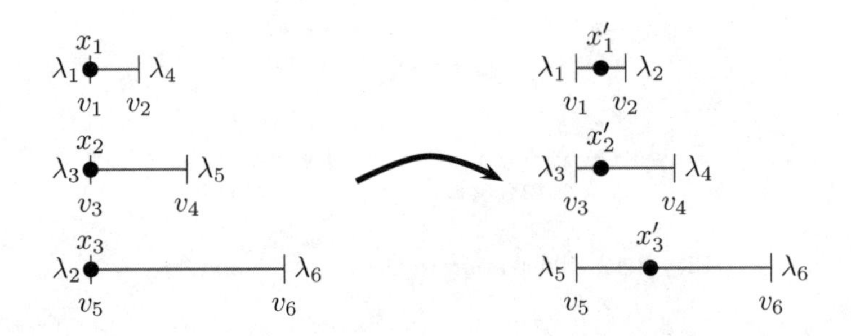

Fig. 8.13. Illustration to Case 2 of Lemma 8.8

This assignment of the λ-weights implies that $d(v_{2i}, x_i) = d(v_{2i+1}, x_{i+1})$ for any $i = 3, \ldots, p - 1$. Indeed, since $\sigma_{2i} = 2i < 2i + 1 = \sigma_{2i+1}$ then $d(v_{2i}, x_i) \leq d(v_{2i+1}, x_{i+1})$, and since $\lambda_{2i+1} > \lambda_{2i+2}$ we deduce that x_{i+1} is located as close as possible to v_{2i+1}, $i = 3, \dots, p-1$. Hence, it implies that $d(v_{2i}, x_i) = d(v_{2i+1}, x_{i+1})$ for any $i = 3, \dots, p-1$.

Moreover, with this assignment of the λ-weights and since, by (8.11), $2\lambda_2 > \lambda_4 + \lambda_5 + \lambda_8$ the optimal location for x_1, x_2 and x_3 must be: $x_1 = v_1$, $x_2 = v_3$, $x_3 = v_5$.

However, this is a contradiction, because we will prove that the above configuration of X_p does not provide an optimal solution of Problem (8.8). Indeed, move x_1, x_2 and x_3 to the new positions x_1', x_2' and x_3', respectively, where $x_1' = v_1 + 1$, $x_2' = v_3 + 1$, $x_3' = v_5 + 3$. Using the condition $d(v_{2i}, x_i) = d(v_{2i+1}, x_{i+1})$ for any $i = 3, \dots, p$, this movement allows us to displace three units length: 1) x_i towards v_{2i-1} for any even index $i = 4, \dots, p$; and 2) x_j towards v_{2j} for any odd index $j = 4, \dots, p$; without any reassignment of the λ-weights corresponding to these nodes. Therefore, these movements produce the following change in the objective function:

$$+\lambda_1 + \lambda_2 + \lambda_3 + \lambda_4 - \lambda_5 - 3\lambda_6 - 3(\lambda_7 - \lambda_8) + 3(\lambda_9 - \lambda_{10}) - 3(\lambda_{11} - \lambda_{12}) + \dots.$$

We prove that this amount is negative. Indeed, by the definition of the λ-weights, see (8.7) and (8.9)-(8.11), they satisfy that $+\lambda_1+\lambda_2+\lambda_3+\lambda_4-\lambda_5-3\lambda_6$ is negative. Besides,

$$-3(\lambda_7 - \lambda_8) + 3(\lambda_9 - \lambda_{10}) - 3(\lambda_{11} - \lambda_{12}) + \dots$$

is negative because the sequence $\lambda_7 - \lambda_8$, $\lambda_9 - \lambda_{10}$, ... is decreasing. This fact contradicts the optimality of X_p since the objective function decreases.

Case 3: $j_o = 4$. The proof is similar to the one in Case 2, and therefore it is omitted.

Case 4: $j_o > 4$ (see Figure 8.14 and 8.15).
Using Remark 8.5 ii), Lemma 8.7 and a similar argument to that used in Case 2.i, λ_1 must be assigned to v_1, λ_2 to v_3, λ_{2i+2} to v_{2i} for $i = 1, \dots, j_o - 2$, λ_{2i-3} to v_{2i-1} for $i = 3, \dots, j_o - 2$, λ_{2j_o-3} to v_{2j_o-3}, λ_{2j_o-5} to v_{2j_o-1}, λ_{2j_o-1} to v_{2j_o-2}, and λ_i to v_i for $i = 2j_o, \dots, 2p$.
Moreover, notice that, $\lambda_{\sigma_{2i-1}} > \lambda_{\sigma_{2i}}$ for $i > 1$. Hence, following a similar argument to the one in Case 2, we obtain that $d(v_{2i}, x_i) = d(v_{2j-1}, x_j)$ whenever $\sigma_{2j-1} - \sigma_{2i} = 1$.

For this assignment of the λ-weights, using (8.7) and (8.9)-(8.11), the optimal location of x_1, x_2, x_3, x_4 and x_5 must be $x_1 = v_1$, $x_2 = v_3$, $x_3 = v_5$, and either

1. $x_4 = v_7 + 4$ and $x_5 = v_9 + 2$ if $j_o = 5$ (Figure 8.14).

or

2. $x_4 = v_7 + 2$ and $x_5 = v_9 + 4$ if $j_o > 5$ (Figure 8.15).

However, this is a contradiction, because we will prove that the above configuration of X_p does not provide an optimal solution of Problem (8.8). Move x_1, x_2, x_3, x_4 and x_5 to the new positions x_1', x_2', x_3', x_4' and x_5', respectively, where $x_1' = v_1 + 1$, $x_2' = v_3 + 1$, $x_3' = v_5 + 4$, and

1. $x_4' = v_7 + 4$ and $x_5' = v_9 + 3$ if $j_o = 5$ (Figure 8.14).
2. $x_4' = v_7 + 3$ and $x_5' = v_9 + 4$ if $j_o > 5$ (Figure 8.15).

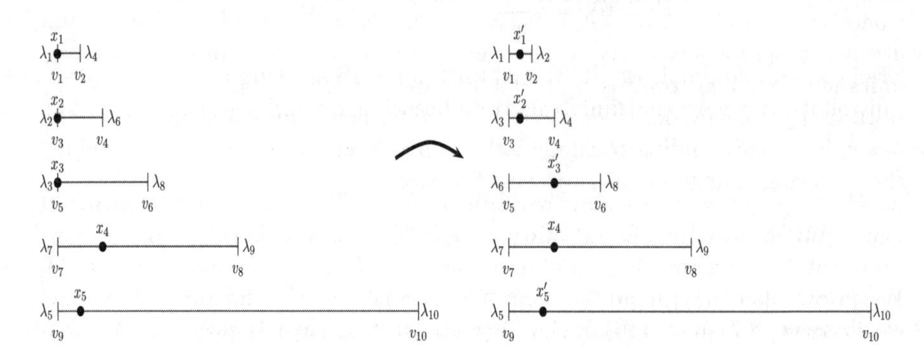

Fig. 8.14. Illustration to Case 4 i) of Lemma 8.8

These displacements permit us to move x_i towards either v_{2i-1} or v_{2i} for $i = 6, \dots, p$ using the condition $d(v_{2i}, x_i') = d(v_{2j-1}, x_j')$ when $\sigma_{2j-1} - \sigma_{2i} = 1$; without any reassignment of their corresponding λ-weights. The change of the objective function is as follows:
i) If $j_o = 5$:

$$+\lambda_1+\lambda_2+\lambda_3+\lambda_4+\lambda_5-4\lambda_8-\lambda_{10}-(\lambda_{11}-\lambda_{12})+(\lambda_{13}-\lambda_{14})-(\lambda_{15}-\lambda_{16})+\dots.$$

By (8.7) and (8.9)-(8.11), we have that $+\lambda_1 + \lambda_2 + \lambda_3 + \lambda_4 + \lambda_5 - 4\lambda_8 - \lambda_{10}$ is negative (by the definition of the λ-weights). Besides, $-(\lambda_{11} - \lambda_{12}) + (\lambda_{13} - \lambda_{14}) - (\lambda_{15} - \lambda_{16}) + \dots$ is negative because the sequence $\lambda_{11} - \lambda_{12}$, $\lambda_{13} - \lambda_{14}$, $\lambda_{15} - \lambda_{16}$, ... is decreasing.

ii) If $j_o > 5$:

$$+\lambda_1+\lambda_2+\lambda_3+\lambda_4+\lambda_5-4\lambda_8-\lambda_{10}+\sum_{\{j\geq 0, 7+6j<2j_o-3\}}(-1)^{j+1}\cdot 0\cdot(\lambda_{7+6j}-\lambda_{12+6j})$$

$$+\sum_{\{j\geq 0,9+6j<2j_o-3\}}(-1)^{j+1}4(\lambda_{9+6j}-\lambda_{14+6j})+\sum_{\{j\geq 0,11+6j<2j_o-3\}}(-1)^{j+1}(\lambda_{11+6j}-\lambda_{16+6j})$$

$$+r(-1)^{j_r}(\lambda_{2j_o-3}-\lambda_{2j_o-1})+\sum_{j=j_o}^{p-1}t(-1)^{j_t+(j-j_o)}(\lambda_{2j+1}-\lambda_{2j+2}),$$

where

$$t = \begin{cases} 0 \text{ if } \exists j_t \text{ s.t. } 2j_o - 3 = 7 + 6j_t \\ 4 \text{ if } \exists j_t \text{ s.t. } 2j_o - 3 = 9 + 6j_t \\ 1 \text{ if } \exists j_t \text{ s.t. } 2j_o - 3 = 11 + 6j_t \end{cases} ; r = \begin{cases} 0 \text{ if } \exists j_r \text{ s.t. } 2j_o + 1 = 7 + 6j_r \\ 4 \text{ if } \exists j_r \text{ s.t. } 2j_o + 1 = 9 + 6j_r \\ 1 \text{ if } \exists j_r \text{ s.t. } 2j_o + 1 = 11 + 6j_r. \end{cases}$$

In case i) we proved that $+\lambda_1 + \lambda_2 + \lambda_3 + \lambda_4 + \lambda_5 - 4\lambda_8 - \lambda_{10}$ is negative.

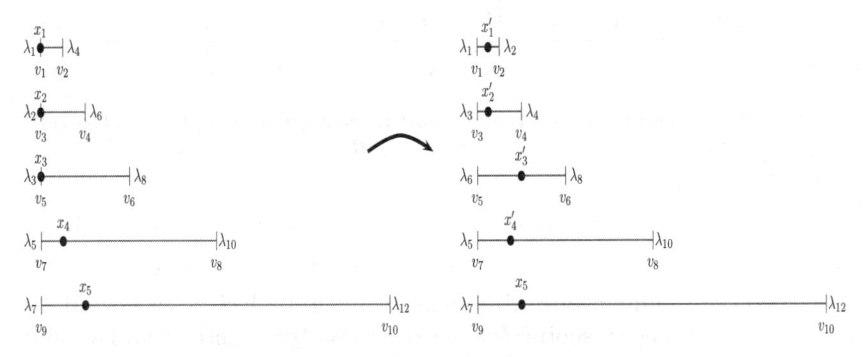

Fig. 8.15. Illustration to Case 4 ii) of Lemma 8.8

Moreover,

$$\sum_{\{j \geq 0, 9+6j < 2j_o - 3\}} (-1)^{j+1} 4(\lambda_{9+6j} - \lambda_{14+6j}) + \sum_{\{j \geq 0, 11+6j < 2j_o - 3\}} (-1)^{j+1} (\lambda_{11+6j} - \lambda_{16+6j})$$

$$+ r(-1)^{j_r}(\lambda_{2j_o-3} - \lambda_{2j_o-1}) + \sum_{j=j_o}^{p-1} t(-1)^{j_t + (j-j_o)}(\lambda_{2j+1} - \lambda_{2j+2})$$

is negative, because we can decompose the expression above in different sums, where each one of them constitutes a decreasing sequence in absolute value with alternate signs and being its first element negative.

Since in all the possible cases we get a contradiction, the initial hypothesis that λ_2 is assigned to a vertex with odd index is inconsistent. Therefore, using Lemma 8.7 we conclude that λ_2 can only be assigned to v_2. □

8.4.2 Main Results

Once the previous results have been stated, we can prove the main results in this section:

Theorem 8.7 *If X_p is an optimal solution of Problem (8.8) then $\lambda_{\sigma_i} = \lambda_i$ for $i = 1, \ldots, 2p$.*

Proof.
First, we prove that λ_1 must be assigned to v_{2i-1} for some i, $i = 1, \ldots, p$.

Indeed, if $\sigma_{2i} = 1$ for some i, $i = 1, \ldots, p$ then, by Remark 8.5 i), $\sigma_{2i-1} < 1$. However, this is impossible because $\sigma_{2i-1} \in \{1, \ldots, 2p\}$.

Second, by lemmas 8.7 and 8.8, we have that every λ_{2k} for $k = 1, \ldots, p$, is assigned to v_{2i} for some i, $i = 1, \ldots, p$. Moreover, Remark 8.5 ii), implies that $\sigma_{2i} = 2i$, (i.e. λ_{2i} is assigned to v_{2i}) for $i = 1, \ldots, p$. Therefore, using a recursive argument and Remark 8.5 i), we obtain that $\sigma_1 = 1$ (i.e. λ_1 is assigned to v_1), $\sigma_3 = 3$ and so on. Thus, the result follows. □

Remark 8.6 *The above result has been proven assuming that* $|v_{2i+1} - v_{2i+2}| = 2|v_{2i-1} - v_{2i}|$. *However, the reader may notice that the result also holds whenever* $|v_{2i+1} - v_{2i+2}| \geq 2|v_{2i-1} - v_{2i}|$.

In order to disprove the polynomial cardinality of any finite dominating set (FDS) for the multifacility ordered median problem, we consider the graph G of Figure 8.8 (assume that p is even). Let $P = \{1, \ldots, p\}$ and $J = \{j_1, j_2, \ldots, j_{\frac{p}{2}-1}, j_{\frac{p}{2}}\} \subseteq P$, such that, $1 = j_1 < j_2 < \ldots < j_{\frac{p}{2}-1} < j_{\frac{p}{2}} = p$. On the graph G we formulate the following $(\frac{p}{2})$-facility ordered median problem:

$$\min_{X_{\frac{p}{2}} \subset A(G)} \sum_{i=1}^{2p} \lambda'_{\sigma_i} w'_i d(v_i, X_{\frac{p}{2}}) \,, \tag{8.12}$$

where $\lambda' = (0, \ldots, 0, \lambda_1, \ldots, \lambda_p)$, such that, λ_i is defined by (8.7) for $i = 1, \ldots, p$, $w'_{2j-1} = w'_{2j} = 1$ for each $j \in J$ and $w'_{2j-1} = w'_{2j} = 0$ for each $j \in P \setminus J$.

Moreover, since $w'_{2j-1} d(v_{2j-1}, X_{\frac{p}{2}}) = w'_{2j} d(v_{2j}, X_{\frac{p}{2}}) = 0 \; \forall j \in P \setminus J$, the first p positions of the ordered sequence of weighted distances between each node and its service facility are given by $w'_{2j-1} d(v_{2j-1}, X_{\frac{p}{2}})$, $w'_{2j} d(v_{2j}, X_{\frac{p}{2}})$ with $j \in P \setminus J$ (indeed, these positions are always zeros). Thus, we can assume without loss of generality that the λ-weights allocated to v_{2j-1} and v_{2j} for any $j \in P \setminus J$ are the first p components of the vector λ', that is, 0.

Notice that the nodes v_{2j-1} and v_{2j} $\forall j \in P \setminus J$ are not really taken into account in the objective value because $w'_{2j-1} = w'_{2j} = 0$. Thus, this problem reduces to locate $\frac{p}{2}$ service facilities on a graph with $\frac{p}{2}$ edges. Indeed, if we consider

$$V' = \bigcup_{i=1, j_i \in J}^{\frac{p}{2}} \{v_{2j_i - 1}, v_{2j_i}\} := \bigcup_{i=1}^{\frac{p}{2}} \{v'_{2i-1}, v'_{2i}\},$$

and the path graph G' induced by the set of nodes V', Problem (8.12) can be reformulated as

$$\min_{X_{\frac{p}{2}} \subset A(G')} \sum_{i=1}^{p} \lambda_{\sigma_i} d(v'_i, X_{\frac{p}{2}}). \tag{8.13}$$

Observe that the components of the vector $\lambda = (\lambda_1, \ldots, \lambda_p)$ coincide with the first p entries of (8.7) and therefore, they satisfy (8.9)-(8.11), $\sigma_i \in \{1, \ldots, p\}$. In addition, $\sigma_i < \sigma_j$ if $d(v'_i, X_{\frac{p}{2}}) \leq d(v'_j, X_{\frac{p}{2}})$ (the w-weights are all equal to one).

Theorem 8.8 *If $X_{\frac{p}{2}}$ is the optimal solution of Problem (8.12) then $x_i = v_{2j_i-1} + z_i$ with $z_1 = 1$ and $z_i = 2^{j_{i-1}} - z_{i-1}$ for $i = 2, \ldots, \frac{p}{2}$.*

Proof.
To prove this result we consider the equivalent formulation of Problem (8.12) given in (8.13). Applying Theorem 8.7 and Remark 8.6, we get that the λ-weight allocated to the node v'_i is λ_i for $i = 1, \ldots, p$.

In addition, the solution $X_{\frac{p}{2}}$ satisfies the relationship

$$d(v'_{2i}, x_i) = d(v'_{2i+1}, x_{i+1}) \text{ for } i = 1, \ldots, \frac{p}{2} - 1.$$

Indeed, since λ_{2i} and λ_{2i+1} are assigned to v'_{2i} and v'_{2i+1} respectively, for $i = 1, \ldots, \frac{p}{2} - 1$, then $d(v'_{2i}, x_i) \leq d(v'_{2i+1}, x_{i+1})$. Moreover, x_{i+1} must be located as close as possible to v'_{2i+1} because $\lambda_{2i+1} > \lambda_{2i+2}$, which in turn implies that $d(v'_{2i}, x_i) = d(v'_{2i+1}, x_{i+1})$.

Next, we prove that $d(v'_1, x_1) = 1$. Notice that, by Remark 8.5, we have that $d(v'_1, x_1) \leq 1$. If $d(v'_1, x_1) < 1$ then we would move x_1 towards v'_2 a small enough amount, ξ. This movement would allow us to move x_i towards v'_{2i-1} for any even index $i = 2, \ldots, \frac{p}{2}$, and x_j towards v'_{2j} for any odd index $j = 2, \ldots, \frac{p}{2}$, the same amount ξ, without any reassignment of the λ-weights. These movements would produce the following change in the objective function:

$$\xi\Big(- (\lambda_2 - \lambda_1) + \sum_{k=2}^{\frac{p}{2}} (-1)^{k-1} (\lambda_{2k-1} - \lambda_{2k})\Big).$$

This amount is negative because $-(\lambda_2 - \lambda_1)$ is negative and $\{\lambda_{2k-1} - \lambda_{2k}\}_{k \geq 2}$ is a decreasing sequence of positive values, that is, $\lambda_{2k-1} - \lambda_{2k} > \lambda_{2k+1} - \lambda_{2k+2} > 0$ for $k = 1, \ldots, \frac{p}{2}$. However, this is not possible since $X_{\frac{p}{2}}$ is optimal. Therefore, we obtain that $x_1 = v'_1 + 1$ and that $X_{\frac{p}{2}}$ is the unique solution satisfying that $d(v_{2i-1}, x_i) \leq d(v_{2i}, x_i)$ for $i = 1, \ldots, p$ (see Remark 8.5). Finally, since $d(v'_{2i}, x_i) = d(v'_{2i+1}, x_{i+1})$ for $i = 1, \ldots, \frac{p}{2} - 1$, the result follows. □

Our next result proves that there is no polynomial size cardinality FDS for the multifacility ordered median problem. The proof consists of building a family of $O(M^M)$ problems on the same graph with different solutions (each solution contains at least one point not included in the remaining), M being the number of nodes.

Theorem 8.9 *There is no polynomial size FDS for the multifacility ordered median problem.*

Proof.
Consider Problem (8.12), by Theorem 8.8 and Remark 8.5, for each choice of the set $J \subseteq P$, we have an unique optimal solution satisfying that $d(v_{2i-1}, x_i) \leq d(v_{2i}, x_i)$ for $i = 1, \ldots, p$, such that, the service facility located on the edge $[v_{2p-1}, v_{2p}]$ has a different location, recall that $j_{\frac{p}{2}} = p$. Thus, since there are $\binom{2p-2}{\frac{p}{2}-2}$ different choices of the set J, any FDS for the considered problem contains at least $\binom{2p-2}{\frac{p}{2}-2}$ elements.

Therefore, we have found a family of problems for which a valid FDS is at least of order $O(M^M)$ where M denotes the number of nodes (recall that for our case $M = 2p$). □

9

The Single Facility Ordered Median Problem on Networks

In this chapter we analyze different aspects of the single facility OMP on networks. We focus mainly on the development of efficient solution algorithms. In addition, we consider geometrical properties that help to understand the underlying structure of these problems on networks. Table 9.1 summarizes the algorithmic results presented in this chapter.

Table 9.1. Summary of Complexity Results for ordered median problems.

Complexity bounds				
	Networks			
	Undirected			Directed
	General	Trees	Line	
General[1] Ordered Median	$O(NM^2 \log M)$	$O(M^3 \log M)$	$O(M^3 \log M)$	$O(NM \log M)$
Concave Ordered Median	$O(M^2 \log M)$	$O(M^2)$ [5]	$O(M^2)$	$O(M^2 \log M)$
Convex Ordered Median	$O(NM^2 \log M)$	$O(M \log^2 M)$	$O(M \log^2 M)$	$O(M^2 \log M)$
k-centrum [2]	$O(NM \log M)$ [3]	$O(M \log M)$	$O(M)$ [4]	$O(M^2 \log M)$

We start by analyzing two examples that indicate the geometrical subdivision induced on the networks by the equilibrium points and the behavior of the OMf for different choices of the λ-weights. We devote a complete section to the single facility k-centrum problem. In this section, we include algorithms to

[1] Arbitrary node-weights.
[2] Nonnegative node-weights.
[3] Unweighted. See Proposition 9.8.
[4] See Corollary 9.3.
[5] See Section 9.3.2.

solve the k-centrum problem on the line, tree graphs and on general networks (see Tamir [187].) Some of these algorithms are nowadays slightly inferior to some others developed later. Nevertheless, we have decided to include them for the sake of completeness and because all of them provide some insight into the structure of the OMP. Needless to say, that we have also included the fastest algorithms in the respective section (the linear time algorithm for the problem on the line [154]).

In the last section, we study, according to Kalcsics et al. [117], the single facility ordered median problem on general undirected and directed networks and present $O(NM^2 \log M)$ and $O(NM \log M)$ algorithms, respectively. For (undirected) trees the induced bound is $O(M^3 \log M)$. We show how to improve this bound to $O(M \log^2 M)$ in the convex case.

9.1 The Single Facility OMP on Networks: Illustrative Examples

In this section we illustrate the geometry of the OMP on networks and state some properties of the single facility ordered median problem. We start recalling the localization result which generalizes the well-known Theorem of Hakimi [84] and gives some insight into the connection between median and center problems: Theorem 8.1 proves that an optimal solution for the single facility OMP can always be found in the set $Cand := \mathsf{EQ} \cup V$.

Theorem 8.1 also gives rise to some geometrical subdivision of the graph G where the problem is solved. As it was indicated in its proof, we can assign to every subgraph $G_i, i = 1, \ldots, k$ a M-tuple giving in the j-th position the j-th nearest vertex to all points in G_i. As an example, we have in Figure 9.1 a graph with 3 nodes and all weights w_i and edge lengths equal to 1. The partition of the graph G into the subgraphs G_i, $i = 1, \ldots, k$, can be seen as a kind of high order Voronoi diagram of G related to the Voronoi partition of networks introduced in [85].

Example 9.1

Consider the network given in Figure 9.2 with $w_1 = w_2 = w_5 = 1$ and $w_3 = w_4 = w_6 = 2$. In Table 9.2 we list the set EQ, where the labels of the rows EQ_{ij} indicate that i, j are the vertices under consideration and the columns indicate the edge $e = [v_r, v_s]$. The entry in the table gives for a point $x = (e, t)$ the value of t (if t is not unique an interval of values is shown).

Having the equilibrium points, to solve the OMP for different choices of λ-parameters, we only have to evaluate the objective function with a given set of λ-values for the elements in EQ and determine the minima. Table 9.3 gives the solutions for some specific choices of λ. To describe the solution set we use the notation EQ_{kl}^{ij} to denote the part of EQ_{ij} which lies on the edge $[v_k, v_l]$.

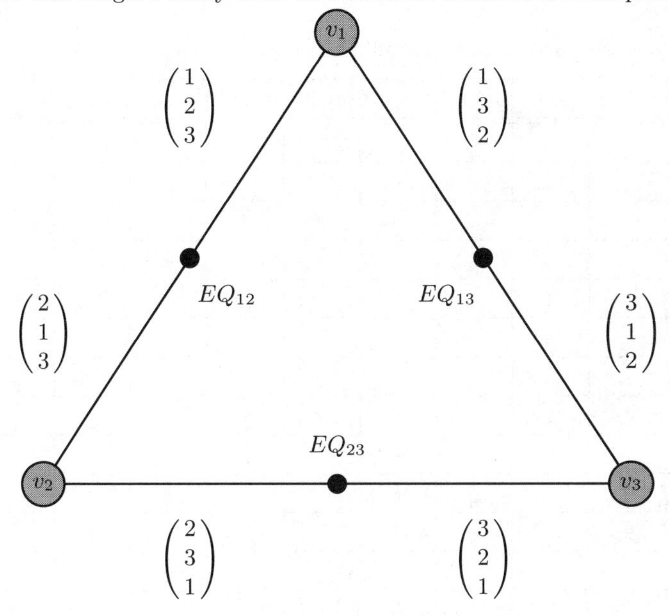

Fig. 9.1. A 3-node network with $\mathsf{EQ} = \{EQ_{12}, EQ_{13}, EQ_{23}, v_1, v_2, v_3\}$ and the geometrical subdivision

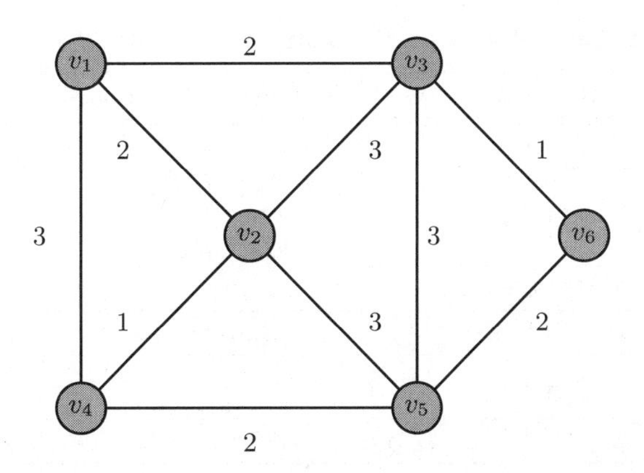

Fig. 9.2. The network used in Example 9.1

We have seen in the previous chapters that $\mathsf{OM}(\lambda)$ is a continuous, piecewise linear, concave function of $\lambda \in \Omega$, being Ω closed and convex; so we can state that $\mathsf{OM}(\lambda)$ always achieves its minimum value at some extreme points of Ω. This fact together with the linearity of $f_\lambda(x)$ for fixed x can be combined to derive an upper bound as well. Let λ_1 be the extreme point of Ω satisfying:

Table 9.2. Equilibrium points in Example 9.1.

	$[v_1, v_2]$	$[v_1, v_3]$	$[v_1, v_4]$	$[v_2, v_3]$	$[v_2, v_4]$	$[v_2, v_5]$	$[v_3, v_5]$	$[v_3, v_6]$	$[v_4, v_5]$	$[v_5, v_6]$
EQ_{12}	$\frac{1}{2}$		$\frac{2}{3}$	$\frac{5}{6}$			$\frac{2}{3}$			$\frac{1}{2}$
EQ_{13}		$\frac{2}{3}$		$\frac{4}{9}$			$\frac{2}{3}$			$\frac{1}{2}$
EQ_{14}	1		$\frac{2}{3}$	0	0	$\frac{8}{9}$	$\frac{8}{9}$			$\frac{1}{6}$
EQ_{15}			$\frac{5}{6}$		$\frac{1}{2}$	$\frac{1}{6}$	$\frac{1}{6}$	$\frac{1}{2}$		
EQ_{16}		1		1		$\frac{8}{9}$	$\frac{8}{9}$	0	$\frac{5}{6}$	
EQ_{23}		$\frac{1}{3}$		$\frac{2}{3}$			$\frac{2}{3}$			$\frac{1}{2}$
EQ_{24}			$\frac{2}{3}$		$\frac{2}{3}$				$\frac{1}{2}$	
EQ_{25}		$[\frac{3}{4}, 1]$		1		$\frac{1}{2}$	0	0	$\frac{1}{4}$	
EQ_{26}		$\frac{2}{3}$		$\frac{8}{9}$			$\frac{1}{3}$			$\frac{1}{6}$
EQ_{34}	$\frac{1}{4}$		$\frac{1}{6}$	$\frac{1}{3}$			$\frac{5}{6}$			$\frac{1}{4}$
EQ_{35}	$\frac{1}{6}$		$\frac{1}{9}$	$\frac{1}{3}$			$\frac{1}{3}$	1		1
EQ_{36}			$[\frac{5}{6}, 1]$		1	$\frac{1}{3}$	$\frac{5}{6}$	$\frac{1}{2}$	0	
EQ_{45}	$\frac{1}{2}$		$\frac{1}{3}$	$\frac{1}{3}$		$\frac{1}{9}$			$\frac{1}{3}$	
EQ_{46}	0	0	0	$\frac{1}{2}$		$[\frac{2}{3}, 1]$	$[\frac{2}{3}, 1]$		1	0
EQ_{56}		$\frac{1}{2}$		$\frac{2}{3}$			$\frac{1}{9}$			$\frac{2}{3}$

Table 9.3. Optimal solutions in Example 9.1.

obj. function	corresponding λ	set of optimal solutions	obj. value
center	$\lambda = (0,0,0,0,0,1)$	EQ_{46}^{23}, EQ_{34}^{35}, EQ_{34}^{56}	5
2-centra	$\lambda = (0,0,0,0,\frac{1}{2},\frac{1}{2})$	$[EQ_{45}^{23}, EQ_{56}^{23}]$, $[EQ_{34}^{35}, EQ_{14}^{35}]$, $[EQ_{14}^{56}, EQ_{12}^{56}]$	5
3-centra	$\lambda = (0,0,0,\frac{1}{3},\frac{1}{3},\frac{1}{3})$	EQ_{26}^{23}	$\frac{40}{9}$
median	$\lambda = (\frac{1}{6},\frac{1}{6},\frac{1}{6},\frac{1}{6},\frac{1}{6},\frac{1}{6})$	$EQ_{16}^{23} = v_3$	3
cent-dian	$\lambda = (\frac{\hat\lambda}{6},\frac{\hat\lambda}{6},\frac{\hat\lambda}{6},\frac{\hat\lambda}{6},\frac{\hat\lambda}{6},\frac{6-5\hat\lambda}{6})$	EQ_{34}^{56}, $0 \le \hat\lambda \le \frac{36}{43}$, v_3 otherwise	$-\frac{17}{12}\hat\lambda + 5, -5\hat\lambda + 8$
noname	$\lambda = (\frac{1}{4},\frac{1}{4},0,0,\frac{1}{4},\frac{1}{4})$	EQ_{16}^{23}, EQ_{12}^{56}	$\frac{13}{4}$

$$\mathsf{OM}(\lambda_1) = \min_{\lambda \in \Omega} \mathsf{OM}(\lambda)$$

and denote by $ext(\Omega)$ the set of extreme points of Ω. Then, it can be easily seen that for any graph G:

$$\mathsf{OM}(\lambda_1) \le f_\lambda(x) \le \max_{\lambda \in ext(\Omega)} f_\lambda(x) \quad \text{for any } x \in A(G).$$

We also include in this section a result on the size of the set of optimal solutions for these problems. Once the parameter λ specifying the objective function $f_\lambda(x)$ is fixed, we denote for any permutation $\sigma \in \mathcal{P}(1\ldots M)$:

$$f_{\lambda,\sigma}(x) := \sum_{i=1}^{M} \lambda_i w_{\sigma(i)} d(x, v_{\sigma(i)}).$$

Lemma 9.1 *Let x_1, x_2 be two different optimal solutions of the single facility ordered median problem, then*

$$d(x_1, x_2) \leq \min_{\sigma \in \mathcal{P}(1\ldots M)} \left(\frac{f_{\lambda,\sigma}(x_1) + f_{\lambda,\sigma}(x_2)}{\sum_{i=1}^{M} \lambda_i w_{\sigma(i)}} \right).$$

Proof.
Using the triangular inequality we have for any $\sigma \in \mathcal{P}(1\ldots M)$:

$$w_{\sigma(j)} d(x_1, x_2) \leq w_{\sigma(j)} d(x_1, v_{\sigma(j)}) + w_{\sigma(j)} d(x_2, v_{\sigma(j)}).$$

Multiplying both sides by λ_j and adding in j:

$$d(x_1, x_2) \sum_{j=1}^{M} \lambda_j w_{\sigma(j)} \leq \sum_{j=1}^{M} \lambda_j w_{\sigma(j)} d(x_1, v_{\sigma(j)}) + \sum_{j=1}^{M} \lambda_j w_{\sigma(j)} d(x_2, v_{\sigma(j)}).$$

Since this inequality holds for any σ, it also holds for the minimum, i.e.:

$$d(x_1, x_2) \leq \min_{\sigma \in \mathcal{P}(1\ldots M)} \left(\frac{f_{\lambda,\sigma}(x_1) + f_{\lambda,\sigma}(x_2)}{\sum_{i=1}^{M} \lambda_i w_{\sigma(i)}} \right).$$

□

It should be noted that this bound is tight. To see this consider the following easy example.

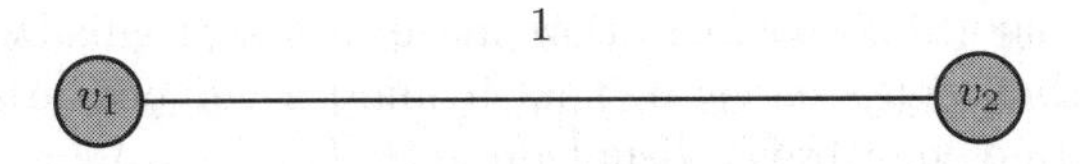

Fig. 9.3. The network used in Example 9.2

Example 9.2
Consider the network given in Figure 9.3 with $w_1 = w_2 = 1$ and $\lambda_1 = \lambda_2 = 1$. In this case the entire edge $[v_1, v_2]$ is optimal and $\mathsf{OM}(\lambda) = 1$. Furthermore, let $x_1 = v_1$ and $x_2 = v_2$ then,

$$\frac{1}{\sum_{j=1}^{M} \lambda_j w_{\sigma(j)}} \Big(f_{\lambda,\sigma}(x_1) + f_{\lambda,\sigma}(x_2) \Big) = \frac{1}{2}(1 + 1) = 1.$$

Since $d(x_1, x_2) = 1$ it also equals the value of the bound of Lemma 9.1.

9.2 The k-Centrum Single Facility Location Problem

This section is devoted to the single facility k-centrum location problem. This problem is a particular case of the OMP for a specific choice of $\bar{\lambda} = (0, \ldots, 0, 1, \overset{(k)}{\ldots}, 1)$, which allows us to develop some special properties. Moreover, the study of the k-centrum single facility problem is interesting because it will be used later to extend some algorithmic results for the general single facility ordered median problem. Most of the results in this section follow the presentation in Tamir [187].

In the single facility k-centrum problem on $G = (V, E)$ we wish to find a point x in $A(G)$, minimizing the sum of the k farthest weighted distances to the nodes from x. Specifically, suppose that each node v_i, $i = 1, ..., M$, is associated with a nonnegative weight w_i. For each vector $(x_1, ..., x_M) \in \mathbb{R}^M$, recall that $r_k(x_1, ..., x_M)$ (see 1.15) is the sum of the k-largest entries in the set of reals $\{x_1, ..., x_M\}$. The objective function of the single facility k-centrum problem is to find a point x of $A(G)$ minimizing

$$f_{\bar{\lambda}}(x) = r_k(w_1 d(x, v_1), ..., w_M d(x, v_M)).$$

In the discrete version of the problem we also require x to be a node. A minimizer of $f_{\bar{\lambda}}(x)$ is called an *absolute k-centrum*, and an optimal solution to the discrete problem is a *discrete k-centrum*. The unweighted single facility k-centrum problem refers to the case where $w_i = 1$, $i = 1, ..., M$.

From the definition it follows that if $k = 1$ the above model reduces to the classical center problem, and when $k = M$ we obtain the classical median problem. Several special cases of the single facility k-centrum problem have been discussed in the literature. To the best of our knowledge, the k-centrum concept was first introduced by Slater [183]. Slater [183] as well as Andreatta and Mason [4] considered the case of the discrete single facility on tree graphs, and observed some properties including the convexity of the objective on tree graphs. There are no algorithms with specified complexity in these papers. Peeters [159] studied the single facility problem on a graph, which he called the upper-k 1-median. He presented an $O(M^2 \log M + NM)$ algorithm for solving only the discrete version of this model, where the solution is restricted to be a node.

This section develops polynomial time algorithms for finding a single facility k-centrum on path graphs, tree graphs and general graphs based on results by Tamir [187]. Some of these results will be improved in the next section analyzing algorithmic approaches to the general OMP.

The Single k-Centrum of a Path

We consider first the case of locating a single facility on a line (a path graph). In this case we view x as a point on the real line, and each node v_i is also regarded as a real point. So we have

$$f_{\bar{\lambda}}(x) = r_k(w_1|x - v_1|, ..., w_M|x - v_M|),$$

with $\bar{\lambda} = (0, \ldots, 0, 1, \overset{(k)}{\ldots}, 1)$. To compute $f_{\bar{\lambda}}(x)$ in the case of the single facility k-centrum problem on the line, we look at the k largest functions in the collection $\{w_i|x - v_i| = \max(w_i(x - v_i), -w_i(x - v_i))\}$, $i = 1, ..., M$. Since $k \leq M$, we may assume that for each i, and a real x, at most one of the 2 values, $\{w_i(x - v_i), -w_i(x - v_i)\}$, i.e., a nonnegative one, belongs to the subcollection of the k largest function values at x. Thus, the single facility k-centrum problem on the line amounts to find the minimum point of the sum of the k largest linear functions in the collection consisting of the above $2M$ linear functions, $\{w_i(x - v_i), -w_i(x - v_i)\}$, $i = 1, ..., M$.

Nowadays, there are several approaches that solve efficiently this problem. For the sake of completeness, we will present three approaches illustrating the different tools which can be used to handle location problems on graphs. Needless to say, the reader only interested in the fastest algorithm is referred to Lemma 9.2 and the remaining material that presents a linear time algorithm for the considered problem.

In our first approach to solve the above problem we consider the following more general problem. Let $[a, b]$ be an interval on the real line, and let k be a positive integer. Given is a collection of $q \geq k$ linear functions $\{a_i x + b_i\}$, $i = 1, ..., q$, defined on $[a, b]$. The objective is to find x^*, a minimum point of $f_{\bar{\lambda}}(x) = r_k(a_1 x + b_1, ..., a_q x + b_q)$ in $[a, b]$.

It is known, that if we have a set of q linear functions, the sum of the k largest functions is a convex, piecewise linear function. Dey [53] showed that the number of breakpoints of this convex function is bounded above by $O(qk^{1/3})$. Katoh (see Eppstein [72]) demonstrated that the number of breakpoints can achieve the bound $\Omega(q \log k)$.

To find the optimum, x^*, we use the general parametric procedure of Megiddo [132] with the modification in Cole [43]. We only note that the master program that we apply is the sorting of the q linear functions, $\{a_i x + b_i\}$, $i = 1, ..., q$, (using x as the parameter).

Proposition 9.1 *The optimum x^* of the function*

$$f_{\bar{\lambda}}(x) = r_k(a_1 x + b_1, ..., a_q x + b_q)$$

in $[a, b]$ can be found in $O(q \log q)$ time

Proof.
The test for determining the location of a given point x' w.r.t. x^* is based on the linear time algorithm for finding the k-th largest element of a set of reals, Blum et al. [22]. Due to the convexity of the function $f_{\bar{\lambda}}(x)$, it is sufficient to determine the signs of the one-sided derivatives at x'. This is done as follows. Consider, for example, $f^+_{\bar{\lambda}}(x')$, the derivative from the right. First, using the algorithm in [22] we compute in linear time, the function value $f_{\bar{\lambda}}(x')$, and the k-th largest linear function at x'. Then, we can have 2 different situations:

1. If there is only one function in the collection of the q functions whose value is equal to the k-th largest value at x', then $f^+_{\bar{\lambda}}(x')$ is the sum of the slopes of the k linear functions corresponding to the k largest values at x'.
2. Otherwise, let $k' < k$, be the number of linear functions whose value at x' is larger than the k-th largest value, and let A be the sum of the slopes of these k' functions. Let I be the subcollection of linear functions whose value at x' is equal to the k-th largest value. Again, using the algorithm in [22] we compute the $k - k'$ largest slopes from the set of slopes of the linear functions in I. Let B be the sum of these $k - k'$ slopes. We then have $f^+_{\bar{\lambda}}(x') = A + B$.

We now conclude that with the above test the parametric approach in [132, 43] will find x^* in $O(q \log q)$ time.

□

Now we can state the following result:

Corollary 9.1 *The algorithm described in the proof of Proposition 9.1 solves the k-centrum problem on the line in $O(M \log M)$.*

Proof.
For this problem it is enough to take $q = 2M$ and the result follows. □

We can also get the result for the discrete k-centrum problem, which is the following.

Corollary 9.2 *The discrete k-centrum can be found in $O(M \log M)$ time.*

Proof.
Because of convexity the solution to the discrete k-centrum problem (the discrete k-centrum) is one of the two consecutive nodes bounding the absolute k-centrum. The discrete solution can also be found directly by a simpler algorithm. First, sort the nodes, and then, using the above linear time procedure to compute the derivatives of $f_{\bar{\lambda}}(x)$, perform a binary search on the nodes. This approach also takes $O(M \log M)$ time. □

Finally the complexity can be improved for the unweighted case, and we can find the solution in linear time.

Proposition 9.2 *The unweighted k-centrum, ($w_i = 1$, for $i = 1, ..., M$), can be found in linear time.*

Proof.
Define $l = \lfloor (k + 1)/2 \rfloor$ (for any real number y, $\lfloor y \rfloor$ denotes the smallest integer which is greater than or equal to y). It is easy to verify that the

midpoint of the interval connecting the l-th smallest point in V with the l-th largest point in V is a local minimum point. By the convexity of the objective function, we conclude that it is an absolute k-centrum. Thus, by using the linear time algorithm, in [22] we can find the unweighted k-centrum of an unsorted sequence of M points on the line in $O(M)$ time. □

Conceptually, the above characterization can be extended to the weighted case. But it is not clear whether this can be used to obtain a linear time algorithm for the weighted case. A necessary and sufficient condition in the weighted case is that the k farthest weighted distances to nodes will be evenly located on both sides of the centrum, i.e., the weight of those nodes on either side of the centrum should be half of the total weight of the k nodes. In the unweighted case the total weight is fixed, and is equal to k, while in the weighted case, the total weight of the k furthest nodes is not known a priori.

Recently, a new approach has been developed to solve the absolute k-centrum in linear time. It is included at the end of this section. However, and for the sake of completeness, we also present another slightly inferior algorithm which solves the weighted model for a fixed value of k, in $O(M)$ time. Specifically, for any fixed value of k, this algorithm will find the (weighted) k-centrum x^* on the line in $O((k^2 \log k)M)$ time.

Again, we consider a general collection of q linear functions $\{a_i x + b_i\}$, $i = 1, ..., q$. Let x^* be a minimum point of the function $f_{\bar{\lambda}}(x)$, defined above, in $[a, b]$.

We first prove the following statement:

Proposition 9.3 *We can remove, in $O(q)$ time, a fixed proportion of the linear functions from the entire collection, without affecting the optimality of x^*.*

Proof.
First divide the collection of q functions into $q/(k+1)$ sets of $k+1$ functions each (for simplicity suppose that $q/(k+1)$ is integer). For each set of $k+1$ functions, at most k of the functions will appear in the objective. Specifically, for each x, the minimum of the $k+1$ functions in the set is ineffective, and therefore can be dropped. The minimum of the $k+1$ functions of a set, called the minimum envelope, is a concave piecewise linear function with k breakpoints. Because of the convexity of the objective, if we know the pair of breakpoints bracketing x^*, we can then omit the minimum function (on the interval connecting the pair) from this collection of $k+1$ linear functions. □

We proceed as follows. For each one of the above $q/(k+1)$ sets of $k+1$ functions, we find in $O(k \log k)$ time its minimum envelope (Kariv and Hakimi [118]). The total effort of this step is $O(q \log k)$.

Next we look at the set P_1, consisting of the $q/(k+1)$ median breakpoints of the $q/(k+1)$ envelopes. We find the median point in P_1, and check its position w.r.t. x^* in $O(q)$ time, as explained above, by implementing the algorithm in [22]. Note that the time to check the location of a given point with respect to x^* is proportional to the number of linear functions in the (current) collection. Thus, for at least half of the $q/(k+1)$ envelopes, we know the position of their medians w.r.t. the centrum. We consider only these $[q/2(k+1)]$ envelopes and continue this binary search. For example, in the next iteration each such envelope contributes its $[k/4]$-th or $[3k/4]$-th breakpoint, depending on the location of the centrum x^* w.r.t. its median. Let P_2 be the set of these $[q/2(k+1)]$ points.

We now check the position of the median element in P_2, w.r.t. x^*. After $O(\log k)$ iterations we identify $O(q/(k(k+1)))$ sets (envelopes) such that for each one of them we have a pair of adjacent breakpoints (segment) bracketing x^*. Thus, from each one of these $O(q/(k(k+1)))$ envelopes, we can eliminate the linear function which coincides with the envelope on this segment.

To summarize, at the end of the first phase, in a total effort of $O(q \log k)$, ($O(\log k)$ iterations of $O(q)$ time each), we eliminate $O(q/k^2)$ functions. We then recursively restart the process with the remaining subcollection of linear functions. Again, it should be emphasized that the time to check the location of a given point with respect to the centrum is proportional to the number of linear functions in the remaining subcollection of linear functions. In particular, the total effort spent in the second phase is $O(q(1-1/k^2)\log k)$.

At the last phase, the subcollection will consist of $O(k^2)$ functions, and therefore, from the discussion above, the time needed to find x^* for this subcollection is $O(k^2 \log k)$.

Now we can calculate the complexity of this algorithm:

Proposition 9.4 *The complexity of the above algorithm is* $O((k^2 \log k)q)$.

Proof.
To evaluate the complexity of the algorithm, suppose that we have a collection of m linear functions. Let $T(q)$ be the total effort to compute x^*, a minimum point of the function, defined as the sum of the k largest functions in the collection (in the single k centrum problem on a path $q = 2M$).

The recursive equation is

$$T(q) \leq c_k q + T(a_k q),$$

where $c_k = c \log k$, for some constant c, and $a_k = 1 - 1/(k+1)^2$. Thus, $T(q) = O((k^2 \log k)q)$. □

We note in passing that with the same effort we can also find the discrete k-centrum, since it is one of the pair of adjacent nodes bracketing the absolute k-centrum.

Recently, a new approach to this problem has been presented by Ogryczak and Tamir [154] who develop an $O(M)$ procedure to minimize the sum of the

k largest linear functions out of M. This technique can also be applied to get a procedure for the $k-$centrum on a path in linear time.

To facilitate the presentation, we first introduce some notation. For any real number z, define $(z)^+ = \max(z, 0)$. Let $y = (y_1, y_2, \ldots, y_M)$ be a vector in $\mathbb{R}^M$. Define $\theta(y) = (\theta_1(y), \theta_2(y), \ldots, \theta_M(y))$ to be the vector in $\mathbb{R}^M$, obtained by sorting the M components in y in non-increasing order, i.e., $\theta_1(y) \geq \theta_2(y) \geq \cdots \geq \theta_M(y)$. $\theta_i(y)$ will be referred to as the i-th largest component of y. Finally, for $k = 1, 2, \ldots, M$, we recall that $r_k(y) = \sum_{i=1}^{k} \theta_i(y)$ is the sum of the k largest components of y.

We will need the following lemma. For the proof, the reader is referred to [154].

Lemma 9.2 *For any vector $y \in \mathbb{R}^M$ and $k = 1, \ldots, M$,*

$$r_k(y) = \frac{1}{M}\left(k\sum_{i=1}^{M} y_i + \min_{t\in\mathbb{R}} \sum_{i=1}^{M}[k(t-y_i)^+ + (M-k)(y_i - t)^+]\right).$$

Moreover, $t^* = \theta_k(y)$ is an optimizer of the above minimization problem.

It follows from the above lemma that $r_k(y)$ can be represented as the solution value of the following linear program:

$$\begin{aligned} r_k(y) = \min\ & \frac{1}{M}\Big(\sum_{i=1}^{M}[k\hat{d}_i + (M-k)\tilde{d}_i + ky_i]\Big) \\ \text{s.t. } & \tilde{d}_i - \hat{d}_i = y_i - t, \quad i = 1, \ldots, M, \\ & \tilde{d}_i, \hat{d}_i \geq 0, \quad i = 1, \ldots, M. \end{aligned}$$

Substituting $\hat{d}_i = \tilde{d}_i - y_i + t$, we obtain

$$\begin{aligned} r_k(y) = \min\ & \left(kt + \sum_{i=1}^{M} \tilde{d}_i\right) \\ \text{s.t. } & \tilde{d}_i \geq y_i - t, \quad \tilde{d}_i \geq 0, \quad i = 1, \ldots, M. \end{aligned}$$

Given the collection of functions, $\{g_i(x)\}_{i=1}^{M}$, and the polyhedral set $Q \in \mathbb{R}^d$, let $g(x) = (g_1(x), \ldots, g_M(x))$. The problem of minimizing $r_k(g(x))$, the sum of the k largest functions of the collection over Q, can now be formulated as

$$\begin{aligned} \min \quad & \left(kt + \textstyle\sum_{i=1}^{M} \tilde{d}_i\right) \\ \text{s.t.} \quad & \tilde{d}_i \geq g_i(x) - t, \qquad \tilde{d}_i \geq 0, \quad i = 1, \ldots, M \\ & x = (x_1, \ldots, x_d) \in Q. \end{aligned}$$

Now, let us consider the linear case. For $i = 1, \ldots, M$, $g_i(x) = a^i(x) + b_i$, where $a^i = (a^i_1, \ldots, a^i_d) \in \mathbb{R}^d$, and $b_i \in \mathbb{R}$. With the above notation, the problem can be formulated as the following linear program

$$\begin{aligned} \min \quad & \left(kt + \textstyle\sum_{i=1}^{M} \tilde{d}_i\right) \\ \text{s.t.} \quad & \tilde{d}_i \geq a^i x + b_i - t \quad i = 1, \ldots, M, \\ & \tilde{d}_i \geq 0, \qquad\qquad\quad i = 1, \ldots, M, \\ & x = (x_1, \ldots, x_d) \in Q. \end{aligned}$$

Note that this linear program has $M + d + 1$ variables, $\tilde{d}_1, \ldots, \tilde{d}_M$, $x_1, \ldots, x_d$, t, and $2M + p$ constraints. This formulation constitutes a special case of the class of linear programs defined as the dual of linear multiple-choice knapsack problems. Therefore, using the results in [215], when d is fixed, an optimal solution can be obtained in $O(p + M)$ time (see also [135]).

Now we are going to solve the rectilinear $k-$centrum problem. For each pair of points $u = (u_1, \ldots, u_d)$, $v = (v_1, \ldots, v_d)$ in $\mathbb{R}^d$ let $d(u, v)$ denote the rectilinear distance between u and v,

$$d(u, v) = \sum_{j=1}^{d} |v_j - u_j|.$$

Given is a set $\{v^1, \ldots, v^M\}$ of M points in $\mathbb{R}^d$. Suppose that v^i, $i = 1, \ldots, M$, is associated with a nonnegative real weight w_i. For each point $x \in \mathbb{R}^d$ define the vector $D(x) \in \mathbb{R}^d$ by $D(x) = (w_1 d_1(x), \ldots, w_M d_M(x))$. For a given $k = 1, \ldots, M$, the single facility rectilinear $k-$centrum problem in $\mathbb{R}^d$ is to find a point $x \in \mathbb{R}^d$ minimizing the objective $r_k(x) = \sum_{i=1}^{k} \theta_i(D(x))$. Note that the case $k = 1$ coincides with the classical (weighted) rectilinear 1-center problem in $\mathbb{R}^d$, while the case $k = M$ defines the classical (weighted) rectilinear 1-median problem, proposed by Hakimi. It is well known that the last two problems can be formulated as linear programs. The center problem ($k = 1$) is formulated as

$$\begin{aligned} \min \quad & t \\ \text{s.t.} \quad & t \geq w_i d_i(x), \qquad i = 1, \ldots, M, \\ & x = (x_1, \ldots, x_d) \in \mathbb{R}^d. \end{aligned}$$

To obtain a linear program, we replace each one of the M nonlinear constraints $t \geq w_i d_i(x)$, by a set of 2^d linear constraints. For $i = 1, \ldots, M$, let $\Delta^i = (\delta^d_1, \ldots, \delta^i_d)$ be a vector all of whose components are equal to +1 or -1. Consider the set of 2^d linear constraints, $t \geq \sum_{j=1}^{d} \delta^i_j w_i(x_j - v^i_j)$, $\delta^i_j \in \{-1, +1\}$, $j = 1, \ldots, d$. This linear program has $d + 1$ variables, $t, x_1, \ldots, x_d$ and $2^d M$ constraints. Therefore, when d is fixed, it can be solved in $O(M)$ time by the algorithm of Meggido [133].

Similarly, the median problem ($k = M$) is formulated as,

$$\begin{aligned}
\min \quad & \textstyle\sum_{i=1}^{M} z_i \\
\text{s.t. } & z_i \geq \textstyle\sum_{j=1}^{d} \delta_j^i w_i(x_j - v_j^i),\ i = 1, \dots, M, \\
& \delta_j^i \in \{-1, +1\}, \qquad j = 1, \dots, d, \quad i = 1, \dots, M, \\
& x = (x_1, \dots, x_d) \in \mathbb{R}^d.
\end{aligned}$$

This linear program has $M + d$ variables, $z_1, \dots, z_M, x_1, \dots, x_d$, and $2^d M$ constraints. This formulation is also a special case of the class of linear programs discussed by Zemel [215]. Therefore, when d is fixed, an optimal solution can be obtained in $O(M)$ time. The above formulation of the median problem can be replaced by another, where the number of variables is $Md + d$ and the number of constraints is only $2dM$.

$$\begin{aligned}
\min \quad & \textstyle\sum_{i=1}^{M} \sum_{j=1}^{d} y_{i,j} \\
\text{s.t.} \quad & y_{i,j} \geq w_i(x_j - v_j^i), \quad j = 1, \dots, d, \quad i = 1, \dots, M, \\
& y_{i,j} \geq -w_i(x_j - v_j^i), \ j = 1, \dots, d, \quad i = 1, \dots, M, \\
& x = (x_1, \dots, x_d) \in \mathbb{R}^d.
\end{aligned}$$

The latter compact formulation is also solvable in $O(M)$ time by the procedure of Zemel when d is fixed. In fact, it is easy to see from this formulation that the $d-$dimensional median problem is decomposable into d 1-dimensional problems. Therefore, it can be solved in $O(dM)$ time.

We note in passing that when d is fixed, even the $1-$centdian objective function, defined by Halpern [87] as a convex combination of the center objective $r_1(x)$ and the median function $r_M(x)$ can be solved in linear time.

Apart from the above algorithm and to the best of our knowledge, for a general value of k no linear time algorithms are reported in the literature even for $d = 1$. In [187] the case $d = 1$ is treated as a special case of a tree network. In particular, this one dimensional problem is solved in $O(M)$ time when k is fixed, and in $O(M \log M)$ time when k is variable. Subquadratic algorithms for any fixed d and variable k are given in [117]. For example, for $d = 2$ the algorithm there has $O(M \log^2 M)$ complexity.

Using the above results, we can now formulate the rectilinear $k-$centrum problem in $\mathbb{R}^d$ as the following optimization problem:

$$\begin{aligned}
\min \quad & \left(kt + \textstyle\sum_{i=1}^{M} \tilde{d}_i\right) \qquad (9.1) \\
\text{s.t.} \quad & \tilde{d}_i \geq w_i d_i(x) - t, \quad \tilde{d}_i \geq 0, \quad i = 1, \dots, M, \\
& x = (x_1, \dots, x_d) \in \mathbb{R}^d.
\end{aligned}$$

As above, to obtain a linear program we replace each one of the M nonlinear constraints $\tilde{d}_i \geq w_i d_i(x) - t$ by a set of 2^d linear constraints. The rectilinear $k-$centrum problem is now formulated as the linear programming problem,

$$
\begin{aligned}
\min \quad & \left(kt + \sum_{i=1}^{M} \tilde{d}_i\right) \\
\text{s.t. } & \tilde{d}_i + t \geq \sum_{j=1}^{d} \delta_i^j w_i(x_j - v_j^i),\ i = 1\ldots, M, \\
& \delta_j^i \in \{-1, 1\}, \qquad j = 1, \ldots, d, \quad i = 1\ldots, M, \\
& \tilde{d}_i \geq 0, \quad i = 1, \ldots, M.
\end{aligned}
$$

Note that the linear program has $M+d+1$ variables, $\tilde{d}_1, \ldots, \tilde{d}_M, x_1, \ldots, x_d$, t and $2^d M + M$ constraints. This formulation is again a special case of the class of linear programs defined as the dual of linear multiple-choice knapsack problems. Therefore, using the results of Zemel [215], when d is fixed, an optimal solution can be found in $O(M)$ time. In the particular case of $d = 1$ the above problem reduces to the k-centrum problem on the line. Therefore, we can state the following result:

Corollary 9.3 *The single facility k-centrum problem on a line is solvable in $O(M)$ time.*

We note in passing that the results for the rectilinear problems can be extended to other polyhedral norms. For example, if we use the ℓ_∞ and let the distance between $u, v \in \mathbb{R}^d$ be defined by $d(u,v) = \max_{j=1,\ldots,d} |v_j - u_j|$, we get the following formulation for the respective $k-$centrum problem:

$$
\begin{aligned}
\min \quad & \left(kt + \sum_{i=1}^{M} \tilde{d}_i\right) \\
\text{s.t. } & \tilde{d}_i + t \geq w_i d_i(x),\ j = 1, \ldots, d, \quad i = 1, \ldots, M, \\
& \tilde{d}_i + t \geq -w_i d_i(x),\ j = 1, \ldots, d, \quad i = 1, \ldots, M.
\end{aligned}
$$

The Single k-Centrum of a Tree

Some of the above algorithms for a path graph (see Corollary 9.3) can be extended to a tree $T = (V, E)$, $V = \{v_1, ..., v_M\}$. We use a 2-phase approach. In the first phase we identify a pair of adjacent nodes (an edge) such that the absolute k-centrum is between them. In the second phase we have to solve the above problem on the edge (segment) containing the absolute k-centrum. The latter subproblem is a special case of the problem on a path discussed in the previous section. Therefore, the total time for the second phase is $O(M)$. We will next show that the first phase can be performed in $O(M \log^2 M)$ time.

From the discussion on path graphs we know that the objective function is convex on each path of the tree. Therefore, we can apply the centroid decomposition search procedure of Kariv and Hakimi [118]. They used this search procedure to find the weighted 1-center of a tree. The procedure is based on searching over a sequence of $O(\log M)$ nodes, which are centroids of some nested sequence of subtrees (the total time needed to generate the sequence is only $O(M)$). For each node in this sequence they have a simple $O(M)$ test to identify whether this node is the 1-center of the tree. Moreover, if the node is not the 1-center, the test identifies the direction from the node towards the 1-center. All we need now to implement their search procedure is an efficient test which finds a direction from a given node v to the absolute k-centrum if v itself is not the absolute k-centrum.

Now we introduce the notation that we will need during the section. Consider an arbitrary node v of the given tree $T = (V, E)$. Let $T_1 = (V_1, E_1), ..., T_q = (V_q, E_q)$, denote the set of all maximal connected components obtained by the removal of v and all its incident edges from T. For $i = 1, ..., q$, let u_i be the closest node to v in $T_i = (V_i, E_i)$. Let r be the k-th largest element in the set of weighted distances of the nodes of T from v. For $i = 1, ..., q$, let V_i^+ ($V_i^=$) be the index set of the nodes in T_i, whose weighted distance from v is greater than (equal to) r. Let $t_i = |V_i^+|$, $s_i = |V_i^=|$, and let T_i^+ denote the subtree obtained by augmenting the edge (v, u_i) to T_i. Also, let $V^+ = \bigcup_{i=1}^{q} V_i^+$ and $V^= = \bigcup_{i=1}^{q} V_i^=$.

With the above notation we can state a simple test to determine whether the k-centrum is in a given T_j^+. Define $t = \sum_{i=1}^{q} t_i$, and $s = \sum_{i=1}^{q} s_i$. Note that $t \leq k \leq t + s$.

Consider a point $x \neq v$, on the edge (v, u_j), which is sufficiently close to v. Then it is easy to check that the difference between the objective value at x and the objective value at v is given by

$$d(v, x)(-A_j + B_j - C_j + D_j),$$

where, $A_j = \sum_{h \in V_j^+} w_h$, $B_j = \sum_{h \in V^+ - V_j^+} w_h$, D_j is the sum of the $m_j = \min(k - t, s - s_j)$ largest weights in the set $\{w_h : h \in V^= - V_j^=\}$, and C_j is the sum of the $k - t - m_j$ smallest weights in the set $\{w_h : h \in V_j^=\}$.

It is now clear, that if v is not an absolute k-centrum, then the latter is contained in the component T_j^+, if and only if the above expression for the difference is negative. v is an absolute k-centrum if and only if $(-A_j + B_j - C_j + D_j)$ is nonnegative for all $j = 1, ..., q$.

Proposition 9.5 *The total time needed to calculate A_j, B_j, C_j for all $j = 1, ..., q$, is $O(M)$. Terms D_j, $j = 1, ..., q$ is $O(M \log M)$.*

Proof.
It is clear from the definitions that the total effort needed to compute A_j, B_j, for all $j = 1, ..., q$, is $O(M)$. For each $j = 1, ..., q$, C_j can be computed in $O(s_j)$ time by using the linear time algorithm in Blum et al. [22]. Thus, the total time to compute C_j for all $j = 1, ..., q$, is $O(s) = O(M)$. The computation of the terms D_j, $j = 1, ..., q$, is more involved.

We first sort the elements of the set $\{w_h : h \in V^=\}$ in decreasing order. Let $L = (w_{h(1)}, w_{h(2)}, ..., w_{h(s)})$ denote the sorted list. We then compute the partial sums $W_p = \sum_{g=1,\dots,p} w_{h(g)}$, $p = 1, \dots, s$. The total effort of this step is $O(s \log s) = O(M \log M)$.

To compute a term D_j, $j = 1, \dots, q$, we now perform the following steps. First, we delete from the list L, the elements $\{w_h : h \in V_j^=\}$, and obtain the reduced sublist L_j. The effort needed is $O(s_j \log s)$. Next, we identify the m_j-th largest element in L_j. Suppose that this is the element $w_{h(y)}$ in the list L. The term D_j is now defined by $D_j = W_y - \sum\{w_{h(z)} : h(z) \in V_j^=, z < y\}$. The effort to compute D_j is $O(s_j \log s)$. Therefore, the total effort to compute all the terms D_j, $j = 1, \dots, q$, is $O(s \log s) = O(M \log M)$. □

And now we can state the following result:

Corollary 9.4 *The absolute k-centrum of a tree can be found in $O(M \log^2 M)$ time.*

Proof.
For a given node v, the total effort to compute the above difference for all $j = 1, \dots, q$, is $O(M \log M)$. Combined with the centroid decomposition, which requires using this step $O(\log M)$ times, we will have an $O(M \log^2 M)$ algorithm to find the absolute k-centrum of a tree. □

As in the former section, the discrete k-centrum is one of the 2 nodes of the edge containing the absolute k-centrum, so it can be calculated with the same effort.

Proposition 9.6 *The unweighted absolute k-centrum of a tree can be found in $O(M + k \log M)$ time.*

Proof.
First, observe that from the discussion in the previous section it follows that the time needed for the second phase is only $O(M)$. The first phase can be performed in $O(M + k \log M)$ time. There are $O(\log M)$ iterations in the first phase.

In the first iteration we find the centroid of the original tree, say v. We compute the terms $\{A_j, B_j, C_j, D_j\}$, $j = 1, \dots, q$, defined above. In the unweighted case $A_j = t_j$, $B_j = t - t_j$, $C_j = k - t - m_j$, and $D_j = m_j$, $j = 1, \dots, q$. The effort of this step is therefore proportional to the number of nodes. If v is not the k-centrum, the k-centrum is in some subtree T_j^+ (since v is a centroid the number of nodes in T_j is at most $M/2$).

Let V' be a subset of nodes in $V - V_j$, corresponding to a set of $\min([(k+1)/2], |V - V_j| - 1)$ largest distances from v to nodes in $V - V_j$. From the discussion on the unweighted k-centrum of a path in the previous section, we conclude that the unweighted k-centrum is determined only by its distances to the nodes of T_j^+ and the nodes in V'.

Let T' denote the tree obtained by augmenting the node set V' to T_j^+, and connecting each node $v_t \in V'$ to the centroid v with an edge of length

$d(v_t, v)$. The number of nodes of T' is at most $(k+1)/2 + 1 + M/2$. In the second iteration T' replaces the original tree T. We find the centroid of T', and proceed as above.

Since there are $O(\log M)$ iterations, the total effort of the first phase is $O(M + k \log M)$. Thus, the time needed to locate a single k-centrum of a tree in the unweighted case is $O(M + k \log M)$. □

We believe that the $O(M \log^2 M)$ bound for the weighted case can be improved to $O(M \log M)$.

The Single k-Centrum of a Graph

Peeters [159] studied the single facility problem on a general graph. He presented an $O(M^2 \log M + NM)$ algorithm for solving the discrete case only. Finding the absolute k-centrum is much more involved. We point out that this latter task can be performed efficiently by using modern computational geometry methods.

As in the 1-center problem on a graph, the solution approach is to find the best local absolute k-centrum on each edge of the graph, and then select the global absolute k-centrum as the best of the N local solutions.

In the preprocessing phase we compute the distances between all pairs of nodes. This can be done in $O(NM + M^2 \log M)$ time, Fredman and Tarjan [80].

Consider an edge of the graph. The edge will be viewed as a line segment. In this case we have a collection of M "roof" functions, and we need the minimum point of the objective function defined as the sum of the k largest roof functions in the collection (each such roof function represents the weighted distance of a given node from the points on the edge, and it is the minimum of 2 linear functions, defined on the edge, Kariv and Hakimi [118]). The objective function is piecewise linear, but it is not convex over the line segment representing the edge. A naive approach to find an optimal point on the segment is to compute the $O(M^2)$ intersections points of the $2M$ linear functions, and evaluate the objective function at each one of them. The effort will be superquadratic in M. We show, how to improve upon this bound and obtain a subquadratic algorithm.

Remark 9.1 *It is clear that a minimum value of this objective is attained at a breakpoint of the k-th largest function of the given collection of M roof functions. Therefore, it will suffice to construct the piecewise linear k-th largest function (a piecewise linear function defined on the real line is represented by its sequence of breakpoints and their respective function values).*

So now, we have that the main result of the section is the following:

Proposition 9.7 *The single k-centrum on a graph can be found in $O(NM\, k^{1/3}\alpha(M/k) \log^2 M)$ time.*

Proof.
Agarwal et al. [1] present results on the complexity bound of the k-th largest function of a collection of "segment" functions (a segment function is defined over the real line. It is linear over some segment of its domain, and is $-\infty$ elsewhere). This bound (combined with Dey [53]) is $O(Mk^{1/3}\alpha(M/k))$ (the function $\alpha(M/k)$ is the inverse Ackermann function, which is a very slow growing function of its argument. See Sharir and Agarwal [181]). This bound is useful in our context, since each roof function can be represented by two "segment" functions.

The k-th largest function of a collection of M segment functions can be constructed in $O(Mk^{1/3}\alpha(M/k)\log^2 M)$ time by the algorithm in Edelsbrunner and Welzl [70], which generates the k-th largest function of a collection of linear functions. As noticed above, the local k-centrum on an edge is a breakpoint of the k-th largest function of the collection of the M roof functions corresponding to that edge.

Therefore, with this approach the total time to find the global single k-centrum will be $O(NMk^{1/3}\alpha(M/k)\log^2 M)$ (this effort dominates the preprocessing time). □

Finally we can state a similar result for the unweighted case:

Proposition 9.8 *The unweighted single k-centrum can be found in $O(NM \log M)$ time.*

Proof.
In the unweighted case the slopes of each roof function are $+1$ and -1, and the k-th largest function of the collection of the M roof functions corresponding to a given edge, has $O(M)$ breakpoints only (this follows from the fact that the increasing (decreasing) part of a roof function can coincide with the piecewise linear k-th largest function on at most one interval connecting a pair of adjacent breakpoints of the k-th largest function). We note that a k-th largest function of an edge can be constructed in $O(M \log M)$ time. For the sake of brevity we skip the details. The total time to find the global unweighted single k-centrum is therefore $O(NM \log M)$ (this effort dominates the preprocessing time). □

9.3 The General Single Facility Ordered Median Problem on Networks

In this section we develop algorithms for models using the single facility ordered median function. We study the ordered median problem on several metric spaces: networks with positive and negative node weights and tree networks. Moreover, we discuss also the problem defined on directed networks.

Then, we present for the problem in general undirected and directed networks $O(NM^2 \log M)$ and $O(NM \log M)$ algorithms, respectively. For (undirected) trees the induced bound is $O(M^3 \log M)$. We show, how to improve this bound to $O(M \log^2 M)$ in the convex case. Further details on the material presented in this section can be found in [117].

9.3.1 Finding the Single Facility Ordered Median of a General Network

To introduce the algorithmic results we first recall some finite dominating sets (FDS) for the OMP.

Theorem 8.1 proved that $V \cup \mathsf{EQ}$ is an FDS for the ordered median problem with nonnegative weights. When some of the node weights are negative, Theorem 8.1 may not hold. However, it is possible to extend the result so that $V \cup \mathsf{EQ} \cup \mathsf{NBN}$ is a finite dominating set for the single facility ordered median problem with general node weights (see 8.2 in Chapter 8).

We next show how to solve the ordered median problem on a general network. We solve the problem independently on each one of the edges, in $O(M^2 \log M)$ time. Restricting ourselves to a given edge $e_i = [v_s, v_t]$, from the above results we know that there exists a best point x^i with respect to the objective function, such that x^i is either a node, an equilibrium point or a negative bottleneck point. Hence, it is sufficient to calculate the objective at the two nodes of e_i, the set K_i of $O(M^2)$ equilibrium points, and the set L_i of $O(M)$ negative bottleneck points on e_i (note that if $w_j \geq 0$, $j = 1, \ldots, M$, we can ignore the bottleneck points).

Generating the bottleneck and equilibrium points on $\mathbf{e_i}$

Let x denote a point on e_i (for convenience x will also denote the distance along e_i, of the point from v_s). For each $v_j \in V$, $d_j(x)$ is a piecewise linear concave function with at most one breakpoint (if the maximum of $d_j(v_j)$ is attained at an interior point, this is a bottleneck point). Assuming that all internodal distances have already been computed, it clearly takes constant time to construct $d_j(x)$ and the respective bottleneck point. If $w_j > 0$, $d_j(x)$ is concave and otherwise $d_j(x)$ is convex. To compute all the equilibrium points on e_i, we calculate in $O(M^2)$ total time the solutions to the equations $d_j(x) = d_k(x)$, where $v_j, v_k \in V$, $v_j \neq v_k$. To conclude, in $O(M^2)$ total time we identify the set L_i^* of $O(M)$ bottleneck points and the set K_i of $O(M^2)$ equilibrium points. Let $N_i = \{v_s, v_t\} \cup L_i^* \cup K_i$.

Computing the objective function at all points in $\mathbf{N_i}$

First we sort in $O(|N_i| \log |N_i|)$ time the points in N_i. For any x in the interior of the sub-edge connecting x_q and x_{q+1}, where x_q, x_{q+1} are two consecutive elements in the sorted list of N_i, the order of $\{d_j(x)\}_{j=1}^M$ does not change. In particular, the objective value at x_{q+1} can be obtained from the objective value at x_q in constant time. The first point in the sorted list is $x = v_s$, and the objective value can be obtained in $O(M \log M)$ time. Therefore, the time needed to compute the objective for all points in N_i is $O(M \log M + |N_i| \log |N_i|)$. To summarize the total effort needed to compute a best solution on e_i is

$$O(M^2 + M \log M + |N_i| \log |N_i|) = O(M^2 + |N_i| \log |N_i|).$$

The time to find a single facility ordered median of a network is therefore $O(NM^2 + NM^2 \log M) = O(NM^2 \log M)$.

The above complexity can be improved for some important special cases discussed in the literature. For example, if $w_j \geq 0$ for all $v_j \in V$, then we can disregard the bottleneck points although this does not improve the complexity. Moreover, if in addition $\lambda_1 \geq \lambda_2 \geq \ldots \geq \lambda_M$, then an optimal solution on e_i is attained either at v_s or v_t, [150]. The objective at each node can be computed in $O(M \log M)$ and therefore, in this case, the optimal single facility ordered median point is obtained in $O(M^2 \log M)$ time.

The ordered median problem on a directed graph can be treated similarly to the undirected case as the following analysis shows. Let $G_D = (V, E)$ be a (strongly connected) directed graph. Following Handler and Mirchandani [92], the weighted distance between a point $x \in A(G_D)$ and $v_j \in V$ is given by $\bar{d}_j(x) := w_j(d(x, v_j) + d(v_j, x))$. In addition, denote $\bar{d}(x) =: (\bar{d}_1(x), \ldots, \bar{d}_M(x))$. The definitions of the other concepts carry over from the undirected case.

The ordered median problem on a directed network can be written as

$$\min_{x \in A(G_D)} f_\lambda(x) := \sum_{j=1}^{M} \lambda_j \bar{d}_{(j)}(x). \tag{9.2}$$

First, we make some observations on the above distance functions.

Let $e = [v_i, v_j] \in$ e be an edge of the directed graph G_D, directed from v_i to v_j, and x a point in the interior of this edge. Then for a node $v_k \in V$, the distance function $\bar{d}_k(.)$ is constant on the interior (v_i, v_j) of the edge $[v_i, v_j]$. Moreover, if $w_k \geq 0$ (respectively $w_k < 0$) then $\bar{d}_k(v_i), \bar{d}_k(v_j) \leq \bar{d}_k(x)$ (respectively $\bar{d}_k(v_i), \bar{d}_k(v_j) \geq \bar{d}_k(x)$) for all $x \in (v_i, v_j)$. With this observation we obtained a finite dominating set for this problem: The ordered median problem on directed networks with nonnegative node weights always has an optimal solution in the node-set V. If in addition $\lambda_1 > 0$ and $w_i > 0$, $\forall i = 1, \ldots, M$, then any optimal solution is in V (see Theorem 8.3 in Chapter 8).

From the above results the case of nonnegative node weights can be solved in $O(M^2 \log M)$ time by evaluating the function at each node of the network.

The case with positive and negative weights can also be solved by evaluating at most $O(N + M) = O(N)$ points in $A(G_D)$ (since for strongly connected graphs $N \geq M$). Indeed, since the functions $\bar{d}_k(x)$ are constant in the interior of an edge, we only need to evaluate the objective function at an arbitrary interior point x_e of each edge $e \in E$. Based on the previous analysis, we can solve the problem with positive and negative weights in $O(NM \log M)$ time.

We summarize the above results in the following Theorem.

Theorem 9.1 *An ordered median of an undirected (directed) network can be found in $O(NM^2 \log M)$ $(O(NM \log M))$ time. Moreover, if $w_j \geq 0$, for all $v_j \in V$, and in addition for the undirected case $\lambda_1 \geq ... \geq \lambda_M \geq 0$, then the complexity bounds reduce to $O(M^2 \log M)$.*

9.3.2 Finding the Single Facility Ordered Median of a Tree

Throughout this section we will use the concept of convexity on trees as defined in Dearing et al. [52]. Rodríguez-Chía et al. [177] showed that when $w_i \geq 0$, for $i = 1, ..., M$, the ordered median function is convex on the plane with respect to any metric generated by norms, if the λ-vector satisfies $0 \leq \lambda_1 \leq \lambda_2 \leq \ldots \leq \lambda_M$. In a previous chapter, Theorem 7.2 proves a similar result for trees.

For trees the same discretization results as for general networks hold, with the additional simplification that we have no bottleneck points on trees and therefore, $V \cup \mathsf{EQ}$ is a finite dominating set for the problem with arbitrary weights.

From the analysis in the previous section (since $N = M - 1$) we can conclude that a best solution on a tree can be found in $O(M^3 \log M)$ time. Improvements are possible for some important cases.

The Unweighted Case: $\mathbf{w_j = w}$ for all $\mathbf{v_j \in V}$

In this case, each pair of distinct nodes, v_j, v_k contributes one equilibrium point. Moreover, such a point is the midpoint of the path $P[v_j, v_k]$, connecting v_j and v_k. Thus, $\sum_{e_i \in E} |N_i| = O(M^2)$. Let $T(M)$ denote the total time needed to find all the equilibrium points on the tree network. Then from the discussion on a general network we can conclude that the total time needed to solve the problem is $O(T(M)+M^2 \log M+\sum_{e_i \in E} |N_i| \log |N_i|) = O(T(M)+M^2 \log M)$.

Next, we show that $T(M) = O(M^2 \log M)$. More specifically, we show that with the centroid decomposition approach, in $O(\log M)$ time we can locate the equilibrium point defined by any pair of nodes $\{v_i, v_j\}$.

In the preprocessing phase we obtain, in $O(M \log M)$ total time a centroid decomposition of the tree T, [136]. In a typical step of this process we are given a subtree T' with q nodes, and we find, in $O(q)$ time an unweighted centroid, say v', of T'. We also compute in $O(q)$ time the distances from v' to all nodes in T'. v' has the property that each one of the connected components obtained from the removal of v' from T', contains at most $\frac{q}{2}$ nodes.

Consider now a pair of nodes v_i and v_j and let α be a positive number satisfying $\alpha \leq d(v_i, v_j)$. The goal is to find a point x on the path $P[v_i, v_j]$ whose

distance from v_i is α (the equilibrium point is defined by $\alpha = d(v_i, v_j)/2$). We use the above centroid decomposition recursively. First, we consider the centroid, say v_k, of the original tree T. Suppose that v_i is in a component T' and v_j is in a component T''. If $T' = T''$ we proceed recursively with T'. Otherwise, v_k is on $P[v_i, v_j]$. If $d(v_i, v_k) \geq \alpha$, x is in T'. We proceed recursively with T', where the goal now is to find a point on the path $P[v_i, v_k]$, whose distance from v_i is α. If $d(v_i, v_k) < \alpha$, x is in T'' and now recursively we proceed with T'' looking for a point on $P[v_k, v_j]$, whose distance from v_j is $d(v_i, v_j) - \alpha$. This process terminates after $O(\log M)$ steps, each consuming constant effort. At the end we locate an edge containing the point x on $P[v_i, v_j]$, whose distance from v_i is exactly α. The point x is found in constant time solving the linear equation $d_i(x) = \alpha$.
A further improvement is possible, if we assume $w_j > 0$ for all $v_j \in V$ and $\lambda_1 \geq \lambda_2 \geq \ldots \geq \lambda_M \geq 0$. From the previous section we know that in this case it is sufficient to compute the objective function at the nodes only, since there is an optimal solution which is a node. Then, we need to compute and sort, for each node v_j, the set of distances $\{w_k d(v_j, v_k)\}_{v_k \in V}$. Provided that G is a tree the total effort needed to obtain the M sorted lists of the distances is $O(M^2)$, [188]. Therefore, in this case the problem is solvable in $O(M^2)$ time.

The Convex Case: $\mathbf{w_j \geq 0}$ for all $\mathbf{v_j \in V}$, and $\mathbf{0 \leq \lambda_1 \leq \ldots \leq \lambda_M}$

We first solve this case on a path graph in $O(M \log^2 M)$ time. For a path graph we assume that the nodes in V are points on the real line. Using the convexity of the objective, we first apply a binary search on V, to identify an edge (a pair of adjacent nodes), $[v_i, v_{i+1}]$, containing the ordered median x^*. Since it takes $O(M \log M)$ time to evaluate the objective at any point v_j, the edge $[v_i, v_{i+1}]$ is identified in $O(M \log^2 M)$ time. Restricting ourselves to this edge we note that for $j = 1, \ldots, M$, $w_j|x - v_j|$ is a linear function of the parameter x.

To find the optimum, x^*, we use the general parametric procedure of Megiddo [132] with the modification in Cole [43]. We only note that the master program that we apply is the sorting of the M linear functions, $\{w_j|x - v_j|\}$, $j = 1, ..., M$ (using x as the parameter). The test for determining the location of a given point x' w.r.t. x^* is based on calculating the objective $f_\lambda(x)$ and determining its one-sided derivatives at x'. This can clearly be done in $O(M \log M)$ time. We now conclude that with the above test the parametric approach in [132, 43] will find x^* in $O(M \log^2 M)$ time.

We now turn to the case of a general tree. As shown above, in this case the objective function is convex on any path of the tree network. We will use a binary search (based on centroid decomposition) to identify an edge containing an optimal solution in $O(\log M)$ phases.

In the first phase of the algorithm we find in $O(M)$ time an unweighted centroid of the tree, say v_k. Each one of the connected components obtained by the removal of v_k contains at most $\frac{M}{2}$ nodes. If v_k is not an optimal ordered median, then due to the convexity of the objective, there is exactly one component, say T^j, such that any optimal solution is either in that component or on the edge connecting the component to v_k. We proceed to search in the subtree induced by v_k and T^j. Since T^j contains at most $\frac{M}{2}$ nodes, this search process will have $O(\log M)$ phases, when at the end an edge containing an optimal solution is identified. To locate the ordered median on an edge, we use the above $O(M \log^2 M)$ algorithm for path trees.

To evaluate the total complexity of the above algorithm, we now analyze the effort spent in each one of the $O(\log M)$ phases. At each such iteration a node (centroid) v_k is given and the goal is to check the optimality of v_k and identify the (unique) direction of improvement if v_k is not optimal. To facilitate the discussion, let $\{v_{k(1)}, \dots v_{k(l)}\}$ be the set of neighbors of v_k. To test optimality, it is sufficient to check the signs of the derivatives of the objective at v_k in each one of the l directions (there is at most one negative derivative).

We first compute and sort in $O(M \log M)$ time the multi-set $\{w_j d(v_j, v_k)\}$ of weighted distances of all nodes of the tree from v_k. Let L^k denote this sorted list (we also assume that the weights $\{w_i\}_{i=1,\dots,M}$ have already been sorted in a list W).

Suppose first, that all the elements in L^k are distinct. We refer to this case as the nondegenerate case. Let σ denote the permutation of the nodes corresponding to the ordering of L^k, i.e.,

$$w_{\sigma(1)} d(v_{\sigma(1)}, v_k) < \dots < w_{\sigma(M)} d(v_{\sigma(M)}, v_k).$$

Then the derivative of the objective at v_k in the direction of its neighbor $v_{k(t)}$ is given by

$$- \sum_{v_{\sigma(i)} \in T^{k(t)}} \lambda_i w_{\sigma(i)} + \sum_{v_{\sigma(i)} \in V \setminus T^{k(t)}} \lambda_i w_{\sigma(i)}.$$

Equivalently, the derivative is equal to

$$\sum_{i=1}^{M} \lambda_i w_{\sigma(i)} - 2 \sum_{v_{\sigma(i)} \in T^{k(t)}} \lambda_i w_{\sigma(i)}.$$

It is therefore clear, that after the $O(M \log M)$ effort needed to find σ, we can compute all l directional derivatives in $O(M)$ time.

Next we consider the case where the elements in L^k are not distinct. Assume without loss of generality that $w_j > 0$, for $j = 1, \dots, M$. In this case we partition the node set into equality classes, $\{U^1, \dots, U^p\}$, such that for each $q = 1, \dots, p$, $w_j d(v_j, v_k) = c^q$ for all $v_j \in U^q$, and $c^1 < c^2 < \dots < c^p$. (Note that $c^1 = 0$ and $v_k \in U^1$.)

Consider an arbitrary perturbation where we add ϵ^j to the length of edge e_j, $j = 1, \dots, N$. Let d' denote the distance function on the perturbed tree network. If ϵ is sufficiently small, all the elements in the set $\{w_j d'(v_j, v_k)\}$

are distinct, and for each pair of nodes, v_s, v_t with $v_s \in U^q$ and $v_t \in U^{q+1}$, $q = 1, ..., p-1$, $w_s d'(v_s, v_k) < w_t d'(v_t, v_k)$. Therefore, as discussed above, in $O(M \log M)$ time we can test whether v_k is optimal with respect to the perturbed problem, and if not find the (only) neighbor of v_k, say $v_{k(t)}$, such that the derivative in the direction of $v_{k(t)}$ is negative.

In the next lemma, we will prove that if v_k is optimal for the perturbed problem, it is also optimal for the original problem. Moreover, if v_k is not optimal for the original problem and $v_i = v_{k(t)}$ is a neighbor defining a (unique) direction of improvement for the original problem, then it also defines the (unique) improving direction for the perturbed problem.

This result will imply that in $O(M \log M)$ time we can test the optimality of v_k for the original problem, and identify an improving direction if v_k is not optimal.

Lemma 9.3 *Let $v_i = v_{k(t)}$ be a neighbor of v_k, and let A_i and $A_i(\epsilon)$ denote the derivatives of the objective $f_\lambda(x)$ at v_k in the direction of v_i for the original and the perturbed problems respectively. Then $A_i(\epsilon) \le A_i$.*

Proof.
Because of the additivity property of directional derivatives it is sufficient to prove the result for the case where V is partitioned into exactly two equality classes, U^1, U^2, where $U^1 = \{v_k\}$, $c^1 = 0$, and $w_j d(v_j, v_k) = c^2$ for all $v_j \ne v_k$.

Consider an arbitrary neighbor $v_i = v_{k(t)}$ of v_k, and let T^i be the component of T, obtained by removing v_k, which contains v_i. Let n_i denote the number of nodes in T^i. Let τ denote the permutation arranging the n_i nodes in T^i in a non-increasing order of their weights, and the remaining $M - n_i$ nodes in a non-decreasing order of their weights. We have $v_{\tau(t)} \in T^i$, $t = 1, ..., n_i$, $v_{\tau(t)} \notin T^i$, $t = n_i + 1, ..., M$, $w_{\tau(1)} \ge w_{\tau(2)} \ge ... \ge w_{\tau(n_i)}$, and $w_{\tau(n_i+1)} \le w_{\tau(n_i+2)} \le ... \le w_{\tau(M)}$. Then, it is easy to verify that

$$A_i = -\sum_{t=1}^{n_i} \lambda_t w_{\tau(t)} + \sum_{t=n_i+1}^{M} \lambda_t w_{\tau(t)}.$$

Next, to compute $A_i(\epsilon)$, consider the perturbed problem and the permutation σ arranging the nodes in an increasing order of their distances $\{w_j d'(v_j, v_k)\}$. Specifically, $0 = w_k d'(v_k, v_k) = w_{\sigma(1)} d'(v_{\sigma(1)}, v_k)$, and

$$w_{\sigma(1)} d'(v_{\sigma(1)}, v_k) < ... < w_{\sigma(M)} d'(v_{\sigma(M)}, v_k).$$

With the above notation it is easy to see that

$$A_i(\epsilon) = -\sum_{j | v_{\sigma(j)} \in T^i} \lambda_j w_{\sigma(j)} + \sum_{j | v_{\sigma(j)} \notin T^i} \lambda_j w_{\sigma(j)}.$$

We are now ready to prove that $A_i(\epsilon) \le A_i$.

We will successively bound $A_i(\epsilon)$ from above as follows: Suppose, that there is an index j such that $v_{\sigma(j)} \in T^i$, and $j > n_i$, i.e., there exists a node $v_t \in T^i$ such that in the expression defining $A_i(\epsilon)$, w_t is multiplied by $-\lambda_j$ for some $j > n_i$. Since $|T^i| = n_i$, there exists a node $v_s \notin T^i$ such that w_s is multiplied by λ_b for some index $b \leq n_i$, in this expression. Thus, from $b \leq n_i < j$, we have $\lambda_b \leq \lambda_j$, which in turn yields

$$-\lambda_j w_t + \lambda_b w_s \leq -\lambda_b w_t + \lambda_j w_s.$$

The last inequality implies that if we swap w_s and w_t, multiplying the first by λ_j and the second by $-\lambda_b$, we obtain an upper bound on $A_i(\epsilon)$. Applying this argument successively and swapping pairs of nodes, as long as possible, we conclude that there is a permutation σ' of the nodes in V, such that each node $v_{\sigma'(j)} \in T^i$ is matched with $-\lambda_j$, for some $1 \leq j \leq n_i$, each node $v_{\sigma'(j)} \notin T^i$ is matched with λ_j, for some $n_i < j \leq M$, and

$$A_i(\epsilon) \leq -\sum_{j=1}^{n_i} \lambda_j w_{\sigma'(j)} + \sum_{j=n_i+1}^{M} \lambda_j w_{\sigma'(j)}.$$

Denote the right hand side of the last inequality by $B_i^{\sigma'}(\epsilon)$.

To conclude the proof, we consider the entire collection of all permutations σ", which assign every node in T^i to a (unique) index $1 \leq j \leq n_i$, and every node which is not in T^i to a (unique) index $n_i < j \leq M$. For each such permutation consider the respective expression

$$B_i^{\sigma"}(\epsilon) = -\sum_{j=1}^{n_i} \lambda_j w_{\sigma"(j)} + \sum_{j=n_i+1}^{M} \lambda_j w_{\sigma"(j)}.$$

From Theorem 368 in Hardy et al.[98], mentioned above, the maximum of the above expression over all such permutations is achieved for the permutation τ, which arranges the n_i nodes in T^i in a non-increasing order of their weights, and the remaining $M - n_i$ nodes not in T^i in a non-decreasing order of their weights. We note that this maximizing permutation is exactly the one defining A_i. Therefore,

$$A_i(\epsilon) \leq B_i^{\sigma'}(\epsilon) \leq B_i^{\tau}(\epsilon) = A_i.$$

□

To summarize, to check the optimality of a node v_k in the original problem, we can use the following procedure:

Compute and sort the elements in the set $\{w_j d(v_j, v_k)\}$. From the sorted sequence define the equality classes $\{U^1, ..., U^p\}$. Arbitrarily select an ordering (permutation) of the nodes in V, such that for any $q = 1, ..., p-1$, and any pair of nodes $v_j \in U^q$, $v_t \in U^{q+1}$, v_j precedes v_t. Let σ denote such a selected permutation. The permutation defines a perturbed problem, which in turn corresponds to a nondegenerate case. Therefore, as noted above, in $O(M \log M)$ time we can check whether v_k is optimal for the perturbed problem. If v_k is optimal for the perturbed problem, i.e., $A_i(\epsilon) \geq 0$ for each neighbor v_i, by the above lemma it is also optimal for the original problem. Otherwise, there is a unique neighbor of v_k, say $v_i = v_{k(t)}$ such that the derivative of the perturbed problem at v_k in the direction of v_i is negative. From Lemma 9.3 to test optimality of v_k for the original problem, it is sufficient to check only the sign of the derivative of the original problem at v_k in the direction of v_i. The latter step can be done in $O(M)$ time.

To conclude, the algorithm has $O(\log M)$ phases, where in each phase we spend $O(M \log M)$ time to test the optimality of some node, and identify the unique direction of improvement if the node is not optimal. At the end of this process an edge of the given tree, which contains an optimal solution, is identified. As explained above, the time needed to find an optimal solution on an edge is $O(M \log^2 M)$ time. Therefore, the total time to solve the problem is $O(M \log^2 M)$.

We note in passing that the above approach is applicable to the special case of the k-centrum problem. Since testing optimality of a node for this model will take only $O(M)$ time, the total time for solving the k-centrum problem will be $O(M \log M)$. This bound improves upon the complexity given in Corollary 9.4 by a factor of $O(\log M)$.

The next theorem summarizes the complexity results for tree graphs we have obtained above.

Theorem 9.2 *An ordered median of an undirected tree can be computed in $O(M^3 \log M)$ time.*

1. *In the unweighted case, i.e., $w_j = w$ for all $v_j \in V$, the time is $O(M^2 \log M)$. If, in addition, $w > 0$, and $\lambda_1 \geq \ldots \geq \lambda_M \geq 0$, the time is further reduced to $O(M^2)$.*
2. *If $w_j \geq 0$, for all $v_j \in V$, and $0 \leq \lambda_1 \leq \ldots \leq \lambda_M$, the ordered median can be found in $O(M \log^2 M)$ time.*

10

The Multifacility Ordered Median Problem on Networks

As we have seen in Chapter 8, there is no general FDS with polynomial cardinality which is valid for all instances of the ordered p-median problem on networks. Of course, this fact does not contradict the existence of polynomial size FDS for particular instances of OMP. This will be exactly the scope of this chapter. We will consider instances of the multifacility ordered median problem where a polynomial size FDS exists. Afterwards, our goal is to design polynomial algorithms for these classes of the OMP.

The chapter is organized as follows. We start studying the multifacility k-centrum location problem. In the presentation we follow the results in Tamir [187]. Then, in the following section we address the ordered p-median problem with a special structure in the λ-modeling weights to present an algorithm to find an optimal solution. Section 3 applies the characterization of the FDS to develop a polynomial time algorithm for solving these problems on tree networks. This polynomial time algorithm for trees is combined with the general approximation algorithms of Bartal [14] and Charikar et al. [34] to obtain an $O(\log M \log\log M)$ approximate solution for the p-facility ordered median problem on general networks.

10.1 The Multifacility k-Centrum Problem

The p-facility k-centrum location problem on a graph G consists of finding a subset X_p of $A(G)$, $|X_p| = p$, minimizing

$$f_{\bar{\lambda}}(X_p) = r_k(w_1 d(X_p, v_1), ..., w_n d(X_p, v_n)) \ ,$$

where $\bar{\lambda} = (0, \overset{(M-k)}{\ldots}, 0, 1, \overset{(k)}{\ldots}, 1)$, and $r_k(x)$ is the sum of the k largest entries of $x \in \mathbb{R}^M$ (recall that r_k was defined in (1.15)).

In the node-restricted (discrete) version of the p-facility k-centrum problem we also require X_p to be a subset of nodes. A minimizer of $f_{\bar{\lambda}}(X_p)$ is called an *absolute k-centrum*, and an optimal solution to the discrete problem is a *discrete k-centrum*. The unweighted p-facility k-centrum problem refers to the case where $w_i = 1$, $i = 1, ..., M$.

From the definition it follows that if $k = 1$ the above model reduces to the classical p-center problem, and when $k = M$ we obtain the classical p-median problem.

Since the p-facility k-centrum problem generalizes both the p-median and the p-center problems, it is NP-hard on general graphs. The recent papers by Charikar et al. [36], Jain and Vazirani [110] and Charikar and Guha [35] provide constant-factor approximation algorithms for the p-median problem. At this point it is not clear whether the techniques in [36, 110, 35] can be modified to yield constant-factor algorithms for the p-facility k-centrum problem.

We next show that the problem is polynomially solvable on path and tree graphs. We note that polynomial algorithms for k-centrum problems on tree graphs can be used to approximate such problems on general graphs. The earlier papers by Bartal [14] and Charikar et al. [34] presented $O(\log p \log\log p)$ approximation algorithms for the p-median problem. These algorithms are based on solving p-median problems on a family of trees (the general network metric is approximated by the tree metrics). The basic approach yields an $O(\log M \log\log M)$ approximation algorithm for the p-median model. The same approach can be applied to the p-facility k-centrum model. Therefore, polynomial time algorithms for the latter model on trees are useful in deriving $O(\log M \log\log M)$ approximating solutions for k-centrum problems defined on general graphs (the improved $O(\log p \log\log p)$ bound for the p-median problem, reported in [14, 34] is derived by using a preprocessing stage to reduce the number of nodes to $O(p)$. It is not yet clear whether this preprocessing stage is also applicable for the p-facility k-centrum model).

To simplify the presentation, we first restrict our discussion to the discrete models where the facilities must be located at nodes of the graph.

Also, we assume that the problem is nondegenerate. Specifically, if $\{v_i, v_j\}$ and $\{v_s, v_t\}$ are two distinct pairs of nodes, then $d(v_i, v_j) \neq d(v_s, v_t)$ and $w_i d(v_i, v_j) \neq w_s d(v_s, v_t)$. We will later show how to resolve degenerate instances.

We add a new parameter, R, to the above problem, and solve a modified problem for each relevant value of this parameter. We look at the *p-facility k-centrum problem with service distance* R, as the problem of minimizing the sum of the k largest weighted distances, provided the k-th largest weighted distance is at least R, and the $(k+1)$-st weighted distance is smaller than R. If there is no such solution, the objective value will be regarded as ∞. Note that if $R = 0$, to get a finite value we must have $k = |V|$, and the problem reduces to the p-median problem. (An $O(pM)$ algorithm for the p-median problem on a path appears in [99]. $O(p^2M^2)$ and $O(pM^2)$ algorithms solving the p-median problem on a tree are given in [119] and [185], respectively.) Thus, we will assume that $R > 0$. Because of the nondegeneracy assumption, the solution to the original problem is the best solution to the parameterized version over all values of the parameter R.

The Discrete p-Facility k-Centrum Problem on a Path

In the following we develop a dynamic programming approach for the discrete case on a path.

We represent the path by its sorted set of M nodes $V = \{v_1, ..., v_M\}$, where v_{i+1} is adjacent to v_i, $i = 1, ..., M-1$. v_i, $i = 1, ..., M$, is also viewed as a real point, and thus, $v_i < v_{i+1}$, $i = 1, ..., M-1$.

For each pair $i < j$, a radius R, and an integer $q \leq k$, suppose that there are facilities at v_i and v_j only, and define $r_q(i,j)$ to be the sum of the q largest weighted distances of the nodes in the subpath $[v_i, v_j]$ to their respective nearest facility. Also, define $r'_q(i,j,R) = r_q(i,j)$ if the q-th largest weighted distance is greater than or equal to R and the $(q+1)$-st largest weighted distance is smaller than R, and $r'_q(i,j,R) = \infty$, otherwise. In particular, with the nondegeneracy assumption, for each R there is a unique value of q, $q(i,j,R)$, such that $r'_q(i,j,R)$ is finite.

For each i, a radius R, an integer $q \leq k$, and an integer $s \leq p$, define $G(i,q,R,s)$ to be the minimum of the sum of the q largest weighted distances of the nodes in the subpath $[v_i, v_M]$, to their respective nearest facilities, provided that there are s facilities in the interval, (v_i is one of them), the q-th largest weighted distance is at least R, and the $(q+1)$-st largest weighted distance is smaller than R. Note that $s \leq p$, $s \leq M - i + 1$, $q \leq k$, and $q \leq M - i + 1$. If there is no such solution with the q-th largest weighted distance being at least R, and the $(q+1)$-st largest distance being smaller than R, define $G(i,q,R,s) = \infty$.

Suppose that R is a positive real number. We have the following recursion:

1. Let $i = M$. Then $G(M,0,R,s) = 0$ and $G(M,1,R,s) = \infty$.
2. Suppose $i < M$.
3. Let $s = 1$. $G(i,q,R,1)$ is equal to the sum of the q largest weighted distances from v_i, if the q-th largest weighted distance is at least R, and the $(q+1)$-st largest weighted distance is less than R. Otherwise, $G(i,q,R,1) = \infty$.
4. Let $s \geq 2$. Then $G(i,q,R,s)$ is finite only if

$$G(i,q,R,s) = \min_{\{j: j>i, q(i,j,R) \leq q\}} \{r'_q(i,j,q(i,j,R),R) + G(j, q-q(i,j,R), R, s-1)\}.$$

We assume without loss of generality, that an optimal solution to the problem on a path has a facility at v_1, otherwise, we can augment a node v_0, left of v_1, with $v_1 - v_0 = \infty$, and increase the number of facilities by 1.

From the above recursion it follows that the total effort to solve the discrete p-facility k-centrum problem with service radius r (i.e., compute $G(1,k,R,p)$), is $O(kpM^2)$. The parameter R can be restricted to values in the set $\{w_i|v_i -$

$v_j| : i, j = 1, ..., M\}$. Therefore, assuming that all the terms $r_q(i,j)$ are already available, the complexity of the above scheme to solve the discrete p-facility k-centrum problem on a path is $O(kpM^4)$.

The preprocessing phase of computing the $r_q(i,j)$ functions is dominated by the above bound. For each pair $i < j$, there are only $O(M)$ distinct values of $r_q(i,j)$ corresponding to the values $R = w_t(v_t - v_i)$ or $R = w_t(v_j - v_t)$ for some $t = i, i+1, ..., j$. Hence, there are only $O(M^3)$ distinct values of all functions $r'_q(i,j,R)$.

For each pair $i < j$, the total time needed to compute the $O(M)$ values of $r_q(i,j)$ is $O(M \log M)$. Suppose, as above, that the nodes of the path are points on the real line with $v_i < v_{i+1}$, for $i = 1, ..., M-1$. We first sort the $O(M)$ elements in the set $\{\min[w_t(v_t - v_i), w_t(v_j - v_t)] : t = i, i+1, ..., j\}$. It then takes $O(M)$ additional time to compute $r_q(i,j)$, for all relevant values of q. Thus, the total preprocessing time is $O(M^3 \log M)$. In the unweighted case we can avoid the sorting and the preprocessing phase takes only $O(M^3)$ time.

The Discrete p-Facility k-Centrum Problem on a Tree

Without loss of generality suppose that the given tree $T = (V, E)$ is binary and its root is v_1, (see [122]). For each node v_i in V, we let V_i denote the subset of all nodes in V that are *descendants* of v_i, i.e., v_i is on the paths connecting them to the root. In particular, for the root v_1, $V_1 = V$. A *child* of a node v_i is a descendant of v_i, which is connected to v_i by an edge.

We now describe a recursive algorithm to solve the discrete p-facility k-centrum problem with service distance R, defined above. The algorithm is a "bottom-up" scheme which starts at the leaves of the rooted tree, and recursively computes solutions for subproblems corresponding to the subtrees defined by the sets V_i.

For each set $(v_i, q, R, s : v_j, v_k)$, where $v_j \in V_i$, $v_k \in V - V_i$, we let $G(v_i, q, R, s : v_j, v_k)$ be the minimum sum of the q largest weighted distances of the nodes in V_i to their respective nearest centers, provided there are s centers in V_i (with the closest to v_i being v_j), there is already at least one center at $V - V_i$, (the closest such center is v_k), the q-th largest weighted distance is at least R, and the $(q+1)$-st largest weighted distance is smaller than R. Note that $q \leq |V_i|$, $q \leq k$, $s \leq |V_i|$, and $s \leq p$. (If $s = 0$, assume that v_j does not exist. By adding a super root, say v_0, and connecting it to the original root by an edge of infinite length, we may assume that there is a center at v_0 and therefore v_k always exists.)

If there is no solution satisfying the requirements, we set the value of the function G to be ∞. If $q = 0$, then $G(v_i, q, R, s : v_j, v_k)$ is finite (and is equal to 0) only if there is a feasible solution with s centers such that the service distances of all nodes in V_i are less than R.

To simplify the recursive equations, for each pair of reals, x, R, define $a(x, R) = 1$ if $x \geq R$, and $a(x, R) = 0$, otherwise.

We now define the function $G(v_i, q, R, s : v_j, v_k)$ recursively.

1. Let v_i be a leaf of the rooted tree.
   Suppose that $s = 0$.
   If $w_i d(v_i, v_k) \geq R$, set $G(v_i, 1, R, 0 : -, v_k) = w_i d(v_i, v_k)$, and $G(v_i, 0, R, 0 : -, v_k) = \infty$. Otherwise, set $G(v_i, 1, R, 0 : -, v_k) = \infty$, and $G(v_i, 0, R, 0 : -, v_k) = 0$.
   Suppose that $s = 1, (v_j = v_i)$.
   Set $G(v_i, 1, R, 1 : v_i, -) = \infty$, and $G(v_i, 0, R, 1 : v_i, -) = 0$.
2. Let v_i be a non leaf node, and let v_{i1} and v_{i2} be its two children.
   Suppose that $s = 0$.
   Let $x = w_i d(v_i, v_k)$. If $a(x, R) = 1$ set $G(v_i, 0, R, 0 : -, v_k) = \infty$. Otherwise, set

$$G(v_i, 0, R, 0 : -, v_k) = G(v_{i1}, 0, R, 0 : -, v_k) + G(v_{i2}, 0, R, 0 : -, v_k).$$

   For $q \geq 1$, set

$$G(v_i, q, R, 0 : -, v_k) = a(x, R)x + \min_{\substack{q_1+q_2=q-a(x,R) \\ |q_1| \leq |V_{i1}| \\ |q_2| \leq |V_{i2}|}} \{G(v_{i1}, q_1, R, 0 : -, v_k) + G(v_{i2}, q_2, R, 0 : -, v_k)\}.$$

   Suppose that $s \geq 1$.
3. If $v_j = v_i$, set

$$G(v_i, q, R, s : v_i, -) = B,$$

   where

$$B = \min_{\substack{q_1+q_2=q \\ s_1+s_2=s-1 \\ |q_1|,|s_1| \leq |V_{i1}| \\ |q_2|,|s_2| \leq |V_{i2}|}} \{\min_{v_t \in V_{i1}} G(v_{i1}, q_1, R, s_1 : v_t, v_i) + \min_{v_s \in V_{i2}} G(v_{i2}, q_2, R, s_2 : v_s, v_i)\}.$$

4. If $v_j \neq v_i$, suppose without loss of generality that $v_j \in V_{i2}$.
5. Suppose that $d(v_i, v_j) < d(v_i, v_k)$. Define $x = w_i d(v_i, v_j)$. If $a(x, R) = 1$, and $q = 0$, set $G(v_i, 0, R, s : v_j, v_k) = \infty$. Otherwise, set

$$G(v_i, q, R, s : v_j, v_k) = a(x, R)x + B,$$

   where

$$B = \min_{\substack{q_1+q_2=q-a(x,R) \\ s_1+s_2=s \\ |q_1|,|s_1| \leq |V_{i1}| \\ |q_2|,|s_2| \leq |V_{i2}| \\ |s_2| \geq 1}} \{ \min_{\substack{v_t \in V_{i1} \\ d(v_t,v_i) > d(v_i,v_j)}} G(v_{i1}, q_1, R, s_1 : v_t, v_j) + G(v_{i2}, q_2, R, s_2 : v_j, v_k)\}.$$

6. Suppose that $d(v_i, v_k) < d(v_i, v_j)$.
   Define $x = w_i d(v_i, v_k)$. If $a(x, R) = 1$ and $q = 0$, set $G(v_i, 0, R, s : v_j, v_k) = \infty$. Otherwise, set

$$G(v_i, q, R, s : v_j, v_k) = a(x, R)x + B,$$

   where

$$B = \min_{\substack{q_1+q_2=q-a(x,R)\\ s_1+s_2=s\\ |q_1|,|s_1|\le|V_{i1}|\\ |q_2|,|s_2|\le|V_{i2}|\\ |s_2|\ge 1}} \{ \min_{\substack{v_t\in V_{i1}\\ d(v_t,v_i)>d(v_i,v_j)}} G(v_{i1}, q_1, R, s_1 : v_t, v_k)$$

$$+G(v_{i2}, q_2, R, s_2 : v_j, v_k)\}.$$

To compute the complexity of the above scheme, note that for a fixed value of R, the total time needed to (recursively) evaluate $G(v_1, k, R, p : v_j, -)$ for all nodes v_j in $V = V_1$, (v_1 is assumed to be the root of the tree), is $O(k^2p^2M^3)$. (The optimal solution value to the respective p-facility k-centrum problem with service distance R is then given by $\min\{G(v_1, k, R, p : v_j, -) : v_j \in V\}$.)

A more careful analysis of the above scheme, using the results in Section 3 of [185], reveals that the complexity of computing $\min\{G(v_1, k, R, p : v_j, -) : v_j \in V\}$ is $O(kp^2M^3)$ if $k \ge p$, and $O(k^2pM^3)$ if $k < p$. Thus, the complexity bound is only $O((\min(k,p))kpM^3)$.

There are only $O(M^2)$ relevant values of the parameter R. These are the elements of the set $\{w_i d(v_i, v_j) : i, j = 1, ..., M\}$. Therefore, the discrete p-facility k-centrum model on a tree can be solved in $O((\min(k,p))kpM^5)$ time.

We note that the above algorithm can easily be modified to solve the generalized discrete model, where there are setup costs for the facilities and the objective is to minimize the sum of the setup costs of the p facilities and the k largest service distances. The complexity will be the same.

The above complexity is significantly higher than the results known for some of the special cases. For the p-median problem ($k = M$), $O(p^2M^2)$ and $O(pM^2)$ algorithms are given in [119] and [185], respectively. For the p-center problem ($k = 1$), an $O(M \log^2 M)$ algorithm is presented in [134, 43].

Solving the Continuous Multifacility k-Centrum Problems on Paths and Trees

It is well-known from Kariv and Hakimi [118] that there is an optimal solution to the multifacility center problem on tree graphs, such that each facility is located either at a node or at a point which is at equal weighted distances from a pair of nodes. It can easily be shown that this result extends to the multifacility k-centrum problem. (Note, however, that unlike the center problem, in the weighted case, the location point of a k-centrum which is at equal weighted distances from a pair of nodes, is not necessarily on the simple path connecting this pair of nodes.)

Consider first the continuous multifacility k-centrum problem on a path. In this case, for each pair of nodes with positive weights, there are at most 2 points on the path that are at equal weighted distances from the two nodes. Therefore, we can discretise the problem and look at $O(M^2)$ potential sites for the facilities.

A similar discretization argument applies to the tree case. For each pair of nodes $\{v_i, v_j\}$, with $w_i \neq w_j$, and each edge of a tree, there is at most one point on the edge which is at equal weighted distances from the two nodes. (If $w_i = w_j$, the pair $\{v_i, v_j\}$ contributes only the middle point of the path connecting them as a potential site for a k-centrum.) Therefore, there are $O(M^3)$ potential sites for the facilities. In the unweighted case this number is only $O(M^2)$.

Solving Degenerate Models

We have assumed above that the problems that we solve satisfy a nondegeneracy assumption on the set of distances. This assumption can be removed by using a standard perturbation on the edge distances. Suppose that the edge set E is defined by $E = \{e_1, ..., e_N\}$. If ϵ denotes a small positive infinitesimal, we add the term ϵ^i to the length of edge e_i, $i = 1, ..., N$. For all values of ϵ, which are sufficiently small, the perturbed problem satisfies the nondegeneracy assumption. Therefore, we can apply the above algorithms symbolically to the perturbed problem, and obtain an optimal solution to the unperturbed problem, defined by $\epsilon = 0$.

10.2 The Ordered p-Median Problem with λ^s-Vector $\lambda^s = (a, \overset{M-s}{\ldots}, a, b, \overset{s}{\ldots}, b)$

In this section we consider the p-facility OMP with a special configuration of the λ-modeling weights: $\lambda^s = (a, \overset{M-s}{\ldots}, a, b, \overset{s}{\ldots}, b)$. For this case, Theorem 8.5 shows that there is always an optimal solution so that one of the facilities (e.g. x_p) is a node or an equilibrium point. From the proof of the theorem it follows that all other solution points are either nodes or pseudo-equilibria with respect to the range of the equilibrium or one of the ranges of the node(s). Hence, in order to solve this problem, we first compute the set of equilibria EQ, then the ranges R and afterwards the r-extreme points for every $r \in \mathsf{R}$. The latter must be saved with a reference to r in a set $\mathsf{PEQ}[r]$. Next, we choose a candidate x_p from the set $V \cup \mathsf{EQ}$. If $x_p \in EQ_{ij}$ is an equilibrium of range $r = d_i(x_p) = d_j(x_p)$, then the objective function value $f_\lambda(X_{p-1} \cup \{x_p\})$ is determined for all $p-1$ subsets $X_{p-1} = \{x_1, \ldots, x_{p-1}\}$ of $V \cup \mathsf{PEQ}[r]$.

If $x_p = v_i$ is a node, then the set R_{v_i} is computed (see proof of Theorem 8.5). Furthermore, for all subsets $X_{p-1} = \{x_1, \ldots, x_{p-1}\}$ of $V \cup$

$\{\mathsf{PEQ}[r] \,|\, r \in \mathsf{R}_{v_i}\}$ the objective function value $f_\lambda(X_{p-1} \cup \{x_p\})$ must be determined.

A summary of the steps required to find an optimal solution of this version of the ordered p-median problem with $\lambda^s \in \mathbb{R}^M_{0+}$, $1 \le s \le M-1$ is given below.

Algorithm 10.1: *Computation of an optimal solution set X_p^**

Data : Network $G = (V, E)$, distance-matrix D, $p \ge 2$ and a vector $\lambda^s = (a, \dots, a, b, \dots, b)$, $1 \le s \le M-1$

Result : An optimal solution set X_p^*

1 Initialization
  Let $X_p^* := \emptyset$, $\quad res := +\infty$;

2 Compute EQ, then the set of ranges R and based on these sets determine for every $r \in \mathsf{R}$ the r-extreme points and save them with a reference to r in a set $\mathsf{PEQ}[r]$;

3 **for** *all* $EQ \in \mathsf{EQ}$ **do**
    Let $x_p := EQ \in EQ_{ij}$ and compute the range r of the equilibrium, i.e. $r := d_i(EQ) = d_j(EQ)$;
    for *all* $X_{p-1} = \{x_1, \dots, x_{p-1}\} \subseteq V \cup \mathit{PEQ}[r]$ **do**
        Compute $f_\lambda(X_p)$, where $X_p := X_{p-1} \cup \{x_p\}$;
        if $f_\lambda(X_p) < res$ **then** $X_p^* := \{X_p\}$, $res := f_\lambda(X_p^*)$
    end
  end

4 **for** *all* $v_i \in V$ **do**
    Let $x_p := v_i$ and compute the set R_{v_i} of all ranges of the node ;
    for *all* $X_{p-1} = \{x_1, \dots, x_{p-1}\} \subseteq V \cup \{\mathit{PEQ}[r] \,|\, r \in R_{v_i}\}$ **do**
        Compute $f_\lambda(X_p)$, where $X_p := X_{p-1} \cup \{x_p\}$;
        if $f_\lambda(X_p) < res$ **then** $X_p^* := \{X_p\}$, $res := f_\lambda(X_p^*)$
    end
  end

5 **return** X_p^*

The above algorithm has complexity $O(pN^{p-1}M^p \log M(K + M^p))$. To show this, observe that the computation of the set of equilibria EQ is possible in $O(N \min[(M+K)\log M, M^2])$ steps where $K < M^2$ (see page 203). If we integrate the computation of the range of an equilibrium and a node in the line-sweep algorithm, then the complexity for obtaining the set R is $O(K + M^2)$, where it is possible to compute PEQ in $O(NM(K + M^2))$. In Step 3 and Step 4 we have to evaluate the objective function for a node or an equilibrium for all subsets of size $p-1$ of $V \cup \mathsf{PEQ}[r]$, respectively $V \cup \{\mathsf{PEQ}[r] \,:\, r \in \mathsf{R}_{v_i}\}$. In the first case we have $O(\binom{M+MN}{p-1}) = O((MN)^{p-1})$ and in the second case $O(\binom{M+(M-1)MN}{p-1}) = O((M^2N)^{p-1})$, different subsets $(|\mathsf{R}_{v_i}| = M-1)$. Since the evaluation of the objective function takes $O(pM \log M)$ time (because it is no longer possible to compute the ordered vectors $d_{\le}(X_p)$ a priori in the line-sweep algorithm), we obtain for Step 3 and Step 4 the complexity

$$O(pM \log M(K(MN)^{p-1} + M(M^2N)^{p-1})) = O(pN^{p-1}M^p \log M(K + M^p)),$$

which is the total complexity of the algorithm. It is clear, that these problems are NP-hard because the p-median and p-center problems are particular instances. Due to this reason we may have to apply approximation algorithms in order to solve the problem.

In the next section we show how to solve the ordered p-median problem with special λ-modeling weights on tree graphs in polynomial time. This development is important because it can be applied to approximate such problems in general graphs. In Bartal [14] and Charikar et al. [34], $O(\log M \log\log M)$ approximation algorithms are given for the p-median problem. These algorithms are based on solving p-median problems on a family of trees. (The general network metric is approximated by the tree metrics.) The same approach can be applied to the ordered p-median problem with λ-modeling weights. Therefore, polynomial time algorithms for solving that problem on trees are useful to derive $O(\log M \log\log M)$ approximating solutions for ordered p-median problems on general networks.

10.3 A Polynomial Algorithm for the Ordered p-Median Problem on Tree Networks with λ^s-Vector, $\lambda^s = (a, \overset{M-s}{\ldots}, a, b, \overset{s}{\ldots}, b)$

In the following we solve the ordered p-median problem with at most two values of λ-weights ($\lambda^s = (a, \overset{M-s}{\ldots}, a, b, \overset{s}{\ldots}, b)$) in polynomial time on a tree. In order to do that, we follow the presentation in [116]. Assume that we are given a tree T with $|V| = M$.

Using the discretization result in Theorem 8.5, it is clear that the optimal ordered p-median set can be restricted w.l.o.g. to the set $Y = \mathsf{PEQ}$.

Once we restrict to trees, the set Y is of cardinality $O(M^4)$ (notice that $|E| = M - 1$). Computing and augmenting these points into the node set of T has complexity $O(M^4)$ by the procedure in Kim et al. [122]. Let T^a denote the augmented tree with the node set Y. Each point in Y is called a *semi-node*. In particular, a node in V is also a semi-node.

Suppose now that the given tree $T = (V, E)$, $|V| = M$ and $|E| = M - 1$, is rooted at some distinguished node, say, v_1. For each pair of nodes v_i, v_j, recall that we say that v_i is a *descendant* of v_j if v_j is on the unique path connecting v_i to the root v_1. If v_i is a descendant of v_j and v_i is connected to v_j with an edge, then v_i is a *child* of v_j and v_j is the (unique) *father* of v_i. If a node has no children, it is called a *leaf* of the tree.

As shown in [116], we can now assume w.l.o.g. that the original tree is binary, where each non-leaf node v_j has exactly two children, $v_{j(1)}$ and $v_{j(2)}$. The former is called the *left child*, and the latter is the *right child*. For each node v_j, V_j will denote the set of its descendants.

Once the tree T^a has been obtained, a second preprocessing phase is performed. For each node v_j we compute and sort the distances from v_j to all semi-nodes in T^a. Let this sequence be denoted by $L_j = \{t_j^1, ..., t_j^m\}$, where $t_j^i \leq t_j^{i+1}$, $i = 1, ..., m-1$, and $t_j^1 = 0$. We can assume w.l.o.g. that there is a one to one correspondence between the elements in L_j and the semi-nodes in Y (Tamir et al. [190]). The semi-node corresponding to t_j^i is denoted by y_j^i, $i = 1, ..., m$.

We note that the total computational effort of this phase is $O(M^6)$ and can be achieved by using the centroid decomposition approach as in [122], or the procedure described in [185]. For each node v_j, an integer $q = 0, 1, ..., p$, $t_j^i \in L_j$, an integer $l = 0, 1, \ldots, s$, and c being a weighted distance from any node to a semi-node, let $G(v_j, q, t_j^i, l, c)$ be the optimal value of the subproblem defined on the subtree T_j, given that a total of at least one and at most q semi-nodes (service centers) can be selected in T_j. Moreover, we assume that at least one of them is located in $\{y_j^1, \ldots, y_j^i\} \cap Y_j$, exactly l vertices are associated to b λ-weights, and that the minimal distance allowed for an element with a b λ-weight is c (in the above subproblem we implicitly assume no interaction between the semi-nodes in T_j, and the rest of the semi-nodes in T). The function $G(v_j, q, t, l, c)$ is computed only for $q \leq |V_j|$, where V_j is the node set of T_j, $l \leq \min(s, |V_j|)$ (notice that a larger l would not be possible) and if $l > 0$ then $c \leq \max\{w_k d(v_k, y) | v_k \in V_j \text{ and } y \in Y_j\}$. Also, for each node v_j we define

$$G(v_j, 0, t, 0, c) = +\infty.$$

Analogously, $G(v_j, q, t, l, c) = +\infty$ for any combination of parameters that leads to an infeasible configuration.

Similarly, for each node v_j, an integer $q = 0, 1, ..., p$, $t_j \in L_j$, an integer $l = 0, 1, 2 \ldots, s$, and c being a weighted distance from any node to a semi-node, we define $F(v_j, q, t_j, l, c)$ as the optimal value of the subproblem defined in T_j satisfying the following conditions:

1. a total of q service centers can be located in T_j;
2. there are already some selected semi-nodes in $Y \backslash Y_j$ and the closest among them to v_j is at a distance t_j;
3. there are exactly $l \leq \min\{|V_j|, s\}$ vertices with b λ-weight in T_j, and
4. c is the minimal weighted distance allowed for a weighted distance with a b λ-weight.

Obviously, the function F is only computed for those t_j^i that correspond to $y_j^i \in Y \setminus Y_j$.

The algorithm computes the function G and F at all the leaves of T, and then recursively, proceeding from the leaves to the root, computes these functions at all nodes of T. The optimal value of the problem will be given by

$$\min_c G(v_1, p, t_1^m, s, c),$$

where v_1 is the root of the tree.

Define

$$f_j(t,l,c) = \begin{cases} at & \text{if } t < c \\ bt & \text{if } t \geq c \text{ and } l < s \\ +\infty & \text{if } t \geq c \text{ and } (l = s \text{ or } l = 0) \end{cases}$$

and

$$g_j(t,l,c) = \begin{cases} at & \text{if } t < c \\ bt & \text{if } t \geq c \text{ and } l > 0 \\ +\infty & \text{if } t \geq c \text{ and } l = 0. \end{cases}$$

Let v_j be a leaf of T. Then,

$$G(v_j, 0, t_j^i, 0, c) = +\infty, \quad i = 1, 2, \ldots, m, \quad c \neq 0$$
$$G(v_j, 1, t_j^1, 0, c) = 0, \quad c \neq 0,$$
$$G(v_j, 1, t_j^i, l, c) = +\infty \text{ otherwise.}$$

For each $i = 1, \ldots, m$ such that $y_j^i \in Y \setminus Y_j$

$$F(v_j, 0, t_j^i, 0, c) = \begin{cases} at_j^i & \text{if } t_j^i < c \\ +\infty & \text{if } t_j^i \geq c \end{cases}$$
$$F(v_j, 0, t_j^i, 1, c) = \begin{cases} bt_j^i & \text{if } t_j^i \geq c \\ +\infty & \text{if } t_j^i < c \end{cases}$$
$$F(v_j, 1, t_j^i, 0, c) = 0$$
$$F(v_j, 1, t_j^i, 1, c) = +\infty.$$

Let v_j be a non-leaf node in V, and let $v_{j(1)}$ and $v_{j(2)}$ be its left and right children, respectively. The element t_j^1 corresponds to v_j. In addition, it corresponds to a pair of elements say $t_{j(1)}^k \in L_{j(1)}$ and $t_{j(2)}^h \in L_{j(2)}$, respectively. Then,

$$G(v_j, q, t_j^1, l, c) = \min \Bigg\{ \min_{\substack{q_1 + q_2 = (q-1)^+ \\ l_1 + l_2 = l}} \{F(v_{j(1)}, q_1, t_{j(1)}^k, l_1, c) + F(v_{j(2)}, q_2, t_{j(2)}^h, l_2, c)\};$$

$$\min_{\substack{q_1 + q_2 = (q-2)^+ \\ l_1 + l_2 = l \\ y_{j(1)}^i \in (v_{j(1)}, v_j)}} \{F(v_{j(1)}, q_1, t_{j(1)}^i, l_1, c) + F(v_{j(2)}, q_2, t_{j(2)}^h, l_2, c)\};$$

$$\min_{\substack{q_1 + q_2 = (q-2)^+ \\ l_1 + l_2 = l \\ y_{j(2)}^i \in (v_{j(2)}, v_j)}} \{F(v_{j(1)}, q_1, t_{j(1)}^k, l_1, c) + F(v_{j(2)}, q_2, t_{j(2)}^i, l_2, c)\} \Bigg\},$$

where for any number a, we denote by $a^+ = \max(0, a)$.

For $i = 2, \ldots, m$ consider t_j^i. If $y_j^i \in Y \setminus Y_j$ then

$$G(v_j, q, t_j^i, l, c) = G(v_j, q, t_j^{i-1}, l, c).$$

If $y_j^i \in Y_{j(1)}$, then it corresponds to $t_{j(1)}^k \in L_{j(1)}$ and to $t_{j(2)}^h \in L_{j(2)}$. If $y_j^i \in Y_{j(1)}$ we can compute G in the following way:

$$G(v_j,q,t_j^i,l,c) = \min\Big\{ G(v_j,q,t_j^{i-1},l,c), g_j(t_j^i,l,c) + \min_{\substack{q_1+q_2=q\\ 1\le q_1\le |V_{j(1)}|\\ q_2\le |V_{j(2)}|\\ l_1+l_2=\begin{cases} l & \text{if } t_j^i<c\\ (l-1)^+ & \text{if } t_j^i\ge c\end{cases}\\ l_i\le \min\{l,|V_{j(i)}|\},\ i=1,2}} \{G(v_{j(1)},q_1,t_{j(1)}^k,l_1,c) + F(v_{j(2)},q_2,t_{j(2)}^k,l_2,c)\}\Big\}.$$

If $y_j^i \in (v_j, v_{j(1)})$ and $y_j^i \neq v_{j(1)}$, then

$$G(v_j,q,t_j^i,l,c) = \min\Big\{ G(v_j,q,t_j^{i-1},l,c), g_j(t_j^i,l,c) + \min_{\substack{q_1+q_2=q-1\\ 1\le q_1\le |V_{j(1)}|\\ q_2\le |V_{j(2)}|\\ l_1+l_2=\begin{cases} l & \text{if } t_j^i<c\\ (l-1)^+ & \text{if } t_j^i\ge c\end{cases}\\ l_i\le \min\{l,|V_{j(i)}|\},\ i=1,2}} \{F(v_{j(1)},q_1,t_{j(1)}^k,l_1,c) + F(v_{j(2)},q_2,t_{j(2)}^k,l_2,c)\}\Big\}.$$

Analogue formulas can be derived for $y_j^i \in Y_{j(2)}$ with the obvious changes.

Once the function G is obtained, we compute the function F. Let y_j^i be a semi-node in $Y \setminus Y_j$. Thus, y_j^i corresponds to some elements, say $t_{j(1)}^k \in L_{j(1)}$ and $t_{j(2)}^h \in L_{j(2)}$. Therefore,

$$F(v_j,q,t_j^i,l,c) = \min\Big\{ G(v_j,q,t_j^i,l,c), f_j(t_j^i,l,c) + \min_{\substack{q_1+q_2=q\\ q_1\le |V_{j(1)}|\\ q_2\le |V_{j(2)}|\\ l_1+l_2=\begin{cases} l & \text{if } t_j^i<c\\ (l-1)^+ & \text{if } t_j^i\ge c\end{cases}\\ l_i\le \min\{l,|V_{j(i)}|\},\ i=1,2}} \{F(v_{j(1)},q_1,t_{j(1)}^k,l_1,c) + F(v_{j(2)},q_2,t_{j(2)}^k,l_2,c)\}\Big\}.$$

Complexity.

It is clear that the complexity required to evaluate the functions G and F depends on the cardinality of the FDS for this problem. If $a \geq b$ then V is a FDS with cardinality $O(M)$, see Theorem 8.4, else if $a < b$ then PEQ is a FDS with cardinality $O(M^4)$. Therefore, it follows directly from the recursive equations that the effort to compute the function G at a given node v_j, for all relevant values of q, t, l and c is $O(p^2Ms^2M^2)$ in case $a \geq b$, and $O(p^2(M^4)s^2M(M^4))$ if $a < b$. Therefore, the overall complexity of the algorithm is clearly

$$\begin{cases} O(p^2 s^2 M^3) & \text{if } a \geq b \\ O(M(psM^4)^2) & \text{if } a < b. \end{cases}$$

However, it is easy to verify that the analysis in [185] can also be applied to the above model to improve the complexity to

$$\begin{cases} O(ps^2 M^3) & \text{if } a \geq b \\ O(pM(sM^4)^2) & \text{if } a < b. \end{cases}$$

It is also worth noting that for $a < b$ and $s = 1$ the algorithm reduces to the one given in [190] and the complexity is $O(pM^6)$. Moreover, if $s = 0$ the problem reduces to the p-median problem and the complexity is $O(pM^2)$ by the algorithm in [185]. Compared with the complexity $O((\min(k,p))kpM^5)$ obtained for the algorithm for the p-facility k-centrum presented in Section 10.1, the results in this section are similar. Note that for the continuous k-centrum, the number of possible candidates is $O(M^3)$ instead of $O(M)$ and therefore, the overall complexity is at least $O((\min(k,p))kpM^8)$.

Since it is already known that the complexity for the p-centdian is $O(pM^6)$ (see [190]), the complexity of the algorithms in this chapter for the more general ordered p-median problems are competitive.

11

Multicriteria Ordered Median Problems on Networks

11.1 Introduction

In previous chapters of Part III we have considered location problems on networks with a single objective function. In this chapter we deal with a different approach to the ordered median problem. Here, we assume that the network is multiobjective in the sense that each edge has associated several lengths or cost functions. These lengths may represent the time needed to cross the edge, the travel cost, the environmental impact, etc. Moreover, we allow multiple weights for the nodes, expressing demands with respect to the different edge lengths or for different products. Our goal is to locate a new facility that considers simultaneously all these criteria using ordered median functions. Usually, these cost functions are opposed, so that decreasing one of them leads to increase some of the others. Therefore, in the multiobjective problem the sense given to optimize must be reconsidered. Among the different (alternative) solution concepts for this kind of problem, we concentrate on finding a particular type of solution set: the set of Pareto optimal or non-dominated solutions.

Scanning the literature of location analysis one can find several references of multiobjective problems on networks. There are excellent research papers dealing with the multiobjective location problem. In particular, [89] and [88] develop efficient algorithms for multiobjective minisum and minisum-minimax problems. The reader interested in an exhaustive review is referred to [152]. In the case of tree networks more is known. Oudjit [156] studied the multi-objective 1-median problem on trees. He proved that the entire set of non-dominated 1-median points of the problem, in the considered tree, is the sub-tree spanned by the union of all the minimum paths between all the pairs of 1-median solutions for each objective function. In fact, this property extends to any problem with convex objective functions. This condition is not true for general networks, as it was proved by Colebrook [44] who has also developed an algorithm to obtain the set of efficient points for the multiobjective median and the multiobjective cent-dian problems in general networks. The OMP on networks with multiple objectives was first introduced by Kalcsics [115]. He showed, how to extend the results in [89] for the classical median problem to the OMP. However, no detailed algorithmic analysis was given.

In order to describe the problem, we assume that we are given a connected and non-directed network $G = (V, E)$, where V is the set of vertices and E is the set of edges. Each edge has associated a vector $(l_e^1, l_e^2, \dots, l_e^Q)$ of Q components that represents Q different distance or cost functions, one per criterion. We denote by l_e^r the length of edge e under the r-th criterion, and $d^r(v_i, v_j)$ is the shortest path distance from v_i to v_j under the r-th criterion (recall that we denote the set of all the points of the graph G by $A(G)$). We will refer to interior points of an edge e through the parameter $t \in [0, 1]$ that gives the point in the parametrization, $t \times l_e^r$, with respect to one of the extreme points of e. Analogously, we will refer to edges and subedges as segments defined by the corresponding interval of the parameter t.

In addition, for each node in V we define the following weight functions:

$$\begin{array}{rcl} w: \quad V & \mapsto & \mathbb{R}^Q \\ v_i \in V & \mapsto & w(v_i) = w_i = (w_i^1, w_i^2, \dots, w_i^Q). \end{array}$$

For each λ, the weight function together with the lengths, induce Q ordered median functions defined as:

$$f_{\lambda^r}^r(x) = \sum_{i=1}^{M} \lambda_i d_{(i)}^r(x) \qquad r = 1, \dots, Q \ ,$$

where

$$d_{(i)}^r(x) = w_{(i)}^r d_{(i)}^r(x, v_i).$$

Finally, for a given vector $\lambda = (\lambda^1, \dots, \lambda^Q)$, such that $\lambda^i = (\lambda_1^i, \dots, \lambda_M^i)$, for $i = 1, \dots, Q$, let $f_\lambda(x) = (f_{\lambda^1}^1(x), f_{\lambda^2}^2(x), \dots, f_{\lambda^Q}^Q(x)) \in \mathbb{R}^Q$. The problem that we deal with throughout this chapter is to find the entire set of points $x_{om} \in A(G)$ such that

$$f_\lambda(x_{om}) \in \mathsf{OM}_{multi}(\lambda) := v - \min_{x \in A(G)} f_\lambda(x).$$

This problem must be understood as searching for the Pareto or non-dominated solutions. For the sake of completeness, we recall the concept of domination which induces the efficient solutions of a multicriteria problem. Let $g = (g^1, g^2, \dots, g^Q)$ and $h = (h^1, h^2, \dots, h^Q)$ be two vectors in $\mathbb{R}^Q$. g is said to dominate h, denoted as $g \prec h$ if and only if $g^i \le h^i$, $\forall i$ and $g^j < h^j$ for at least one $j = 1, \dots, Q$.

This domination structure is applied to the problem in the following way. Let $U = \{(f_\lambda^1(x), f_\lambda^2(x), \ \dots, f_\lambda^Q(x)) : \ \forall x \in A(G)\}$ be the range of values of f_λ on G. A vector $g \in U$ is *non-dominated* or *Pareto-optimal* if $\nexists\, h \in U$ such that $h \prec g$. The set of all non-dominated vectors is denoted by $\mathsf{OM}_{multi}(\lambda)$.

Based on the set U, we define the set $L = \{x \in A(G) : (f^1_\lambda(x), f^2_\lambda(x), \ldots, f^Q_\lambda(x)) \in \mathsf{OM}_{multi}(\lambda)\}\}$. A point $x \in L$ is called *non-dominated* or *efficient.* Our goal in this chapter is to find the set $\mathsf{OM}_{multi}(\lambda)$ and thus, the set of efficient location points L on $A(G)$.

11.2 Examples and Remarks

The structure of the solution set of the multicriteria ordered median problem (MOMP from now on) can be rather unexpected and of course differs from the structure of the set of ordered median solutions. In order to gain insights into the problem, we present several examples of MOMP.

Example 11.1
Consider the network in Figure 11.1, with two lengths per edge and node-weights equal to one:

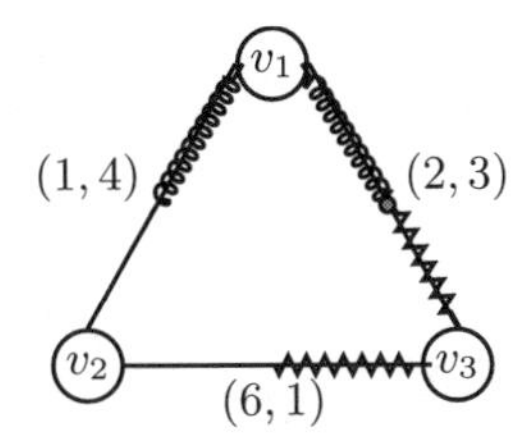

Fig. 11.1. The graph of Example 11.1

We want to solve the MOMP with $\lambda = ((0,1,1),(0,1,1))$, i.e., the 2-centrum bicriteria problem. First, we look for the optimal solutions of the 2 single objective problems induced by the two distances defined on the edges. Applying the results of previous chapters we find the optimal solution sets.

The optimal solutions for the first problem are given by the union of two subedges: $([v_1,v_2],(0,0.5)) \cup ([v_1,v_3],(0,0.5))$. For the second problem, the optimal solution set is again the union of two subedges: $([v_1,v_3],(0.5,1)) \cup ([v_2,v_3],(0.5,1))$.

Let $a_1 = ([v_1,v_2],0.5)$, $a_2 = ([v_2,v_3],0.5)$ and $a_3 = ([v_1,v_3],0.5)$. The objective values of the vertices v_1 and v_3 are $(3,7)$ and $(8,4)$, respectively; and the objective values of a_1 and a_2 are $(3,5)$ and $(7,4)$, respectively. In this example, a_3 minimizes both objective functions simultaneously. Hence, this is the unique non-dominated solution.

The example shows several interesting facts.

1. Some optimal solutions of the single objective problems are not efficient for the multiobjective problem.

2. We also observe that the efficient set for the multiobjective problem is smaller than the optimal solution sets of the single objective problems. This is somehow surprising since usually the efficient points are large sets.

Example 11.2
Let us now consider the graph depicted in Figure 11.2 with node-weights equal to one. In this example we want to solve the MOMP for $\lambda = ((0,0,0,1,1), (0,0,0,1,1))$. This choice of λ parameters amounts to consider the 2-centrum bicriteria problem.

The optimal solutions with respect to the single objective problem, using the first distance induced in the graph, is the entire edge $[v_4, v_5]$. The solution of the second problem is the union of the following two subedges: $([v_3, v_4], (0.75, 1)) \cup ([v_4, v_5], (0, 0.875))$.

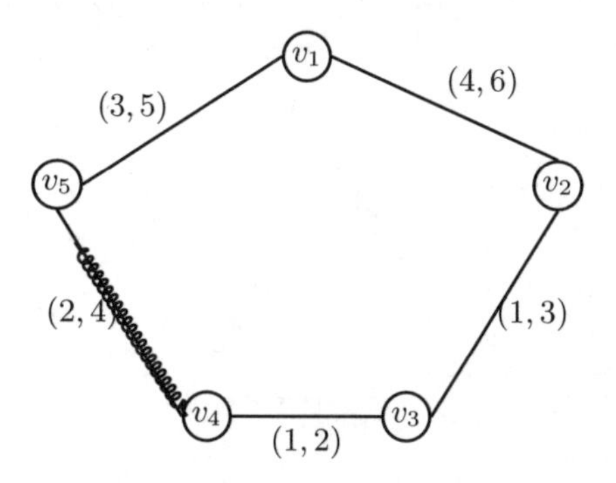

Fig. 11.2. The graph of Example 11.2

In this example, the intersection of the 2 solution sets for the single objective problems is the segment, $([v_4, v_5], (0, 0.875))$ which is the solution set for the two-objective problem as it minimizes both criteria simultaneously.

Example 11.3
Consider the network described in Figure 11.3. In this case we present a different situation.
Take the following node-weight vectors for the 2 criteria: $w^1 = (1, 2, 1, 2, 2, 1)$ and $w^2 = (2, 1, 2, 2, 2, 1)$.

We want to solve the 2-centrum bicriteria problem on the network of Figure 11.3 with $\lambda = ((0, 0, 0, 0, 1, 1), (0, 0, 0, 0, 1, 1))$. The optimal solutions for the first single objective problem are the union of following two subedges: $S_1 = ([v_4, v_6], (0.833, 1)) \cup ([v_5, v_6], (0.667, 1))$. Analogously, the optimal solutions for the second single objective problem are: $S_2 = ([v_3, v_4], (0, 0.333)) \cup ([v_3, v_5], (0, 0.167))$.

In this example the sets of optimal solutions of the single objective problems have empty intersection, so there are no points minimizing both criteria

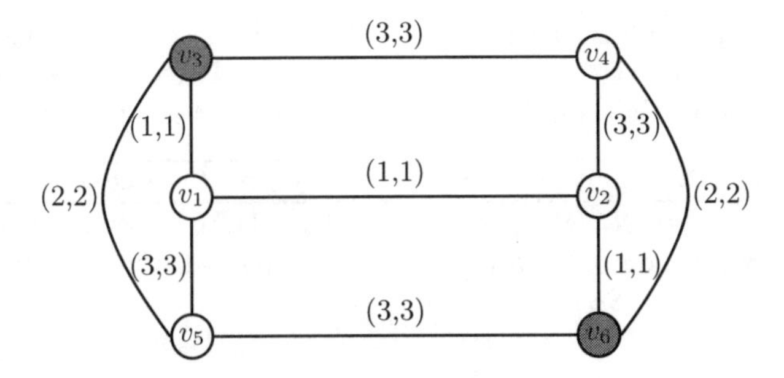

Fig. 11.3. The graph of Example 11.3

simultaneously. To identify the set of efficient (non-dominated) solutions, we use a domination argument.

Let $c_1 = ([v_3, v_4], 0.333)$, $c_2 = ([v_3, v_5], 0.167)$, $c_3 = ([v_5, v_6], 0.667)$ and $c_4 = ([v_4, v_6], 0.833)$. The points c_1 and c_2 have objective value $(14, 10)$ and $(13.33, 10)$, respectively, while the node v_3 has an objective value of $(12, 10)$. Hence, c_1 and c_2 are dominated by v_3. It can be checked that v_3 also dominates all the optimal solutions of the first single objective problem.

On the other hand, c_3 and c_4 have objective values $(10, 14)$ and $(10, 13.33)$, respectively; while the node v_6 has an objective value of (10,12). The reader can check that v_6 also dominates all the points in the solution set for the second single objective problem.

If there would exist an efficient point outside of the two sets S_1 and S_2, it should have the objective values with respect to both criteria less than or equal to 12. It is straightforward to check that this situation cannot happen in any edge of the graph, apart from the above mentioned solution sets of the single objective problems.

11.3 The Algorithm

This section presents a procedure to find the entire set of efficient points of the multiobjective ordered median location problem on networks. In our presentation we extend the approach in [44] which was proven to be valid for the multicriteria p-median and p-centdian problems.

We first observe that on an edge $e \in E$ the ordered median function f_λ^q, for $1 \leq q \leq Q$ is neither concave nor convex. Therefore, to have an useful representation of these functions, we resort to split them according to their breakpoints (Bottleneck and equilibrium points.)

Using the linear representation of all the ordered median functions within the identified subedges, in the Algorithm 11.1, we present the computation of the set of non-dominated points and non-dominated segments. The algorithm computes edge by edge all the locally non-dominated points and segments.

Once these tasks are performed, the locally efficient solutions must be compared to discard those that are globally dominated. These tasks are done using specific procedures that are described in algorithms 11.2, 11.3 and 11.6.

Algorithm 11.1: *Computation of P_{ND} and S_{ND} for the MOMP*

Data : Network N=(V,E), distance-matrix D, parameters Q, λ.
Result: Set of non-dominated points P_{ND} and set of non-dominated segments S_{ND}.

1 Initialization;
Let $P := \emptyset$ be the set of candidate points to be non-dominated;
Let $S := \emptyset$ be the set of candidate segments to be non-dominated;
2 **for** *all edges* $e := (v_s, v_t) \in E$ **do**
  Calculate $b_1, b_2, \ldots, b_j$ the breakpoints (equilibrium points) of the Q ordered median functions;
  Sort these points in increasing order with respect to the first length;
  Let $v_s = x_0, x_1, x_2, \ldots, x_j, x_{j+1} = v_t$ be the sorted sequence of these points including the endnodes;
  for $i := 0$ *to* j **do**
    Let $[x_i, x_{i+1}]$ be a segment of edge e ;
    for $q := 1$ *to* Q **do**
      | Compute $f_\lambda^r(x)$;
    end
    Let $f_\lambda(x) = (f_\lambda^1(x), f_\lambda^2(x), \ldots f_\lambda^Q(x))$;
    if $f_\lambda(x_i) \prec f_\lambda(x_{i+1})$, **then**
      | $P := P \cup \{x_i\}$
    else
      if $f_\lambda(x_{i+1}) \prec f_\lambda(x_i)$, **then**
        | $P := P \cup \{x_{i+1}\}$
      else
        | $P := P \cup \{x_i\} \cup \{x_{i+1}\}$ and $S := S \cup \{[x_i, x_{i+1}]\}$
      end
    end
  end
end
3 Compare the points in P using Algorithm 11.2 and include in P_{ND} the non-dominated points ;
4 Compare the segments in S using Algorithm 11.6 and include in S_{ND} non-dominated segments;
5 Compare the elements in P_{ND} and S_{ND} using Algorithm 11.3, removing from the corresponding sets the dominated elements ;
6 **return** P_{ND} *and* S_{ND}.

On each edge, the complexity of Algorithm 11.1 can be calculated in the following way. The process of splitting and finding the expression of the Q ordered median functions according to their breakpoints is done in at most

$O(QM^2 \log M)$ steps since there are at most $O(M^2)$ breakpoints on each of the Q functions. The number of segments and points generated for all the edges is $O(NM^2Q)$. Comparing pairwise all these elements takes $O(N^2M^4Q^2)$ steps and each comparison step takes $O(Q)$ time. Thus, the multicriteria ordered median algorithm runs in $O(N^2M^4Q^3 + QNM^3 \log M) = O(N^2M^4Q^3)$ time.

11.4 Point Comparison

Algorithm 11.1 presented above refers to three additional procedures to compare points, segments as well as points and segments. In this section we present the point-to-point and the point-to-segment comparison. The point comparison is a very intuitive algorithm that consists of just a pairwise comparison to determine the non-dominated points.

Algorithm 11.2: *Point-to-point comparison to compute the non-dominated points*

Data : Set P.
Result : Set of non-dominated points, P_{ND}.
1 Initialization;
Let $\{x_1, x_2, \ldots, x_p\}$ be the points belonging to P, and P_{ND} the set of non-dominated points ;
$P_{ND} = \emptyset$;
2 **for** $i := 1$ *to* p **do**
  Let x_i be a point in P;
  if $\nexists x_j \in P_{ND} : \ x_j \prec x_i$ **then**
    $P_{ND} := P_{ND} \cup \{x_i\}$
  end
  if $\exists x_k \in P_{ND} : \ x_i \prec x_k$ **then**
    $P_{ND} := P_{ND} \backslash \{x_k\}$
  end
end
3 **return** P_{ND}.

The second task that we describe in this section is the procedure to compare points and segments. Algorithm 11.3 is specifically designed to perform this comparison. To simplify the presentation of this algorithm we define the dominance relationship between a point and a segment.

Definition 11.1 *Let $X = [x_0, x_1]$ be an interval. We say that a point z dominates X (at least a part of X) ($z \prec X$) if we can find two points x'_0 and x'_1, with $x_0 \le x'_0$, $x_1 \ge x'_1$ and $x'_0 \le x'_1$ such that, for every $x \in [x'_0, x'_1] \subset X$, $z \prec x$.*

Algorithm 11.3: *Point-to-segment comparison*

```
  Data   : Set P, Set S.
  Result : Set of non-dominated points, P_ND, Set of non-dominated segments,
           S_ND.
1 Initialization;
  P_ND := P, S_ND := S;
2 for all points z ∈ P_ND do
      for all segments X := [x_0, x_1] ∈ S_ND do
          if z ≺ X then
              Let [x_min, x_max] ⊂ X be the interval dominated by point z;
              X := X\[x_min, x_max]
          end
          if X ≺ z then
              P_ND := P_ND\{z}
          end
      end
  end
3 return P_ND and S_ND.
```

11.5 Segment Comparison

The segment-to-segment comparison is also needed to calculate the set of efficient points. In order to describe such a process, we recall some important facts. First, within a segment, the ordered median function may not have a unique linear representation. To avoid this inconvenience, we break the segment into smaller subsegments where the objective function is linear.

Let us consider the segment $[x_i, x_{i+1}]$. We divide it in such a way that each of the Q single objective functions we work with, is linear within its new subintervals. To do this, we order all the breakpoints of the Q objective functions, $b_j, b_k, \ldots, b_p$, with respect to the first objective to get a collection of subintervals $[x_i, b_j] \cup [b_j, b_k] \cup \cdots \cup [b_p, x_{i+1}]$ in which the objective functions have a linear representation. Notice that this is always possible just considering consecutive equilibrium and bottleneck points. Let S denote the set of candidate segments to be non dominated. By construction, each function f_λ^r, for all $r = 1, \ldots, Q$ has a unique linear representation within each segment in S.

To facilitate the description of the algorithm we introduce the dominance relationship between segments.

Definition 11.2 *Let $X = [x_0, x_1], Y = [y_0, y_1]$ be two intervals. We say that X dominates Y ($X \prec Y$) if we can find four points x'_0, x'_1, y'_0, y'_1, with $x_0 \leq x'_0$, $x_1 \geq x'_1$ and $x'_0 \leq x'_1$ and $y_0 \leq y'_0$, $y_1 \geq y'_1$ and $y'_0 \leq y'_1$ such that, for every $x \in [x'_0, x'_1] \subseteq X$, and every $y \in [y'_0, y'_1] \subseteq Y$ we have that $x \prec y$.*

Let $X = [x_0, x_1]$ and $Y = [y_0, y_1]$ be two segments of the set S. Let $x \in X$ and $y \in Y$ be two inner points, then the Q objective functions have the following expressions:

$$f^r_{\lambda,X}(x) = f^r_{\lambda,X}(x_0) + m^r_X(x - x_0), \forall r = 1, \dots Q,$$
$$f^r_{\lambda,Y}(y) = f^r_{\lambda,Y}(y_0) + m^r_Y(y - y_0), \forall r = 1, \dots Q.$$

Now, if segment X dominates segment Y ($X \prec Y$), then, the following inequalities have to be satisfied for some $x \in X$ and some $y \in Y$:

$$f^r_{\lambda,X}(x) \leq f^r_{\lambda,Y}(y), \ r = 1, \dots, Q$$
$$\Downarrow$$
$$f^r_{\lambda,X}(x_0) + m^r_X(x - x_0) \leq f^r_{\lambda,Y}(y_0) + m^r_Y(y - y_0), \text{ for } r = 1, \dots, Q.$$

This implies that

$$y \geq \frac{f^r_{\lambda,X}(x_0) - f^r_{\lambda,Y}(y_0) - m^r_X x_0 + m^r_Y y_0}{m^r_Y} + \frac{m^r_X}{m^r_Y} x \quad \text{for some } x \in X, \ y \in Y. \tag{11.1}$$

Now, let

$$p^r = \frac{f^r_{\lambda,X}(x_0) - f^r_{\lambda,Y}(y_0) - m^r_X x_0 + m^r_Y y_0}{m^r_Y} \text{ and } q^r = \frac{m^r_X}{m^r_Y}, \ r = 1, \dots, Q.$$

Then, Equation (11.1) can be rewritten as $y \geq p^r + q^r x, \ x \in X, \ y \in Y$. Using this representation, we classify the different types of inequalities that can appear according to the values of m^r_X, q^r, p^r and m^r_Y. The classification gives rise to eight different types of inequalities.

- a $y \leq p^r + q^r x$ with $q^r > 0$.
- b $y \geq p^r + q^r x$ with $q^r > 0$.
- c $y \leq p^r + q^r x$ with $q^r < 0$.
- d $y \geq p^r + q^r x$ with $q^r < 0$.
- e $y \leq p^r$ with $m^r_X = 0$ and then $q^r = 0$.
- f $y \geq p^r$ with $m^r_X = 0$ and then $q^r = 0$.
- g $y \leq u^r$ with $m^r_y = 0$ and then $u^r = \frac{f^r_{\lambda,X}(x_0) - f^r_{\lambda,Y}(y_0) - m^r_X x_0}{m^r_X}$.
- h $y \geq u^r$ with $m^r_y = 0$ and then $u^r = \frac{f^r_{\lambda,X}(x_0) - f^r_{\lambda,Y}(y_0) - m^r_X x_0}{m^r_X}$.

Let us denote by R the region where $X \prec Y$, and let T be the set of inequalities understood in the following way. T can be divided into different subsets according to the above classification: $T = T_a \cup T_b \cup \dots \cup T_h$. We will refer to the inequalities by the letter of the type they belong to; for instance, $a \in T_a$ will represent the line defining the corresponding inequality. Also, $a(x)$ will be the value of the equation at the point x. It is clear that if two inequalities $c \in T_c$ and $d \in T_d$ can be found so that they satisfy $c(x) < d(x)$ $\forall \ x \in [x_0, x_1]$, then R is an empty region, hence $X \not\prec Y$. This result can be stated in a more general way through the following lemma:

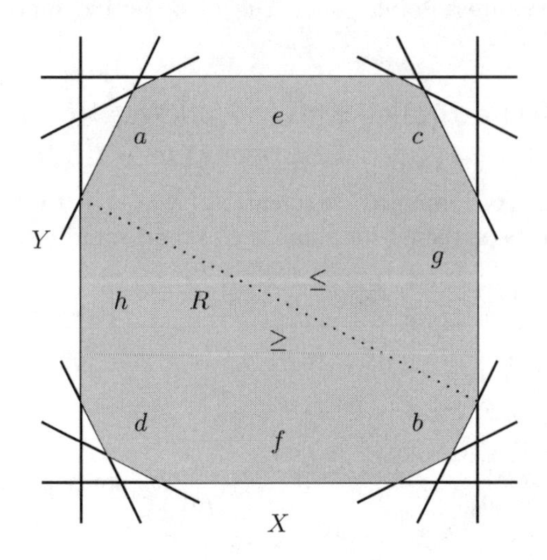

Fig. 11.4. The eight different types of inequalities with the domination region R

Lemma 11.1 *If there are inequalities* $a \in T_a, b \in T_b, \ldots, h \in T_h$, *such that* $a(x) < b(x)$ *or* $c(x) < d(x)$ *or* $e(x) < f(x)$ *or* $g(y) < h(y)$, *for all points* $x \in X$ *and* $y \in Y$, *then* $X \nprec Y$.

Proof.
Any of these conditions lead to an empty feasible region R of points $x \in X$ and $y \in Y$ satisfying (11.1) and hence $X \nprec Y$. □

In fact, the segment-to-segment comparison problem can be reformulated as a feasibility two-variable linear programming problem, so that it could be solved applying the simplex algorithm. This approach will be presented in the last section of this chapter. Here, the segment-to-segment comparison will be solved in a different way using computational geometry techniques.

In the following we show how to find the points in X that dominate points in Y, that is, we look for four points, that we denote x', $x'' \in X$ and $y_{\min}$, $y_{\max} \in Y$ so that $[x', x''] \prec [y_{\min}, y_{\max}]$, or equivalently, that for every $x \in [x', x'']$ and for every $y \in [y_{\min}, y_{\max}]$ the relationship $x \prec y$ holds. Moreover, $x_{\min}$ ($x_{\max}$) will denote the corresponding x-value in the equation for which $y_{\min}$ ($y_{\max}$) is attained.

In the following lemma, $y_{\max}$ is found and later, by means of a similar analysis $y_{\min}$ will be also found.

Lemma 11.2 *If* $T_a = \emptyset$ *and* $T_c = \emptyset$ *then* $y_{\max} = y_1$. *When* $T_a = \emptyset$ *then* $y_{\max} = \min_{c \in T_c} c(x_0)$, *with* $x_{\max} = x_0$. *Likewise, if* $T_c = \emptyset$, $y_{\max} = \min_{a \in T_a} a(x_0)$, *with* $x_{\max} = x_1$.

Proof.
The proof is straightforward. □

In the case in which $T_a \neq \emptyset$ and $T_c \neq \emptyset$ we can find the value $y_{\max}$ as the intersection point of the equations defining two inequalities, one of T_a and one of T_c. In order to find this intersection, we will need to define some other sets. Let $I(a, c) \in X$ be the intersection between $a \in T_a$ and $c \in T_c$; and let $I = \{I(a, c) : \forall a \in T_a, \forall c \in T_c\}$ be the set of all the intersection points between inequalities in T_a and in T_c.

We start by assuming that $I \neq \emptyset$, i.e., we can find at least an intersection point between an inequality $a \in T_a$ and another $c \in T_c$. Otherwise, all the inequalities in T_a are below the inequalities of T_c or vice versa, and this means that we can find $y_{\max}$ using the same procedure as in Lemma 11.2.

We will also suppose that the intersection point is below y_1, if not, we can state the value of $(y_{\min}, y_{\max})$ using the following lemma:

Lemma 11.3 *If all the intersections between the inequalities in* T_a *and* T_c *take place above* y_1, *then* $y_{\max} = y_1$, *being* $[x^0_{\max}, x^1_{\max}]$ *the interval where this maximum value is achieved, with* $x^0_{\max} = \max\{x \in X : t(x) = y_1, t \in T_a\}$ *and* $x^1_{\max} = \min\{x \in X : t(x) = y_1, t \in T_c\}$.

Proof.
The proof is straightforward. □

Taking into account this latter assumption, there must be a point $z \in I$ such that $\widehat{a}(z) = \min_{x \in I, a \in T_a} a(x)$. Therefore, $x_{\max} = z$ and $y_{\max} = \widehat{a}(z)$. We denote by $\widehat{a}$ and $\widehat{c}$ the inequalities attaining the minimum. The next lemma proves this result:

Lemma 11.4 $y_{\max} = \widehat{a}(z)$ *and hence* $x_{\max} = z$.

Proof.
Being R a convex region, the maximal value $y_{\max}$ is attained at the intersection (extreme point) of two inequalities with opposite sign slope. Any other intersection point must have its height above $\widehat{a}(z)$, i.e., $a(z) \geq \widehat{a}(z)$, $\forall a \in T_a$, and $c(z) \geq \widehat{c}(z)$ $\forall c \in T_c$. If there is some $a^* \in T_a$, with $a^* \neq \widehat{a}$, such that $a^*(z) < \widehat{a}(z)$, then $x^* = I(a^*, \widehat{c})$ and hence, $\widehat{a}(x^*) < \widehat{a}(z)$, which contradicts that $\widehat{a}(z)$ is the minimal value. The same analysis can be done for any inequality $c \in T_c$, and so the result follows. □

There is an immediate consequence of this lemma:

Corollary 11.1 *Provided that all the intersection points lie inside $X \times Y$, let $a \in T_a$ and $c \in T_c$, with $x = I(a,c)$ and $y = \widehat{a}(x)$, $y' = \widehat{c}(x)$. If $a(x) < y$ then $\widehat{a}(I(\widehat{a},c)) < y$, and if $\widehat{c}(x) < y'$ then $\widehat{c}(I(a,\widehat{c})) < y'$.*

This result can be used to speed up the computation of $(x_{\max}, y_{\max})$. Once we have obtained the value of $y_{\max}$, we can state that if $y_{\max} < y_0$ then R is an empty region and that would mean $X \not\prec Y$.

Searching for $y_{\min}$ is done analogously as for $y_{\max}$. $y_{\min}$ can be found applying lemmas 11.2, 11.3 and 11.4 on inequalities T_d and T_b, and using that $\min(y) = -\max(-y)$. Now, let $\widehat{d}$ and $\widehat{b}$ be the inequalities that determine $x_{\min} = I(\widehat{d},\widehat{b})$ with $y_{\min} = \widehat{d}(x_{\min})$.

Once $y_{\max}$ and $y_{\min}$ have been obtained, if we have $y_{\min} > y_{\max}$, then $X \not\prec Y$. We have to study now whether these maximal and minimal values can be improved by searching the intersections between other types of inequalities, for instance T_a and T_b or T_c and T_d.

If it happens that $x_{\min} < x_{\max}$ then we can have one of the two situations in Figure 11.5.

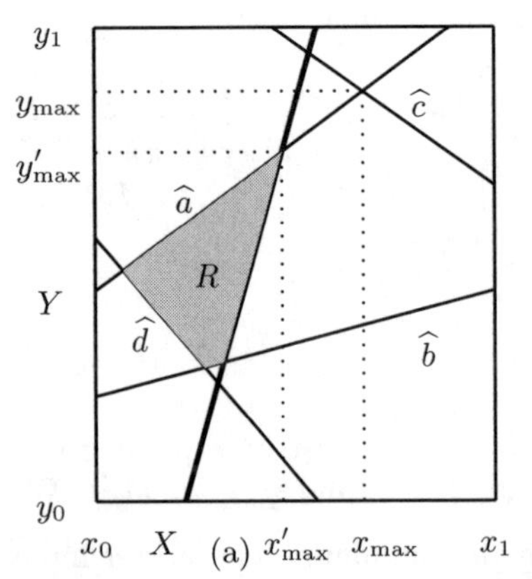

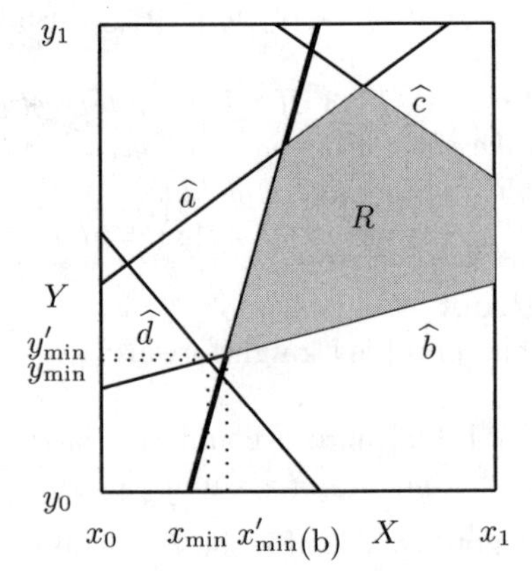

Fig. 11.5. Two examples where new maximum and minimum values have to be computed

In these cases, it is necessary to recompute new values for $y_{\min}$ and $y_{\max}$ that will be denoted $y'_{\min}$ and $y'_{\max}$. Before calculating these new values, there is another case that can be simplified because it leads to $X \not\prec Y$.

Lemma 11.5 *If $x_{\min} < x_{\max}$ and $\widehat{a}(x_{\min}) < y_{\min}$ and $\widehat{b}(x_{\max}) > y_{\max}$, then $X \not\prec Y$.*

Proof.
The proof follows since $\widehat{a}$ is completely below $\widehat{b}$ and then region R is empty.

□

In order to search for the values of $y'_{\min}$ and $y'_{\max}$, we redefine the set T_b to a new set T'_b eliminating the useless inequalities. The new set is:

$$T'_b = T_b \backslash \{b \in T_b : b(x_{max}) \leq y_{\max}\}.$$

If $T'_b = \emptyset$, then there is no $b \in T_b$ such that $b(x_{\max}) > y_{\max}$ and so, the maximum point $(x_{\max}, y_{\max})$ remains unchanged.

Otherwise, we use a similar argument as in Lemma 11.4 to get a new maximum point $(x'_{\max}, y'_{\max})$. Let us define the set $I' = \{I(a,b) : \forall a \in T_a, \forall b \in T'_b, \text{slope}(a) < \text{slope}(b)\}$. This set contains all the intersections between inequalities in T_a and in T'_b satisfying that the slope of a is smaller than the slope of b. The condition slope$(a) <$ slope(b) is very important, because we are interested in those intersections with larger y value. Since we suppose $T'_b \neq \emptyset$, there exists $z' \in I'$ such that $\widehat{a}(z') = \min_{x \in I'} a(x)$ and so we can state the following lemma.

Lemma 11.6 $y'_{\max} = \widehat{a}(z')$ *and hence* $x'_{\max} = z'$.

Proof.
As $T'_b \neq \emptyset$, then $\exists b^* \in T'_b$ with $x^* = I(\widehat{a}, b^*)$ such that $\widehat{a}(x^*) < y_{\max}$ (See Figure 11.5). Any other inequality $b \in T'_b$ verifying $b(x^*) > \widehat{a}(x^*)$, or any inequality $a \in T_a$ with $a(x^*) < \widehat{a}(x^*)$ can improve the value of $a(x^*)$. Then, $\exists \widehat{a'} \in T_a$ and $\exists \widehat{b'} \in T'_b$ with slope$(\widehat{a'}) <$ slope$(\widehat{b'})$ and $z' = I(\widehat{a'}, \widehat{b'})$ such that $y'_{\max} = \widehat{a'}(z')$ with $x'_{\max} = z'$. □

As in Corollary 11.1, we can improve the search of $(x'_{\max}, y'_{\max})$ by using the proof of the preceding lemma.

Corollary 11.2 *Assuming that all intersection points fall inside* $X \times Y$, *let* $a \in T_a$ *and* $b \in T'_b$, *with* $x = I(a,b)$ *and* $y = \widehat{a'}(x)$. *If* $\widehat{a'}(x) < y$ *then* $\widehat{a'}(I(\widehat{a'}, b)) < y$, *and if* $\widehat{b'}(x) > y$ *then* $\widehat{b'}(I(a, \widehat{b'})) < y$.

Next we proceed to tighten $y_{\min}$ by searching for a new value $y'_{\min}$ in the intersection points between inequalities of T_a and T_b (see Figure 11.5). In order to do that, we redefine the set T_a by eliminating all useless inequalities

$$T'_a = T_a \backslash \{a \in T_a : a(x_{\min}) \geq y_{\min}\}.$$

In this case, it is not necessary to keep the set T_a because it will not be used later. If $T'_a = \emptyset$, there is no inequality in T_a that can improve $y_{\min}$. Otherwise, the new minimum value $y'_{\min}$ can be obtained in a similar way to Lemma 11.6 along with the fact that $\min(y) = -\max(-y)$.

Finally, when $x_{\min} > x_{\max}$ we obtain analogous situations to those in Figure 11.5, with the consideration that the new situations take place on the right-hand side.

Lemma 11.7 *If* $x_{\min} > x_{\max}$ *and* $\widehat{c}(x_{\min}) < y_{\min}$ *and* $\widehat{d}(x_{\max}) > y_{\max}$*, then* $X \not< Y$.

Proof.
The proof follows since $\widehat{c}$ is completely below $\widehat{d}$, and hence the region R is empty. □

If we are not in the situation of Lemma 11.7, we must obtain the new maximum value $y'_{\max}$. As before, we remove the useless inequalities below $y_{\max}$

$$T'_d = T_d \backslash \{d \in T_d : d(x_{\max}) \leq y_{\max}\}.$$

Now, if $T'_d \neq \emptyset$, we look for $y'_{\max}$ in the intersection points between inequalities in T_c and T'_d. Once this is done, we eliminate all the useless equations in T_c

$$T'_c = T_c \backslash \{c \in T_c : c(x_{\min}) \geq y_{\min}\}$$

and continue searching for $y'_{\min}$ in the intersection points of T'_d and T'_c.

Algorithm 11.5 performs all the process above described. This algorithm calls for the procedure *ComputeMaximum*, which corresponds to Algorithm 11.4. The complexity of Algorithm 11.4 can be stated by the following theorem.

Theorem 11.1 *The algorithm ComputeMaximum runs in* $O(Q)$ *time.*

Proof.
Each set of inequalities has at most Q elements. According to Lemma 11.8, each iteration of the "while" loop removes at least $|P|/2$ inequalities from P. Thus, the number of inequalities processed within this loop is

$$Q + \frac{Q}{2} + \frac{Q}{4} + \cdots + \frac{Q}{2^l} = Q\left(\frac{2^l + 2^{l-1} + \cdots + 1}{2^l}\right) = \frac{Q}{2^l}\sum_{i=0}^{l} 2^i = \frac{Q}{2^l}(2^{l+1} - 1).$$

The loop keeps on until only two inequalities remain. Then $(Q/2^l) = 2$ which implies $Q = 2^{l+1}$, and hence $(Q/2^l) \cdot (2^{l+1} - 1) = 2(Q-1) < 2Q \in O(Q)$. □

Megiddo [131] and Dyer [67] proposed $O(Q)$ algorithms for calculating respectively, the minimal and maximal values of a two-variable linear programming problem. The time complexity of their methods is bounded by $4Q$, whereas this approach is bounded by $2Q$.

We describe in Algorithm 11.4 the procedure *ComputeMaximum*. In this algorithm T_L and T_R denote any inequality set with $L, R \in \{a, b, \ldots, h\}$.

***Algorithm 11.4**: ComputeMaximum : maximum dominated value inside R*

Data : Set of Inequalities T_L, Set of Inequalities T_R, Value y_{limit}.
Result: limit value (x_m, y_m) and the inequalities l_m and r_m.

1 Initialization;
Choose any $l_m \in T_L$ and $r_m \in T_R$ that intersect inside $X \times Y$. ;
$x_m := I(l_m, r_m)$, $y_m := F(x_m)$;
2 **while** $T_L \neq \emptyset$ *and* $T_R \neq \emptyset$ **do**
  Let $P := \{(l, r) : l \in T_L, r \in T_R\}$ be a matching between inequalities in T_L and T_R such that $|P| := \max\{|T_L|, |T_R|\}$;
  for *all the pairs* $(l, r) \in P$ **do**
    $x := (l, r)$, $y := F(x)$
    Check if $X \not\prec Y$ using Lemma 11.1 ;
    Try to improve value y by using Corollary 11.1 or 11.2 ;
    if $F(I(l_m, r)) < y$ **then** $l := l_m$ **if** $F(I(l, r_m)) < y$ **then** $r := r_m$
    if *y has been improved* **then** recompute $x := (l, r)$ and $y := F(x)$ Check if intersection is below lower limit;
    if $y < y_{\text{limit}}$ **then return** $X \not\prec Y$ Store the minimum value found so far.;
    if $y < y_m$ **then**
      $y_m := y$, $x_m := x$;
      $l_m := l$, $r_m := r$
    end
  end
  Look for the lower inequalities under y_m ;
  for *all* $l \in T_L$ **do**
    if *l is below* l_m **then** $l_m := l$
  end
  for *all* $r \in T_R$ **do**
    if *r is below* r_m **then** $r_m := r$
  end
  Check if $X \not\prec Y$ using Lemma 11.1 ;
  $x_m := I(l_m, r_m)$, $y_m := F(x_m)$;
  for *all* $l \in T_L$ **do**
    if *l is over* y_m **then** $T_L := T_L \backslash \{l\}$
  end
  for *all* $r \in T_R$ **do**
    if *r is over* y_m **then** $T_R := T_R \backslash \{r\}$
  end
end
3 **return** (x_m, y_m) *and* l_m, r_m.

Computing the maximal and minimal points inside a region R satisfying $X \prec Y$ is based on the comparison of values over the intersection points among the inequalities.

The set I has been defined as the set of the intersection points. If we have an inequality for each of the objectives $q = 1, \ldots, Q$, there are at most

$O(Q)$ inequalities in R. Therefore, $|I| \leq Q^2$. But it can be proved that both $(x_{\min}, y_{\min})$ and $(x_{\max}, y_{\max})$ can be calculated in $O(Q)$ time.

We can start by analyzing the computation of $(x_{\max}, y_{\max})$. Let $P = \{(a,c) : \text{for some } a \in T_a, c \in T_c\}$ be a set of pairs of inequalities of T_a and T_c such that $|P| = \max\{|T_a|, |T_c|\}$ with $|P| \leq |I|$.

Lemma 11.8 *In each iteration m of the search for the maximal (minimal) point $(x_{\max}, y_{\max})$ we can remove at least $|P|/2$ inequalities from P.*

Proof.
Assume that at iteration m of the searching process the current set of inequalities is $P = \{(a,c) : \text{for some } a \in T_a, c \in T_c\}$. Each pair of inequalities $(a,c) \in P$ yields a point $x = I(a,c)$ and a value $y = a(x)$. Let $x_m \in X$ be a point verifying that $a(x_m) = \min_{(a,c) \in P} a(I(a,c))$. This point x_m is a candidate to be maximal, with $x_{\max} = x_m$, $y_{\max} = y_m = F(x_m)$ and a_m and c_m being the inequalities defining $(x_{\max}, y_{\max})$. Accordingly, all inequalities in T_a and T_c will be deleted. Otherwise, there may still be some inequalities below a_m and c_m.

Let $a^* \in T_a$ such that $a^*(x_m) = \min_{a \in T_a} a(x_m)$ and $c^* \in T_c$ such that $c^*(x_m) = \min_{c \in T_c} c(x_m)$ be the lowest inequalities underneath $a_m(x_m)$. Let $x'_m = I(a^*, c^*)$ and $y'_m = a^*(x_m)$. This value is the new optimal point. Furthermore, we can now eliminate, in the worst case, one inequality a or c from each pair in $(a,c) \in P$. Indeed, each pair may have one single inequality under y'_m, that is, either $a(x_m) < y'_m$ or $c(x_m) < y'_m$. Both inequalities cannot be below since it contradicts the fact that (x'_m, y'_m) is the minimal point. Therefore, at least $|P|/2 = \max\{|T_a|, |T_c|\}/2$ inequalities are deleted.

□

Algorithm 11.5: *Checking $X \prec Y$*

Input: Set of inequalities T, Interval $X = [x_0, x_1]$, Interval $Y = [y_0, y_1]$.
Output: $[y_{\min}, y_{\max}]$, the points in Y where $X \prec Y$.

1 Initialization
Classify all inequalities in $T = T_a \cup T_b \cup \cdots \cup T_h$
Bound X and Y
$X := [\max\{x_0, \max_{t \in T_h} t\}, \min\{x_1, \max_{t \in T_g} t\}]$
$Y := [\max\{y_0, \max_{t \in T_f} t\}, \min\{y_1, \max_{t \in T_e} t\}]$

2 **if** $x_0 > x_1$ *or* $y_0 > y_1$ **then return** $X \not\prec Y$

3 $y_{\max} := y_1$
if $T_a \neq \emptyset$ *and* $T_c \neq \emptyset$ **then**
  if *there is an intersection point between* T_a *and* T_b *below* y_1 **then**
    if *Lemma 11.2 holds* **then** store solution in $(x_{\max}, y_{\max})$
    else $(x_{\min}, y_{\min}) := ComputeMaximum(T_a, T_b, y_0)$
  else apply Lemma 11.3 to get $[x^0_{\max}, x^1_{\max}]$
end

4 $y_{\min} := y_0$
if $T_d \neq \emptyset$ *and* $T_b \neq \emptyset$ **then**
  $T'_d := \{-d(x) : \forall d \in T_d\}$, $T'_b := \{-b(x) : \forall b \in T_b\}$
  if *there is an intersection point between* T'_d *and* T'_b *below* $-y_0$ **then**
    if *Lemma 11.2 holds for* T'_d *and* T'_b **then** store solution in $(x_{\min}, y_{\min})$
    else
      $(x_{\min}, y_{\min}) := ComputeMaximum(T'_d, T'_b, -y_1)$
      $y_{\min} := -y_{\min}$
    end
  else apply Lemma 11.3 on T'_b and T'_d to get $[x^0_{\min}, x^1_{\min}]$
end

5 **if** $y_{\min} > y_{\max}$ **then return** $X \not\prec Y$
continued next page

Algorithm 11.5 *continued*

6 **if** $x_{\min} < x_{\max}$ **then**
  Check Lemma 11.5
  if $\widehat{a}(x_{\min}) < y_{\min}$ *and* $\widehat{b}(x_{\max}) > y_{\max}$ **then return** $X \nprec Y$
  else
    $T'_b = T_b \backslash \{b \in T_b : b(x_{\max}) \leq y_{\max}\}$
    if $T'_b \neq \emptyset$ **then** $(x_{\max}, y_{\max}) = ComputeMaximum(T_a, T'_b)$
    $T_a := T_a \backslash \{a \in T_a : a(x_{\min}) \geq y_{\min}\}$
    if $T_a \neq \emptyset$ **then**
      $T'_b := \{-b(x) : \forall b \in T_b\}$, $T'_a := \{-a(x) : \forall a \in T_a\}$
      $(x_{\min}, -y_{\min}) := ComputeMaximum(T'_b, T'_a)$
    end
  end
else
  if $x_{\min} > x_{\max}$ **then**
    Check Lemma 11.7
    if $\widehat{c}(x_{\min}) < y_{\min}$ *and* $\widehat{d}(x_{\max}) > y_{\max}$ **then return** $X \nprec Y$ **else**
      $T'_d := T_d \backslash \{d \in T_d : d(x_{\max}) \leq y_{\max}\}$
      if $T'_d \neq \emptyset$ **then** $(x_{\max}, y_{\max}) = ComputeMaximum(T_c, T'_d)$
      $T_c := T_c \backslash \{c \in T_c : c(x_{\min}) \geq y_{\min}\}$
      if $T_c \neq \emptyset$ **then**
        $T'_d := \{-d(x) : \forall d \in T_d\}$, $T'_c := \{-c(x) : \forall c \in T_c\}$
        $(x_{\min}, -y_{\min}) := ComputeMaximum(T'_d, T'_c)$
      end
    end
  end
end

7 **return** $[y_{\min}, y_{\max}]$.

To conclude this section, we present an algorithm for the comparison between segments. This procedure finalizes the entire description of all the ingredients needed in Algorithm 11.1. Hence, Algorithm (11.1) is able to compute the entire set of efficient points of the MOMP.

Algorithm 11.6: *Segment-to-segment comparison*

Data : Set S.
Result : Set S_{ND}.
1 Initialization ;
$S_{ND} := S$;
2 **for** *all segments* $X := [x_0, x_1] \in S_{ND}$ **do**
  for *all segments* $Y := [y_0, y_1] \in S_{ND}$ *successors in* S_{ND} *to* X **do**
    for $r := 1$ *to* q **do**
      Create inequality $y(x)$;
      $T := T \cup y(x)$
    end
    Apply Algorithm 11.5 a (T, X, Y);
    if $X \prec Y$ **then** $Y := Y \backslash [y_{\min}, y_{\max}]$ Change inequalities $y(x)$ to $x(y)$ defining the complementary region $\overline{R}$;
    Apply Algorithm 11.5 a (T, Y, X);
    if $Y \prec X$ **then** $X := X \backslash [x_{\min}, x_{\max}]$
  end
end
3 **return** S_{ND}.

11.6 Computing the Set of Efficient Points Using Linear Programming

The solution method based on the Linear Programming approach consists of a pairwise comparison of subedges, as in the former case. The idea is to partition the network into subedges where the objective functions are linear. The points where the piecewise linear functions change in slope are in fact the bottleneck and equilibrium points. We then make a pairwise comparison of all the subedges and delete the dominated parts. The result is the complete set of efficient solutions $\mathrm{OM}_{multi}(\lambda)$. In the presentation we follow the approach in [89] that was developed for median problems.

It is important to note that a partial subedge may be efficient, starting at a point which is not a node or an edge-breakpoint. For each comparison of two subedges we will construct a linear program to detect dominated points (segments), that can be found in linear time by the method in Megiddo [131].

Let $z^q(t) = f^q_\lambda(x_t)$, for $q = 1, \ldots, Q$; where $x_t = (e, t)$. These Q functions are all piecewise linear and their breakpoints correspond to the equilibrium and bottleneck points. Assume that, for a given edge, there are $P + 1$ breakpoints including the two nodes. We then have P subedges. Let the breakpoints on (e, t) be denoted by t_j, $j = 0, 1, \ldots, P$ $(1 \leq P \leq \binom{M}{2} Q)$, with $t_0 = v_i$, $t_P = v_j$ and $t_{j-1} < t_j$, $\forall j = 1, 2, \ldots, P$. For $t \in [t_{j-1}, t_j]$, $z^q(t)$, $q = 1, \ldots, Q$, are linear functions of the form

$$z^q(t) = m^q_j t + b^q_j, \qquad \forall q = 1, 2, \ldots, Q.$$

Let us now compare the subedge $A = (e_A, [t_{j-1}, t_j])$ on edge e_A, with the subedge $B = (e_B, [s_{p-1}, s_p])$ on e_B. A point $(e_A, t) \in (e_A, [t_{j-1}, t_j])$ is dominated by some point $(e_B, s) \in (e_B, [s_{p-1}, s_p])$ if and only if

$$m_p^q s + b_p^q \le m_j^q t + b_j^q, \qquad \forall q = 1, 2, \ldots, Q, \tag{11.2}$$

where at least one of those inequalities is strict.

Let us define T as the set of points where the inequalities (11.2) hold (for these particular subedges):

$$T = \{(s,t) | m_j^q t - m_p^q s \ge b_p^q - b_j^q, \forall q \in Q\} \cap ([s_{p-1}, s_p] \times [t_{j-1}, t_j]).$$

If $T = \emptyset$, $(e_B, [s_{p-1}, s_p])$ does not contain any point dominating any point in $(e_A, [t_{j-1}, t_j])$. Otherwise $T \neq \emptyset$ is taken as the feasible set of the following 2-variable linear programs:

$$LB = \min\{t | (s,t) \in T\} \quad \text{and} \quad UB = \max\{t | (s,t) \in T\}.$$

Using the algorithm in Megiddo [131], LB and UB can be calculated in $O(Q)$ time. We now check, if we have only *weak dominance*. This means that none of the inequalities need to be strict as we require in our definition of dominance. Note that points with weak dominated objective function may be efficient. Let s_{LB} and s_{UB} be the optimal values of s corresponding to LB and UB. These $s-$values are not necessarily unique. In the case where s_{LB} (and/or s_{UB}) is not unique ($s_{LB} \in [s_a, s_b]$), we choose $s_{LB} = \frac{1}{2}(s_a + s_b)$ to avoid problems with weak dominance in the subedge endnodes. To check for weak dominance, we examine the subedge endnodes. If $m_p^q s_{LB} + b_p^q = m_j^q LB + b_j^q, \forall q \in Q$, then the solution corresponding to LB is only weakly dominated and can therefore still be efficient. Similarly, if $m_p^q s_{UB} + b_p^q = m_j^q UB + b_j^q, \forall q \in Q$, then the solution corresponding to UB is only weakly dominated. If both the solutions corresponding to LB and UB are only weakly dominated, the entire subedge $(e_A, [t_{j-1}, t_j])$ is only weakly dominated by $(e_B, [s_{p-1}, s_p])$. This means that all the inequalities in T are in fact equalities. Otherwise, the dominated part of the subedge is deleted. If both the solutions corresponding to LB and UB are dominated, then

$$(e_A, [t_{j-1}, t_j]) = (e_A, [t_{j-1}, t_j]) \setminus (e_A, [LB, UB])$$

and if, say, the solution corresponding to LB is only weakly dominated, then

$$(e_A, [t_{j-1}, t_j]) = (e_A, [t_{j-1}, t_j]) \setminus (e_A, (LB, UB])$$

leaving the solution corresponding to LB. This comparison can also be done in linear time. The approach is simplified if one or both subedges consists of a single point (e_A, t') (or (e_B, s'')). If $(e_A, [t_{j-1}, t_j]) = (e_A, t') = x$, then $LB = UB = t'$ and

$$T' = \{s| - m_p^q s \geq b_p^q - f^q(x), \forall q \in Q\} \cap [s_{p-1}, s_p].$$

If $(e_B, [s_{p-1}, s_p]) = (e_B, s'') = y$, then

$$T'' = \{t | m_j^q t \geq f^q(y) - b_j^q, \forall q \in Q\} \cap [t_{j-1}, t_j]$$

and

$$LB = \min\{t | t \in T''\} \qquad UB = \max\{t | t \in T''\}.$$

Since we are removing a connected piece of $(e_A, [t_{j-1}, t_j])$, three things can happen. First, $(e_A, [t_{j-1}, t_j])$ can be completely deleted if $t_{j-1} = LB$ and $t_j = UB$ are both dominated. Second, a piece of $(e_A, [t_{j-1}, t_j])$ that includes one of the endpoints t_{j-1} or t_j can be deleted, in which case, one connected subedge remains, say $(e_A, [t_{j-1}, LB))$ or $(e_A, [t_{j-1}, LB])$. The third case is when an interior part of $(e_A, [t_{j-1}, t_j])$ is deleted, so we end up with two new subedges $(e_A, [t_{j-1}, LB))$ and $(e_A, (UB, t_j])$, possibly including one of the points LB or UB.

In order to complete the comparison, we simply perform an ordered subedge comparison. First, we compare $(e_1, [t_0, t_1])$ with all the other subedges, possibly dividing $(e_1, [t_0, t_1])$ into new subedges. Then we compare the second subedge $(e_1, [t_1, t_2])$ with all the remaining subedges. If $(e_1, [t_0, t_1])$ is not completely dominated, we also compare with this subedge. This comparison continues until we have compared the last subedge $(e_N, [s_{P-1}, s_P])$ with all the remaining subedges.

Notice that we can still use the entire subedge $(e_A, [t_{j-1}, t_j])$ to compare with the other subedges, even though a part of it is dominated. It is only for the set of efficient points $\mathsf{OM}_{multi}(\lambda)$, that we have to remember what part of $(e_A, [t_{j-1}, t_j])$ is efficient. But if the whole subedge $(e_A, [t_{j-1}, t_j])$ is dominated, we should delete it from further consideration, also in the comparison process.

Assume that edge $e_i \in E$ is divided into P_i breakpoint subedges. Each of the N edges may have up to $O(QM^2)$ equilibrium plus bottleneck points, giving at most $O(QM^2N)$ subedges. If we make the global pairwise comparison on the $O(QM^2N)$ breakpoint subedges, each taking $O(Q)$ time, we get for Algorithm 11.7 a complexity bound of $O(Q^3M^4N^2)$ time.

Algorithm 11.7: *Computation of the solution set* $OM_{multi}(\lambda)$

Data : Network $G = (V, E)$, distance-matrix D, parameters Q, λ.
Result: $\mathrm{OM}_{multi}(\lambda)$ the solution set to the multicriteria ordered median problem.

```
1 Initialization ;
  OM_multi(λ) = G(V,E);
2 for i := 1 to N do
      for x := 1 to P_i do
          for j := 1 to N do
              for r := 1 to P_j do
                  compare (e_i, [t_{x-1}, t_x]) with (e_j, [t_{y-1}, t_y]);
                  OM_multi(λ) unchanged if no points are dominated ;
                  OM_multi(λ) = OM_multi(λ) \ (e_i, [LB, UB]) if LB and UB
                  are dominated;
                  OM_multi(λ) = OM_multi(λ) \ (e_i, (LB, UB]) if only UB is
                  dominated;
                  OM_multi(λ) = OM_multi(λ) \ (e_i, [LB, UB)) if only LB is
                  dominated;
              end
          end
      end
  end
3 return OM_multi(λ).
```

12

Extensions of the Ordered Median Problem on Networks

In the previous chapters we consider location problems where the objective is to locate a specified number of servers optimizing some criterion, which usually depends on the distances between the demand points and their respective servers (represented by points). In this chapter we study the location of connected structures, which cannot be represented by isolated points in the network. These problems are motivated by concrete decision problems related to routing and network design. For instance, in order to improve the mobility of the population and reduce traffic congestion, many existing rapid transit networks are being updated by extending or adding lines. These lines can be viewed as new facilities. Studies on location of connected structures (which are called *extensive* facilities) appeared already in the early eighties (see [145], [100], [183], [140], [139], [19]).

Almost all the location problems of extensive facilities that have been discussed in the literature and shown to be polynomially solvable, are defined on tree networks (see e.g., [3], [18], [23], [86], [160], [161], [173], [182], [183], [186], [189], [192], [201], [202], and the references therein). In this chapter, following the presentation in Puerto and Tamir [173], we study the location of a tree-shaped facility S on a tree network, using the ordered median function of the weighted distances to represent the total transportation cost objective. According to a common classification in logistic models, e.g. [28], two settings are considered: *tactical* and *strategic*. In the tactical model, there is an explicit bound L on the length of the subtree, and the goal is to select a subtree of size L, which minimizes the cost function. In the strategic model the length of the subtree is variable, and the objective is to minimize the sum of the transportation cost and the length of the subtree. We consider both discrete and continuous versions of the tactical and the strategic models. If the leaves of a subtree are restricted to be nodes of the tree, the subtree facility is called *discrete*, otherwise it is *continuous*. We note that the discrete tactical problem is NP-hard since Knapsack is a special case, and we solve the continuous tactical problem in polynomial time using a Linear Programming (LP) approach. We also use submodularity properties for the strategic problem. These properties allow us to solve the discrete strategic version in strongly polynomial time. Moreover, the continuous version is also solved via LP. For the special case of the k-centrum objective we obtain improved algorithmic results using a Dynamic Programming (DP) algorithm and discretization results.

It is worth noting that there is a fundamental difference between the analysis of problems with the ordered median objective and with other classical functions in location analysis (median, center, centdian, or even k-centrum). In spite of its inherent similarity the ordered median objective does not possess any clear separability properties. Therefore, standard DP approaches, commonly used to solve location problems with median, center and even k-centrum objectives [186], [187], [191], [192], are not applicable here. As a result, we discuss alternative approaches that rely on different tools. We apply submodularity properties and linear programming formulations that turn out to be instrumental in solving discrete and continuous versions of the problems, respectively.

In Table 12.1 we present the complexity of the algorithms discussed in the chapter.

Table 12.1. New results in the chapter

Complexity of Tree-shaped facilities [(1)]				
	Tactical		Strategic	
	Discrete	Continuous	Discrete	Continuous
Convex ordered median	NP-hard	$O(M^7 + M^6 I)$	$O(M^6 \log M)$	$O(M^7 + M^6 I)$
k-centrum	NP-hard	$O(M^3 + M^{2.5} I)$ [(2)]	$O(kM^3)$ [(2)]	$O(kM^7)$ [(2)]

For the sake of readability the chapter is organized as follows. In Section 12.1 we define the models that we deal with, and state the notation used throughout the chapter. Sections 12.2 and 12.3 analyze the tactical and strategic models of the convex ordered median subtree problem, respectively. In the next two sections we restrict ourselves to the special case of the subtree k-centrum problem. There we prove nestedness properties and develop a DP algorithm that solves both the strategic continuous and discrete subtree k-centrum problems. The chapter ends with some concluding remarks.

12.1 Notation and Model Definitions

We follow the notation already introduced in Section 7.1. Specifically, let $T = (V, E)$ be an undirected tree network with node set $V = \{v_1, ..., v_M\}$ and

[1] I denotes the input size of the problem.

[2] A nestedness property with respect to the point solution holds.

edge set $E = \{e_2, ..., e_M\}$. Each edge e_j, $j = 2, 3, \ldots, M$, has a positive length l_j, and is assumed to be rectifiable. In particular, an edge e_j is identified as an interval of length l_j, so that we can refer to its interior points. We assume that T is embedded in the Euclidean plane. Let $A(T)$ denote the continuum set of points on the edges of T. We view $A(T)$ as a connected and closed set which is the union of $M - 1$ intervals. Let $P[v_i, v_j]$ denote the unique simple path in $A(T)$ connecting v_i and v_j. Suppose that the tree T is rooted at some distinguished node, say v_1. For each node v_j, $j = 2, 3, \ldots, M$, let $p(v_j)$, the *parent* of v_j, be the node $v \in V$, closest to v_j, $v \neq v_j$ on $P[v_1, v_j]$. v_j is a *child* of $p(v_j)$. e_j is the edge connecting v_j with its parent $p(v_j)$. S_j will denote the set of all children of v_j. A node v_i is a *descendant* of v_j if v_j is on $P[v_i, v_1]$. V_j will denote the set of all descendants of v_j. The path $P[x, y]$ is also viewed as a collection of edges and at most two subedges (partial edges). $P(x, y)$ will denote the open path obtained from $P[x, y]$, by deleting the points x, y, and $P(x, y]$ will denote the half open path obtained from $P[x, y]$, by deleting the point x. A subset $Y \subseteq A(T)$ is called a *subtree* if it is closed and connected. Y is also viewed as a finite (connected) collection of partial edges (closed subintervals), such that the intersection of any pair of distinct partial edges is empty or is a point in V. We call a subtree *discrete* if all its (relative) boundary points are nodes of T. For each $i = 1, \ldots, M$, we denote by T_i the subtree induced by V_i. T_i^+ is the subtree induced by $V_i \cup \{p(v_i)\}$. If Y is a subtree we define the *length* or size of Y, $L(Y)$, to be the sum of the lengths of its partial edges. We also denote by $D(Y)$ the *diameter* of Y (the diameter of a graph is the maximum value of the minimum distance between any two nodes). In our location model the nodes of the tree are the demand points (customers), and each node $v_i \in V$ is associated with a nonnegative weight w_i. The set of potential servers consists of subtrees. There is a transportation cost function of the customers (assumed to be at the nodes of the tree) to the serving facility. We use the ordered median objective. Given a real L, the *tactical* subtree problem with an ordered median objective is the problem of finding a subtree (facility) Y of length smaller than or equal to L, minimizing the ordered median objective. When $L = 0$, we will refer to the problem as the *point problem* instead of the subtree problem, and call its solution a *point solution.* Given a positive real α, the *strategic* subtree model is the problem of finding a subtree (facility) Y minimizing the sum of the ordered median objective and the setup cost of the facility (the setup cost will be represented by $\alpha L(Y)$). We say that a tactical (strategic) model is *discrete* when the endnodes (leaves) of Y must be nodes in V. If the endnodes of Y may be anywhere in $A(T)$, we call the model *continuous.*

12.2 Tactical Subtree with Convex Ordered Median Objective

In this section we consider the problem of selecting a subtree of a given length which minimizes the convex ordered median objective. We recall, that the discrete model is NP-hard even for the median function, since the Knapsack problem is a special case. Thus, we consider the continuous version. First, we note that if a continuous subtree does not contain a node it is properly contained in an edge. Hence, to solve the continuous model, it is sufficient to solve $O(M)$ continuous subproblems where the subtree is restricted to contain a distinguished node, and in addition, $M - 1$ continuous subproblems where the subtree is restricted to be a subedge of a distinguished edge. We show, how to find an optimal subedge of a given edge by converting the problem into a minimization of a single variable, convex and piecewise linear function. Later, we find an optimal subtree containing a distinguished point (node) with a different machinery.

12.2.1 Finding an Optimal Tactical Subedge

We show here, how to find an optimal subedge for the tactical model. As mentioned above, these are subproblems that we need to consider in the process of finding optimal continuous subtrees. Consider an arbitrary edge of the tree $e_k = (v_s, v_t)$. We show that an optimal subedge of e_k with respect to the tactical model can be found in $O(M \log^2 M)$ time. A point on e_k is identified by its distance, x, from the node v_s. A subedge of length L, incident to x is identified by its two endpoints: x and $x + L$. Let V^s be the set of nodes in the component of T, obtained by removing e_k, which contains v_s. For each node $v_i \in V^s$, its weighted distance from the subedge is given by $h_i(x, L) = w_i(d(v_i, v_s) + x)$. For each node $v_i \in V - V^s$, the respective distance is $h_i(x, L) = w_i(d(v_i, v_t) + d(v_s, v_t) - (x + L))$. In the tactical model L is fixed, and the problem of finding an optimal subedge of e_k of length L reduces to minimizing the single variable, x, ordered median function of a collection of the linear functions $\{h_i(x, L)\}$ (the domain is defined by the constraints $x \geq 0$, and $x + L \leq d(v_s, v_t) = l_k$). The latter task can be performed in $O(M \log^2 M)$ time, [117].

Remark 12.1 We note in passing that instead of minimizing over each edge separately, we can use the convexity of the objective and save by using global minimization. Globally, the problem reduces to finding a continuous subpath of the tree of length smaller than or equal to L, which minimizes the ordered median objective. (In this tactical continuous subpath model, the objective can be expressed in a way that makes it convex over each path of the tree. Specifically, restricting ourselves to a path of T, and using x as a single variable point along the path, each function $h_i(x, L)$ becomes a single variable, x, piecewise linear convex function with three pieces. The respective slopes are $\{-w_i, 0, w_i\}$.)

12.2.2 Finding an Optimal Tactical Continuous Subtree Containing a Given Node

We assume that the selected subtree must contain v_1, the root of the tree T. For each edge e_j of the rooted tree, connecting v_j with its parent, $p(v_j)$, assign a variable x_j, $0 \leq x_j \leq l_j$. The interpretation of x_j is as follows: Suppose that $x_j > 0$, and let $x_j(e_j)$ be the point on edge e_j, whose distance from $p(v_j)$, the parent of v_j, is x_j. The only part of e_j, included in the selected subtree rooted at v_1 is the subedge $P[p(v_j), x_j(e_j)]$. In order for the representation of a subtree by the x_j variables to be valid, we also need the following condition to be satisfied:

$$x_j(l_i - x_i) = 0, \text{ if } v_i = p(v_j), v_i \neq v_1, \text{ and } j = 2, \ldots, M. \tag{12.1}$$

Equation (12.1) ensures the connectivity of the subtree represented by the variables x_j. Unfortunately, these constraints are non-linear. For each node v_t, the weighted distance of v_t from a subtree (represented by the variables x_j), is

$$y_t = w_t \sum_{v_k \in P[v_t, v_1)} (l_k - x_k).$$

We call a set $X = \{x_2, ..., x_M\}$ *admissible* if $x_2, ..., x_M$, satisfy $0 \leq x_j \leq l_j$, $j = 2, \ldots, M$, and

$$\sum_{j=2}^{M} x_j \leq L. \tag{12.2}$$

Note that an admissible solution induces a closed subset of $A(T)$ of length less than or equal to L, which contains v_1.

Proposition 12.1 *Let $X = \{x_2, ..., x_M\}$ be admissible. For each $v_t \in V$ define $y_t = w_t \sum_{v_k \in P[v_t, v_1]} (l_k - x_k)$. Then, there exists an admissible set $X' = \{x'_2, ..., x'_M\}$, satisfying (12.1), such that for each $v_t \in V$*

$$y'_t = w_t \sum_{v_k \in P[v_t, v_1)} (l_k - x'_k) \leq y_t.$$

Proof.
The positive components of X induce a collection of connected components (subtrees) on the rooted tree T. Each such subtree is rooted at a node of T. One of these components contains v_1 (if no positive variable is associated with an edge incident to v_1, then $\{v_1\}$ will be considered as a connected component). If there is only one component in the collection X itself satisfies

(12.1). Suppose that there are at least two connected components. A connected component T' is called minimal if there is no other component T'', such that the path from T'' to v_1 passes through T'. Let T' be a minimal component. Let v_i be the root of T', and let $x_j > 0$ induce a leaf edge of T'. In other words, v_i is the closest node to v_1 in T', and there is a node v_j, possibly in T', such that $x_t = 0$ for each descendant v_t of v_j. Let T_1 be the connected component containing v_1. Consider the point of T_1, which is closest to v_i. This point is either v_1 or it is a leaf of T_1. It is uniquely identified by a node v_k on $P[v_i, v_1]$, and its associated variable x_k. (It is the point on the edge $(v_k, p(v_k))$ whose distance from $p(v_k)$ is exactly x_k.) Let T^*, be the closest component to v_k, amongst all the components which intersect $P[v_i, v_k]$ and do not contain v_1. Let v_q be the root of T^*. Set

$$\epsilon = \min[x_j, d(v_q, v_k) + l_k - x_k].$$

If $\epsilon = d(v_q, v_k) + l_k - x_k$, define an admissible solution $Z = \{z_2, ..., z_M\}$ by $z_p = l_p$ for each node v_p on $P[v_q, v_k]$, $z_j = x_j - \epsilon$, and $z_s = x_s$ for any other node v_s, $v_s \neq v_j$, v_s not on $P[v_q, v_k]$. It is easy to check that $w_t \sum_{v_k \in P[v_t, v_1)} (l_k - z_k) \leq y_t$, for each $v_t \in V$. Note that the number of connected components induced by Z is smaller than the number of connected components induced by X. We can replace X by Z and proceed. Suppose that $\epsilon = x_j < d(v_q, v_k) + l_k - x_k$. We similarly define an admissible solution $Z = \{z_2, ..., z_M\}$. Specifically, let v_m be a node on $P[v_i, v_k]$ such that $d(v_m, v_k) + l_k - x_k > x_j \geq d(v_m, v_k) - l_m + l_k - x_k$. Define $z_p = l_p$ for each node v_p, $v_p \neq v_m$, on $P[v_m, v_k]$, $z_m = x_j + x_k + l_m - d(v_m, v_k) - l_k$, $z_j = 0$, and $z_s = x_s$ for any other node v_s, $v_s \neq v_j$, v_s not on $P[v_m, v_k]$. It is easy to check that $w_t \sum_{v_k \in P[v_t, v_1)} (l_k - z_k) \leq y_t$, for each $v_t \in V$. Note that the number of leaf edges of minimal connected components in the solution induced by Z is smaller than the respective number in the solution induced by X. We can replace X by Z and proceed. To conclude in $O(M)$ steps we will identify X' as required. □

The representation of subtrees by the variables x_j is not linear. In spite of that fact, Proposition 12.1 proves that looking for optimal subtrees with respect to an isotone function of the weighted distances, one obtains connectivity without imposing it explicitly. (A function is isotone if it is monotone non-decreasing in each one of its variables.) This argument is formalized in the following corollary.

Corollary 12.1 *Consider the problem of selecting a subtree of total length L, containing v_1, and minimizing an isotone function, $f(y_2, ..., y_M)$, of the weighted distances $\{y_t\}$, $v_t \in V$, of the nodes in V from the selected subtree. Then the following is a valid formulation of the problem:*

$$
\begin{aligned}
&\min f(y_2, \ldots, y_M), \\
&s.t.\ y_t = w_t \sum_{v_k \in P[v_t, v_1)} (l_k - x_k), \ \textit{for each } v_t \in V, \\
&\quad 0 \le x_j \le l_j, \qquad\qquad j = 2, \ldots, M, \\
&\quad \textstyle\sum_{j=2}^{M} x_j \le L.
\end{aligned}
$$

We can now insert the LP formulations, for the k-centrum and the convex ordered median objectives, presented in (9.1). (See [154] for more details.) This will provide a compact LP formulation for finding a continuous subtree rooted at a distinguished point, whose length is at most L, minimizing a convex ordered median objective. For convenience we define $\lambda_{n+1} = 0$.

$$
\begin{aligned}
\min\ & \sum_{k=1}^{M} (\lambda_k - \lambda_{k+1})(kt_k + \sum_{i=1}^{M} d_{i,k}^{+}) \\
\text{s.t. } & d_{i,k}^{+} \ge y_i - t_k,\ d_{i,k}^{+} \ge 0, \qquad i = 1, \ldots, M,\ k = 1, \ldots, M, \\
& y_t = w_t \sum_{v_k \in P[v_t, v_1)} (l_k - x_k), \qquad \text{for each } v_t \in V, \\
& 0 \le x_j \le l_j, \qquad j = 2, \ldots, M, \\
& \textstyle\sum_{j=2}^{M} x_j \le L.
\end{aligned}
$$

(This LP formulation is later used to formulate the continuous strategic model, when the objective is to minimize the sum of the total length of the selected subtree and the ordered median function of the weighted distances to the subtree. Specifically, we add the linear function $\sum_{j=2}^{M} x_j$ to the objective, and remove the length constraint $\sum_{j=2}^{M} x_j \le L$.) The above formulation uses $p = O(M^2)$ variables and $q = O(M^2)$ constraints. Assuming integer data, let I denote the total number of bits needed to represent the input. Then, by [197], the LP can be solved by using only $O(M^6 + M^5 I)$ arithmetic operations. We also note that in the unweighted model, where the distances of all demand points are equally weighted, all the entries of the constraint matrix can be assumed to be $0, 1$ or -1. Therefore, by [193], the number of arithmetic operations needed to solve the unweighted version is strongly polynomial, i.e., it is bounded by a polynomial in M, and is independent of the input size I.

To evaluate the overall time complexity of this algorithm for the continuous tactical problem, we note that we solve M linear programs, and find $M - 1$ optimal subedges. Therefore, the total complexity is $O(M^7 + M^6 I)$.

12.3 Strategic Subtree with Convex Ordered Median Objective

Unlike the tactical model, we will show in this section that the strategic discrete subtree problem is solvable in polynomial time. Specifically, we will formulate this discrete model as a minimization problem of a submodular function over a lattice. But first we consider the continuous version and show, how to solve it using LP techniques. We assume that the objective function is the sum of the convex ordered median function and the total length of the selected subtree. The argument used above for the tactical continuous subtree problem proves that it is sufficient to consider only two types of (strategic) subproblems. The first type corresponds to a subproblem where the optimal subtree is a subedge of a given edge. In the second type the optimal subtree must contain a distinguished node.

12.3.1 Finding an Optimal Strategic Subedge

We use the same notation as in the subsection on finding the tactical subedge. In the case of the strategic model L is variable, and the problem of finding an optimal subedge of e_k reduces to minimizing the 2-variable, $\{x, L\}$, ordered median function of a collection of the linear functions $\{h_i(x, L)\}$. The feasible set for this problem is defined by the constraints $0 \leq x$, $0 \leq L$, and $x + L \leq d(v_s, v_t) = l_k$. This 2-dimensional minimization can be performed in $O(M \log^4 M)$ time, [117].

12.3.2 Finding an Optimal Strategic Continuous Subtree Containing a Given Node

We use the notation in the above section dealing with the tactical continuous subtree problem. In particular, we assume that the optimal subtree must contain v_1, the root of the tree T. From the discussion above we conclude that this restricted version of the strategic continuous problem can be formulated as the following compact linear program:

$$
\begin{aligned}
\min \ & \sum_{k=1}^{M} (\lambda_k - \lambda_{k+1})(kt_k + \sum_{i=1}^{M} d^+_{i,k}) + \alpha \sum_{j=2}^{M} x_j \\
s.t. \ & d^+_{i,k} \geq y_i - t_k, \ d^+_{i,k} \geq 0, && i = 1, \ldots, M, \ k = 1, \ldots, M, \\
& y_t = w_t \sum_{v_k \in P[v_t, v_1]} (l_k - x_k), && \text{for each } v_t \in V, \\
& 0 \leq x_j \leq l_j, && j = 2, \ldots, M.
\end{aligned}
$$

By the analysis used above for the tactical model, it is easy to see that the overall time complexity of the algorithm for the continuous strategic problem is also $O(M^7 + M^6 I)$.

12.3.3 Submodularity of Convex Ordered Median Functions

To solve the strategic discrete subtree problem we use the Submodularity Theorem 1.4 proved in Chapter 1. We use the submodularity property to claim that the strategic discrete subtree ordered median problem is solvable in strongly polynomial time. First note that the set of all subtrees containing a distinguished point of the tree forms a lattice. For each subtree S of T, consider the vector $a(S) = (a_i(S))$ in $\mathbb{R}^M$, where $a_i(S) = w_i d(v_i, S)$, $i = 1, ..., M$. Define

$$f_\lambda(S) = \sum_{i=1}^{M} \lambda_i a_{(i)}(S).$$

($a_{(i)}(S)$ denotes the i-th largest element in the set $\{a_1(S), ..., a_M(S)\}$.) We now demonstrate that $f_\lambda(S)$ is submodular over the above lattice. Let S_1 and S_2 be a pair of subtrees with nonempty intersection. Then, from [184] we note that

$$w_i d(v_i, S_1 \cup S_2) = w_i \min[d(v_i, S_1), d(v_i, S_2)],$$

$$w_i d(v_i, S_1 \cap S_2) = w_i \max[d(v_i, S_1), d(v_i, S_2)].$$

Then, $\min[a_i(S_1), a_i(S_2)] = w_i d(v_i, S_1 \cup S_2)$ and $\max[a_i(S_1), a_i(S_2)] = w_i d(v_i, S_1 \cap S_2)$. Applying Theorem 1.4, we conclude that $f_\lambda(S)$, the convex ordered median objective, defined on this lattice of subtrees, is submodular, i.e.,

$$f_\lambda(S_1 \cup S_2) + f_\lambda(S_1 \cap S_2) \leq f_\lambda(S_1) + f_\lambda(S_2).$$

From [184] we note that the length function $L(S)$ is modular and the diameter function $D(S)$ is submodular. Thus, any linear combination with nonnegative coefficients of the functions $f_\lambda(S), L(S)$ and $D(S)$ is submodular. Therefore, we can also add the diameter function to the objective and preserve the submodularity. Specifically, we now extend the objective function of the discrete strategic model to

$$f_\lambda(S) + \alpha L(S) + \beta D(S),$$

where β is a nonnegative real. In addition to the ellipsoidal algorithms [83], there are now combinatorial strongly polynomial algorithms to minimize a submodular function over a lattice [130]. The best complexity for a strongly polynomial algorithm is $O(M^4 OE)$ in [83], (quoted from [174]), where OE is the time to compute the submodular function. It is easy to check that for the strategic discrete subtree problem with a convex ordered median function $OE = O(M \log M)$. Therefore, minimizing over each lattice family (subtrees containing a distinguished node) takes $O(M^5 \log M)$. Solving the discrete strategic subtree ordered median problem will then take $O(M^6 \log M)$!!! (For comparison purposes, the special case of the discrete k-centrum objective is solved in the next section in $O(kM^3) = O(M^4)$ time.)

The problem above can be generalized by replacing the sum of weighted distances by the sum of transportation cost functions of the nodes (demand points) to the subtree (facility). Indeed, let f_i, $i = 1, \ldots, M$, be real isotone functions. We consider S_1, S_2 two subtrees of the given lattice (subtrees containing a distinguished node). Then, it is clear that

$$\begin{aligned} f_i(S_1 \cup S_2) &:= f_i(d(v_i, S_1 \cup S_2)) = f_i(d(v_i, S_1) \wedge d(v_i, S_2)) \\ &= \min[f_i(d(v_i, S_1)), f_i(d(v_i, S_2))], \\ f_i(S_1 \cap S_2) &:= f_i(d(v_i, S_1 \cap S_2)) = f_i(d(v_i, S_1) \vee d(v_i, S_2)) \\ &= \max[f_i(d(v_i, S_1)), f_i(d(v_i, S_2))]. \end{aligned}$$

These inequalities allow us to apply Theorem 1.4 to the function:

$$g(S) = \sum_{i=1}^{M} \lambda_i f_{(i)}(S),$$

where $f^*(S) = (f_i(S))_{i=1,\ldots,M}$, $f_i(S) = f_i(d(v_i, S))$; $i = 1, \ldots, M$. ($f_{(i)}(S)$ is the i-th largest element in the set $\{f_1(S), \ldots, f_M(S)\}$.) Thus, we conclude that $g(S)$ is submodular on the considered lattice of subtrees. A special case of the above extension is the *conditional* version of this model, defined as follows. It is assumed that there are already facilities established at some closed subset Y' of $A(T)$. If we select a subtree S as the new facility to be setup, then each demand point v_i will be served by the closest of S and Y'. Hence, the respective transportation cost function is given by $f_i(S) = w_i \min[d(v_i, S), d(v_i, Y')]$. We can extend further the result on submodularity to the case where customers are represented by subtrees instead of nodes. Let us assume that we are given a family of subtrees $T^i = (V^i, E^i)$, $i = 1, \ldots, t$, such that $\bigcup_{i=1}^{t} V^i = V$. For each subtree S of T consider the vector $b(S) = (b_i(S)) \in \mathbb{R}^t$, where $b_i(S) = w_i d(T^i, S)$, $i = 1, \ldots, t$ (in particular, when each $T^i = (\{v_i\}, \emptyset)$, $i = 1, \ldots, M$, we get the customer-node model). Define

$$g(S) = \sum_{i=1}^{t} \lambda_i b_{(i)}(S).$$

($b_{(i)}(S)$ is the i-th largest element in the set $\{b_1(S), \ldots, b_t(S)\}$.) If S_1, S_2 is a pair of subtrees with non-empty intersection, then we note that

$$\begin{aligned} w_i d(T^i, S_1 \cup S_2) &= w_i \min[d(T^i, S_1), d(T^i, S_2)] = \min[b_i(S_1), b_i(S_2)], \\ w_i d(T^i, S_1 \cap S_2) &= w_i \max[d(T^i, S_1), d(T^i, S_2)] = \max[b_i(S_1), b_i(S_2)]. \end{aligned}$$

Once more, we can apply the submodularity Theorem 1.4 to conclude that $g(S)$ is submodular over the considered lattice of subtrees.

12.4 The Special Case of the Subtree k-Centrum Problem

In this section we focus on the special case of the k-centrum objective. Recall that the k-centrum objective value for a subtree S is given by the sum of the k

largest elements in $X(S) = \{w_1 d(v_1, S), ..., w_M d(v_M, S)\}$, the set of weighted distances from S. We consider the strategic discrete and continuous subtree k-centrum problems, and show how to solve them polynomially using dynamic programming techniques. In [173] it is proven that the discrete version is solvable in cubic time for any fixed k, and the same recursive approach it is also extended to the continuous version of the problem. As noted above the tactical discrete model is NP-hard. The tactical continuous model is polynomially solvable, as a special case of the convex ordered median problem discussed above. However, one can do better for the k-centrum objective case. Following [173], we will show in the next subsections that it is sufficient to solve only one subproblem where the selected subtree must contain a point k-centrum of the tree (recall that a point k-centrum is a solution to the tactical model when the length of the subtree is zero).

12.4.1 Nestedness Property for the Strategic and Tactical Discrete k-Centrum Problems

We are interested in studying whether there exists a distinguished point, e.g. an optimal solution to the point k-centrum problem, that must be included in some optimal solution to the subtree k-centrum problem. We call this property *nestedness.* Before we study the nestedness property for the strategic discrete model, we first observe that the discrete tactical k-centrum problem does not have the nestedness property with respect to the point solution. Indeed, it can be seen that even for the regular median objective, M-centrum, the property fails to be true, as shown by the following example on the line, taken from [173]: $v_1 = 0, v_2 = 2, v_3 = v_2 + 1/4$, and $v_4 = v_3 + 1$. $w_1 = 2$ and $w_i = 1$ for $i = 2, 3, 4$. The unique solution for the tactical discrete problem with $L = 0$ is v_2, and the unique solution for the tactical discrete problem with $L = 1$ is the edge (v_3, v_4). This negative result leads us to investigate the strategic discrete k-centrum problem. The next theorem the proof of which can be found in [173] shows that a nestedness property holds for this model.

Theorem 12.1 *Let v' be an optimal solution for the continuous point k-centrum problem. If v' is a node, there is an optimal solution to the strategic discrete subtree k-centrum problem which contains v'. If v' is not a node, there is an optimal solution which contains one of the two nodes of the edge containing v'.*

It is easy to verify that the above result also validates the following nestedness property of the continuous model.

Theorem 12.2 *Let v' be an optimal solution for the continuous point k-centrum problem. There is an optimal solution to the strategic continuous subtree k-centrum problem which contains v'.*

Following the above theorems we note that an optimal solution to the continuous point k-centrum problem can be found in $O(M \log M)$ time by the algorithm in page 274, taken from [117].

We described above that the strategic discrete subtree k-centrum problem possesses nestedness property which does not extend to the tactical version of the model. We also include the following result taken from [173] which ensures that the tactical continuous subtree k-centrum problem has this property.

Theorem 12.3 *Let v' be an optimal solution to the continuous point k-centrum problem. Then there exists an optimal solution to the tactical continuous subtree k-centrum problem which contains v'.*

Remark 12.2 A nestedness property for the continuous tactical model implies the result for the continuous case of the strategic version. Thus, the last theorem provides an alternative proof of Theorem 12.2.

From the above nestedness result we can assume that an optimal subtree to the tactical continuous k-centrum problem is rooted at some known point of the tree. Without loss of generality suppose that this point is the node v_1. Therefore, using the formulation in Section 12.2.2, the problem reduces to a single linear program. Moreover, since $\lambda_i = 1$ for $i = M - k + 1, \ldots, M$, and $\lambda_i = 0$ for $i = 1, \ldots, M - k$, the formulation has only $O(M)$ variables and $O(M)$ constraints:

$$\begin{aligned}
\min\ & kt_k + \sum_{i=1}^{M} d^+_{i,k} \\
\text{s.t. } & d^+_{i,k} \geq y_i - t_k,\ d^+_{i,k} \geq 0, && i = 1, ..., M, \\
& y_t = w_t \sum_{v_q \in P[v_t, v_1)} (l_q - x_q), && \text{for each } v_t \in V, \\
& 0 \leq x_j \leq l_j, && j = 2, ..., M, \\
& \textstyle\sum_{j=2}^{M} x_j \leq L.
\end{aligned}$$

We conclude that the overall time complexity to solve the continuous tactical k-centrum problem is $O(M^3 + M^{2.5} I)$.

12.4.2 A Dynamic Programming Algorithm for the Strategic Discrete Subtree k-Centrum Problem

In this section we present a dynamic programming bottom-up algorithm to solve the strategic discrete subtree k-centrum problem when the selected subtree is restricted to contain a distinguished node, say v_1. From the nestedness results proved above, we know that solving at most two such restricted subproblems will suffice for the solution of the unrestricted problem. We follow the approach in [173], and assume without loss of generality that the tree T is binary and rooted at v_1. (If a node v_i is not a leaf, its two children are denoted by $v_{i(1)}$ and $v_{i(2)}$). Also, following the arguments in [173], we assume without loss of generality that all weighted distances between pairs of nodes are distinct. For the sake of readability, we recall that V_i is the set of descendants of v_i, T_i is the subtree induced by V_i, and T_i^+ is the subtree induced by $V_i \cup \{p(v_i)\}$ (see Section 12.1).

- Define $G_i(q, r)$ to be the optimal value of the objective of the subproblem with a subtree rooted at v_i restricted to T_i, (we take the sum of the q largest weighted distances), when the q-th largest weighted distance is exactly r (if there is no feasible solution set $G_i(q, r) = +\infty$).
- Define $G_i^+(q, r)$ to be the optimal value of the objective of the subproblem with a subtree rooted at v_i restricted to T_i, when the q-th largest weighted distance is greater than r, and the $(q+1)$-th largest weighted distance is smaller than r (if there is no feasible solution set $G_i^+(q, r) = +\infty$).
- Define $A_i(q, r)$ to be the sum of the q largest weighted distances of nodes in $T \setminus T_i$ to v_i, when the q-th largest is exactly r (in case there is no such set, let $A_i(q, r) = +\infty$).
- Define $A_i^+(q, r)$ as above, with the condition that the q-th largest is greater than r, and the $(q+1)$-th is smaller than r (if there is no feasible solution set $A_i^+(q, r) = +\infty$).
- Define $B_i(1, q, r)$ to be the sum of the q largest weighted distances of nodes in $V_{i(1)} \cup \{v_i\}$ to v_i, when the q-th largest is exactly r (in case there is no such set, let $B_i(1, q, r) = +\infty$).
- Define $B_i(2, q, r)$ to be the sum of the q largest weighted distances of nodes in $V_{i(2)} \cup \{v_i\}$ to v_i, when the q-th largest is exactly r (in case there is no such set, let $B_i(2, q, r) = +\infty$).
- Define $B_i^+(1, q, r)$ to be the sum of the q largest weighted distances of nodes in $V_{i(1)} \cup \{v_i\}$ to v_i, with the condition that the q-th largest is greater than r, and the $(q+1)$-th is smaller than r (in case there is no such set, let $B_i^+(1, q, r) = +\infty$).
- Define $B_i^+(2, q, r)$ to be the sum of the q largest weighted distances of nodes in $V_{i(2)} \cup \{v_i\}$ to v_i, with the condition that the q-th largest is greater than r, and the $(q+1)$-th is smaller than r (in case there is no such set, let $B_i^+(2, q, r) = +\infty$).
- Remark: For convenience, the 0-th largest weighted distance is $+\infty$, and for each i, in the definition of G_i, G_i^+, B_i and B_i^+, for any $q > |V_i|$, the $(q+1)$-th largest distance is $-\infty$. A similar convention is used for A_i and A_i^+.

In the above definitions the parameter r is restricted to the set $R^* = \{w_i d(v_i, v_j)\}$, $v_i, v_j \in V$ (note that $r = 0$ is an element of R^*). With the above notation, the optimal objective value for the strategic discrete subtree problem, when the subtree is rooted at v_1, is

$$OV_1 = \min_{r \in R^*} G_1(k,r).$$

It is convenient to consider the case $r = 0$ separately. $G_1(k,0)$ is the optimal value when the k-th largest weighted distance is equal to 0. Hence, for any $p \geq k$, the p-th largest weighted distance is equal to 0. Therefore, the problem reduces to finding a subtree, rooted at v_1, which contains at least $M+1-k$ nodes, and minimizes the sum of weighted distances of all nodes to the subtree and the length of the subtree. This problem can be phrased as a special case of the model considered in [113] (see also [40, 73, 76]). In particular, using the algorithm in [113], $G_1(k,0)$ can be computed in $O(kM)$ time.

Recursive Equations for $G_i(q,r)$ and $G_i^+(q,r)$ when $r > 0$

$G_i(q,r)$: Without loss of generality suppose that $r > 0$ is the weighted distance of some node $v_j \in V_{i(1)}$ from the selected subtree.
If the subtree does not include the edges $(v_i, v_{i(1)})$ and $(v_i, v_{i(2)})$, then the best value is

$$C_i = \min_{1 \leq q_1 \leq q} \{B_i(1, q_1, r) + B_i^+(2, q - q_1, r)\}.$$

If the subtree includes the edge $(v_i, v_{i(1)})$ but not $(v_i, v_{i(2)})$, then the best value is

$$D_i = \alpha d(v_i, v_{i(1)}) + \min_{1 \leq q_1 \leq q} \{G_{i(1)}(q_1, r) + B_i^+(2, q - q_1, r)\}.$$

If the subtree includes the edge $(v_i, v_{i(2)})$ but not $(v_i, v_{i(1)})$, then the best value is

$$E_i = \alpha d(v_i, v_{i(2)}) + \min_{1 \leq q_1 \leq q} \{B_i(1, q_1, r) + G_{i(2)}^+(q - q_1, r)\}.$$

If the subtree includes the edges $(v_i, v_{i(1)})$ and $(v_i, v_{i(2)})$, then the best value is

$$F_i = \alpha(d(v_i, v_{i(1)}) + d(v_i, v_{i(2)})) + \min_{1 \leq q_1 \leq q} \{G_{i(1)}(q_1, r) + G_{i(2)}^+(q - q_1, r)\}$$

$$G_i(q,r) = \min\{C_i, D_i, E_i, F_i\}.$$

$G_i^+(q,r)$:

First of all, we define $G_i^+(0,r)$ to be the length of the minimum subtree, T_i', rooted at v_i, ensuring that the weighted distance to each node in V_i is smaller than r (for each node v_j in V_i, define $v_{k(j)}$ to be the node on $P[v_i, v_j]$, closest to v_i, satisfying $w_j d(v_j, v_{k(j)}) < r$. T_i' is the subtree induced by v_i, and $v_{k(j)}, v_j \in V_i$).

If the subtree does not include the edges $(v_i, v_{i(1)})$ and $(v_i, v_{i(2)})$, then the best value is

$$C_i^+ = \min_{0 \le q_1 \le q} \{B_i^+(1, q_1, r) + B_i^+(2, q - q_1, r)\}.$$

If the subtree includes the edge $(v_i, v_{i(1)})$ but not $(v_i, v_{i(2)})$, then the best value is

$$D_i^+ = \alpha d(v_i, v_{i(1)}) + \min_{0 \le q_1 \le q} \{G_{i(1)}^+(q_1, r) + B_i^+(2, q - q_1, r)\}.$$

If the subtree includes the edge $(v_i, v_{i(2)})$ but not $(v_i, v_{i(1)})$, then the best value is

$$E_i^+ = \alpha d(v_i, v_{i(2)}) + \min_{0 \le q_1 \le q} \{B_i^+(1, q_1, r) + G_{i(2)}^+(q - q_1, r)\}.$$

If the subtree includes the edges $(v_i, v_{i(1)})$ and $(v_i, v_{i(2)})$, then the best value is

$$F_i^+ = \alpha(d(v_i, v_{i(1)}) + d(v_i, v_{i(2)})) + \min_{0 \le q_1 \le q} \{G_{i(1)}^+(q_1, r) + G_{i(2)}^+(q - q_1, r)\}$$

$$G_i^+(q, r) = min\{C_i^+, D_i^+, E_i^+, F_i^+\}.$$

Preprocessing:

To compute all the functions A_i, A_i^+, B_i and B_i^+, it is actually sufficient to find and sort the largest k weighted distances in V_i and $V \setminus V_i$ for all v_i. The latter task will consume only $O(M^2)$ time by the methods presented recently in Tamir [188].

Initialization:

If v_i is a leaf, then by the general definitions and conventions above, for $r > 0$, $G_i^+(1, r) = +\infty$ and $G_i^+(0, r) = 0$. As noted above, the best objective value attained by a subtree containing v_1 is

$$OV_1 = \min_{r \in R^*} G_1(k, r).$$

To evaluate the total complexity of the algorithm we note that for each node v_i, the functions $G_i(q, r)$ and $G_i^+(q, r)$ are computed for k values of q and $O(M^2)$ values of r. Hence, the total complexity, including the preprocessing, is $O(k^2M^3)$ (we note in passing that since the parameter q is bounded by $\min\{k, |V_i|\}$, the analysis in [185] is applicable also to the above algorithm. In particular, the actual complexity is only $O(kM^3)$).

12.5 Solving the Strategic Continuous Subtree k-Centrum Problem

The nestedness result shown above implies that it is sufficient to solve a restricted model where the continuous selected subtree contains a point k-centrum of the tree (as noted above the latter can be found in $O(M \log M)$

by the algorithm presented in page 274 which is taken from [117]). Without loss of generality suppose that the subtree must contain the root v_1. We now discretize the continuous problem as follows. Let

$$\mathsf{R} = \{r \in \mathbb{R}^+ : \exists x \in A(T), v_i, v_j \in V, w_i \neq w_j, w_i d(v_i,x) = w_j d(v_j,x) = r, \\ \text{or } \exists v_i, v_j \in V,\ r = w_i d(v_i,v_j)\}.$$

For each node v_i and $r \in \mathsf{R}$, let $x_i(r)$ be the point on $P[v_i, v_1]$ satisfying $w_i d(v_i, x_i(r)) = r$. (If there is no such point set $x_i(r) = v_i$.) Next define

$$\mathsf{PEQ} = \{x_i(r) \in A(T) : \text{ for any } r \in \mathsf{R} \text{ and } i = 1, 2, \dots, M\}.$$

Theorem 12.4 *The set PEQ is a Finite Dominating Set (FDS) for the strategic continuous subtree k-centrum problem.*

Proof.
Let Y be a given subtree containing v_1, and suppose that no leaf of Y is a node (the leaves are interior points of the edges). The objective function is the sum of the function $\alpha L(Y)$ and the sum of the k-largest weighted distances to the leaves of Y. Consider the change in the objective, due to small perturbations of the leaves of Y. Since the change in $L(\cdot)$ is linear in these perturbations, an FDS for the continuous k-centrum subtree location problem depends only on the change in the sum of the k-largest weighted distances from Y. It is known that the k-centrum objective is reduced to the convex ordered median problem with only two different λ values (namely $0, 1$). Moreover, if Y has p leaves then the subtree problem can now be reduced to a p-facility problem. Indeed, since the tree can be considered rooted at v_1, the distance of each node v_j from Y is the distance between v_j and the closest point to v_j in $Y \cap P[v_1, v_j]$ (if v_j is in Y, then $d(v_j, Y) = 0$). Thus, the p leaves of Y can be identified with the p facilities (points) to be located in the p-facility problem. Therefore, an FDS for the p-facility convex ordered median problem with only two different λ values is also an FDS for the strategic k-centrum subtree problem. Finally, by (8.1) the set PEQ is an FDS for this special p-facility location problem. Hence, it is also an FDS for the strategic continuous k-centrum subtree problem. □

To summarize, we note that the strategic continuous k-centrum subtree problem can be discretized by introducing $O(M^3)$ additional nodes, with zero weight, and solving the respective discrete model as above. The $O(M^3)$ augmented nodes do not contribute to the k-centrum objective, and each one of them has only one child of the rooted augmented tree. Therefore, it is easy to check that the implementation of the above algorithm to the problem defined on the augmented tree consumes only $O(kM^7)$ and not $O(kM^9)$ time.

Remark 12.3 Notice that the same FDS is also valid for the modified continuous strategic subtree problem in which the sum of the k-largest weighted distances from a subtree T' is replaced by:

$$a \sum_{l=1}^{k} d_{(l)}(T') + b \sum_{l=k+1}^{M} d_{(l)}(T'); \quad a > b > 0, \tag{12.3}$$

($d_{(l)}(T')$ is the l-th largest element in the set $\{w_1 d(v_1, T'), \ldots, w_M d(v_M, T')\}$). We note in passing that the DP algorithm of Section 12.4.2 can easily be adapted to deal with the more general objective function described in (12.3). Thus, it also solves the corresponding strategic discrete and continuous subtree problems.

12.6 Concluding Remarks

We have presented above a variety of polynomial algorithms to solve problems of locating tree-shaped facilities using the convex ordered median objective. For the special case of the k-centrum objective we proved nestedness results, which led to improved algorithms. At the moment we do not know whether these nestedness results extend to the ordered median objective. It is an interesting research question to investigate the case of path-shaped facilities, using the convex ordered median objective. The goal is to extend the recent subquadratic algorithms for locating path-shaped facilities with the median and center objectives reported in [3, 191, 201, 202]. Note that unlike the case of tree-shaped facilities, there are only $O(M^2)$ topologically different path-shaped facilities on a tree. Therefore, the models with path-shaped facilities are clearly polynomial. The goal is to determine whether low order algorithms exist also for the ordered median objective.

Part IV

The Discrete Ordered Median Location Problem

13

Introduction and Problem Statement

In contrast to the previous two parts of the book, we now have the discrete case and not necessarily a metric environment anymore. On the other hand, the 1-facility case is quite simple, and we can focus on the multifacility case.

Discrete location problems have been widely studied due to their importance in practical applications. A number of survey articles and textbooks have been written on these problems, see e.g. Daskin [49], Drezner & Hamacher [61], Mirchandani & Francis [142], and references therein. Discrete location models typically involve a finite set of *sites* at which facilities can be located, and a finite set of *clients*, whose demands have to be satisfied by allocating them to the facilities. Whilst many problem variations have been considered in the literature, we will focus on problems in which a fixed number of facilities must be located at sites chosen from among the given set of candidate sites. Moreover, a given client can only be supplied from a single facility. For each client-site pair, there is a given cost for meeting the total demand of the client from a facility located at the site.

As in the continuous and network case a variety of objective functions has been considered in the literature. The *median* objective is to minimize the sum of the costs of supplying all clients from the facilities at the selected sites. The *center* objective is to minimize the maximum cost of supplying a client, from amongst the sites chosen, over all clients. The *centdian* objective is a convex combination of the median and center objectives which aims at keeping both the total cost and largest cost low. These are the three objectives most frequently encountered in the literature. It is worth noting that every problem has its own solution method, including its own algorithmic approach.

In the last part of this monograph we will first give a nonlinear mixed-integer formulation of the discrete OMP and explain how to linearize this initial formulation. A chapter is then devoted to exact and heuristic solution procedures. The last chapter treats related problems and gives an outlook to further research.

In the following we introduce the discrete version of the OMP and present some illustrative examples. In addition, we point out many modeling possibilities to obtain classical as well as new discrete facility location problems as particular cases of the OMP. Finally, we describe a quadratic integer programming formulation originally proposed by Nickel [149].

13.1 Definition of the Problem

Let A denote a given set of M sites, identified with the integers $1, \ldots, M$, i.e. $A = \{1, \ldots, M\}$.

Let $C = (c_{kj})_{k,j=1,\ldots,M}$ be a given nonnegative $M \times M$ cost matrix, where c_{kj} denotes the cost of satisfying the total demand of client k from a facility located at site j. We may also say c_{kj} denotes the cost of allocating client k to facility j.

As being usual in discrete facility location problems, we assume without loss of generality that the number of candidate sites is identical to the number of clients. Let $1 \leq N \leq M$ be the desired number of facilities to be located at the candidate sites. A solution to the facility location problem is given by a set $X \subseteq A$ of N sites, i.e. $|X| = N$. We assume, that each new facility has unlimited capacity. Therefore, each client k will be allocated to a facility located at site j of X with lowest cost, i.e.

$$c_k(X) := \min_{l \in X} c_{kl} = c_{kj}. \tag{13.1}$$

Ties are broken arbitrarily.

What distinguishes this problem from the classical uncapacitated N-median problem is its objective function. As in the previous parts, the entries of the cost vector for allocating the clients to the new facilities $c(A, X) = (c_1(X), \ldots, c_M(X))$ are sorted in non-decreasing order and we get the corresponding sorted cost vector

$$sort_M(c(A, X)) = (c_{(1)}(X), \ldots, c_{(M)}(X)) = c_{\leq}(A, X)\,. \tag{13.2}$$

We call any permutation transforming $c(A, X)$ into $sort_M(c(A, X)) = c_{\leq}(A, X)$ a *valid permutation* for X.

Then, in accordance with Definition 1.1, the objective function applies a linear cost factor, with coefficient $\lambda_i \geq 0$, to the ith lowest cost of supplying a client, $c_{(i)}(X)$, for each $i = 1, \ldots, M$. Let $\lambda = (\lambda_1, \ldots, \lambda_M)$ with $\lambda_i \geq 0$, $i = 1, \ldots, M$, be a given vector, which will be essential to model different discrete facility location problems. The discrete OMP is defined as

$$\min_{X \subseteq A\,,\, |X| = N} f_\lambda(X) = \langle \lambda, sort_M(c(A, X)) \rangle = \sum_{i=1}^{M} \lambda_i \, c_{(i)}(X). \tag{13.3}$$

Next, an example is given to illustrate the calculation of the discrete ordered objective function value.

Example 13.1
Let $A = \{1, \ldots, 5\}$ be the set of sites and assume that we are interested in building $N = 2$ new facilities. Let the cost matrix C be as follows:

$$C = \begin{pmatrix} 0 & 4 & 5 & 3 & 3 \\ 5 & 0 & 6 & 2 & 2 \\ 7 & 2 & 0 & 5 & 6 \\ 7 & 4 & 3 & 0 & 5 \\ 1 & 3 & 2 & 4 & 0 \end{pmatrix}.$$

Choose $\lambda = (0, 0, 1, 1, 0)$, which leads to the so-called $(2, 1)$-trimmed mean problem (see Section 1.3) and evaluate the function at $X = \{1, 4\}$. It turns out, that sites 1 and 5 are allocated to Facility 1, while the remaining sites are allocated to Facility 4, see Figure 13.1. As a result, the associated cost vector is

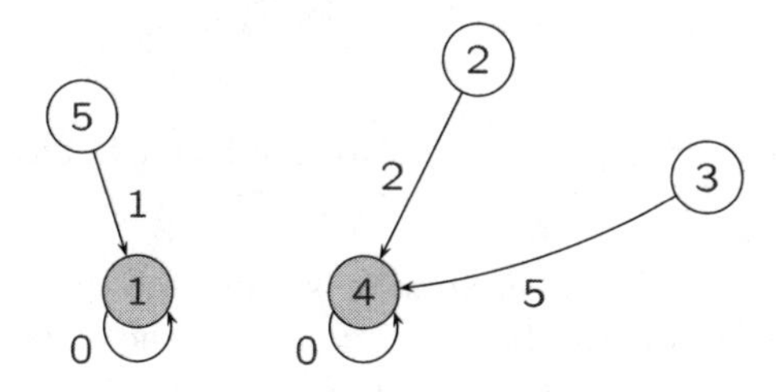

Fig. 13.1. Optimal solution of Example 13.1

$c(A, X) = (0, 2, 5, 0, 1)$. Thus, the sorted cost vector is $c_{\leq}(A, X) = (0, 0, 1, 2, 5)$ and the optimal objective function value is equal to:

$$f_\lambda(X) = \langle \lambda, sort_M(c(A, X)) \rangle = 0 \times 0 + 0 \times 0 + 1 \times 1 + 1 \times 2 + 0 \times 5 = 3.$$

Next, we show that the choice of the objective function has a tremendous impact on the optimal locations of new facilities.

Example 13.2
Let $A = \{1, \ldots, 5\}$ be the set of sites and assume as before that we are interested in building $N = 2$ new facilities. But we consider now the following cost matrix:

$$C = \begin{pmatrix} 2 & 20 & 2 & 20 & 20 \\ 20 & 2 & 20 & 2 & 5 \\ 4 & 20 & 3 & 20 & 20 \\ 20 & 5 & 20 & 4 & 5 \\ 6 & 20 & 20 & 11 & 5 \end{pmatrix},$$

and assume that it represents distances. Then, depending on the selection of the λ vector, we obtain different optimal locations for the new facilities and objective function values. The choice of the objective function to be minimized is determined by the purpose of the new facilities to be built. Therefore, if for example the two new facilities represent schools, then the 2-median problem should be solved to find the best locations, i.e. those which minimize the total sum of the distances. In this case, by setting $\lambda = (1, 1, 1, 1, 1)$ in the OMP, we get that the optimal solution is formed by the facilities $X_M = \{1, 4\}$, see Figure 13.2, with an objective function value equal to 18. If the two new

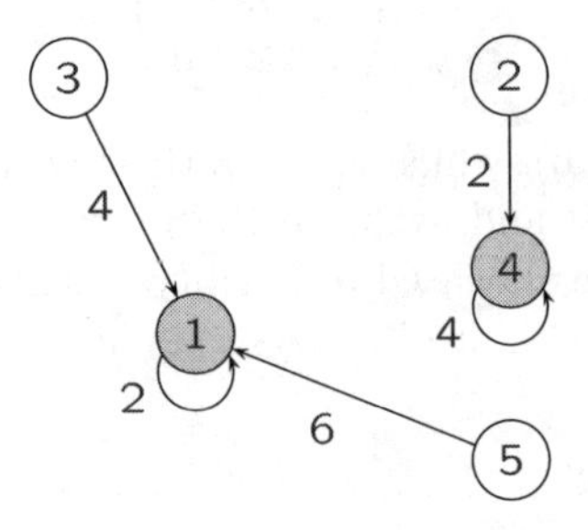

Fig. 13.2. 2-median solution of Example 13.2

facilities represent emergency services, it is more suitable to consider the 2-center problem to obtain the best locations, i.e. those which minimize the maximal distance. In this case, we set $\lambda = (0, 0, 0, 0, 1)$ and observe that the optimal solution is formed either by the facilities $X_C = \{1, 5\}$ or by $X_C = \{3, 5\}$, see Figure 13.3, both with an objective function value equal to 5.

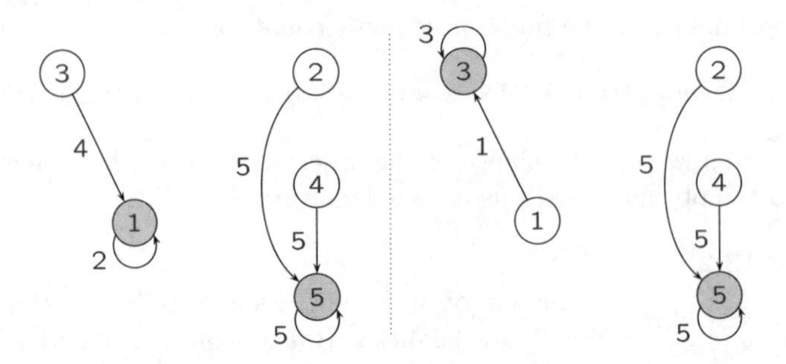

Fig. 13.3. 2-center solution of Example 13.2

Nevertheless, if the purpose of locating two new facilities is to establish new leisure centers, then the optimal solution can ignore those clients that are either very far away or too closely located in order to give a better service to those clients who are at a middle distance. Thus, we may consider the $(1, 1)$-trimmed mean problem. By setting $\lambda = (0, 1, 1, 1, 0)$ in the OMP, the optimal solution is formed by the facilities $X_T = \{3, 4\}$, see Figure 13.4, with an objective function value equal to 9.

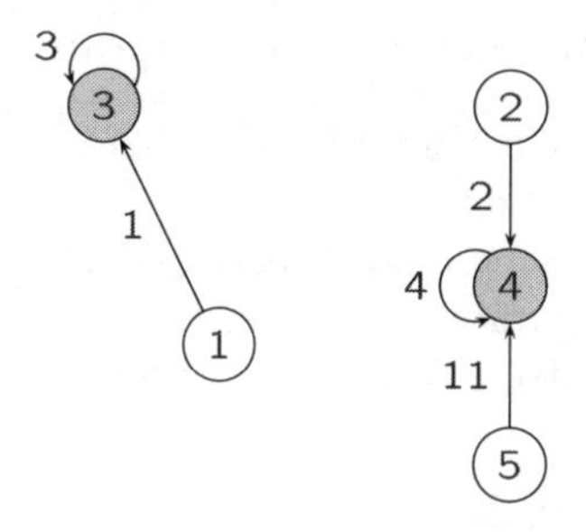

Fig. 13.4. Solution corresponding to the $(1, 1)$-trimmed mean problem of Example 13.2

The above example refers to a case where the (k_1, k_2)-trimmed mean problem yields an optimal solution different to the corresponding ones given by the N-median and the N-center problems.

Before dealing with solution approaches, we quickly discuss the complexity status of the discrete OMP.

Theorem 13.1 *The discrete OMP belongs to the class of NP-hard problems.*

Proof.
The discrete N-median problem, which is NP-hard (see Kariv & Hakimi [119]), is a particular case of the discrete OMP. Hence, clearly the discrete OMP belongs to the class of NP-hard problems. □

13.2 A Quadratic Formulation for the Discrete OMP

The main issue when dealing with discrete optimization problems is to find a compact formulation as a mathematical program. These formulations help in finding solution methods and gaining insights into the structure of the problems. The goal of this section is to provide a first formulation for the discrete OMP. The OMP can be seen as a (location) problem containing two types of subproblems: one corresponds to the sorting of the cost vector, while the other one corresponds to the location of the new facilities and the allocation of the clients to each service center. Therefore, the OMP can be formulated as a combination of an integer linear program which models the sorting subproblem, and a mixed-integer linear programming approach which solves the location-allocation subproblem. Unfortunately, this combination will lead to a problem with quadratic constraints and objective function. The presentation follows the approach originally proposed in [149].

13.2.1 Sorting as an Integer Linear Program (ILP)

In order to develop a formulation for the OMP, it is necessary first to represent the sorting process (according to a *valid permutation*) as an ILP. Since any permutation can be represented by an assignment problem, we only need to add some additional constraints to obtain the correct permutation as shown below. The matching model is displayed in Figure 13.5.

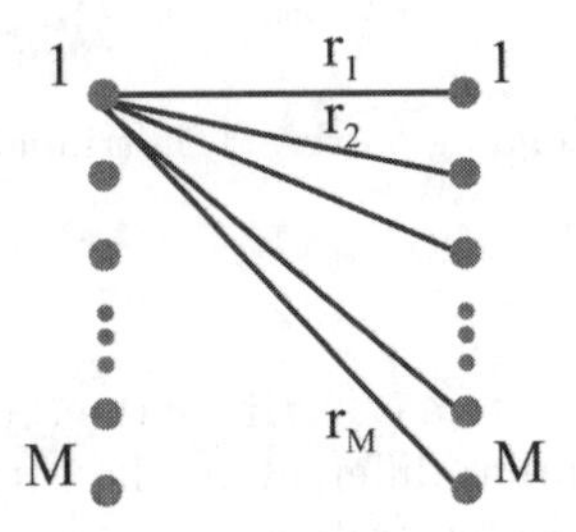

Fig. 13.5. Graph to illustrate the integer sorting subproblem (*SORT*)

In the sorting subproblem, a given set of real numbers $r_1, \ldots, r_M$ must be sorted in non-decreasing order: $r_{\sigma(1)} \leq r_{\sigma(2)} \leq \ldots \leq r_{\sigma(M)}$, $\sigma \in \mathcal{P}(1...M)$. To this aim, we define the following decision variables:

$$s_{ik} = \begin{cases} 1 \text{ if } \sigma(i) = k, \text{ i.e. if } r_k \text{ is the } i\text{th smallest value} \\ 0 \text{ otherwise} \end{cases} \qquad i, k = 1, \ldots, M. \tag{13.4}$$

Then, any solution of the following integer linear programming formulation yields an (optimal) solution to the sorting subproblem, i.e. a realization of the $sort_M$ function.

$$\text{(SORT)} \quad \min \quad 1$$

$$\text{s.t.} \quad \sum_{i=1}^{M} s_{ik} = 1 \qquad \forall\, k = 1, \ldots, M \tag{13.5}$$

$$\sum_{k=1}^{M} s_{ik} = 1 \qquad \forall\, i = 1, \ldots, M \tag{13.6}$$

$$\sum_{k=1}^{M} s_{ik} r_k \leq \sum_{k=1}^{M} s_{i+1,k} r_k \qquad \forall\, i = 1, \ldots, M-1 \tag{13.7}$$

$$s_{ik} \in \{0,1\} \qquad \forall\, i,k = 1,\ldots,M. \tag{13.8}$$

The first group of constraints (13.5) ensures that each given real number is placed at only one position. Analogously, by constraints (13.6) each position contains only one given real number. Constraints (13.7) guarantee the non-decreasing order of the sorted real numbers. Finally, the integrality constraints (13.8) are a natural consequence of the definition of the variables s_{ik}, $i,k = 1,\ldots,M$, given in (13.4).

(SORT) is just a feasibility problem, since the interest of this formulation lies in the feasible region, i.e. the construction of feasible solutions. Hence, by setting

$$r_{\sigma(i)} := \sum_{k=1}^{M} s_{ik}\, r_k,\ i = 1,\ldots,M$$

we obtain the desired non-decreasing sequence.

Next, we present a small example that illustrates the formulation (SORT).

Example 13.3

Let $r_1 = 6$, $r_2 = 4$, $r_3 = 5$ be given. Then, the unique feasible values for the variables $s_{ik} \in \{0,1\}$, $i,k = 1,\ldots,3$, are obtained by setting $s_{12} = s_{23} = s_{31} = 1$ and the remaining variables equal to zero.

Clearly, constraints (13.5) and (13.6) are fulfilled. Furthermore, constraints (13.7) are also satisfied:

$$\begin{aligned} 4 = 6s_{11} + 4s_{12} + 5s_{13} &\le 6s_{21} + 4s_{22} + 5s_{23} = 5 \\ 5 = 6s_{21} + 4s_{22} + 5s_{23} &\le 6s_{31} + 4s_{32} + 5s_{33} = 6. \end{aligned}$$

Notice that the constraints presented above are fundamental to fix the values of the variables $s_{ik} \in \{0,1\}$, $i,k = 1,\ldots,3$. Therefore,

$s_{12} = 1 \Rightarrow \sigma(1) = 2$, i.e. $r_2 = 4$ is placed at the 1st position $(r_{\sigma(1)} = r_2)$;
$s_{23} = 1 \Rightarrow \sigma(2) = 3$, i.e. $r_3 = 5$ is placed at the 2nd position $(r_{\sigma(2)} = r_3)$;
$s_{31} = 1 \Rightarrow \sigma(3) = 1$, i.e. $r_1 = 6$ is placed at the 3rd position $(r_{\sigma(3)} = r_1)$.

Obviously, if the given real numbers are not pairwise different, then there exist multiple feasible solutions of (SORT). Thus, as in (13.1) ties are broken arbitrarily.

Next, a well-known formulation of the location-allocation subproblem will be introduced.

13.2.2 Formulation of the Location-Allocation Subproblem

Observe that by choosing $\lambda = (1,1,\ldots,1)$, the sorting permutation σ has no effect on the objective function and therefore can be omitted. In this case, we obtain the classical discrete N-median problem which coincides with the location-allocation subproblem (see Daskin [49], Labbé et al. [127] and Mirchandani [141]). We use the standard variables

$$x_j = \begin{cases} 1 \text{ if a new facility is built at site } j \\ 0 \text{ otherwise} \end{cases} \quad j = 1, \ldots, M, \tag{13.9}$$

and

$$y_{kj} = \begin{cases} 1 \text{ if client } k \text{ is allocated to facility } j \\ 0 \text{ otherwise} \end{cases} \quad k, j = 1, \ldots, M. \tag{13.10}$$

Therefore, the location-allocation subproblem can be formulated using the following mixed-integer linear program proposed by Revelle & Swain [176]:

$$(N\text{-MED}) \quad \min \quad \sum_{k=1}^{M} \sum_{j=1}^{M} c_{kj}\, y_{kj}$$

$$\text{s.t.} \quad \sum_{j=1}^{M} x_j = N \tag{13.11}$$

$$\sum_{j=1}^{M} y_{kj} = 1 \qquad \forall\, k = 1, \ldots, M \tag{13.12}$$

$$x_j \geq y_{kj} \qquad \forall\, j, k = 1, \ldots, M \tag{13.13}$$

$$x_j \in \{0, 1\}, \quad y_{kj} \geq 0 \qquad \forall\, j, k = 1, \ldots, M. \tag{13.14}$$

The objective function is to minimize the sum of the costs of satisfying the total demand of the clients.

By constraint (13.11), N new facilities are built. Constraints (13.12) guarantee that each client is allocated to one facility. Constraints (13.13) ensure that a client i may only be allocated to a facility j if this facility is open.

Observe that the requirement of variable y to be binary (see (13.10)) is relaxed. Notice that the x_j, $j = 1, \ldots, M$, are the strategic variables. Therefore, once a set of variables x_j satisfying the integrality constraints and (13.11) is specified, it is simple to determine a set of allocation variables y_{kj}, $j, k = 1, \ldots, M$, that solves the resulting linear program for the fixed x_j, $j = 1, \ldots, M$, since the facilities have unlimited capacity. Therefore, by dropping the integrality requirement of variable y, an optimal set of y_{kj} is given by $y_{kj^*} = 1$ and $y_{kj} = 0$, $j \neq j^*$ where $c_{kj^*} = min\{c_{kj} : x_j = 1\}$. Thus, no integrality constraints on the variables y_{kj} are necessary, and we simply set $y_{kj} \geq 0$, see Cornuejols et al. [47].

Next, we present an example of a location-allocation problem.

Example 13.4
Consider the data presented in Example 13.1, and assume that we are interested in building $N = 2$ new facilities in order to minimize the sum of the total allocation costs. The optimal solution of the 2-median problem is formed by the facilities $X = \{1, 2\}$. Therefore, the non-zero values of the decision variables x_j are $x_1 = x_2 = 1$. Moreover, the decision variables y_{kj} are all equal to zero except for

$y_{11} = 1$, i.e. the total demand of client 1 is satisfied by itself;
$y_{22} = 1$, i.e. the total demand of client 2 is satisfied by itself;
$y_{32} = 1$, i.e. the total demand of client 3 is satisfied by facility 2;
$y_{42} = 1$, i.e. the total demand of client 4 is satisfied by facility 2;
$y_{51} = 1$, i.e. the total demand of client 5 is satisfied by facility 1.

Hence, the associated cost vector is $c(A, X) = (0, 0, 2, 4, 1)$, see Figure 13.6, and the optimal objective function value is equal to 7.

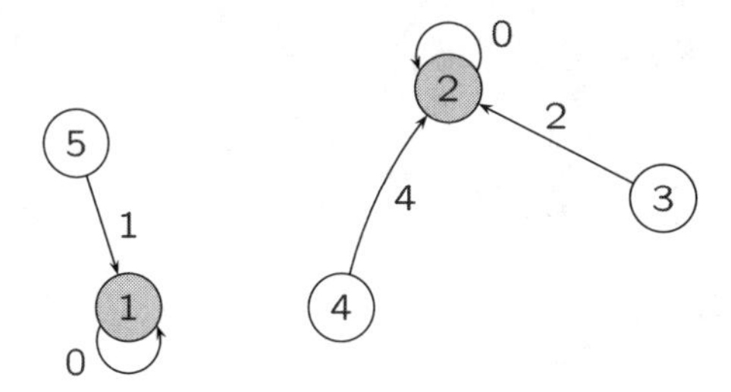

Fig. 13.6. Optimal solution corresponding to the 2-median problem of Example 13.4

13.2.3 Quadratic Integer Programming Formulation for OMP

Now, we combine the ILP-formulation of the sorting process, (SORT), proposed in Section 13.2.1, with the formulation (N-MED) introduced in Section 13.2.2. The combination of both models is not a difficult task. The elements r_k that have to be sorted are the costs $c_k(X)$ which are represented in the (N-MED) formulation by

$$r_k = c_k(X) = \sum_{j=1}^{M} c_{kj}\, y_{kj}, \; k = 1, \ldots, M.$$

Hence, using the decision variables s_{ik} defined in (13.4), the objective function of the OMP can be written as

$$\sum_{i=1}^{M} \lambda_i \, c_{(i)}(X) = \sum_{i=1}^{M} \lambda_i \sum_{k=1}^{M} s_{ik} \, c_k(X) = \sum_{i=1}^{M} \lambda_i \sum_{k=1}^{M} s_{ik} \sum_{j=1}^{M} c_{kj} \, y_{kj}.$$

Furthermore, we consider the decision variables y_{kj} and x_j, $j,k = 1,\ldots,M$, defined by (13.9) and (13.10), respectively, and add the constraints corresponding to the sorting and the location-allocation subproblems. The OMP can be formulated as follows:

$$\text{(OMP)} \quad \min \sum_{i=1}^{M} \lambda_i \sum_{k=1}^{M} s_{ik} \sum_{j=1}^{M} c_{kj} \, y_{kj} \tag{13.15}$$

$$\text{s.t.} \quad \sum_{i=1}^{M} s_{ik} = 1 \qquad \forall \, k = 1,\ldots,M$$

$$\sum_{k=1}^{M} s_{ik} = 1 \qquad \forall \, i = 1,\ldots,M$$

$$\sum_{k=1}^{M} s_{ik} \sum_{j=1}^{M} c_{kj} \, y_{kj} \le \sum_{k=1}^{M} s_{i+1,k} \sum_{j=1}^{M} c_{kj} \, y_{kj} \quad \forall \, i = 1,\ldots,M-1 \tag{13.16}$$

$$\sum_{j=1}^{M} x_j = N$$

$$\sum_{j=1}^{M} y_{kj} = 1 \qquad \forall \, k = 1,\ldots,M$$

$$x_j \ge y_{kj} \qquad \forall \, j,k = 1,\ldots,M$$

$$s_{ik}, x_j \in \{0,1\}, \quad y_{kj} \ge 0 \quad \forall \, i,j,k = 1,\ldots,M.$$

If constraint (13.11) is dropped, and (13.15) is modified in such a way that fixed costs associated to each facility (f_j for all $j \in A$) are part of the objective function as follows:

$$\min \sum_{j=1}^{M} f_j \, x_j + \sum_{i=1}^{M} \lambda_i \sum_{k=1}^{M} s_{ik} \sum_{j=1}^{M} c_{kj} \, y_{kj},$$

we obtain the Uncapacitated Facility Location (UFL) problem (see Cornuejols et al. [47]) with sorted supply costs.

Observe that (*OMP*) is a mixed-integer formulation containing a quadratic objective function and furthermore, a group of quadratic constraints (see constraints (13.16)). Consequently, the formulation of the OMP is very hard to be straightly handled. On the other hand, the presented formulation is a natural consequence of combining the two ingredients that have to be used in ordered multifacility discrete location models: ordering and location-allocation.

14

Linearizations and Reformulations

In the previous chapter we defined the discrete OMP and proposed a mixed-integer nonlinear program to model it. This formulation, (OMP), has a quadratic objective function as well as a quadratic group of constraints. In this chapter we propose equivalent mixed-integer linear programs to handle (OMP), see Nickel [149] and Boland et al. [25].

In other words, we look for linear reformulations (linearizations) of (OMP) which provide the same feasible solutions and the same optimal objective function value as the original program. In this chapter we propose three linearizations that are motivated on linearizations developed for the Quadratic Assignment Problem, see Burkard [30] and Rendl [175]. In addition, we show how properties of optimal solutions can be used to strengthen these models, and discuss relationships between them. Furthermore, computational results are presented.

14.1 Linearizations of (OMP)

There are many approaches (linearizations) to avoid the quadratic terms in the objective function (13.15) and in the constraints (13.16) of (OMP). The underlying idea of all these approaches is to define a group of new variables as an expression based on the quadratic terms in order to substitute them in the complete formulation. Additional constraints corresponding to these new variables are often included to keep the feasibility of the solutions and their objective function value. In this section three possible linearizations are developed.

14.1.1 A First Linearization: (OMP^1)

A first mixed-integer linear formulation for the OMP can be derived analogously to the linearizations for the Quadratic Assignment Problem proposed by Lawler [128]. Hence, we use the following binary variables

$$z_{ikj} = s_{ik}\, y_{kj} \quad i,j,k = 1,\dots,M. \tag{14.1}$$

Thus, $z_{ikj} = 1$ if client k is supplied by a facility located at site j and this allocation corresponds to the ith lowest cost, and $z_{ikj} = 0$, otherwise.

Observe that we do not consider any capacity constraints and therefore, variable y does not need any integrality requirement, see Section 13.2.2. Hence, we can also drop the integrality constraint corresponding to variable z.

The first linearization is as follows:

$$
\begin{array}{llll}
(\mathsf{OMP}^{1'}) & \min & \displaystyle\sum_{i=1}^{M}\lambda_i \sum_{k=1}^{M}\sum_{j=1}^{M} c_{kj}\, z_{ikj} & \\
& \text{s.t.} & \displaystyle\sum_{i=1}^{M} s_{ik} = 1 & \forall\, k = 1,\dots,M \\
& & \displaystyle\sum_{k=1}^{M} s_{ik} = 1 & \forall\, i = 1,\dots,M \\
& & \displaystyle\sum_{k=1}^{M}\sum_{j=1}^{M} c_{kj}\, z_{ikj} \le \sum_{k=1}^{M}\sum_{j=1}^{M} c_{kj}\, z_{i+1,kj} & \forall\, i = 1,\dots,M-1 \quad (14.2) \\
& & \displaystyle\sum_{j=1}^{M} x_j = N & \\
& & \displaystyle\sum_{j=1}^{M} y_{kj} = 1 & \forall\, k = 1,\dots,M \\
& & x_j \ge y_{kj} & \forall\, j,k = 1,\dots,M \\
& & \displaystyle\sum_{i=1}^{M} z_{ikj} = y_{kj} & \forall\, k,j = 1,\dots,M \quad (14.3) \\
& & \displaystyle\sum_{j=1}^{M} z_{ikj} = s_{ik} & \forall\, i,k = 1,\dots,M \quad (14.4) \\
& & \displaystyle\sum_{k=1}^{M}\sum_{i=1}^{M}\sum_{j=1}^{M} z_{ikj} = M & \quad (14.5) \\
& & s_{ik} + y_{kj} - 2 z_{ikj} \ge 0 & \forall\, i,j,k = 1,\dots,M \quad (14.6) \\
& & s_{ik}, x_j \in \{0,1\}, \quad z_{ikj}, y_{kj} \ge 0 & \forall\, i,j,k = 1,\dots,M.
\end{array}
$$

This mixed-integer linear program is equivalent to the quadratic formulation of the OMP as shown in the following theorem.

Theorem 14.1 *The formulations (OMP) and (OMP$^{1'}$) are equivalent, i.e. a solution (x', y', s') is feasible to (OMP) if and only if (x', y', s', z') is feasible to (OMP$^{1'}$) by setting $z'_{ikj} = s'_{ik}\, y'_{kj}$. Moreover, their corresponding objective function values coincide.*

Proof.
Let (x', y', s') be a feasible solution to (OMP). Then, constraints (13.5), (13.6), (13.11), (13.12) and (13.13) are also satisfied in (OMP$^{1'}$).

Now, we show that by setting $z'_{ikj} := s'_{ik}\, y'_{kj}$, the remaining constraints also hold.

First, from constraints (13.16) we obtain the validity of constraints (14.2), since

$$\sum_{k=1}^{M}\sum_{j=1}^{M} c_{kj}\, z'_{ikj} = \sum_{k=1}^{M}\sum_{j=1}^{M} c_{kj}\, s'_{ik}\, y'_{kj} \le \sum_{k=1}^{M}\sum_{j=1}^{M} c_{kj}\, s'_{i+1,k}\, y'_{kj} = \sum_{k=1}^{M}\sum_{j=1}^{M} c_{kj}\, z'_{i+1,kj}$$

for every $i = 1, \dots, M-1$.

Furthermore, taking into account the definition of z'_{ikj} and summing over the index i we have

$$\sum_{i=1}^{M} z'_{ikj} = \sum_{i=1}^{M} s'_{ik}\, y'_{kj} = y'_{kj} \underbrace{\sum_{i=1}^{M} s'_{ik}}_{=1 \text{ by } (13.5)} = y'_{kj}$$

and therefore, constraints (14.3) are also fulfilled.

Analogously, summing over the index j we can show the validity of constraints (14.4) as follows

$$\sum_{j=1}^{M} z'_{ikj} = \sum_{j=1}^{M} s'_{ik}\, y'_{kj} = s'_{ik} \underbrace{\sum_{j=1}^{M} y'_{kj}}_{=1 \text{ by } (13.12)} = s'_{ik}.$$

In addition, computing as before, we get

$$\sum_{k=1}^{M}\sum_{i=1}^{M}\sum_{j=1}^{M} z'_{ikj} = \sum_{k=1}^{M}\sum_{i=1}^{M}\sum_{j=1}^{M} s'_{ik}\, y'_{kj} = \sum_{k=1}^{M} \underbrace{\sum_{i=1}^{M} s'_{ik}}_{=1 \text{ by } (13.5)} \overbrace{\sum_{j=1}^{M} y'_{kj}}^{=1 \text{ by } (13.12)} = M$$

and therefore, constraint (14.5) is also satisfied.

Finally, in order to show that constraints (14.6) are valid, we distinguish two cases depending on the value of the binary variable s. On the one hand, if $s'_{ik} = 0$, we obtain as a consequence that $z'_{ikj} = 0$ and therefore, $s'_{ik} + y'_{kj} - 2z'_{ikj} = y'_{kj} \geq 0$ from constraints (13.14). On the other hand, if $s'_{ik} = 1$ then $z'_{ikj} = y'_{kj}$ and hence, $s'_{ik} + y'_{kj} - 2z'_{ikj} = 1 + y'_{kj} - 2y'_{kj} = 1 - y'_{kj} \geq 0$ using constraints (13.13) and (13.14). Therefore, in both cases, constraints (14.6) hold.

Thus, (x', y', s', z') is feasible to ($\mathsf{OMP}^{1'}$). Furthermore, by definition of z', (x', y', s') provides the same objective function value in (OMP) as (x', y', s', z') in ($\mathsf{OMP}^{1'}$).

Now, let (x', y', s', z') be a feasible solution to ($\mathsf{OMP}^{1'}$). Then, as before, constraints (13.5), (13.6), (13.11), (13.12) and (13.13) are also valid for (OMP).

If we show that the equivalence $z'_{ikj} = s'_{ik}\, y'_{kj}$ holds, the implication is proven.

Variable z is nonnegative. Therefore, using constraints (14.4) and taking into account that $s'_{ik} \in \{0, 1\}$ (from constraints (13.14)), we obtain that $0 \leq z'_{ikj} \leq 1$. Hence, we have to consider three cases. If $z'_{ikj} = 1$ we get from constraints (14.6) that $s'_{ik} + y'_{kj} - 2 \geq 0$ and therefore, $s'_{ik} = 1$ and $y'_{kj} = 1$. In particular, we have $z'_{ikj} \leq s'_{ik}\, y'_{kj}$. If $z'_{ikj} = 0$ we obtain directly from the non-negativity constraints of variables s and y that $z'_{ikj} \leq s'_{ik}\, y'_{kj}$ also holds. Finally, we only have to investigate the case when $0 < z'_{ikj} < 1$. On the one hand, by constraints (14.4), we have that $0 < z'_{ikj} \leq s'_{ik}$ and, since s is a binary variable, we obtain $s'_{ik} = 1$. On the other hand, using constraints (14.3) and the non-negativity of variable z, we have that $z'_{ikj} \leq y'_{kj}$. Therefore, as $s'_{ik} = 1$, we have that $z'_{ikj} \leq s'_{ik}\, y'_{kj}$. Thus, in either case, $z'_{ikj} \leq s'_{ik}\, y'_{kj}$.

Now assume that there exist i^*, k^*, j^* with $z'_{i^*k^*j^*} < s'_{i^*k^*}\, y'_{k^*j^*}$. Hence,

$$\sum_{k=1}^{M} \sum_{i=1}^{M} \sum_{j=1}^{M} z'_{ikj} < \sum_{k=1}^{M} \sum_{i=1}^{M} \sum_{j=1}^{M} s'_{ik}\, y'_{kj} = \sum_{k=1}^{M} \sum_{i=1}^{M} s'_{ik} \sum_{j=1}^{M} y'_{kj} = M \,,$$

which contradicts constraint (14.5). Therefrom it follows that

$$z'_{i^*k^*j^*} = s'_{i^*k^*}\, y'_{k^*j^*}.$$

Thus, (x', y', s') is feasible to (OMP) and the objective function values for (x', y', s', z') in ($\mathsf{OMP}^{1'}$) and for (x', y', s') in (OMP) coincide. □

Constraints (14.3) and (14.4) show a tight link between variables s and z, and between y and z, respectively. Thus, we can substitute variables s and y by expressions on variable z, as the next corollary shows.

Corollary 14.1 *The formulation ($\mathsf{OMP}^{1'}$) can be simplified and written as follows:*

$$
\begin{array}{llll}
(\mathit{OMP}^{1}) & \min & \sum_{i=1}^{M}\sum_{k=1}^{M}\sum_{j=1}^{M} \lambda_i \, c_{kj} \, z_{ikj} & \\
 & s.t. & \sum_{i=1}^{M}\sum_{j=1}^{M} z_{ikj} = 1 & \forall\, k = 1,\dots,M \quad (14.7)\\
 & & \sum_{k=1}^{M}\sum_{j=1}^{M} z_{ikj} = 1 & \forall\, i = 1,\dots,M \quad (14.8)\\
 & & \sum_{k=1}^{M}\sum_{j=1}^{M} c_{kj} z_{ikj} \le \sum_{k=1}^{M}\sum_{j=1}^{M} c_{kj} z_{i+1,kj} & \forall\, i = 1,\dots,M-1\\
 & & \sum_{j=1}^{M} x_j = N & \\
 & & x_j \ge \sum_{i=1}^{M} z_{ikj} & \forall\, k,j = 1,\dots,M \quad (14.9)\\
 & & z_{ikj}, x_j \in \{0,1\} & \forall\, i,j,k = 1,\dots,M.
\end{array}
$$

Before we prove this result, we remark on some aspects of this new formulation, (OMP^1).

Remark 14.1

- *The constraint* (14.5) *is redundant, since it is simply an aggregation of either constraints* (14.7) *or* (14.8).
- *To obtain the equivalence between the linearizations* $(\mathit{OMP}^{1'})$ *and* (OMP^{1}), *the binary character of variable* z *is necessary.*

In the following we proceed by showing the validity of (OMP^1).

Proof.
To prove that variable z can be considered as a binary variable, we only have to show that each feasible solution (x', y', s', z') to $(\mathsf{OMP}^{1'})$ can be transformed into a solution with z'' binary. On the one hand, if z' takes a Boolean value, then we just have to consider $z'' = z'$. On the other hand, if z' is not binary, since $z'_{ikj} = s'_{ik}\, y'_{kj}$ with s'_{ik} binary, then the behavior of z' depends only on y'. As there are no capacity constraints, client k can be allocated to a facility j such that $c_{kj} = \min\{c_{kl} : l \in X\}$ where $X = \{l \in A : \; x'_l = 1\}$ and therefore, we can consider that $y''_{kj} = 1$. Otherwise, if $x'_j = 0$ we obtain that $y''_{kj} = 0$. Therefore, defining $z''_{ikj} = s'_{ik}\, y''_{kj}$ there exists a solution (x', y'', s', z'') to $(\mathsf{OMP}^{1'})$ with z'' binary.

Then, substituting y''_{kj} and s'_{ik} by their expressions involving z''_{ikj} as shown by constraints (14.3) and (14.4), respectively, we get the validity of constraints (14.7), (14.8) and (14.9).

Hence, (x', z'') is feasible to (OMP^1). Moreover, we also get the same objective function value in $(\mathsf{OMP}^{1'})$ for (x', y', s', z') and in (OMP^1) for (x', z'').

Let (x', z') be a feasible solution to (OMP^1). As $z'_{ikj} \in \{0,1\}$, let i^*, k^*, j^* be such that $z'_{i^*k^*j^*} = 1$. By constraints (14.7) we obtain that $z'_{i^*k^*j} = 0$, $\forall j \neq j^*$. Therefore, by setting $s'_{ik} := \sum_{j=1}^{M} z'_{ikj} \in \{0,1\}$ and $y'_{kj} := \sum_{i=1}^{M} z'_{ikj} \geq 0$ we have that

$$s'_{ik} + y'_{kj} - 2z'_{ikj} = \sum_{l=1}^{M} z'_{ikl} + \sum_{l=1}^{M} z'_{lkj} - 2z'_{ikj} = \sum_{l=1, l\neq j}^{M} z'_{ikl} + \sum_{l=1, l\neq i}^{M} z'_{lkj} \geq 0.$$

Thus, constraints (14.6) hold and (x', y', s', z') is feasible to $(\mathsf{OMP}^{1'})$. Furthermore, we obtain the same objective function value for both linearizations.

□

From now on, we will consider (OMP^1) instead of $(\mathsf{OMP}^{1'})$ as our first linearization of (OMP).

Note that (OMP^1) has $O(M^3)$ variables and $O(M^2)$ constraints. In the next section we present an alternative linearization of the quadratic formulation (OMP).

14.1.2 A Linearization Using Less Variables: (OMP^2)

In this section a second linearization for the OMP, which uses $O(M^2)$ variables rather than the $O(M^3)$ variables of (OMP^1), is developed. This linearization is inspired in the mixed-integer linear programming formulation proposed by Kaufman & Broeckx [121] for the Quadratic Assignment Problem. Variables are introduced which take on values of *costs* of supply; in a sense they absorb the objective function, see (14.10).

We define M^2 new nonnegative continuous variables w_{ik} by

$$w_{ik} = s_{ik} \sum_{j=1}^{M} c_{kj}\, y_{kj} \quad i, k = 1, \ldots, M, \tag{14.10}$$

so w_{ik} is the cost of supplying client k if k is the ith client in non-decreasing linear order of supply cost, and zero otherwise.

Therefore, we obtain the following mixed-integer linear problem:

$$
\begin{aligned}
(\mathsf{OMP}^2) \quad \min \quad & \sum_{i=1}^{M}\sum_{k=1}^{M} \lambda_i w_{ik} \\
\text{s.t.} \quad & \\
& \sum_{i=1}^{M} s_{ik} = 1 && \forall\, k = 1,\dots,M \\
& \sum_{k=1}^{M} s_{ik} = 1 && \forall\, i = 1,\dots,M \\
& \sum_{k=1}^{M} w_{ik} \le \sum_{k=1}^{M} w_{i+1,k} && \forall\, i = 1,\dots,M-1 \qquad (14.11) \\
& \sum_{j=1}^{M} x_j = N && \\
& \sum_{j=1}^{M} y_{kj} = 1 && \forall\, k = 1,\dots,M \\
& x_j \ge y_{kj} && \forall\, j,k = 1,\dots,M \\
& \sum_{i=1}^{M} w_{ik} = \sum_{j=1}^{M} c_{kj}\, y_{kj} && \forall\, k = 1,\dots,M \qquad (14.12) \\
& w_{ik} \ge \sum_{j=1}^{M} c_{kj}\, y_{kj} - \sum_{j=1}^{M} c_{kj}\,(1 - s_{ik}) && \forall\, i,k = 1,\dots,M \qquad (14.13) \\
& s_{ik}, x_j \in \{0,1\}, \quad y_{kj}, w_{ik} \ge 0 && \forall\, i,j,k = 1,\dots,M.
\end{aligned}
$$

The following theorem shows the equivalence between this mixed-integer linear program and the quadratic formulation of the OMP.

Theorem 14.2 *The formulation (OMP^2) is a valid linearization of (OMP), i.e. a solution (x', y', s') is feasible to (OMP) if and only if (x', y', s', w') is feasible to (OMP^2) by setting $w'_{ik} = s'_{ik} \sum_{j=1}^{M} c_{kj}\, y'_{kj}$. Moreover, their corresponding objective function values coincide.*

Proof.
Let (x', y', s') be a feasible solution to (OMP). Then, constraints (13.5), (13.6), (13.11), (13.12) and (13.13) are immediately satisfied for (OMP^2).

The validity of the remaining constraints will be shown by setting

$$w'_{ik} := s'_{ik} \sum_{j=1}^{M} c_{kj}\, y'_{kj}.$$

First, from constraints (13.16) we get

$$\sum_{k=1}^{M} w'_{ik} = \sum_{k=1}^{M} s'_{ik} \sum_{j=1}^{M} c_{kj}\, y'_{kj} \leq \sum_{k=1}^{M} s'_{i+1,k} \sum_{j=1}^{M} c_{kj}\, y'_{kj} = \sum_{k=1}^{M} w'_{i+1,k},$$

for every $i = 1, \dots, M-1$, and hence, constraints (14.11) hold.

In addition, we obtain

$$\sum_{i=1}^{M} w'_{ik} = \underbrace{\sum_{i=1}^{M} s'_{ik}}_{=1 \text{ by } (13.5)} \sum_{j=1}^{M} c_{kj}\, y'_{kj} = \sum_{j=1}^{M} c_{kj}\, y'_{kj},$$

and therefore, constraints (14.12) are also fulfilled.

Finally, we can prove the validity of constraints (14.13) using the binary character of variable s. Thus, we distinguish two cases. On the one hand, if $s'_{ik} = 0$ then $w'_{ik} = 0$. From constraints (13.13) and (13.14) we obtain that $y'_{kj} \leq 1$ and therefore, since $c_{kj} \geq 0$,

$$c_{kj}\, y'_{kj} \leq c_{kj} \Leftrightarrow \sum_{j=1}^{M} c_{kj}\, y'_{kj} \leq \sum_{j=1}^{M} c_{kj} \Leftrightarrow \sum_{j=1}^{M} c_{kj}\, y'_{kj} - \sum_{j=1}^{M} c_{kj} \leq 0 = w'_{ik}. \tag{14.14}$$

On the other hand, if $s'_{ik} = 1$ then $w'_{ik} = \sum_{j=1}^{M} c_{kj}\, y'_{kj}$. Hence, in both cases, constraints (14.13) hold.

Summarizing, we obtain that (x', y', s', w') is feasible to (OMP^2). Moreover, by construction of w', the objective function of (OMP) for (x', y', s') provides the same value as the objective function of (OMP^2) for (x', y', s', w').

Now, let (x', y', s', w') be a feasible solution to (OMP^2). Then, constraints (13.5), (13.6), (13.11), (13.12) and (13.13) are also fulfilled for (OMP).

To show that the remaining constraints hold, we just need to prove the validity of the equivalence $w'_{ik} = s'_{ik} \sum_{j=1}^{M} c_{kj}\, y'_{kj}$.

Variable s is binary from constraints (13.8), so we consider two cases. On the one hand, if $s'_{ik} = 0$ we get by constraints (14.13)

$$\sum_{j=1}^{M} c_{kj}\, y'_{kj} - \sum_{j=1}^{M} c_{kj} \leq w'_{ik} \quad \forall\, i = 1, \dots, M.$$

In addition, by the same argument reasoning that in (14.14) we obtain that

$$\sum_{j=1}^{M} c_{kj}\, y'_{kj} - \sum_{j=1}^{M} c_{kj} \leq 0 \leq w'_{ik}.$$

On the other hand, if $s'_{ik} = 1$ we get directly from constraints (14.13) that

$$w'_{ik} \geq \sum_{j=1}^{M} c_{kj}\, y'_{kj} \ .$$

In summary, in both cases we obtain that

$$w'_{ik} \geq s'_{ik} \sum_{j=1}^{M} c_{kj}\, y'_{kj} \ .$$

Now assume that there exist i^*, k^* such that

$$w'_{i^* k^*} > s'_{i^* k^*} \sum_{j=1}^{M} c_{k^* j}\, y'_{k^* j} \ .$$

Hence, summing over the index i, we get

$$\sum_{i=1}^{M} w'_{ik} > \underbrace{\sum_{i=1}^{M} s'_{ik}}_{=1 \text{ by } (13.5)} \sum_{j=1}^{M} c_{kj}\, y'_{kj} = \sum_{j=1}^{M} c_{kj}\, y'_{kj} \ ,$$

which contradicts constraints (14.12). Thus, it follows that

$$w'_{i^* k^*} = s'_{i^* k^*} \sum_{j=1}^{M} c_{k^* j}\, y'_{k^* j} \ .$$

Therefore, (x', y', s') is also feasible to (OMP). Furthermore, the objective function values for (x', y', s', w') in (OMP^2) and for (x', y', s') in (OMP) coincide. □

In the following section we present a third linearization of (OMP) based on the variables required by (OMP^2) but strongly simplified.

14.1.3 Simplifying Further: (OMP^3)

Looking at the variables defined for (OMP^2), we see that for every $i = 1, \ldots, M$, $w'_{ik} = s_{ik} \sum_{j=1}^{M} c_{kj}\, y_{kj}$ can only take a non-zero value for a unique $k^* = 1, \ldots, M$ corresponding to $s_{ik^*} = 1$. (Note that s is a binary variable and $\sum_{k=1}^{M} s_{ik} = 1$ from constraints (13.8) and (13.6), respectively.) Therefore, it is not necessary to include variables to record the cost of supplying client k. It suffices to record the ith cheapest supply cost. Thus, the formulation is simplified by introducing M new nonnegative continuous variables w_i representing the cost of supplying the client with the ith cheapest supply cost:

$$w_i = \sum_{k=1}^{M} w'_{ik} = \sum_{k=1}^{M} s_{ik} \sum_{j=1}^{M} c_{kj}\, y_{kj}.$$

Using these variables, we get the following linearization of the formulation (OMP):

$$
\begin{array}{llll}
(\mathsf{OMP}^3) & \min & \displaystyle\sum_{i=1}^{M} \lambda_i w_i & \\
 & \text{s.t.} & & \\
 & & \displaystyle\sum_{i=1}^{M} s_{ik} = 1 & \forall\, k = 1,\dots,M \\
 & & \displaystyle\sum_{k=1}^{M} s_{ik} = 1 & \forall\, i = 1,\dots,M \\
 & & w_i \le w_{i+1} & \forall\, i = 1,\dots,M-1 \quad (14.15) \\
 & & \displaystyle\sum_{j=1}^{M} x_j = N & \\
 & & \displaystyle\sum_{j=1}^{M} y_{kj} = 1 & \forall\, k = 1,\dots,M \\
 & & x_j \ge y_{kj} & \forall\, j,k = 1,\dots,M
\end{array}
$$

$$\sum_{i=1}^{M} w_i = \sum_{j=1}^{M}\sum_{k=1}^{M} c_{kj}\, y_{kj} \qquad (14.16)$$

$$w_i \ge \sum_{j=1}^{M} c_{kj}\, y_{kj} - \sum_{j=1}^{M} c_{kj}(1 - s_{ik}) \;\; \forall\, i,k = 1,\dots,M \qquad (14.17)$$

$$s_{ik}, x_j \in \{0,1\}, \quad y_{kj}, w_i \ge 0 \quad \forall\, i,j,k = 1,\dots,M.$$

The quadratic formulation of the OMP is equivalent to this mixed-integer linear program as the following theorem shows.

Theorem 14.3 *The formulations (OMP) and (OMP^3) are equivalent, i.e. a solution (x', y', s') is feasible to (OMP) if and only if (x', y', s', w') is feasible to (OMP^3) by setting $w'_i = \sum_{k=1}^{M} s'_{ik} \sum_{j=1}^{M} c_{kj}\, y'_{kj}$. Moreover, their corresponding objective function values coincide.*

Proof.
Let (x', y', s') be a feasible solution to (OMP). Then, it satisfies immediately constraints (13.5), (13.6), (13.11), (13.12) and (13.13). Thus, they are also fulfilled in (OMP^3).

By setting $w'_i := \sum_{k=1}^{M} s'_{ik} \sum_{j=1}^{M} c_{kj}\, y'_{kj}$ we will see that the remaining constraints also hold.

First, we obtain constraints (14.15) using constraints (13.16), since

$$w'_i = \sum_{k=1}^{M} s'_{ik} \sum_{j=1}^{M} c_{kj}\, y'_{kj} \le \sum_{k=1}^{M} s'_{i+1,k} \sum_{j=1}^{M} c_{kj}\, y'_{kj} \le w'_{i+1},$$

for every $i = 1, \ldots, M-1$.

Furthermore, from the definition of w' we get

$$\sum_{i=1}^{M} w'_i = \sum_{i=1}^{M}\sum_{k=1}^{M} s'_{ik} \sum_{j=1}^{M} c_{kj}\, y'_{kj} = \sum_{k=1}^{M} \underbrace{\sum_{i=1}^{M} s'_{ik}}_{=1 \text{ by } (13.5)} \sum_{j=1}^{M} c_{kj}\, y'_{kj} = \sum_{k=1}^{M}\sum_{j=1}^{M} c_{kj}\, y'_{kj},$$

and therefore, constraint (14.16) is also fulfilled.

Finally, we will show the validity of constraints (14.17). From the definition of variable s and constraints (13.6), we obtain that for each index $i \in \{1, \ldots, M\}$ there exists an index $k^* = 1, \ldots M$ such that $s'_{ik^*} = 1$ and $s'_{ik} = 0$, $\forall k \neq k^* = 1, \ldots, M$. Therefore,

$$\begin{aligned} w'_i = \sum_{k=1}^{M} s'_{ik} \sum_{j=1}^{M} c_{kj}\, y'_{kj} &= s'_{ik^*} \sum_{j=1}^{M} c_{k^*j}\, y'_{k^*j} \\ &= \sum_{j=1}^{M} c_{k^*j}\, y'_{k^*j} = \sum_{j=1}^{M} c_{k^*j}\, y'_{k^*j} - \sum_{j=1}^{M} c_{k^*j}(1 - s'_{ik^*}). \end{aligned}$$

In addition, because x is a binary variable and by constraints (13.13) we get that $y'_{kj} \le 1$. Hence, since $c_{kj} \ge 0$,

$$c_{kj}\, y'_{kj} \le c_{kj} \Leftrightarrow \sum_{j=1}^{M} c_{kj}\, y'_{kj} \le \sum_{j=1}^{M} c_{kj} \Leftrightarrow \sum_{j=1}^{M} c_{kj}\, y'_{kj} - \sum_{j=1}^{M} c_{kj} \le 0.$$

Thus, for $k \neq k^*$ we have

$$\sum_{j=1}^{M} c_{kj}\, y'_{kj} - \sum_{j=1}^{M} c_{kj}(1 - s'_{ik}) = \sum_{j=1}^{M} c_{kj}\, y'_{kj} - \sum_{j=1}^{M} c_{kj} \le 0 \le w'_i.$$

Therefore, for all $i, k = 1, \ldots, M$, constraints (14.17) are satisfied.

Summarizing, we obtain that (x', y', s', w') is feasible to (OMP^3). Furthermore, by the definition of w' we also get the same objective function value in (OMP) for (x', y', s') as in (OMP^3) for (x', y', s', w').

Now, let (x', y', s', w') be a feasible solution to (OMP^3). Then, constraints (13.5), (13.6), (13.11), (13.12) and (13.13) of (OMP) are fulfilled.

We will show that the equivalence $w'_i = \sum_{k=1}^{M} s'_{ik} \sum_{j=1}^{M} c_{kj}\, y'_{kj}$ holds and therefore, the implication is proven.

From the binary character of variable s and constraints (13.6) we know that for every $i = 1, \ldots, M$ there exists $k^* = 1, \ldots, M$ such that $s'_{ik^*} = 1$ and $s'_{ik} = 0$ for $k \neq k^*$. Then, we obtain by constraints (14.17) that for k^*

$$w'_i \geq \sum_{j=1}^{M} c_{k^*j}\, y'_{k^*j} = s'_{ik^*} \sum_{j=1}^{M} c_{k^*j}\, y'_{k^*j} = \sum_{k=1}^{M} s'_{ik} \sum_{j=1}^{M} c_{kj}\, y'_{kj} \quad \forall\, i = 1, \ldots, M.$$

Now assume that there exists i^* with $w'_{i^*} > \sum_{k=1}^{M} s'_{i^*k} \sum_{j=1}^{M} c_{kj}\, y'_{kj}$. Thus,

$$\sum_{i=1}^{M} w'_i > \sum_{k=1}^{M} \underbrace{\sum_{i=1}^{M} s'_{ik}}_{=1 \text{ by } (13.5)} \sum_{j=1}^{M} c_{kj}\, y'_{kj} = \sum_{k=1}^{M} \sum_{j=1}^{M} c_{kj}\, y'_{kj},$$

which is a contradiction to constraint (14.16). Hence, it follows that

$$w'_{i^*} = \sum_{k=1}^{M} s'_{i^*k} \sum_{j=1}^{M} c_{kj}\, y'_{kj}\ .$$

Therefore, (x', y', s') is feasible to (OMP). Moreover, the objective function values for (x', y', s', w') in (OMP^3) and for (x', y', s') in (OMP) coincide. □

Since (OMP^2) and (OMP^3) are tightly related, a theoretical comparison between their feasible regions is presented in following section.

14.1.4 Comparison Between (OMP^2) and (OMP^3)

We will show in the following that the feasible region described by $(\mathsf{OMP}^2_{\mathsf{LR}})$ (the linear programming relaxation of (OMP^2)) is a subset of the feasible region of $(\mathsf{OMP}^3_{\mathsf{LR}})$ (the linear programming relaxation of (OMP^3)). Moreover, we show that this inclusion is strict. This is somewhat surprising, since (OMP^3) seems to be the more compact formulation.

Theorem 14.4 *Let (x', y', s', w') be a feasible solution of $(\mathsf{OMP}^2_{\mathsf{LR}})$. Then $(x'$,y',s',$w'')$ is a feasible solution of $(\mathsf{OMP}^3_{\mathsf{LR}})$, by setting $w''_i = \sum_{k=1}^{M} w'_{ik}$ for all $i = 1, \ldots, M$.*

Proof.
Since (x', y', s', w') is feasible to $(\mathsf{OMP}^2_{\mathsf{LR}})$, constraints (13.5), (13.6), (13.11), (13.12) and (13.13) are immediately fulfilled by (x', y', s', w'') with

$$w''_i = \sum_{k=1}^{M} w'_{ik} \quad \forall\, i = 1, \ldots, M\ .$$

For the remaining constraints we proceed as follows.

Constraints (14.15) are fulfilled using $w''_i = \sum_{k=1}^{M} w'_{ik}$ and (14.11):

$$w''_i = \sum_{k=1}^{M} w'_{ik} \le \sum_{k=1}^{M} w'_{i+1,k} = w''_{i+1}.$$

From constraints (14.12) and summing over the index k we have

$$\sum_{i=1}^{M} w''_i = \sum_{i=1}^{M}\sum_{k=1}^{M} w'_{ik} = \sum_{j=1}^{M}\sum_{k=1}^{M} c_{kj}\, y'_{kj}$$

and hence, constraint (14.16) is also satisfied.

Finally, because (x', y', s', w') fulfills constraints (14.13) and $w'_{ik} \ge 0$, we obtain that (x', y', s', w'') verifies constraints (14.17) as follows

$$w''_i = \sum_{l=1}^{M} w'_{il} \ge w'_{ik} \ge \sum_{j=1}^{M} c_{kj}\, y'_{kj} - \sum_{j=1}^{M} c_{kj}\,(1 - s'_{ik}) \quad \forall\, k = 1, \ldots, M.$$

Hence, (x', y', s', w'') is a feasible solution to $(\mathsf{OMP}^3_{\mathsf{LR}})$. □

In the following example we present a point that is feasible to $(\mathsf{OMP}^3_{\mathsf{LR}})$ which cannot be transformed into a feasible point to $(\mathsf{OMP}^2_{\mathsf{LR}})$. Therefore, the inclusion shown in Theorem 14.4 is strict.

Example 14.1
Let $M = 3$ and $N = 2$, i.e. we have three sites, and we would like to locate two service facilities. Let $C = \begin{pmatrix} 0\,1\,3 \\ 1\,0\,3 \\ 1\,3\,0 \end{pmatrix}$ be the 3×3 cost matrix. Then consider the point (x', y', s', w'') defined as follows:

$$x' = \begin{pmatrix} 0.5 \\ 0.75 \\ 0.75 \end{pmatrix}, y' = \begin{pmatrix} 0 & 0.5 & 0.5 \\ 0.25 & 0 & 0.75 \\ 0.25 & 0.75 & 0 \end{pmatrix}, s' = \begin{pmatrix} 0 & 0.5 & 0.5 \\ 0.5 & 0 & 0.5 \\ 0.5 & 0.5 & 0 \end{pmatrix}, w'' = \begin{pmatrix} 0.5 \\ 3 \\ 3.5 \end{pmatrix}.$$

It satisfies constraints (13.5), (13.6), (13.11), (13.12), (13.13), (14.15), (14.16) and (14.17). Thus, this point belongs to the feasible region of $(\mathsf{OMP}^3_{\mathsf{LR}})$.

To obtain a feasible point to $(\mathsf{OMP}^2_{\mathsf{LR}})$ we only need to find values for w'_{ik} which satisfy constraints (14.11), (14.12) and (14.13) such that $\sum_{k=1}^{M} w'_{ik} = w''_i$ for all $i = 1, 2, 3$. However, this is not possible, since on the one hand, by constraints (14.13), we obtain

$$w'_{11} \geq 0,\ w'_{12} \geq 0.5 \text{ and } w'_{13} \geq 0.5,$$

but on the other hand, $\sum_{k=1}^{M} w'_{1k} = 0.5 = w''_1$. Thus, there is no solution to the above system of equations and therefore, (x', y', s', w'') cannot be transformed into a feasible solution of $(\mathsf{OMP}^2_{\mathsf{LR}})$.

The improvement of the three presented linearizations of (OMP) is the focus of the next section. Observe that Theorem 14.4 implies that any improvements found for linearization (OMP^3) easily carry over to (OMP^2), see Section 14.2.3.

14.2 Reformulations

The goal of this section is to improve the three linearizations (OMP^1), (OMP^2) and (OMP^3). These improvements are obtained adding constraints, strengthened forms of original constraints, and preprocessing steps, such as fixing some variables to zero, or relaxing integrality requirements on some others. By doing so, we reduce the computing time required to solve the discrete OMP, either by reducing the gap between the optimal objective function value and the relaxed linear programming (LP) solution (*integrality gap*), or by reducing the number of variables for which integrality must be ensured. For some of these properties, we will need free self-service, i.e. we assume $c_{ii} = 0$, for all $i \in A$. For short, we will denote this property by (FSS). Also note that, for most improvements, λ is required to be nonnegative. However, the initial formulations are also valid for general λ.

The first strengthening approach is based on any upper bound, z_{UB}, on the optimal objective function value of the OMP which we assume to be given. Upper bounds are, of course, easy to come by; any set of N locations yields one. Since *good* upper bounds are in general more difficult to obtain, we give in Section 15.2 different heuristic procedures to be used to get upper bounds. Consider a client k: Either the cost of allocating k is among the highest $M - m$

such costs, or its cost is among the cheapest m costs. In the latter case the objective value of the problem must be at least $\sum_{i=m}^{M} \lambda_i$ multiplied by the cost of allocating client k; this value is clearly a lower bound. Obviously we are only interested in solutions with objective value less than or equal to z_{UB}, i.e. solutions with either client k's cost ranked $m+1$ or higher, or with k allocated to some location j with $c_{kj} \leq \frac{z_{\text{UB}}}{\sum_{i=m}^{M} \lambda_i}$. To simplify the presentation, recall that

$$f_\lambda(X) := \sum_{i=1}^{M} \lambda_i c_{(i)}(X)$$

denotes the total cost of a solution X. We write $\mathsf{rk}(k)$ to denote the rank of client k in the cost ordering, so if $c_{(i)}(X) = c_k(X)$ for some $i, k = 1, \ldots, M$ then $\mathsf{rk}(k) = i$. Note that we adopt the convention that if $\sum_{i=m}^{M} \lambda_i = 0$ then $\frac{z_{\text{UB}}}{\sum_{i=m}^{M} \lambda_i} = \infty$ for all $j, k = 1, \ldots, M$.

Proposition 14.1 *If $f_\lambda(X) \leq z_{UB}$ for some z_{UB}, then $\mathsf{rk}(k) \geq m+1$ for all $k = 1, \ldots, M$ and all ranks $m = 1, \ldots, M-1$ which fulfill $c_k(X) > \frac{z_{UB}}{\sum_{i=m}^{M} \lambda_i}$.*

Proof.
Suppose that there exist $k \in \{1, \ldots, M\}$ and $m \in \{1, \ldots, M-1\}$ such that $c_k(X) > \frac{z_{\text{UB}}}{\sum_{i=m}^{M} \lambda_i}$ and $\mathsf{rk}(k) \leq m$. By construction,

$$c_{(M)}(X) \geq \cdots \geq c_{(m)}(X) \geq c_{(r)}(X) = c_k(X).$$

Since C and λ are both nonnegative, we have

$$f_\lambda(X) = \sum_{i=1}^{M} \lambda_i c_{(i)}(X) \geq \sum_{i=m}^{M} \lambda_i c_{(i)}(X) \geq \sum_{i=m}^{M} \lambda_i c_k(X) = c_k(X) \sum_{i=m}^{M} \lambda_i > z_{\text{UB}},$$

which yields a contradiction, and therefore, the result follows. □

The following proposition states a general property which will be used in the next subsections.

Proposition 14.2 *Given X with $|X| = N$. If (FSS) holds and $\lambda \geq 0$, then $c_i(X) = 0$ for all $i \in X$. Furthermore, there exists a valid permutation for X, such that*

$$\{(i) \; : \; i = 1, \ldots, N\} = X. \tag{14.18}$$

Proof.
For each client $k = 1, \ldots, M$ there exists an open facility $j(k) \in \operatorname{argmin}_{j \in X} c_{kj}$,

where $X = \{j \in A : x_j = 1\}$, such that client k causes a cost equal to $c_k(X) = c_{kj(k)}$.

Now since $c_{ij} \geq 0$, for all $i, j = 1, \ldots, M$ and by (FSS) $c_{kk} = 0$, for all $k = 1, \ldots, M$, in each optimal allocation we have that $c_k(X) = 0$ for all $k \in X$ and $c_k(X) \geq 0$ for all $k \in A \setminus X$. Observe that for each $k \in X$, by (FSS), k itself is a minimizer of $\min_{j \in X} c_{kj}$, thus, we can set $j(k) = k$ for all $k \in X$.

Therefore, there exists a valid permutation for X such that the first N elements are the open facilities. This permutation satisfies

$$0 = c_{(1)}(X) = \cdots = c_{(N)}(X) \leq c_{(N+1)}(X) \leq \cdots \leq c_{(M)}(X). \tag{14.19}$$

Furthermore, if there would exist some $h > 0$ such that

$$\begin{aligned} 0 = c_{(1)}(X) = \ldots = c_{(N)}(X) & \\ = c_{(N+1)}(X) = \cdots = c_{(N+h)}(X) & \leq \cdots \leq c_{(M)}(X), \end{aligned}$$

it is possible to permute zeros without increasing the total cost such that (14.18) is satisfied, i.e. the first N entries of $c_{\leq}(X)$ (see (13.2)) correspond to ones with $c_{kk} = 0$, for all $k \in X$.

Thus, in either case, (14.18) is fulfilled. □

In the following subsections we use Proposition 14.2 to get improvements for the different linearizations of Section 14.1.

All equivalent formulations of the OMP use different variables but induce the same solution X for the OMP. Instead of writing the solution value $f_\lambda(X)$ induced by the variables u of a specific reformulation we allow to write $f_\lambda(u)$ directly. As a consequence, in the following we will not distinguish between variables and their values in order to simplify the notation.

14.2.1 Improvements for (OMP^1)

First, we use an upper bound in order to fix some variables z_{ikj} to zero, as the next lemma shows.

Lemma 14.1 *Let z_{UB} be an upper bound on the optimal value of the OMP. Then any optimal solution (x, z) to (OMP^1) must satisfy*

$$z_{mkj} = 0 \quad \forall\, m, k, j \text{ with } c_{kj} > \frac{z_{UB}}{\sum_{i=m}^{M} \lambda_i}. \tag{14.20}$$

Proof.
Suppose (x, z) is an optimal solution to (OMP^1). Assume further that z_{UB} is an upper bound on the value of the OMP and there exist m, j^*, k^* such that $c_{k^*j^*} > \dfrac{z_{\text{UB}}}{\sum_{i=m}^{M} \lambda_i}$ and $z_{mk^*j^*} = 1$.

By the definition of variable z, see (14.1), it follows from $z_{mk^*j^*} = 1$ that site j^* is in the solution set X induced by (x, z) (i.e. $x_{j^*} = 1$), and moreover, client k^* is ranked at position m (i.e. $\mathsf{rk}(k^*) = m$). Furthermore, client k^* is allocated to site j^*, thus $c_{k^*}(X) = c_{k^*j^*} > \frac{z_{\text{UB}}}{\sum_{i=m}^{M} \lambda_i}$.

In addition, $f_\lambda(X)$ is the objective value of (x, z) in (OMP^1), and therefore $f_\lambda(X) \leq z_{\text{UB}}$.

Summarizing, by Proposition 14.1, we have that $\mathsf{rk}(k^*) \geq m + 1$. But $\mathsf{rk}(k^*) = m$. This contradiction completes the proof. □

Assuming that property (FSS) holds, it is possible to strengthen the original formulation of (OMP^1). Notice that it is natural to consider that the allocation cost of one client to itself is 0. Therefore, this assumption is often fulfilled in actual applications. To simplify the presentation, we will assume in the rest of this section that (FSS) holds.

Lemma 14.2 *There exists an optimal solution (x, z) to (OMP^1) satisfying*

$$\sum_{i=1}^{N} z_{ikk} = x_k, \qquad \forall\ k = 1, \ldots, M. \tag{14.21}$$

Proof.
Let (x, z) be an optimal solution to (OMP^1). Let X be the set of facilities induced by x, i.e. $X = \{j \in A : \ x_j = 1\}$. We have $|X| = N$ by constraints (13.11).

Consider a valid permutation for X. By (FSS) and Proposition 14.2, we may assume that this permutation also satisfies (14.18), and thus $\mathsf{rk}(k) \in \{1, \ldots, N\}$ for each $k \in X$.

For each $k \in A \setminus X$, let $j(k)$ be a minimizer of $\min_{j \in X} c_{kj}$. Observe that by (FSS) $j(k) = k$ for all $k \in X$.

Then, define z' by $z'_{ikj} = 1$ if $j = j(k)$ and $k = (i)$, and $z'_{ikj} = 0$ otherwise, for each $i, j, k = 1, \ldots, M$. By Theorem 14.1 and Corollary 14.1, (x, z') is an optimal solution to (OMP^1). We distinguish two cases to prove constraints (14.21).

Case 1: $k \in X$. In this case $j(k) = k$ and $\mathsf{rk}(k) \in \{1, \ldots, N\}$, so $z'_{\mathsf{rk}(k)kk} = 1$ and $z'_{ikk} = 0$ for all $i = 1, \ldots, N$, $i \neq \mathsf{rk}(k)$. Thus, $\sum_{i=1}^{N} z'_{ikk} = z'_{\mathsf{rk}(k)kk} = 1 = x_k$ as required.

Case 2: $k \notin X$. In this case $j(k) \neq k$, since $j(k) \in X$ by definition. Therefore, $z'_{ikk} = 0$ for all $i = 1, \ldots, M$. Thus, $\sum_{i=1}^{N} z'_{ikk} = 0 = x_k$ as required.

In either case, constraints (14.21) are fulfilled by (x, z'), which is an optimal solution to (OMP^1). □

The following corollary unifies the two results presented above.

Corollary 14.2 *There exists an optimal solution to* (OMP^1) *satisfying constraints* (14.20) *and* (14.21).

Proof.
By Lemma 14.2 there exists an optimal solution (x, z) satisfying constraints (14.21). By Lemma 14.1, every optimal solution fulfills constraints (14.20), so (x, z) does. Thus, there exists an optimal solution (x, z) satisfying both constraints (14.20) and (14.21). □

Finally, the integrality requirement on some variables and the existence of an optimal solution to (OMP^1) which satisfies all the results presented in this section is shown in the next theorem.

In the next step of our analysis we show that there are optimal solutions of (OMP^1) which satisfy the reinforcements defined by (14.20) and (14.21) even if some z_{ikj} are relaxed. Thus, we get a better polyhedral description of (OMP^1).

Let $(\mathsf{OMP}^1_{\mathsf{LR(z)}})$ be the problem that results from (OMP^1) after relaxing the integrality constraints for z_{ikj}, $i = 1, \ldots, N$, $j, k = 1, \ldots, M$.

Theorem 14.5 *Let* (x, z) *be an optimal solution of* $(\mathsf{OMP}^1_{\mathsf{LR(z)}})$ *reinforced with the additional constraints* (14.20) *and* (14.21). *Then* x *induces an optimal solution to the OMP.*

Proof.
Let (x, z) be an optimal solution to $(\mathsf{OMP}^1_{\mathsf{LR(z)}})$ satisfying constraints (14.20) and (14.21).

We will proceed by showing that there exists z' such that (x, z') is an optimal solution to (OMP^1). By Theorem 14.1 and Corollary 14.1, we conclude that x must induce an optimal solution to OMP.

Let X be the set of facilities induced by x. By Proposition 14.2 there exists a valid permutation (14.18). By (14.18), we have

$$x_{(1)} = x_{(2)} = \cdots = x_{(N)} = 1,$$

i.e. $\mathsf{rk}(j) \in \{1, \ldots, N\}$ for all $j \in X$, and $x_j = 0$ for all $j = 1, \ldots, M$ with $\mathsf{rk}(j) > N$. We now define z' by

$$z'_{ikj} := \begin{cases} z_{ikj}, & i > N \\ 1, & i \leq N, k = j = (i) \\ 0, & \text{otherwise} \end{cases}$$

for each $i, k, j = 1, \ldots, M$.

In the following we show that z' is feasible for (OMP^1).

First, we prove that z' satisfies constraints (14.2). On the one hand, for all $i > N$, $z'_{ikj} = z_{ikj}$ for all $j, k \in A$, so the result follows since z already fulfills constraints (14.2). On the other hand, for all $i \leq N$,

$$\sum_{k=1}^{M}\sum_{j=1}^{M} c_{kj}\, z'_{ikj} = c_{(i)(i)} = 0$$

since (FSS) holds. Therefore, constraints (14.2) are fulfilled.

For the constraints of type (14.7) we distinguish two cases.

Case 1: $k \in X$. Since z satisfies constraints (14.21) and in this case $x_k = 1$, we have that $\sum_{i=1}^{N} z_{ikk} = 1$. Furthermore, z fulfills constraints (14.7), and therefore, $\sum_{i=N+1}^{M}\sum_{j=1}^{M} z_{ikj} = 0$. Hence, and from the definition of z', we have

$$\begin{aligned}\sum_{i=1}^{M}\sum_{j=1}^{M} z'_{ikj} &= \sum_{i=1}^{N}\sum_{j=1}^{M} z'_{ikj} + \sum_{i=N+1}^{M}\sum_{j=1}^{M} z_{ikj} \\ &= z'_{\mathsf{rk}(k)kk} + 0 \\ &= 1\end{aligned}$$

as required.

Case 2: $k \notin X$. We start by showing that $\sum_{j\in X} z_{ijj} = 1$ for all $i = 1,\ldots,N$. From constraints (14.21) we have $\sum_{i=1}^{N} z_{ijj} = 1$ for all $j \in X$. Then, by constraints (14.8), we know that $\sum_{j\in X} z_{ijj} \leq 1$ for all $i = 1,\ldots,N$. Suppose that there exists an $i \in \{1,\ldots,N\}$ with $\sum_{j\in X} z_{ijj} < 1$. Then

$$\sum_{i=1}^{N}\sum_{j\in X} z_{ijj} < N = \sum_{j\in X}\sum_{i=1}^{N} z_{ijj}$$

since $|X| = N$. This is a contradiction and therefore $\sum_{j\in X} z_{ijj} = 1$ for all $i = 1,\ldots,N$.

Using constraints (14.8) again, we obtain that $z_{ik^*j} = 0$ for all $i = 1,\ldots,N$, and all $j, k^* = 1,\ldots,M$ with $k^* \neq j$ or $j \notin X$. Thus, applying constraints (14.7), and since $k \notin X$, we have that

$$\sum_{i=N+1}^{M} \sum_{j=1}^{M} z_{ikj} = 1.$$

To conclude, we observe that by the definition of z', and since $k \notin X$, we have that $z'_{ikj} = 0$ for all $i = 1, \ldots, N$ and all $j = 1, \ldots, M$, so

$$\begin{aligned}\sum_{i=1}^{M} \sum_{j=1}^{M} z'_{ikj} &= \sum_{i=1}^{N} \sum_{j=1}^{M} z'_{ikj} + \sum_{i=N+1}^{M} \sum_{j=1}^{M} z_{ikj} \\ &= \sum_{i=1}^{N} \sum_{j=1}^{M} 0 + 1 \\ &= 1\end{aligned}$$

as required.

Therefore, in either case z' fulfills constraints (14.7).

Showing that z' satisfies constraints (14.8) is easier. On the one hand, for $i > N$, $z'_{ikj} = z_{ikj}$ for all $j, k = 1, \ldots, M$, so

$$\sum_{k=1}^{M} \sum_{j=1}^{M} z'_{ikj} = \sum_{k=1}^{M} \sum_{j=1}^{M} z_{ikj} = 1$$

since z fulfills constraints (14.8). On the other hand, for $i \leq N$, $z'_{ikj} = 1$ if $k = j = (i)$, and $z'_{ikj} = 0$, otherwise. Therefore, z' fulfills constraints (14.8).

Proving that z' satisfies constraints (14.9) is similar to showing that z' verifies constraints (14.7), using the above arguments. Again, we distinguish the same two cases.

Case 1: $k \in X$. Using the above arguments, we have that $\displaystyle\sum_{i=N+1}^{M} \sum_{j=1}^{M} z_{ikj} = 0.$

Hence, $\displaystyle\sum_{i=N+1}^{M} z_{ikj} = 0$ for all $j = 1, \ldots, M$. Now, from the definition of z', $z'_{ikj} = 1$ for $i = 1, \ldots, N$ if and only if $j = k = (i)$. Therefore, we have that if $j \neq k$,

$$\sum_{i=1}^{M} z'_{ikj} = \sum_{i=1}^{N} z'_{ikj} + \sum_{i=N+1}^{M} z_{ikj} = 0 + 0 = 0 \leq x_j,$$

while if $j = k$, then $x_j = 1$ since $j = k \in X$, and we obtain

$$\sum_{i=1}^{M} z'_{ikj} = \sum_{i=1}^{N} z'_{ikj} + \sum_{i=N+1}^{M} z_{ikj} = z'_{\mathsf{rk}(k)kk} + 0 = 1 = x_j$$

as required.

Case 2: $k \notin X$. In this case, it follows from the definition of z' that $z'_{ikj} = 0$ for all $i = 1, \ldots, N$ and all $j = 1, \ldots, M$. Thus

$$\sum_{i=1}^{M} z'_{ikj} = \sum_{i=1}^{N} z'_{ikj} + \sum_{i=N+1}^{M} z_{ikj} = 0 + \sum_{i=N+1}^{M} z_{ikj} \le \sum_{i=1}^{M} z_{ikj} \le x_j$$

since z satisfies constraints (14.9), as required.

Therefore, in either case constraints (14.9) hold.

Since z' is binary we obtain that (x, z') is feasible for (OMP^1).

We now show that the objective value of (x, z') in (OMP^1) is less than or equal to that of (x, z). Consider the function

$$\zeta_i(z) = \sum_{k=1}^{M} \sum_{j=1}^{M} c_{kj}\, z_{ikj}$$

for each $i = 1, \ldots, M$. Then we can write the objective function of (OMP^1) as

$$\sum_{i=1}^{M} \lambda_i \zeta_i(z).$$

On the one hand, for $i > N$, $z'_{ikj} = z_{ikj}$ for all k, j, so

$$\zeta_i(z') = \sum_{k=1}^{M} \sum_{j=1}^{M} c_{kj}\, z'_{ikj} = \sum_{k=1}^{M} \sum_{j=1}^{M} c_{kj}\, z_{ikj} = \zeta_i(z).$$

On the other hand, for $i \le N$, $z'_{ikj} = 1$ if and only if $j = k = (i)$, and $z'_{ikj} = 0$ otherwise, thus

$$\zeta_i(z') = \sum_{k=1}^{M} \sum_{j=1}^{M} c_{kj}\, z'_{ikj} = c_{(i)(i)} = 0$$

by (FSS). This implies

$$\zeta_i(z') = 0 \le \zeta_i(z)\ ,\ i = 1, \ldots, M\ .$$

Since $\lambda \ge 0$ is assumed, the objective value of (x, z') in (OMP^1) is

$$f_\lambda(x, z') = \sum_{i=1}^{M} \lambda_i \zeta_i(z') \le \sum_{i=1}^{M} \lambda_i \zeta_i(z) = f_\lambda(x, z), \tag{14.22}$$

i.e. it is smaller than or equal to the objective value of (x, z).

To conclude, notice that (OMP^1) and (OMP^1) reinforced by constraints (14.20) and (14.21) have the same optimal objective value (see Corollary 14.2). Since (x, z) is an optimal solution to $(\mathsf{OMP}^1_{\mathsf{LR(z)}})$, a relaxation of the latter problem, we have

$$f_\lambda(x,z) \le \mathsf{val}(\mathsf{OMP}^1)\ .$$

(Recall that val(P) denotes the optimal value of Problem P.)

In addition, we have proven that (x, z') is feasible for (OMP^1) and therefore,

$$\mathsf{val}(\mathsf{OMP}^1) \le \mathrm{f}_\lambda(\mathrm{x}, \mathrm{z}')\ .$$

Thus, $f_\lambda(x,z) \le f_\lambda(x,z')$. Moreover, by (14.22) they are equal and hence, (x, z') is optimal. □

As a consequence of the above results we define (OMP^{1^*}) as $(\mathsf{OMP}^1_{\mathsf{LR(z)}})$ reinforced by constraints (14.20) and (14.21). In Section 14.3 computational experiments will show the advantages of the strengthened formulation (OMP^{1^*}). After dealing with the first linearization we show how to strengthen the other linearizations in the following.

14.2.2 Improvements for (OMP^3)

In the following, we will prove properties which are fulfilled by optimal solutions to (OMP^3). These results will help to strengthen the original formulation. Using the results of Section 14.1.4, our findings will easily carry over to (OMP^2).

Lemma 14.3 *If (x, y, s, w) is an optimal solution of (OMP^3), then*

$$\sum_{j\,:\,c_{kj} \le \frac{z_{UB}}{\sum_{i=m}^{M} \lambda_i}} x_j \ge \sum_{i=1}^{m} s_{ik} \tag{14.23}$$

is fulfilled for all $k, m = 1, \ldots, M$.

Proof.
Let (x, y, s, w) be an optimal solution to (OMP^3), and let z_{UB} be an upper bound on the optimal objective. Let X be the set of facilities induced by x and let σ be the permutation of $\{1, \ldots, M\}$ induced by s, i.e. let $\sigma(i) = j$ if and only if $s_{ij} = 1$.

Then, by Theorem 14.3, $|X| = N$, σ is a valid permutation for X and $f_\lambda(X)$ is the objective value of (x, y, s, w) in (OMP^3), so $f_\lambda(X) \le z_{\mathrm{UB}}$.

Take any arbitrary $k, m \in \{1, \ldots, M\}$. From constraints (13.5) and the binary character of variable s, $\sum_{i=1}^{m} s_{ik} \in \{0, 1\}$. Since $x \ge 0$, (14.23) trivially holds when $\sum_{i=1}^{m} s_{ik} = 0$. Thus, let us assume that $\sum_{i=1}^{m} s_{ik} = 1$. Therefore, $s_{rk} = 1$ for some $1 \le r \le m$, and so $\sigma(r) = k$ and $\mathsf{rk}(k) = r \le m$. Hence, by Proposition 14.1, we obtain that $c_k(X) \le \frac{z_{\mathrm{UB}}}{\sum_{i=m}^{M} \lambda_i}$. By the definition of

$c_k(X)$, see (13.1), there exists a $j \in X$ with $c_{kj} = c_k(X)$. Thus, $x_j = 1$ and $c_{kj} \leq \frac{z_{\mathrm{UB}}}{\sum_{i=m}^{M} \lambda_i}$, and therefore

$$\sum_{j \,:\, c_{kj} \leq \frac{z_{\mathrm{UB}}}{\sum_{i=m}^{M} \lambda_i}} x_j \geq 1 = \sum_{i=1}^{m} s_{ik}$$

as required. □

In the following, we present a result which strengthens constraints (14.17) from the original formulation.

Lemma 14.4 *Let $\overline{c}_k$ be the N-th largest entry of the cost matrix at row k, i.e. let $\overline{c}_k$ be the N-th largest element of the set $\{c_{kj} \mid j = 1, \ldots, M\}$. Then the following inequality*

$$w_i \geq \sum_{j=1}^{M} c_{kj}\, y_{kj} - \overline{c}_k(1 - s_{ik}), \quad \forall i, k = 1, \ldots, M \tag{14.24}$$

is satisfied by any optimal solution (x, y, s, w) to (OMP^3).

Proof.
Let (x, y, s, w) be an optimal solution to (OMP^3) and $X = \{j \in A : x_j = 1\}$ be the set of facilities induced by x. From the binary character of variable s we distinguish two cases.

Case 1: $s_{ik} = 1$. By constraints (14.17), $w_i \geq \sum_{j=1}^{M} c_{kj}\, y_{kj}$, as required.

Case 2: $s_{ik} = 0$. We observe that $y_{kj(k)} = 1$ for some $j(k) \in X = \{j \in A : x_j = 1\}$, and in fact since (x, y, s, w) is optimal, it must be that $j(k) \in \mathrm{argmin}_{j \in X} c_{kj}$. Since $|X| = N$, $c_{kj(k)}$ cannot exceed the value of the N-th largest element in $\{c_{kj} \mid j = 1, \ldots, M\}$, denoted by $\overline{c}_k$. By constraints (13.12), we have that $y_{kj} = 0$ for all $j \neq j(k)$, thus

$$\sum_{j=1}^{M} c_{kj}\, y_{kj} = c_{kj(k)} \leq \overline{c}_k,$$

and therefore

$$\sum_{j=1}^{M} c_{kj}\, y_{kj} - \overline{c}_k(1 - s_{ik}) = c_{kj(k)} - \overline{c}_k \leq 0$$

in this case. By the non-negativity of variable w we obtain that $w_i \geq c_{kj(k)} - \overline{c}_k$, as required.

Hence, in either case, constraints (14.24) hold. □

Constraints (14.24) can even be further strengthened as the next corollary shows.

Corollary 14.3 *Any optimal solution (x, y, s, w) to (OMP^3) satisfies*

$$w_i \geq \sum_{j=1}^{M} c_{kj}\, y_{kj} - \overline{c}_k \left(1 - \sum_{l=1}^{i} s_{lk}\right), \quad \forall\, i, k = 1, \ldots, M \tag{14.25}$$

where $\overline{c}_k$ is the N-th largest entry of the cost matrix at row k.

Proof.
Let (x, y, s, w) be an optimal solution or (OMP^3) and let $i, k = 1, \ldots, M$. Then, from constraints (14.15) and applying Lemma 14.4 we have for any $l \in \{1, \ldots, i\}$ that

$$w_i \geq w_l \geq \sum_{j=1}^{M} c_{kj}\, y_{kj} - \overline{c}_k (1 - s_{lk}). \tag{14.26}$$

By (13.5), $\sum_{l=1}^{M} s_{lk} = 1$. Since s is binary we get either $s_{lk} = 0$ for all $l = 1, \ldots, i$ or there exists $1 \leq l' \leq i$ such that $s_{l'k} = 1$ and $s_{lk} = 0$ for all $l \neq l' = 1, \ldots, i$. In both cases, $\sum_{l=1}^{i} s_{lk} = s_{l'k}$ for some $l' \in \{1, \ldots, i\}$, and by (14.26) taking $l = l'$ constraints (14.25) are satisfied. □

Note that constraints (14.25) are a strengthening of constraints (14.17), and so the former can replace the latter in the formulation of (OMP^3).

In the following, we define an (OMP^3)-*feasible point* induced by a set of facilities $X \subseteq \{1, \ldots, M\}$.

Definition 14.1 *Assume that (FSS) holds. We say that (x, y, s, w) is an* (OMP^3)-feasible point induced by $X \subseteq A$ *if x, y, s and w satisfy the following:*

- *Let $x_j := 1$ if and only if $j \in X$, for each $j = 1, \ldots, M$ and 0 otherwise.*
- *For each $k = 1, \ldots, M$, let $j(k) \in \arg\min_{j \in X} c_{kj}$, and set $y_{kj} := 1$ if and only if $j = j(k)$, for each $j = 1 \ldots, M$ and 0 otherwise.*
- *Let $s_{ik} := 1$ if and only if $c_{(i)}(X) = c_k(X)$ $(\sigma(i) = k)$, for all $i, k = 1, \ldots, M$ and 0 otherwise.*
- *Set $w_i := c_{(i)}(X)$ for all $i = 1, \ldots, M$.*

Of course, to apply the definition, we require $|X| = N$. In this case the name is justified, and (x, y, s, w) is indeed a feasible point of (OMP^3).

The following example illustrates the construction of an (OMP^3)-feasible point induced by X.

Example 14.2
Let $A = \{1,\dots,5\}$ be the set of sites and assume that we are interested in building $N = 2$ new facilities. We use the same cost matrix as in Example 13.1:

$$C = \begin{pmatrix} 0 & 4 & 5 & 3 & 3 \\ 5 & 0 & 6 & 2 & 2 \\ 7 & 2 & 0 & 5 & 6 \\ 7 & 4 & 3 & 0 & 5 \\ 1 & 3 & 2 & 4 & 0 \end{pmatrix}.$$

Set $\lambda = (0,0,1,1,0)$, which leads to the so-called $(2,1)$-trimmed mean problem. The optimal solution is $X = \{1,4\}$ and the associated cost vector is $c(X) = (0,2,5,0,1)$. Therefore, by Definition 14.1, X induces a feasible point to (OMP^3) as follows:

- Set $x = (1,0,0,1,0)$.
- To determine y we have

$$\begin{aligned}
k = 1 : j(1) &= \arg\min_{j=1,4} c_{1j} = 1 \Rightarrow y_{11} = 1 \text{ and } y_{1j} = 0 \text{ for all } j \neq 1;\\
k = 2 : j(2) &= \arg\min_{j=1,4} c_{2j} = 4 \Rightarrow y_{24} = 1 \text{ and } y_{2j} = 0 \text{ for all } j \neq 4;\\
k = 3 : j(3) &= \arg\min_{j=1,4} c_{3j} = 4 \Rightarrow y_{34} = 1 \text{ and } y_{3j} = 0 \text{ for all } j \neq 4;\\
k = 4 : j(4) &= \arg\min_{j=1,4} c_{4j} = 4 \Rightarrow y_{44} = 1 \text{ and } y_{4j} = 0 \text{ for all } j \neq 4;\\
k = 5 : j(5) &= \arg\min_{j=1,4} c_{5j} = 1 \Rightarrow y_{51} = 1 \text{ and } y_{5j} = 0 \text{ for all } j \neq 1\,.
\end{aligned}$$

- Notice that $c_{\leq}(X) = (0,0,1,2,5)$ and a valid permutation is (1,4,5,2,3).
- Therefore, s is defined by $s_{11} = s_{24} = s_{35} = s_{42} = s_{53} = 1$, and the remaining values are set to zero.
- Finally, we set $w = (0,0,1,2,5)$.

Another family of inequalities for (OMP^3) is described in our next results.

Lemma 14.5 *If (FSS) holds, then there exists an optimal solution (x,y,s,w) to (OMP^3) which satisfies*

$$w_1 = w_2 = \cdots = w_N = 0, \tag{14.27}$$

$$\sum_{i=1}^{N} s_{ij} = x_j, \qquad \forall\, j = 1,\dots,M, \tag{14.28}$$

and

$$w_{N+1} \geq \underline{c} x_{j^*} + \underline{d}(1 - x_{j^*}), \tag{14.29}$$

where $\underline{c} = \min_{\substack{k,j=1,\dots,M,\\ k\neq j}} c_{kj} = c_{k^*j^*}$, *and* $\underline{d} = \min_{\substack{k,j=1,\dots,M,\\ k\neq j,\ j\neq j^*}} c_{kj}$.

Proof.
Let X be an optimal solution to the OMP and let (x, y, s, w) be an (OMP^3)-feasible point induced by X. By construction, we have

$$\sum_{i=1}^{M} \lambda_i w_i = \sum_{i=1}^{M} \lambda_i c_{(i)}(X) = f_\lambda(X) .$$

Since X is optimal for the OMP (x, y, s, w) is optimal to (OMP^3) as well (see Theorem 14.3).

From (14.19), and since $w_i = c_{(i)}(X)$ for all $i = 1, \ldots, M$, we have that $w_1 = w_2 = \cdots = w_N = 0$, and therefore, constraint (14.27) holds.

Furthermore, by (14.18), and the definition of s, we have, for each $j = 1, \ldots, M$, that if $x_j = 1$ then $s_{ij} = 1$ for a unique $i \in \{1, \ldots, N\}$ $(j \in X)$, whilst if $x_j = 0$, then $j \notin X$, and therefore $s_{ij} = 0$ for all $i = 1, \ldots, N$. Hence, it follows that constraints (14.28) are valid.

Finally, by (14.18), and since $|X| = N$, the client whose cost is ranked $N + 1$ is not an open facility $((N + 1) \notin X)$. Now $c_{(N+1)}(X) = c_{(N+1),j}$ for some $j \in X$ and therefore, $(N + 1) \neq j$. Thus, $c_{(N+1)}(X) \geq \underline{c}$, with $\underline{c}$ being the minimal off-diagonal entry of the cost matrix C. In addition, $w_{N+1} = c_{(N+1)}(X)$, by definition of w, so we have $w_{N+1} \geq \underline{c}$. Thus, constraint (14.29) holds in the case $x_{j^*} = 1$. Otherwise, if $x_{j^*} = 0$, we have $j^* \notin X$, and $c_{(N+1)}(X) = c_{(N+1),j}$ for some $j \neq j^*$, where of course $j \neq (N + 1)$ as well. Thus, $w_{N+1} = c_{(N+1)}(X) \geq \underline{d}$, as required. Summarizing, we obtain that constraint (14.29) is satisfied. □

Finally, the following theorem unifies the results presented throughout this section and moreover, shows the existence of an optimal solution even if the integrality of s is relaxed.

Let $(\mathsf{OMP}^3_{\mathsf{LR(s)}})$ be the problem that results from (OMP^3) after relaxing the integrality constraints for s_{ik}, $i = 1, \ldots, N$, $k = 1, \ldots, M$.

Theorem 14.6 *Let (x, y, s, w) be an optimal solution of $(\mathsf{OMP}^3_{\mathsf{LR(s)}})$ reinforced with the additional constraints* (14.27), (14.28) *and* (14.29). *Then x induces an optimal solution to the OMP.*

Proof.
Let (x, y, s, w) be an optimal solution to $(\mathsf{OMP}^3_{\mathsf{LR(s)}})$, satisfying (14.27), (14.28) and (14.29). Let X be the set of facilities induced by x. Also let $\hat{y}$, $\hat{s}$ and $\hat{w}$ be such that $(x, \hat{y}, \hat{s}, \hat{w})$ is an (OMP^3)-feasible point induced by X.

Now for $i = N + 1, \ldots, M$, if $s_{ik} = 1$ it follows that $k \notin X$, because otherwise if $k \in X$ then $x_k = 1$, then by (14.28), $\sum_{i=1}^{N} s_{ik} = 1$, and hence, $\sum_{i=N+1}^{M} s_{ik} = 1 - \sum_{i=1}^{N} s_{ik} = 0$, by constraints (13.5).

Furthermore, if $k \notin X$, then $x_k = 0$, so by (14.28): $\sum_{i=1}^{N} s_{ik} = 0$; and hence, from constraints (13.5), $\sum_{i=N+1}^{M} s_{ik} = 1 - \sum_{i=1}^{N} s_{ik} = 1$, i.e. there exists some $i \in \{N+1, \dots, M\}$ such that $s_{ik} = 1$.

Therefore, for any client $k \notin X$, $s_{ik} = 1$ if and only if $k \geq N+1$. Moreover, by (14.18), $c_{(i)}(X) = c_k(X)$ for all $i = 1, \dots, N$ and $k \in X$.

Thus, setting

$$s'_{ik} = \begin{cases} s_{ik}, \ i \geq N+1 \\ 1, \quad i \leq N \text{ and } c_{(i)}(X) = c_k(X) \\ 0, \quad \text{otherwise} \end{cases}$$

for each $i, k = 1, \dots, M$ constraints (13.5) and (13.6) hold, and moreover, $s' \in \{0, 1\}$. Let σ be the permutation of $\{1, \dots, M\}$ induced by s'. Then for $i \in \{1, \dots, N\}$, $\sigma(i) \in X$, and for $i \in \{N+1, \dots, M\}$, $\sigma(i) = k$ if and only if $s_{ik} = 1$, in which case $k \notin X$. Now for each $i = N+1, \dots, M$, we have

$$\begin{aligned} w_i &\geq \sum_{j=1}^{M} c_{\sigma(i)j}\, y_{\sigma(i)j} && \text{(from (14.17) and } s_{i\sigma(i)} = 1) \\ &= \sum_{j \in X} c_{\sigma(i)j}\, y_{\sigma(i)j} && (y_{kj} \leq x_j \text{ for all } j,\, k, \text{ and } x_j = 0 \text{ if } j \notin X) \\ &\geq c_{\sigma(i)}(X) \sum_{j \in X} y_{\sigma(i)j} && (\text{since } c_{\sigma(i)}(X) \leq c_{\sigma(i)j} \text{ for all } j \in X) \\ &= c_{\sigma(i)}(X) && (\text{from (13.12), and } y_{\sigma(i)j} = 0 \text{ for all } j \notin X) \\ &= \hat{w}_{\mathsf{rk}(\sigma(i))} && (\text{by the definition of } \hat{w}.) \end{aligned}$$

Also, for each $i = 1, \dots, N$, we have $w_i = 0$, from constraint (14.27). In addition, since (FSS) holds and $\sigma(i) \in X$, $i = 1, \dots, N$ we obtain $c_{\sigma(i)}(X) = 0$. Thus

$$w_i = 0 = c_{\sigma(i)}(X) = \hat{w}_{\mathsf{rk}(\sigma(i))},$$

again by the definition of $\hat{w}$. Hence,

$$w_i \geq \hat{w}_{\mathsf{rk}(\sigma(i))}, \quad \forall\, i = 1, \dots, M.$$

Observe that w and $\hat{w}$ satisfy the hypothesis of Lemma 1.2. We deduce that

$$\hat{w}_i \leq w_i, \quad \forall\, i = 1, \dots, M.$$

Therefore, $(x, \hat{y}, \hat{s}, \hat{w})$ is feasible for (OMP^3), and its objective value is

$$f_\lambda(x, \hat{y}, \hat{s}, \hat{w})) = \sum_{i=1}^{M} \lambda_i \hat{w}_i \leq \sum_{i=1}^{M} \lambda_i w_i = f_\lambda(x, y, s, w)\,. \tag{14.30}$$

To conclude, notice that (OMP^3) and (OMP^3) reinforced with the inequalities (14.27), (14.28) and (14.29) have the same optimal value (by Lemma 14.5).

Since (x, y, s, w) is an optimal solution to $(\mathsf{OMP}^3_{\mathsf{LR(s)}})$, a relaxation of the latter problem, we have

$$f_\lambda(x, y, s, w) \leq \mathsf{val}((\mathsf{OMP}^3)) .$$

In addition, we have proven that $(x, \hat{y}, \hat{s}, \hat{w})$ is feasible for (OMP^3), which yields

$$\mathsf{val}((\mathsf{OMP}^3)) \leq f_\lambda(x, \hat{y}, \hat{s}, \hat{w}) .$$

Thus, by (14.30), $(x, \hat{y}, \hat{s}, \hat{w})$ induces an optimal solution to OMP. □

From the results presented above, we conclude that if (FSS) holds we can include constraints (14.23), (14.27), (14.28) and (14.29) to (OMP^3), and replace constraints (14.17) by (14.25). Moreover, we can relax the binary variables $s_{ik} \in \{0, 1\}$, to $0 \leq s_{ik} \leq 1$, for all $i = 1, \ldots, N$ and all $k = 1, \ldots, M$. We denote this formulation by (OMP^{3^*}). Some computational experiments will be presented in Section 14.3 to compare (OMP^3) and (OMP^{3^*}).

All these results carry over for (OMP^2) as the following section will show.

14.2.3 Improvements for (OMP^2)

In this section, we will mainly show how properties shown in the last section for (OMP^3) can be transformed to be also valid for (OMP^2). To this aim, the result given by Theorem 14.4 will be essential throughout this section.

For the sake of readability the results corresponding to (OMP^2) are presented in a similar way as those for (OMP^3) in Section 14.2.2.

First, we derive new inequalities for (OMP^2) using a given upper bound z_{UB} of the optimal value of OMP.

Lemma 14.6 *If (x, y, s, w) is an optimal solution to (OMP^2) then*

$$\sum_{j \,:\, c_{kj} \leq \frac{z_{UB}}{\sum_{i=m}^{M} \lambda_i}} x_j \geq \sum_{i=1}^{m} s_{ik}$$

holds for all $k, m = 1, \ldots, M$.

Proof.
Let (x, y, s, w) be an optimal solution to (OMP^2). Applying Theorem 14.4, we conclude that (x, y, s, w') is a feasible solution to (OMP^3) with $w'_i = \sum_{k=1}^{M} w_{ik}$ for all $i = 1, \ldots, M$. Furthermore, by Theorems 14.2 and 14.3, we have that (x, y, s, w') is also optimal to (OMP^3).

Thus, following the proof of Lemma 14.3 we obtain the result. □

The following lemma shows how constraints (14.13) can be strengthened.

Lemma 14.7 *Any optimal solution (x,y,s,w) for (OMP^2) satisfies*

$$\sum_{l=1}^{M} w_{il} \geq \sum_{j=1}^{M} c_{kj}\, y_{kj} - \overline{c}_k(1 - \sum_{l=1}^{i} s_{lk}), \quad \forall\, i,k = 1,\ldots,M. \tag{14.31}$$

where $\overline{c}_k$ is the N-th largest component of the cost matrix at row k.

Proof.
Let (x,y,s,w) be an optimal solution to (OMP^2). Hence, as above we can conclude that (x,y,s,w') is also optimal to (OMP^3) with $w'_i = \sum_{l=1}^{M} w_{il}$ for all $i = 1,\ldots,M$. Thus, by applying Lemma 14.4 and Corollary 14.3 we obtain

$$\sum_{l=1}^{M} w_{il} = w'_i \geq \sum_{j=1}^{M} c_{kj}\, y_{kj} - \overline{c}_k \left(1 - \sum_{l=1}^{i} s_{lk} \right), \quad \forall\, i,k = 1,\ldots,M,$$

as required. □

Analogously to Definition 14.1 we define an (OMP^2)-*feasible point* induced by a set of facilities $X \subseteq \{1,\ldots,M\}$, $|X| = N$ as follows.

Definition 14.2 *Assume that (FSS) holds. Then, (x,y,s,w) is an* (OMP^2)-feasible point induced by *$X \subseteq A$ if x, y, s and w satisfy the following:*

- *Let $x_j := 1$ if and only if $j \in X$, for each $j = 1,\ldots,M$ and 0 otherwise.*
- *For each $k = 1,\ldots,M$, let $j(k) \in \arg\min_{j\in X} c_{kj}$, and set $y_{kj} := 1$ if and only if $j = j(k)$, for each $j = 1\ldots,M$ and 0 otherwise.*
- *Let $s_{ik} := 1$ if and only if $c_{(i)}(X) = c_k(X)$ $(\sigma(i) = k)$, for all $i,k = 1,\ldots,M$ and 0 otherwise.*
- *Set $w_{ik} := c_{(i)}(X)$ if $c_{(i)}(X) = c_k(X)$ and $w_{ik} = 0$ otherwise for all $i,k = 1,\ldots,M$.*

Again, to apply the definition, we require $|X| = N$. In this case the name is justified, and (x,y,s,w) is indeed a feasible point of (OMP^2).

The following example illustrates the construction of an (OMP^2)-feasible point induced by X.

Example 14.3
Consider the data from Example 14.2, and assume that we are interested in building $N = 2$ new facilities in order to minimize the $(2,1)$-trimmed mean problem, i.e. $\lambda = (0,0,1,1,0)$. The optimal solution is given by $X = \{1,4\}$ and the associated cost vector is $c(X) = (0,2,5,0,1)$. Therefore, by Definition 14.2, X induces a feasible point to (OMP^2) as follows:

- Set $x = (1,0,0,1,0)$.
- Set $y_{11} = y_{24} = y_{34} = y_{44} = y_{51} = 1$, and the remaining values to zero.

- The sorted cost vector $c_{\leq}(X) = (0, 0, 1, 2, 5)$ and therefore, a valid permutation is $(1, 4, 5, 2, 3)$.
- Therefore, s is defined by $s_{11} = s_{24} = s_{35} = s_{42} = s_{53} = 1$, and the remaining entries are equal to zero.
- Finally, w is defined by $w_{11} = 0$, $w_{24} = 0$, $w_{35} = 1$, $w_{42} = 2$, $w_{53} = 5$, and the remaining entries are set equal to zero.

The following lemma constructs an optimal solution to (OMP^2) which fulfills some additional constraints.

Lemma 14.8 *If (FSS) holds then there exists an optimal solution (x, y, s, w) to (OMP^2) which satisfies*

$$\sum_{k=1}^{M} w_{1k} = \sum_{k=1}^{M} w_{2k} = \cdots = \sum_{k=1}^{M} w_{Nk} = 0, \tag{14.32}$$

$$\sum_{i=1}^{N} s_{ij} = x_j, \qquad \forall\, j = 1, \ldots, M,$$

and

$$\sum_{k=1}^{M} w_{N+1,k} \geq \underline{c} x_{j^*} + \underline{d}(1 - x_{j^*}), \tag{14.33}$$

where $\underline{c} = \min\limits_{\substack{k,j=1,\ldots,M,\\ k \neq j}} c_{kj} = c_{k^* j^*}$, *and* $\underline{d} = \min\limits_{\substack{k,j=1,\ldots,M,\\ k \neq j,\ j \neq j^*}} c_{kj}$.

Proof.
Let X be an optimal solution to the OMP and let (x, y, s, w) be an (OMP^2)-feasible point induced by X. By Theorem 14.4, we obtain that (x, y, s, w') is also a (OMP^3)-feasible point induced by X with $w'_i = \sum_{k=1}^{M} w_{ik}$. Hence, from Lemma 14.5 the result follows. □

Finally, we present a result which summarizes the properties presented above and moreover, shows the existence of an optimal solution to (OMP^2) even after the relaxation of the integrality on s_{ik}.

Let $(\mathsf{OMP}^2_{\mathsf{LR(s)}})$ be the problem that results from (OMP^2) after relaxing the integrality constraints for s_{ik}, $i = 1, \ldots, N$, $k = 1, \ldots, M$.

Theorem 14.7 *Let (x, y, s, w) be an optimal solution of $(\mathsf{OMP}^2_{\mathsf{LR(s)}})$ reinforced with the additional constraints* (14.28), (14.32) *and* (14.33). *Then x induces an optimal solution to the OMP.*

The proof is completely analogous to the one of Theorem 14.6 and is therefore omitted here.

Again, we conclude if (FSS) holds that we can add constraints (14.23), (14.28), (14.32), and (14.33) to (OMP^2), and replace constraints (14.13) by (14.31). Moreover, we can relax the binary variables $s_{ik} \in \{0,1\}$, to $0 \le s_{ik} \le 1$, for all $i = 1,\dots,N$ and all $k = 1,\dots,M$. We denote this formulation as (OMP^{2^*}).

In the following section we will investigate the relationship between the feasible regions of the formulations (OMP^{2^*}) and (OMP^{3^*}).

14.2.4 Comparison Between (OMP^{2^*}) and (OMP^{3^*})

Analogous to Theorem 14.4, we compare now the strengthened linearization (OMP^{2^*}) and (OMP^{3^*}).

Theorem 14.8 *There exists a surjective function, π, which maps each feasible solution (x,y,s,w) of the linear programming relaxation of (OMP^{2^*}), denoted by $(\mathsf{OMP}^{2^*}_{\mathsf{LR}})$, into a feasible solution (x,y,s,w') of $(\mathsf{OMP}^{3^*}_{\mathsf{LR}})$.*

Proof.
First, we have to prove that π is well defined. Let (x,y,s,w) be a feasible solution to $(\mathsf{OMP}^{2^*}_{\mathsf{LR}})$. Then, taking $w'_i = \sum_{k=1}^{M} w_{ik}$ for all $i = 1,\dots,M$, and following the proof of Theorem 14.4, we obtain that (x,y,s,w') is a feasible solution for $(\mathsf{OMP}^{3^*}_{\mathsf{LR}})$.

Secondly, we have to show that π is surjective. Given a feasible solution (x,y,s,w') to $(\mathsf{OMP}^{3^*}_{\mathsf{LR}})$, we have to prove that there exists at least one feasible solution to $(\mathsf{OMP}^{2^*}_{\mathsf{LR}})$ which can be mapped onto (x,y,s,w').

We directly get that constraints (13.5), (13.6), (13.11), (13.12), (13.13), (14.23) and (14.28) are fulfilled.

In addition, from constraint (14.27) we have that $w'_1 = \ldots = w'_N = 0$. Therefore, by taking $w_{ik} = 0$ for all $i = 1,\dots,N$ and $k = 1,\dots,M$, we conclude that constraints (14.32) also hold.

Hence, to obtain constraints (14.11), (14.12), (14.31), and (14.33) we just have to solve the following system of equations

$$\left\{ \begin{array}{lll} \displaystyle\sum_{k=1}^{M} w_{ik} & = w'_i & \forall\, i = N+1,\dots,M \\ \displaystyle\sum_{i=N+1}^{M} w_{ik} & = \displaystyle\sum_{j=1}^{M} c_{kj}\, y_{kj} & \forall\, k = 1,\dots,M \end{array} \right. . \tag{14.34}$$

We prove that this system of equations is solvable and furthermore, has no unique solution in many cases.

To show that (14.34) is solvable, the rank of the matrix of coefficients, M_c, must be equal to the rank of the augmented matrix, M_a. This rank cannot be maximal, since by constraints (14.16) and (14.27) we have

$$\sum_{i=N+1}^{M} w_i' = \sum_{j=1}^{M}\sum_{k=1}^{M} c_{kj}\, y_{kj}\,.$$

We will see that the matrix of coefficients, M_c, has rank equal to $2M-N-1$, i.e. equal to the number of linearly independent equations, and therefore, its rank coincides with that of the augmented matrix, M_a. The matrix of coefficients can be divided into blocks as follows

$$M_c = \begin{pmatrix} B^1 & B^2 & \dots & B^{M-N} \\ I_M & I_M & \dots & I_M \end{pmatrix}, \tag{14.35}$$

where for each $i = 1, \dots, M-N$, B^i is the $(M-N) \times M$ matrix with entries at ith row equal to one and the rest equal to zero, and I_M is the $M \times M$ identity matrix.

One minor of M_c that has rank equal to $2M - N - 1$ can be written as follows

$$\begin{pmatrix} 0_{(M-N-1)\times M} & I_{M-N-1} \\ I_M & B_c^1 \end{pmatrix}, \tag{14.36}$$

where B_c^1 is the $M \times (M-N-1)$ matrix with the first row equal to one and the remaining entries equal to zero.

In addition, (14.34) has no unique solution when the number of unknowns w_{ik} $(M(M-N))$ is larger than the number of linearly independent equations $(M + (M-N) - 1 = 2M - N - 1)$.

Hence, the system of equations presented in (14.34) is solvable and it provides a feasible solution (x, y, s, w) to $(\mathsf{OMP}^{2*}_{\mathsf{LR}})$. Therefore, the function π is surjective. □

With this result we can conclude that the feasible region described by $(\mathsf{OMP}^{3*}_{\mathsf{LR}})$ is a projection of the feasible region described by $(\mathsf{OMP}^{2*}_{\mathsf{LR}})$. Furthermore, both linearizations provide the same integrality gap solving the OMP, i.e. the same gap between the optimal objective function value and the relaxed LP solution. Therefore, the computing time required to solve the OMP directly depends on the number of variables and constraints of the linearization. Since (OMP^{3*}) contains less number of variables and constraints from now on we will only take into account this linearization.

In the next section we show computational comparisons between (OMP^{1*}) and (OMP^{3*}). For the sake of readability we will simply denote the formulations by (OMP^{1}) and (OMP^{3}). A more detailed analysis can be found in [55].

14.3 Computational Results

The different linearizations (OMP^{1}) and (OMP^{3}) are both valid for solving OMP. The results in the previous section show some relationships between them. Nevertheless, there is no simple way to asses the two formulations so that one can conclude the superiority of one formulation over the other. Despite this difficulty, it is important to have some ideas about the performance of these two approaches. A way to get such a comparison is by means of statistical analysis. In this regard, we consider the experimental design proposed in [25, 55]. It consists of the following factors and levels:

- Size of the problem: The number of sites, M, determines the dimensions of the cost matrix (C) and the λ-vector. Moreover, it is an upper bound of the number of facilities (N) to be located. Therefore, M is considered as a factor in this design which has four levels: $M = 8, 10, 12, 15$.
- New facilities: N is the second factor with four levels. To obtain comparable levels, they are taken as proportions of the M values: $\lceil \frac{M}{4} \rceil$, $\lceil \frac{M}{3} \rceil$, $\lceil \frac{M}{2} \rceil$, and $\lceil \frac{M}{2} + 1 \rceil$. Therefore, for $M = 8$ we consider $N = 2, 3, 4, 5$, for $M = 10$, $N = 3, 4, 5, 6$, for $M = 12$, $N = 3, 4, 6, 7$ and finally, for $M = 15$, $N = 4, 5, 8, 9$.
- Type of problem: Each λ-vector is associated with a different objective function. Also λ is considered as a factor. Its levels are designed depending on the value of M as follows:
  1. T1: λ-vector corresponding to the N-median problem, i.e. $\lambda = (\underbrace{1, \dots, 1}_{M})$.
  2. T2: λ-vector corresponding to the N-center problem, i.e. $\lambda = (\underbrace{0, \dots, 0}_{M-1}, 1)$.
  3. T3: λ-vector corresponding to the k-centra problem, i.e. $\lambda = (0, \dots, 0, \underbrace{1, \dots, 1}_{k})$, where $k = \lfloor \frac{M}{3} \rfloor$.
  4. T4: λ-vector corresponding to the (k_1, k_2)-trimmed mean problem, i.e. $\lambda = (\underbrace{0, \dots, 0}_{k_1}, 1, \dots, 1, \underbrace{0, \dots, 0}_{k_2})$, where $k_1 = N + \lceil \frac{M}{10} \rceil$ and $k_2 = \lceil \frac{M}{10} \rceil$.
  5. T5: λ-vector with binary entries alternating both values and finishing with an entry equal to 1, i.e. $\lambda = (0, 1, 0, 1, \dots, 0, 1, 0, 1)$ if M is even and $\lambda = (1, 0, 1, \dots, 0, 1, 0, 1)$ if M is odd.
  6. T6: As T5, but finishing with an entry equal to 0, i.e. $\lambda = (1, 0, 1, 0, \dots, 1, 0, 1, 0)$ if M is even and $\lambda = (0, 1, 0, \dots, 1, 0, 1, 0)$ if M is odd.
  7. T7: λ-vector generated by the repetition of the sequence $0, 1, 1$ from the end to the beginning, i.e. $\lambda = (\dots, 0, 1, 1, 0, 1, 1)$.
  8. T8: λ-vector generated by the repetition of the sequence $0, 0, 1$ from the end to the beginning, i.e. $\lambda = (\dots, 0, 0, 1, 0, 0, 1)$.
- Linearization: This factor has two levels corresponding to the two linearizations we would like to test: (OMP^1) and (OMP^3).

The above description corresponds to a linear model with four factors (M with four levels, N with four levels, λ with eight levels and two linearizations) yielding 256 different combinations. For each combination 15 replications are evaluated, thus this design leads to 3840 problems to be solved. Two dependent variables in the model are considered: the integrality gap and the computing time. All the factors were tested to be simultaneously meaningful performing a multivariate analysis of the variance applied to the model. Additionally, principal component analysis was also applied in order to check whether some factors may be removed from consideration. The conclusion was that none of them can be neglected.

The computational results for solving the OMP using either (OMP^1) or (OMP^3) are given in Tables 14.1–14.9. The OMP was solved with the commercial package ILOG CPLEX 6.6 using the C++ modeling library ILOG Planner 3.3 (see [109]). These computational results were obtained using a Pentium III 800 Mhz with 1 GB RAM. The upper bounds used were derived by a heuristic method based on variable neighborhood search (see [56]).

We first comment briefly on the effect of the improvements presented in Section 14.2 on the linear formulations. Both linearizations (OMP^1) and (OMP^3) are greatly affected. Some small examples ($M = 8$) were tested to compare the performance of each linearization, with and without the strengthening provided by results in Section 14.2. The results are reported in Table 14.1. For all types of problems (except type T1, for which the bound was unaffected) the integrality gap provided by the improved linearizations was significantly smaller than that given by the original formulation. The effect was more pronounced for (OMP^3), which had weaker bounds to begin with. The gap for this formulation was reduced by a factor of up to more than twenty (which was achieved on type T2 problems), with the average reduction being a factor of around 4.6. For (OMP^1) the greatest reduction was observed on type T4 problems, with the gap given by the improved formulation just over one eighth of the gap reported by the original formulation, i.e. an improvement by a factor of almost eight. The average gap reduction for (OMP^1) across all problem types was by a factor of about 2.6. Improvements in computing time were even more dramatic. Even for problems of type T1, for which no improvement in root node bound was reported, the computing time was decreased by an order of magnitude for (OMP^1) and cut in about four for (OMP^3). Indeed, for every problem type, the computing time for (OMP^1) was reduced by at least one order of magnitude, with a reduction of two orders of magnitude for two of the eight problem types. For (OMP^3), the computing time on all types of problems except T1 was reduced by at least one order of magnitude, with two problem types showing reductions of two orders of magnitude, and three types having computing time around three orders of magnitude less. Thus, we see that the strengthening provided by results in Section 14.2 are highly effective in improving both formulations.

Table 14.1. Gaps and computing times for the (OMP^1) and (OMP^3) formulations, with and without the improvements provided by Section 14.2, on problems with $M = 8$. Each row of the table represents an average taken over problems with $N = 2, 3, 4$ and 5, where for each value of N, fifteen instances of the stated problem type were solved and the results averaged.

Formulation	Problem Type	Gap (%) original	Gap (%) improved	Ratio original/ improved	Computing Time (s) original	Computing Time (s) improved	Ratio original/ improved
(OMP^1)	T1	0.71	0.71	1.00	0.63	0.07	9.00
	T2	68.44	39.27	1.74	24.31	0.10	243.10
	T3	63.05	13.88	4.54	15.93	0.10	159.30
	T4	58.40	7.46	7.83	3.19	0.07	45.57
	T5	20.90	15.92	1.31	1.69	0.12	14.08
	T6	21.55	13.82	1.56	0.79	0.09	8.78
	T7	18.36	9.22	1.99	1.10	0.13	8.46
	T8	32.72	26.75	1.22	1.97	0.14	14.07
(OMP^3)	T1	0.71	0.71	1.00	0.40	0.11	3.64
	T2	68.44	3.22	21.25	169.88	0.05	3397.60
	T3	63.05	17.89	3.52	125.04	0.08	1563.00
	T4	100	38.81	2.58	158.96	0.19	836.63
	T5	20.90	12.69	1.65	6.71	0.15	44.73
	T6	100	30.71	3.26	73.52	0.17	432.47
	T7	18.36	11.65	1.58	2.98	0.15	19.87
	T8	32.72	15.79	2.07	14.46	0.12	120.50

The next step is to compare the two linearizations. To this end, the number of nodes in the branch and bound tree, the gap between the optimal solution and the linear relaxation objective function value, and the computing time in seconds are shown. Each row represents the result of 15 replications of each combination (M, N). For this reason, the average, minimum and maximum of the number of nodes, integrality gap and computing time are reported. First, there are detailed results of the experiments with problems having $M = 15$, and then describe the results of the analysis for all 3840 test problems.

In Tables 14.2-14.9 we summarize the results of the experiments on problems with $M = 15$. We denote as L1 and L2 the results corresponding to the linearization (OMP^1) and (OMP^3), respectively.

Finally, the statistical analysis for all 3840 test problems is as follows. The comparison between (OMP^1) and (OMP^3) is made in two steps. First, the results within each type of problem are compared to test whether the integrality gap and/or the computing time show a statistically similar behavior in (OMP^1) and (OMP^3). To perform this test, we use the Mann-Whitney U

Table 14.2. Computational results corresponding to the T1 type problem.

Example			# nodes			gap(%)			CPU(s)		
Lin	M	N	aver	min	max	aver	min	max	aver	min	max
L1	15	4	697.53	166	1577	2.39	0.00	18.80	8.40	2.42	18.94
		5	311.87	6	894	1.19	0.00	10.81	2.99	0.36	7.19
		8	12.53	0	121	0.00	0.00	0.00	0.31	0.19	1.06
		9	3.80	0	19	0.14	0.00	2.08	0.22	0.16	0.31
L2	15	4	2210.33	164	9350	2.39	0.00	18.80	25.41	5.03	114.16
		5	727.40	92	2570	1.19	0.00	10.81	8.79	2.83	21.41
		8	33.07	2	103	0.00	0.00	0.00	0.95	0.28	2.17
		9	19.93	0	63	0.14	0.00	2.08	0.63	0.17	1.61

statistic. With respect to the gap, all types of problems but $T1$ are different under (OMP^1) and (OMP^3) with a confidence level above 90%. For the computing time, all types of problems are different with a confidence level above 88%. Secondly, it is checked which linearization performs better in each problem using robust statistics (trimmed mean) and confidence intervals for the mean value. The conclusions, calculated using all four levels of M, are the following.

From Table 14.10 we observe that (OMP^1) provides stronger lower bounds for (OMP), for problem types T3, T4, T6, T7 and (OMP^3) bounds are stronger for T2, T5, T8.

The behavior of the computing time is substantially different. From Table 14.11 we can observe that the computing time needed for solving the problem using (OMP^1) is significantly shorter than using (OMP^3). Only for T2 does (OMP^3) solve the problem faster.

Table 14.3. Computational results corresponding to the T2 type problem.

Example			# nodes			gap(%)			CPU(s)		
Lin	M	N	aver	min	max	aver	min	max	aver	min	max
L1	15	4	5107.20	548	12181	50.55	31.60	66.16	58.85	4.98	117.09
		5	2242.07	114	8196	53.26	28.00	73.75	17.81	2.00	74.00
		8	42.73	10	129	40.50	26.19	57.14	0.41	0.28	0.70
		9	31.53	0	150	39.65	25.00	60.00	0.38	0.22	0.88
L2	15	4	1.20	0	7	16.21	0.00	40.72	1.03	0.61	1.59
		5	2.67	0	15	16.61	0.00	59.92	0.84	0.23	1.69
		8	0.00	0	0	0.00	0.00	0.00	0.26	0.08	0.41
		9	0.40	0	2	12.06	0.00	45.45	0.29	0.09	0.61

Table 14.4. Computational results corresponding to the T3 type problem.

	Example		# nodes			gap(%)			CPU(s)		
Lin	M	N	aver	min	max	aver	min	max	aver	min	max
L1	15	4	2026.87	405	7079	32.25	16.86	50.00	22.54	3.72	96.09
		5	976.93	138	4620	25.44	16.37	45.90	7.26	1.08	27.08
		8	31.13	0	129	12.63	0.00	22.08	0.41	0.19	0.94
		9	11.87	0	47	6.21	0.00	13.04	0.28	0.17	0.47
L2	15	4	544.40	91	1091	33.79	26.95	45.26	9.41	4.08	14.25
		5	467.27	69	1570	28.90	20.37	43.48	6.51	2.39	14.50
		8	100.60	11	358	13.55	4.76	19.19	1.55	0.53	3.86
		9	64.73	5	188	8.11	4.44	13.62	0.98	0.33	1.95

Table 14.5. Computational results corresponding to the T4 type problem.

	Example		# nodes			gap(%)			CPU(s)		
Lin	M	N	aver	min	max	aver	min	max	aver	min	max
L1	15	4	369.00	28	1054	12.47	1.08	17.29	4.37	1.13	10.92
		5	108.73	12	385	12.77	5.10	19.20	1.70	0.53	5.17
		8	1.67	0	7	10.32	0.00	26.67	0.32	0.20	0.44
		9	1.60	0	5	10.83	0.00	33.33	0.30	0.19	0.39
L2	15	4	41956.87	271	178289	82.65	70.83	91.25	578.05	7.36	2168.08
		5	4386.73	151	34872	76.62	53.85	87.76	63.86	6.22	472.94
		8	34.47	2	209	50.25	25.00	76.92	1.55	0.28	3.72
		9	8.53	0	26	34.51	0.00	75.00	0.62	0.19	1.89

The improvement achieved by the reformulations is tremendous. However, the computational experiments show clearly the limitations of using a standard solver for solving the OMP. Therefore, we will develop in the next chapter specifically tailored solution routines for the OMP.

Table 14.6. Computational results corresponding to the T5 type problem.

	Example		# nodes			gap(%)			CPU(s)		
Lin	M	N	aver	min	max	aver	min	max	aver	min	max
L1	15	4	2357.20	593	5758	13.26	7.46	26.15	17.35	3.66	45.47
		5	1459.87	230	6206	12.15	4.41	26.67	9.04	1.59	37.64
		8	36.67	0	125	8.33	0.00	18.18	0.37	0.22	0.94
		9	37.73	0	147	12.32	0.00	29.17	0.36	0.17	0.70
L2	15	4	15986.47	166	46286	13.39	7.46	26.15	108.06	4.78	358.00
		5	4810.87	85	28586	12.15	4.41	26.67	28.99	2.64	150.23
		8	80.13	4	357	8.33	0.00	18.18	1.42	0.42	3.78
		9	108.60	3	435	12.19	0.00	27.19	1.29	0.33	3.16

Table 14.7. Computational results corresponding to the T6 type problem.

	Example		# nodes			gap(%)			CPU(s)		
Lin	M	N	aver	min	max	aver	min	max	aver	min	max
L1	15	4	1097.73	51	3414	11.96	4.41	23.61	10.63	1.45	28.84
		5	413.80	68	1727	12.03	4.29	20.97	3.62	0.98	10.20
		8	32.73	0	160	12.32	0.00	29.17	0.58	0.25	1.27
		9	5.67	0	13	8.06	0.00	21.43	0.26	0.14	0.41
L2	15	4	29733.13	894	93126	82.36	75.00	87.95	266.90	14.45	719.80
		5	19748.87	612	115575	78.00	64.29	85.55	201.05	7.20	1304.75
		8	261.80	37	785	62.63	25.00	81.25	4.62	1.52	12.42
		9	35.47	0	149	44.73	0.00	70.00	1.37	0.08	3.02

Table 14.8. Computational results corresponding to the T7 type problem.

	Example		# nodes			gap(%)			CPU(s)		
Lin	M	N	aver	min	max	aver	min	max	aver	min	max
L1	15	4	1564.67	320	3960	11.96	7.63	23.61	12.87	2.45	30.77
		5	835.73	142	1744	12.15	5.21	20.83	5.04	1.22	9.66
		8	36.20	0	108	9.95	0.00	15.48	0.37	0.16	0.72
		9	22.13	0	93	9.41	0.00	21.48	0.31	0.19	0.56
L2	15	4	9981.33	225	86922	12.01	8.15	23.81	54.38	4.75	413.48
		5	1692.80	205	9298	11.85	5.21	20.83	12.12	3.84	56.42
		8	106.80	3	399	9.30	0.00	14.96	1.52	0.30	3.20
		9	55.87	0	172	9.33	0.00	21.67	0.90	0.23	2.17

Table 14.9. Computational results corresponding to the T8 type problem.

	Example		# nodes			gap(%)			CPU(s)		
Lin	M	N	aver	min	max	aver	min	max	aver	min	max
L1	15	4	5000.20	493	21037	21.81	13.77	32.14	46.95	2.52	162.47
		5	1400.93	233	3861	21.13	7.84	34.34	7.75	1.22	30.31
		8	66.33	0	263	16.88	0.00	29.17	0.50	0.20	1.22
		9	43.40	0	142	19.13	0.00	39.74	0.37	0.22	0.70
L2	15	4	18360.47	235	134087	21.85	14.49	31.91	104.12	4.31	654.41
		5	2366.33	131	10157	20.66	7.84	31.31	18.40	3.20	68.56
		8	59.87	6	264	15.54	0.00	22.22	1.39	0.42	3.33
		9	43.53	0	301	17.50	0.00	39.74	0.76	0.13	2.64

Table 14.10. p-values and trimmed means obtained for the integrality gap.

		T1	T2	T3	T4	T5	T6	T7	T8
L1 vs L2	p-value	1.000	0.000	0.001	0.000	0.097	0.000	0.091	0.000
L1	trimmed mean	0.443	43.922	16.736	9.987	13.875	12.743	10.484	23.558
L2	trimmed mean	0.443	1.892	20.700	49.266	12.511	52.763	11.413	18.639

Table 14.11. p-values and trimmed means obtained for the computing time.

		T1	T2	T3	T4	T5	T6	T7	T8
L1 vs L2	p-value	0.000	0.000	0.011	0.000	0.000	0.000	0.000	0.115
L1	trimmed mean	0.519	5.481	0.951	0.348	1.020	0.688	0.977	1.461
L2	trimmed mean	1.147	0.223	1.020	6.173	2.453	6.799	1.602	2.092

[illegible]

for each $i = 1, \ldots, M$ [illegible] $= 1$, and hence constraints (14.11) hold.

In addition, we obtain

[illegible]

and therefore, constraints (14.12) are also fulfilled.

Finally, we can prove the validity of constraints (14.16) using the binary character of variable s_{ij}. Thus, we distinguish two cases. On the one hand, if $s'_{ij} = 0$ then $w_{ij} = 0$. From constraints (13.13) and (13.14) [illegible] and therefore, since $c_{ik} \geq 0$,

[illegible]

On the other hand, if $s'_{ij} = 1$ then [illegible]. Hence, in both cases, constraints (14.16) hold.

Summarizing, we obtain that [illegible] is feasible to (OMP[2]). Moreover, by construction of [illegible], the objective function of (OMP) for [illegible] provides the same value as the objective function of (OMP[2]) for [illegible].

Now, let [illegible] be a feasible solution to (OMP[2]). Then, constraints (13.5), (13.7), (13.11), (13.12) and (13.13) are also fulfilled for (OMP).

To show that the remaining constraints hold, we just need to prove the validity of the equivalence [illegible].

Variable s is binary from constraints (13.5), so we consider two cases. On the one hand, if [illegible] $= 0$ we get by constraints (14.9)

[illegible]

In addition, by the same arguments [illegible], we obtain that

[illegible]

15

Solution Methods

Having seen the limitations of standard solver based solution approaches, we will present in this chapter specifically designed algorithms in order to be able to solve larger OMP instances. Since we deal with a NP-hard problem, it should be clear that also specialized exact solution algorithms will only be able to solve medium sized problems. Therefore, we will look at heuristic approaches to tackle large problems. The price we have to pay for using a heuristic is, that we loose the possibility to measure the deviation from the optimal solution. But, with computational experiments, we are able to make empirical comparisons showing the weaknesses and strengths of the investigated heuristics. For the exact method we decided to utilize the most commonly used method in integer programming: Branch and Bound (see [147]). Other exact methods might be successfully applied as well, yielding a better understanding of the structure of the OMP. For the heuristics, some of the most promising modern developments have been used (see [82]): Evolution Program and Variable Neighborhood Search.

15.1 A Branch and Bound Method

The Branch and Bound method (B&B) is the most widely used method for solving integer and mixed-integer problems. Basically, this method solves a model by breaking up its feasible region into successively smaller subsets (branching), calculating bounds on the objective value over each corresponding subproblem (bounding), and using them to discard some of the subproblems from further consideration (fathoming). The bounds are obtained by replacing the current subproblem by an easier model (relaxation), such that the solution of the latter yields a bound for the former one. The procedure ends when each subproblem has either produced an infeasible solution, or has been shown to contain no better solution than the one already at hand. The best solution found during the procedure is an optimal solution. In the worst case, the Branch and Bound method becomes a complete enumeration procedure. The amount of subproblems which can be eliminated depends heavily on the quality of the bounds. For more details on the Branch and Bound method the reader is referred to [147]. We will now explain how the building blocks of the Branch and Bound Methods are adapted to the discrete OMP following the presentation in [55, 25].

The driving variables for the OMP are the binary x_j variables, indicating which sites have been selected for facility location. Once these are known, the objective value is easy to calculate. All the other variables are in the integer linear programming formulations to enable the costs to be calculated. It thus makes sense to build a Branch and Bound (B&B) method based entirely on the x_j variables, i.e. on decisions of whether or not a site is selected for facility location.

In this section, we present a B&B method in which each node represents a disjoint pair of sets of sites: a set of sites at which facilities will be opened and a set of sites at which facilities will *not* be opened. We refer to these as the set of *open* and *closed* sites, respectively (recall that $A = \{1, \ldots, M\}$ denotes the set of sites). For a given node, we may let $F \subseteq A$ denote the set of open sites and $\overline{F} \subseteq A \setminus F$ denote the set of closed sites. We refer to the sites indexed by $A \setminus (F \cup \overline{F})$ as *undecided*. The node in the B&B tree is represented by the pair $(F, \overline{F})$. Of course, a node is a leaf node if either $|F| \geq N$ or $|\overline{F}| \geq M - N$. Otherwise, we calculate a lower bound on the cost function of the problem defined by $(F, \overline{F})$. The lower bound is relatively simple to calculate; it does not require solution of any linear program. We discuss the details of this lower bound in Section 15.1.1.

Because of the nature of the cost function, opening a facility gives us very little information about its impact on a lower bound. It is only making the decision *not* to open a facility which restricts choices and forces the objective up. We thus use a branching rule which is strong, and which ensures that on each branch some site is closed. We discuss the branching rule in detail in Section 15.1.2.

In Section 15.1.3 we present computational experiments to compare the performance of the B&B method with that of the respective best integer linear programming formulation from Chapter 14.

15.1.1 Combinatorial Lower Bounds

A node in the B&B tree is represented by the disjoint pair of sets $(F, \overline{F})$ of sites open and closed, respectively. At each node which is not a leaf node of the B&B tree, we need to calculate a lower bound on the value of the cost function. Let $\hat{F} = A \setminus \overline{F}$ denote the set of sites which are either open, or undecided.

For any set of sites $R \subseteq \{1, \ldots, M\}$ define $f_\lambda(R)$ to be the cost of having open facilities at precisely those sites. So we define

$$c_i(R) = \min_{j \in R} c_{ij} \tag{15.1}$$

to be the minimum assignment cost for a customer at site i to a facility in R. Then we have

$$f_\lambda(R) = \sum_{i=1}^{M} \lambda_i \, c_{(i)}(R). \tag{15.2}$$

In the following we present two lower bounds, based on different strategies. The second one only applies in the case where self-service is cheapest or free. The other one may be stronger, depending on the problem data. As both bounds are very easy to calculate, the lower bound used in the B&B method is the maximum of the two.

In the following proposition we present a lower bound on the objective function value of any feasible solution having facilities in $\overline{F} \subseteq A$ closed.

Proposition 15.1 *Given a set of closed sites $\overline{F} \subseteq A$ with $|\overline{F}| < M - N$, let S be any set of N facilities not in $\overline{F}$, i.e. let $S \subseteq \hat{F}$ and $|S| = N$, where $\hat{F} = A \setminus \overline{F}$. Then*

$$f_\lambda(S) \geq f_\lambda(\hat{F}) =: LB_1(\hat{F}) \, .$$

Proof.

From (15.2) we have that $f_\lambda(S) = \sum_{i=1}^{M} \lambda_i \, c_{(i)}(S)$ and $f_\lambda(\hat{F}) = \sum_{i=1}^{M} \lambda_i \, c_{(i)}(\hat{F})$. Since we assume $\lambda \geq 0$, we just have to show that $c_{(i)}(S) \geq c_{(i)}(\hat{F})$ for all $i = 1, \ldots, M$ to conclude that $f_\lambda(S) \geq f_\lambda(\hat{F})$.

Observe that since $S \subseteq \hat{F}$ by (15.1), $c_i(S) \geq c_i(\hat{F})$ for all $i = 1, \ldots, M$. Taking $r = s = M$, $p_i = c_i(S)$ and $q_i = c_i(\hat{F})$ for all $i = 1, \ldots, M$, we can apply Theorem 1.1 to deduce that $c_{(i)}(S) \geq c_{(i)}(\hat{F})$ for all $i = 1, \ldots, M$, as required. □

Remark 15.1 *$LB_1(\hat{F})$ is a lower bound on the objective function value of any feasible solution having facilities in $\overline{F}$ closed.*

Observe that if the self-service is free (FSS), i.e. $c_{ii} = 0$, for all $i \in A$, this lower bound is likely to be weak, unless $|\hat{F}|$ is not too much greater than N, i.e. unless a relatively large number of facilities have been closed by branching. This will not occur unless the algorithm is relatively deep in the B&B tree, which closes one more site at each level (see Section 15.1.2 for details of the branching rule). Thus, if (FSS) holds, we should consider another lower bound, hopefully more effective higher in the tree. In fact, this lower bound applies more generally to the case that self-service is cheapest, i.e. $c_{ii} \leq c_{ij}$, for all $i, j = 1, \ldots, M$ with $j \neq i$. This condition, which we refer to as *cheapest self-service*, or *(CSS)*, can be assumed in most of the facility location problems without any capacity constraint.

The idea is that for any feasible set of columns $S \subseteq \hat{F}$, the feasible cost vector will consist of N diagonal elements and $M-N$ off-diagonal row minima, taken over S. Hence, using the vector consisting of the N smallest diagonal elements and the $M-N$ smallest off-diagonal row minima, taken over $\hat{F} \supseteq S$, we obtain a sorted cost vector which is in every component no more than the sorted cost vector given by S, and therefore provides a valid lower bound.

To introduce this lower bound, we define $D^{\hat{F}}$ to be the vector of diagonal elements of the cost matrix, taken over columns in $\hat{F}$, and let $d_1, \dots, d_N$ be the N smallest elements in $D^{\hat{F}}$, ordered so that

$$d_1 \leq \cdots \leq d_N. \tag{15.3}$$

Furthermore, we define the vector $H^{\hat{F}} \in \mathbb{R}^M$ via

$$h_i^{\hat{F}} = \min_{j \in \hat{F},\ j \neq i} c_{ij}, \quad \forall\, i \in A, \tag{15.4}$$

to be the vector of cheapest off-diagonal elements in each row, over the columns in $\hat{F}$, and let $h_1, \dots, h_{M-N}$ be the $M-N$ smallest elements in $H^{\hat{F}}$ ordered so that

$$h_1 \leq \cdots \leq h_{M-N}. \tag{15.5}$$

Finally, we define the vector $K^{\hat{F}} = (d_1, \dots, d_N, h_1, \dots, h_{M-N})$ and let $k_1, \dots, k_M$ be the M elements of $K^{\hat{F}}$ ordered so that

$$k_1 \leq \cdots \leq k_M. \tag{15.6}$$

Now we define

$$LB_2(\hat{F}) = \sum_{i=1}^{M} \lambda_i \, k_i. \tag{15.7}$$

Note that if self-service is in fact free, i.e. if (FSS) holds, then $d_1 = \cdots = d_N = 0$, $k_i = 0$ for $i = 1, \dots, N$ and $k_i = h_{i-N}$ for $i = N+1, \dots, M$, and thus

$$LB_2(\hat{F}) = \sum_{i=1}^{M-N} \lambda_{N+i} \, h_i. \tag{15.8}$$

In the proposition below, we prove that $LB_2(\hat{F})$ is a valid lower bound on the objective value $f_\lambda(S)$ for any feasible set $S \subseteq \hat{F}$ with $|S| = N$, if (CSS) holds. Note that the proof relies on a relatively simple, general result concerning sorted vectors: a vector of r real numbers that is componentwise no less than r elements chosen from a vector Q of s real numbers, $s \geq r$, is, when sorted, no less than (componentwise) the vector of the r smallest real numbers in Q, sorted. This general result is proven in Lemma 1.3.

Proposition 15.2 *Given $\overline{F} \subseteq A$ a set of closed sites with $|\overline{F}| < M - N$, let S be any set of N facilities in $\overline{F}$, i.e. let $S \subseteq \hat{F}$ and $|S| = N$, where $\hat{F} = A \setminus \overline{F}$. Then, if (CSS) holds,*

$$f_\lambda(S) \geq LB_2(\hat{F}) .$$

Proof.

From (15.2) we know that $f_\lambda(S) = \sum_{i=1}^{M} \lambda_i \, c_{(i)}(S)$. Thus, since $\lambda \geq 0$, to show that $LB_2(\hat{F})$ is a lower bound on the objective function value of any feasible solution having facilities in $\overline{F}$ closed, we only need to show that $c_{(i)}(S) \geq k_i$, for all $i = 1, \ldots, M$. To do this, we need to consider diagonal and off-diagonal costs separately.

Observe that the cost $c_i(S)$ is the diagonal cost matrix element in row i if $i \in S$; otherwise it is an off-diagonal element in a column in S, in row i. Thus, the vector of all costs $c_i(S)$, $i = 1, \ldots, M$, which we denote by $c(S)$, has N elements which are diagonal cost matrix elements and $M - N$ elements which are off-diagonal cost matrix elements, and where every element is taken from a column of the cost matrix which is in S. Let $D^S \in \mathbb{R}^N$ be the vector consisting of the N diagonal cost elements of $c(S)$, sorted in increasing cost order, i.e. chosen so that

$$d_1^S \leq \cdots \leq d_N^S.$$

Since $S \subseteq \hat{F}$, for all $j = 1, \ldots, N$ there exists a unique $i(j) \in \hat{F}$ such that $d_j^S = d_{i(j)}^{\hat{F}}$. Then by Lemma 1.3 we have that $d_j^S \geq d_j$ for all $j = 1, \ldots, N$.

Similarly, if we let $H^S \in \mathbb{R}^{M-N}$ be the vector consisting of the $M - N$ off-diagonal cost elements of $c(S)$, sorted in increasing cost order, i.e. chosen so that

$$h_1^S \leq \cdots \leq h_{M-N}^S,$$

then for each $j = 1, \ldots, M - N$ there must exist a unique $i(j) \in \hat{F}$ such that $h_j^S \geq h_{i(j)}^{\hat{F}}$. To see this, we note that for each $j = 1, \ldots, M - N$ there must exist a unique $i(j) \in A \setminus S$ such that $h_j^S = c_{i(j)}(S)$, by the definition of H^S. Now since $S \subseteq \hat{F}$ and $i(j) \notin S$ we know that $c_{i(j)}(S) = \min_{i' \in S} c_{i(j)i'} \geq \min_{i' \in \hat{F},\, i' \neq i(j)} c_{i(j)i'} = h_{i(j)}^{\hat{F}}$ by the definition of $H^{\hat{F}}$; and so $h_j^S \geq h_{i(j)}^{\hat{F}}$ as required. Then by Lemma 1.3 we have that $h_j^S \geq h_j$ for all $j = 1, \ldots, M - N$.

Now define $K^S = (d_1^S, \ldots, d_N^S, h_1^S, \ldots, h_{M-N}^S)$ and observe that we have shown that $K_j^S \geq K_j^{\hat{F}}$ for all $j = 1, \ldots, M$. (This is obvious, since if $j \leq N$ then $K_j^S = d_j^S$ and $K_j^{\hat{F}} = d_j$, and we have shown above that $d_j^S \geq d_j$. Similarly if $j > N$ then $K_j^S = h_{j-N}^S$ and $K_j^{\hat{F}} = h_{j-N}$, and we have shown above that $h_{j-N}^S \geq h_{j-N}$.) Note also that K^S is simply a permutation of $c(S)$ and hence, the ith component of K^S when sorted must be $c_{(i)}(S)$. Thus, by Theorem 1.1 we have $c_{(i)}(S) \geq k_i$ for all $i = 1, \ldots, M$, (recall that k_i is, by definition, the ith component of $K^{\hat{S}}$ when sorted), as required.

Thus, as a consequence, since we assume $\lambda \geq 0$, we have that $LB_2(\hat{F})$ is a lower bound on the cost of any feasible solution having facilities in $\overline{F}$ closed. □

Remark 15.2 *$LB_2(\hat{F})$ is a lower bound on the objective function value of any feasible solution having facilities in $\overline{F}$ closed.*

The following example illustrates the two lower bounds, and demonstrates that their relative strength depends not only on how many sites have been closed by branching, but also on the value of λ.

Example 15.1
Let $A = \{1, \ldots, 5\}$ be the set of sites and assume that we need $N = 2$ new facilities. Let the cost matrix be:

$$C = \begin{pmatrix} 0 & 4 & 5 & 6 & 4 \\ 5 & 0 & 6 & 2 & 2 \\ 7 & 1 & 0 & 5 & 1 \\ 7 & 4 & 3 & 0 & 5 \\ 1 & 3 & 5 & 4 & 0 \end{pmatrix}.$$

Suppose, we are at a node of the Branch and Bound tree represented by the pair $(F, \overline{F})$ with $F = \emptyset$ and $\overline{F} = \{1\}$. So site 1 has been closed by branching and $\hat{F} = \{2, 3, 4, 5\}$. Observe that (FSS), and hence (CSS), holds in this example, since $c_{ii} = 0$ for all $i \in A$. Therefore, both $LB_1(\hat{F})$ and $LB_2(\hat{F})$ are lower bounds on the value of the (OMP) at this node.

To calculate the bound $LB_1(\hat{F})$, we determine the vector of row minima over columns in $\hat{F}$, ($c_i(\hat{F})$ for all $i \in A$), to be $(4, 0, 0, 0, 0)$ yielding the sorted vector $(0, 0, 0, 0, 4)$.

To calculate $LB_2(\hat{F})$, we have to determine the off-diagonal cost matrix row minima over columns in $\hat{F}$, i.e. we calculate $H^{\hat{F}} = (4, 2, 1, 3, 3)$. Thus, the $M - N = 3$ smallest off-diagonal row minima are $h_1 = 1$, $h_2 = 2$ and $h_3 = 3$, and, since the diagonal costs are all zero, we get a lower bound based on the cost vector $k = (0, 0, 1, 2, 3)$.

Which one of $LB_1(\hat{F})$ or $LB_2(\hat{F})$ yields the best bound depends on the value of λ. For instance, $\lambda = (1, 0, 0, 0, 1)$ means that $LB_1(\hat{F}) = 0 + 4 = 4$ is better than $LB_2(\hat{F}) = 0 + 3 = 3$. However, $\lambda = (0, 0, 1, 1, 1)$ implies $LB_2(\hat{F}) = 1 + 2 + 3 = 6$ which is better than $LB_1(\hat{F}) = 0 + 0 + 4 = 4$.

Notice that both lower bounds can easily be computed. Hence, when (CSS) holds, and in particular when (FSS) holds, we use the lower bound given by

$$\max\{LB_1(\hat{F}), LB_2(\hat{F})\} \tag{15.9}$$

at a node in the B&B tree identified by sets F and $\overline{F}$, with $\hat{F} = A \setminus \overline{F}$.

Observe that this lower bound is not trivial at the root node, represented by the pair $(\emptyset, \emptyset)$. For $\hat{F} = A$, we can write the gap between the optimal objective function value (z^*) and $\max\{LB_1(A), LB_2(A)\}$ at the root node as follows:

$$\text{gap at root node} = \frac{z^* - \max\{LB_1(A), LB_2(A)\}}{z^*} \times 100. \tag{15.10}$$

In calculating the lower bound, we may have an opportunity to find a feasible solution of the same value, and so be able to prune the node. In calculating $LB_1(\hat{F})$, the row minimum was found for each row, over columns in $\hat{F}$: let $m(i) \in \hat{F}$ denote the column in which the row minimum for row i was found. In case of a tie with a row minimum occurring in a column in F, $m(i)$ is chosen to be in F. Let $V(F, \overline{F}) = \{m(i) \ : \ i \in A\} \setminus F$ be the set of columns in which the selected row minima occur, outside of F. Now if $|V(F, \overline{F})| + |F| \leq N$, then any set $S \subseteq \hat{F}$ with $|S| = N$ and $S \supseteq V(F, \overline{F}) \cup F$ must be an optimal set for the subproblem at that node, and the lower bound $LB_1(\hat{F})$ will be equal to the upper bound obtained from S. In this case, either the value of S is better than the current upper bound, which can thus be updated to the value of S, or the lower bound (equal to the value of S) is not better than the current upper bound; in either case the node can be pruned.

Similarly, in calculating $LB_2(\hat{F})$ in the case that (CSS) holds, the off-diagonal row minimum was found for each row, over columns in $\hat{F}$: let $o(i) \in \hat{F}$ denote the column in which the off-diagonal row minimum for row i was found. In case of a tie with a row minimum occurring in a column in F, $o(i)$ is chosen to be in F. Let $V'(F, \overline{F}) = \{o(i) \ : \ i \in A \setminus F\} \setminus F$ be the set of columns in which off-diagonal row minima occur, outside of F, for rows not in F. Now if $|V'(F, \overline{F})| + |F| \leq N$, then any set $S \subseteq \hat{F}$ with $|S| = N$ and $S \supseteq V'(F, \overline{F}) \cup F$ must be an optimal set for the subproblem at that node, and the lower bound $LB_2(\hat{F})$ will be equal to the upper bound obtained from S, the node can be pruned, and if the value $LB_2(\hat{F})$ is better than the current upper bound, the upper bound can be set to this value.

15.1.2 Branching

The lower bounds presented in the previous section are based on row minima of the cost matrix calculated over columns corresponding to open or undecided sites. Thus, making the decision to open a site will not affect the lower bound. Closing a site, however, would be likely to increase the lower bound. Hence, we consider a branching rule so that a (different) site is closed on each branch. We also ensure that the branching is strong, i.e. it partitions the solution space.

The generic form of the branching rule assumes that the undecided sites will be closed following a given order.

Consider a node of the Branch and Bound tree, defined by the pair of sets $(F, \overline{F})$ with $|F| < N$ and $|\overline{F}| < M - N$. We may also assume that $|V(F, \overline{F})| >$

$N - |F|$ and, if (CSS) holds, that $|V'(F,\overline{F})| > N - |F|$; otherwise the node would have been pruned, as discussed at the end of the previous section. Set $\hat{F} = A \setminus \overline{F}$. Suppose an order is given for the undecided sites, defined to be $U = \hat{F} \setminus F$. Let $\beta : \{1, \ldots, |U|\} \to U$ define the given order, so $\beta(i)$ is the index of the ith undecided site in the order. Note that $|U| = M - |\overline{F}| - |F| > N - |F|$ since $|\overline{F}| < M - N$. The branching rule creates child nodes with the ith child node having site $\beta(i)$ closed and sites $\beta(1), \ldots, \beta(i-1)$ open. A node with more than N sites open is infeasible, so at most $N - |F| + 1$ child nodes need to be created. Furthermore, $|U| > N - |F|$ and thus, $N - |F| + 1$ child nodes can be created. In other words, the child nodes are defined by the pairs of sets $(F^i, \overline{F}^i)$ for $i = 1, \ldots, N - |F| + 1$, where $\overline{F}^i = \overline{F} \cup \{\beta(i)\}$ and $F^i = F \cup \{\beta(1), \ldots, \beta(i-1)\}$, with $F^1 = F$.

For the branching we consider two different orders. The first one is simply the site index order, i.e. we take β so that

$$\beta(1) \leq \cdots \leq \beta(N - |F| + 1).$$

We refer to the resulting branching rule as the *index-order* branching rule. The second order attempts to maximize the impact of the branching on the lower bound, and is more sophisticated. Recall $|V(F,\overline{F})| > N - |F|$. In order to branch having the highest impact on the lower bound, we can eliminate the column with the greatest impact on a row minimum.

Arguably, this will be a column containing the smallest row minimum. Thus, we define for each $j \in V(F,\overline{F})$ the set of rows which have their row minimum in column j to be $W(j) = \{i \in A \; : \; m(i) = j\}$, (where $m(i)$ was defined at the end of the previous section), and define the smallest row minimum in column j to be

$$v_j = \min_{i \in W(j)} c_{ij}.$$

Let σ^V denote a permutation of $V(F,\overline{F})$ which sorts the vector V in increasing order, i.e. such that

$$v_{\sigma^V(1)} \leq v_{\sigma^V(2)} \leq \cdots \leq v_{\sigma^V(|V(F,\overline{F})|)}.$$

Observe, that when the self-service is free, i.e. (FSS) holds, there is little or nothing to differentiate the v values unless a relatively large number of facilities have been closed by branching. This will not occur until relatively deep in the B&B tree. Thus, a secondary key could be used in sorting, such as the second-smallest row costs. For each row i, let u_i denote the second-smallest cost over columns in $\hat{F}$, and let w_j denote the largest difference between the second-smallest and smallest row element in $W(j)$, i.e. set

$$w_j = \max_{i \in W(j)} (u_i - c_{ij})$$

for each $j \in V(F,\overline{F})$. Now we may choose σ^V so that whenever $v_{\sigma^V(j)} = v_{\sigma^V(j')}$ for $j < j'$ then $w_{\sigma^V(j)} \geq w_{\sigma^V(j')}$.

Hence, for the second order we take $\beta(i) = \sigma^V(i)$ for $i = 1, \ldots, N - |F| + 1$. A similarly order can be used if the lower bound was achieved by $LB_2(\hat{F})$, provided that (CSS) holds. The only difference would be that everything should be based on off-diagonal row minima rather than row minima.

The branching rule resulting from this second order can be viewed as closing sites in order to decrease the "maximum regret", i.e. maximizing the cost impact of the decision. Thus, we refer to it as the *max-regret* branching rule. The example below illustrates the use of this rule.

Example 15.2

Consider the data presented in Example 15.1. Assume that the current node is defined by $F = \{4\}$ and $\overline{F} = \{1\}$, and so $\hat{F} = \{2, 3, 4, 5\}$. Note that we expect to have $N - |F| + 1 = 2 - 1 + 1 = 2$ branches from this node.

On the one hand, if the lower bound is given by $LB_1(\hat{F})$, we have to focus on the row minima. These are achieved in columns $m(1) = 2$, $m(2) = 2$, $m(3) = 3$, $m(4) = 4$, $m(5) = 5$, so $V(F, \overline{F}) = \{2, 3, 5\}$, $W(2) = \{1, 2\}$, $W(3) = \{3\}$ and $W(5) = \{5\}$, with $v_2 = v_3 = v_5 = 0$. Note that, in this case, we need the secondary key for sorting. The second-smallest cost over columns in $\hat{F}$, for rows not in F, are $u_1 = 4$, $u_2 = 2$, $u_3 = 1$ and $u_5 = 3$. Then $w_2 = \max\{u_1 - c_{12}, u_2 - c_{22}\} = \max\{4-4, 2-0\} = 2$, $w_3 = u_3 - c_{33} = 1 - 0 = 1$ and $w_5 = u_5 - c_{55} = 3 - 0 = 3$. Therefore $\sigma^V(1) = 5$, $\sigma^V(2) = 2$ and $\sigma^V(3) = 3$. Thus, the two child nodes are defined by the pairs $(\{4\}, \{1, 5\})$ and $(\{4, 5\}, \{1, 2\})$.

On the other hand, if the lower bound is achieved by $LB_2(\hat{F})$, we have to focus on the off-diagonal row minima. These are achieved in columns $o(1) = 2$, $o(2) = 4$, $o(3) = 2$, $o(4) = 3$, $o(5) = 2$, so $V'(F, \overline{F}) = \{2, 3\}$, $W(2) = \{1, 3, 5\}$ and $W(3) = \{4\}$, with $v_2 = 1$ and $v_3 = 3$. Therefore, $\sigma^V(1) = 2$ and $\sigma^V(2) = 3$ and the secondary key is not required. Thus, the two child nodes are defined by the pairs $(\{4\}, \{1, 2\})$ and $(\{4, 2\}, \{1, 3\})$.

15.1.3 Numerical Comparison of the Branching Rules

We present computational results of the B&B method based on a implementation in C++ (see [55, 25]). The upper bound is initialized by a heuristic algorithm based on variable neighborhood search (see [56]). The experiments were performed on a Pentium III 800 Mhz with 1 GB RAM. The structure of the test problem instances is described in Section 14.3. As will be reported in more detail in Section 15.1.4, the B&B method performed very well, and was able to solve much larger instances than the standard solver with the formulations presented in Chapter 14. Here we show the results of running the B&B algorithm using each branching rule on problems with $M = 30$, averaged over fifteen instances for each value of $N = 8, 10, 15, 16$. All cost matrices were randomly generated so that (FSS) holds. The results are presented in Table 15.1.

Table 15.1. Numbers of B&B nodes and computing times for the B&B method using either the index-order or max-regret branching rule on problems with $M = 30$ and $N = 8, 10, 15, 16$, for which (FSS) holds. Results are averages taken over 15 problem instances for each value of N.

Problem Type	# of B&B Nodes index-order	# of B&B Nodes max-regret	Ratio i.-ord./ max-r.	Computing Time (s) index-order	Computing Time (s) max-regret	Ratio i.-ord./ max-r.
T1	1727594.63	578787.85	2.99	912.34	235.92	3.87
T2	156211.15	17841.50	8.76	82.51	7.44	11.09
T3	772448.15	265769.78	2.91	417.62	107.04	3.90
T4	2774061.28	840401.43	3.30	1640.40	339.68	4.83
T5	1306657.1	428017.95	3.05	704.83	179.73	3.92
T6	2093979.28	633977.18	3.30	1143.49	256.07	4.47
T7	1388014.55	463225.33	3.00	730.87	190.62	3.83
T8	981157.38	310314.05	3.16	517.87	129.53	4.00
Average	1400015.44	442291.88	3.81	768.74	180.75	4.99

As can be seen from the table, using the max-regret branching rule reduces the number of B&B nodes by a factor of about 3.8 on average and reduces the computing time by a factor of about 5. The effect was more pronounced for problems of type T2 (i.e. N-center problems) for which the number of nodes required by the B&B algorithm with the max-regret rule was less than one eighth of the number of nodes provided by the algorithm with the index-order branching rule. Furthermore, for this type of problems the computing time was decreased by an order of magnitude.

It is clear that the more sophisticated max-regret rule is more effective than the simple index-order rule. Furthermore, the computing time for solving the instances with $M = 30$ is 180.75 seconds on average, and so problems with even more sites could be expected to be solved. However, after attempting problems with $M = 35$, we found that although all problems could be solved to optimality, the computing times were in most cases around a factor of ten more than for $M = 30$. This indicates that the B&B method is not going to be particularly effective for problems much larger than those with $M = 35$.

15.1.4 Computational Results

In this section we compare the computational performance of the B&B method described in Section 15.1, with the max-regret branching rule, with that of the best linearization (for type T2 problems this is (OMP^2) and for all other problem types it is (OMP^1)). The same upper bounds were used, where needed, in constraints defining the linearizations.

The performance of the B&B method was consistently better than that of the best linearization, with the former out-performing the latter by a significant margin. To illustrate the type of performance, we give results for problems with $M = 18$ and $N = 5$, as well as $N = 10$, in Table 15.2. We report results for these extreme values of N as the performance of the linearizations generally improved as N increased; the performance of the B&B method was, by contrast, relatively consistent. Results for intermediate values of N can be simply interpolated.

Table 15.2. Numerical results for problems with $M = 18$ and the extreme values of N tested. All results are averages taken over 15 problem instances.

	Problem	Best Linearization			B&B Method		
N	Type	gap(%)	# nodes	CPU(s)	gap(%)	# nodes	CPU(s)
5	T1	2.8	1999.1	40.77	48.4	2102.5	0.57
	T2	24.6	3.8	3.37	54.4	473.0	0.13
	T3	34.2	17469.6	370.32	51.7	1477.3	0.40
	T4	12.1	3568.2	53.76	39.5	916.4	0.25
	T5	11.1	19169.1	187.03	49.0	2054.3	0.56
	T6	10.9	12797.7	190.53	44.8	1419.5	0.38
	T7	10.5	24350.6	289.50	49.2	1723.7	0.56
	T8	16.0	36343.93	408.60	49.0	1723.7	0.47
10	T1	0.0	42.1	0.75	20.4	1395.2	0.33
	T2	17.0	0.73	0.95	29.3	222.4	0.05
	T3	11.3	90.8	1.16	22.3	1030.3	0.24
	T4	9.8	11.1	0.64	14.3	1760.5	0.42
	T5	10.2	318.9	1.75	22.7	968.8	0.23
	T6	7.8	38.2	0.99	14.3	1124.5	0.27
	T7	7.9	63.1	0.81	21.2	1107.0	0.26
	T8	17.8	421.1	1.99	23.6	819.5	0.19

We observe that the B&B method always requires less CPU time than the best linearization, and for some of the problems with $N = 5$ more than two orders of magnitude less. The B&B method shows less variation in the number of nodes needed for the different problem types, and also for different values of N. We report the average (root node) gap, but note that whilst this may be indicative of the quality of an integer programming formulation, it is less meaningful for the B&B method, where the bounds are very weak high in the tree but improve rapidly deeper in the tree.

To demonstrate that the B&B method can solve larger problems, we also provide detailed numerical results for problems with $M = 30$, in Table 15.3.

As in Sections 14.3 and 15.1.3, each row of the tables contains results averaged over fifteen instances having the same parameter values.

Table 15.3. Computational results for the B&B method with the max-regret branching rule for problems with $M = 30$.

Problem		# nodes			gap(%)			CPU(s)		
Type	N	aver	min	max	aver	min	max	aver	min	max
T1	8	376661.7	90639	920625	50.8	40.4	58.3	167.19	40.92	408.73
	10	698401.1	50974	1678543	43.1	31.8	51.8	303.11	21.75	762.08
	15	710577.9	71558	1444065	28.2	15.2	38.2	274.74	28.03	562.25
	16	529510.7	107189	1112917	25.0	12.9	36.7	198.65	41.19	417.09
T2	8	30463.8	5043	100375	55.9	36.4	66.7	13.54	2.14	44.19
	10	15012.3	2565	38291	47.0	22.2	62.5	6.39	1.06	16.59
	15	12473.1	0	37005	37.2	0.0	50.0	4.81	0.00	14.84
	16	13416.8	275	35905	35.9	0.0	60.0	5.03	0.09	14.23
T3	8	195498.0	87519	447017	54.1	43.5	65.8	86.44	39.11	197.08
	10	275646.3	58988	867397	45.6	28.3	55.6	117.02	25.20	370.39
	15	326233.7	38961	630901	32.3	13.8	44.8	125.19	15.36	246.95
	16	265701.1	56589	585759	28.7	11.5	42.3	99.51	21.83	217.92
T4	8	124932.5	11396	275910	44.9	30.5	54.3	56.37	5.06	125.25
	10	354196.3	5306	1108572	36.7	20.5	47.8	154.46	2.31	476.23
	15	1606801.0	52119	3474800	21.3	7.1	33.3	649.56	20.81	1457.81
	16	1275675.9	110028	3970039	18.6	6.3	33.3	498.32	42.88	1460.81
T5	8	351434.6	98789	729440	51.8	42.0	60.0	160.75	45.42	332.02
	10	550405.8	58142	1364222	43.8	30.6	52.6	242.00	25.72	601.64
	15	533272.8	38551	1279337	28.9	14.3	40.9	209.68	15.55	502.16
	16	276958.6	47709	606337	26.4	12.5	40.0	106.49	18.88	239.69
T6	8	292640.9	34600	597933	48.3	35.7	54.3	131.82	15.64	272.22
	10	582688.3	29094	1265889	40.7	28.1	48.6	253.11	12.58	561.84
	15	839755.2	65764	1927705	26.4	12.5	40.0	327.44	26.34	777.64
	16	820824.3	50294	2022796	22.0	9.1	35.7	311.89	19.88	792.27
T7	8	380210.3	89852	825887	51.8	41.8	58.3	170.57	40.69	368.23
	10	625290.4	36296	1621612	43.1	30.4	53.3	269.65	15.44	712.63
	15	466872.1	31681	1160957	28.9	13.0	40.0	179.30	12.47	442.06
	16	380528.5	75834	809368	25.8	13.6	38.1	142.95	29.02	313.39
T8	8	337418.9	85926	710147	52.0	41.7	59.0	151.38	38.84	328.09
	10	436277.5	26389	1174336	42.5	28.0	51.7	188.97	11.33	503.84
	15	263030.8	2277	1139116	28.9	11.1	40.0	101.19	0.91	436.17
	16	204529.0	47135	479781	25.8	9.1	36.4	76.59	18.30	173.09

15.2 Two Heuristic Approaches for the OMP

After having seen the potential, but also the limitations, of an exact B&B method we continue with two heuristic approaches which enable us to tackle much larger problems. In our presentation we follow [56].

15.2.1 An Evolution Program for the OMP

The Evolution Program (EP) proposed to solve the OMP is essentially based on a genetic algorithm developed by Moreno Vega [144] for p-median and p-center problems. First, it should be noted that both of these problems can be solved as particular cases of the OMP. Second, the feasible solutions of these problems and those of the OMP have a similar structure. These are the reasons to adapt the procedure in [144] to our problem. In addition, evolution strategies are used to improve the performance of the EP in each iteration.

In the next section we introduce some general concepts of genetic algorithms which are necessary to present the EP.

Genetic Algorithms

Genetic algorithms use a vocabulary taken from natural genetics. We talk about individuals in a population, in the literature these individuals are also called chromosomes. A chromosome is divided into units - genes; see Dawkins [51]. These genes contain information about one or several attributes of the individual.

The evolution in genetic algorithms can be implemented by two processes which mimic nature: natural selection and genetic change in the chromosomes or individuals. Natural selection consists of selecting those individuals that are better adapted to the environment, i.e. those who survive. Genetic changes (produced by genetic operators) can occur either when there exists a crossover between two individuals or when an individual undergoes a kind of mutation. The crossover transformation creates new individuals by combining parts from several (two or more) individuals. The mutations are unary transformations which create new individuals by a small change in a single individual. After some generations the procedure converges — it is expected that the best individual represents a near-optimal (reasonable) solution.

In addition, several extensions of genetic algorithms have been developed (evolutionary algorithms, evolution algorithms and evolutive algorithms). These extensions mainly consist of using new data structures for representing the population members and including different types of genetic operators and natural selection (see Michalewicz [137]).

In the next section we introduce an EP to solve the OMP.

Evolution Program

Classical genetic algorithms use a binary codification to define the chromosomes. But sometimes this representation is very difficult to handle and therefore, some authors decided not to use it (see [50] and [137]). Genetic algorithms, which use codifications different from the binary one and genetic operators adapted to these particular codifications, are called evolution programs (see [137]). In the following we will use a non-binary representation scheme for the individuals of the population.

An EP is a probabilistic algorithm which maintains a population of H individuals, $P(t) = \{x_1^t, \ldots, x_H^t\}$ in each iteration t. Each individual stands for a potential solution to the problem at hand, and is represented by some data structure. Some members of the population undergo transformations (alteration step) by means of genetic operators to form new solutions. Each solution is evaluated to give some measure of its "fitness". Then, a new population (iteration $t + 1$) is formed by selecting the fittest individuals (selection step). The program finishes after a fixed number of iterations where the best individual of the last population is considered as the approximative solution of the problem. We present a scheme of an EP as follows:

***Algorithm 15.1**: Generic evolution program*

$t \leftarrow 0$
initialize $P(t)$
while *(not termination-condition)* **do**
  modify $P(t)$
  evaluate $P(t)$
  $t \leftarrow t + 1$
  generate $P(t)$ from $P(t-1)$
end

An EP for a particular problem must have the following six components (see [137]):

- a genetic representation of potential solutions of the problem,
- a way to create an initial population of potential solutions,
- an evaluation function that plays the role of the environment, rating solutions in terms of their "fitness",
- genetic operators (crossover and mutation) that alter the composition of children,
- a selection criterion that determines the survival of every individual, allowing an individual to survive or not in the next iteration,
- values for various parameters that the genetic algorithm uses (population size, probabilities of applying genetic operators, number of generations, etc.).

All these components will be described in the following sections. First of all, we introduce a codification in order to have an appropriate representation of the individuals, that is, of the feasible solutions of the OMP.

Codification of the Individuals

Taking into account that the set of existing facilities is finite, we can assume that it is indexed. Thus, the feasible solutions of a discrete facility location problem can be represented by an M-dimensional binary vector with exactly N entries equal to 1 (see Hosage & Goodchild [105] and Jaramillo et al. [111]). An i-th entry with value 1 means that facility i is open, the value 0 means that it is closed. The advantage of this codification is that the classical genetic operators (see [137]) can be used. The disadvantages are that these operators do not generate, in general, feasible solutions and that the $M - N$ positions containing a zero also use memory while not providing any additional information.

Obviously, the classical binary codification can be used for the OMP. But the disadvantage of the inefficiently used memory is especially clear for examples with $N \ll M$. For this reason, we represent the individuals as N-dimensional vectors containing the indices of the open facilities, as [143] and [144] proposed for the p-median and p-center problems, respectively. In addition, the entries of each vector (individual) are sorted in increasing order. The sorting is to assure that under the same conditions the crossover operator, to be defined in Section 15.2.1, always yields the same children solutions; see [144]. We illustrate this representation of the individuals with a small example: if $M = 7$ and $N = 5$, the feasible solution $X = (0, 1, 0, 1, 1, 1, 1)$ is codified as $(2, 4, 5, 6, 7)$.

To start an evolution, an initial population is necessary. The process to generate this population is described in the following subsection.

Initial Population

Two kinds of initial populations were considered. The first one is completely randomly generated (denoted by *random* $P(0)$). The individuals of the second one (denoted by *greedy* $P(0)$) are all but one randomly generated, while the last one is a greedy solution of the OMP. The number of individuals of the population in every generation, denoted by H, is constant. In this way, the population *random* $P(0)$ is made up of H randomly generated individuals, and the *greedy* $P(0)$ of $H - 1$ randomly generated and one solution of the OMP constructed with the Greedy Algorithm.

A greedy solution of the OMP is obtained as follows: the first chosen facility is the one that minimizes the ordered median objective function assuming that we are interested in the 1-facility case. After that, at every step we choose the facility with minimal objective function value, taking into account the facilities already selected. This procedure ends after exactly N facilities have been chosen.

Each individual of the population has an associated "fitness" value. In the following subsection the evaluation function that defines the "fitness" measure is described.

Evaluation Function

In order to solve the OMP using an EP, the "fitness" of an individual is determined by its corresponding ordered median function value. Therefore, an individual will be better adapted to the environment than another one if and only if it yields a smaller ordered median function value. Thus, the best adapted individual of a population will be one that provides the minimal objective function value of the OMP among all the individuals of this population.

Genetic Operators

The genetic operators presented in this section are replicas of the classical crossover and mutation operators (see [105]). These operators are adapted to the actual codification (see Section 15.2.1), and they always provide feasible solutions, i.e. vectors of size N in which all entries are different. There are two of them:

- Crossover Operator
  In order to present the crossover operator we define the breaking position as the component where the two parent individuals break to generate two children. The crossover operator interchanges the indices placed on the right-hand side of the breaking position (randomly obtained). When the breaking position has been generated, the output of this operator depends only on the two parent individuals, i.e. their crossing always provides the same children. This is possible because of the sorting in the individual codification, as we mentioned in Section 15.2.1. Moreover, to ensure feasibility during the crossing procedure, the indices of the parent individuals that should be interchanged (i.e. those indices which are common for both parent individuals) are marked. Observe that feasibility of an individual will be lost if it contains the same index more than once.
  The breaking position is randomly chosen. The indices placed to the right-hand side of the breaking position are called cross positions. Then the children are obtained as follows:
  1. both parents are compared, the indices presented in both vectors are marked;
  2. the non-marked indices are sorted (in increasing order) and moved to the left;
  3. the indices of the transformed parents that lie on the right-hand side of the breaking position are interchanged;
  4. the marks are eliminated and both children codifications are sorted.

- Mutation Operator
  The mutation operator is defined as the classical one derived for the binary codification but guaranteeing the feasibility of the solution: interchange one index of the individual with another not presented in the individual. After the interchange the indices of the new individual are sorted.

  These two operators are illustrated in the following example.

Example 15.3
Assume that there are seven sites and five new facilities should be open, i.e. $M = 7$ and $N = 5$. Let us consider two feasible solutions: $(2, 3, 4, 6, 7)$ and $(1, 2, 4, 5, 7)$.

Assume that the breaking position is 1. Then:

parents	marks	sorted non-marked indices	interchange	children
$(2,3,4,6,7)$	$(2^*,3,4^*,6,7^*)$	$(3,\underline{6},\underline{2}^*,\underline{4}^*,\underline{7}^*)$	$(3,5,2,4,7)$	$(2,3,4,5,7)$
$(1,2,4,5,7)$	$(1,2^*,4^*,5,7^*)$	$(1,\underline{5},\underline{2}^*,\underline{4}^*,\underline{7}^*)$	$(1,6,2,4,7)$	$(1,2,4,6,7)$

A mutation of the feasible solution $(2, 3, 4, 6, 7)$ can be originated by the interchange between any index of this individual and an index of the set $\{1, 5\}$. Then, the indices of the new individual are sorted.

A constant probability for all individuals along all iterations is associated with each genetic operator (crossover and mutation). The determination of this probability is based on empirical results, as shown in Subsection 15.2.1.

After the generation of the children, a selection criterion is applied to mimic the natural selection in genetics. This selection depends on the evaluation function ("fitness") presented in Subsection 15.2.1. Our selection criterion is described in the following subsection.

Selection Criterion

The algorithm uses evolution strategies in order to ensure a kind of convergence in each generation, i.e. to avoid the new population being worse than the original one (see Bäck et al. [10] and Schwefel [180]). Hence, we include in the original population all the children generated by crossover and all the mutated individuals. Obviously, the number of individuals in the population after these transformations is normally larger than H. Thus, the selection criterion consists of dropping the worst individuals (i.e. those individuals with the largest objective function values) until the population contains again exactly H individuals. Clearly, this selection criterion ensures that the population size is constant at each iteration.

This method of replacing the population is called *incremental replacement*, since the child solutions will replace "less fit" members of the population, see [111]. Figure 15.1 illustrates one of the advantages of this method. After a few generations (100), we obtain a population containing a set of different good solutions, all of them at most a modest percentage away from the best solution (1.79%). Figure 15.1 shows the best and the worst solution found at each generation as well as the optimal solution.

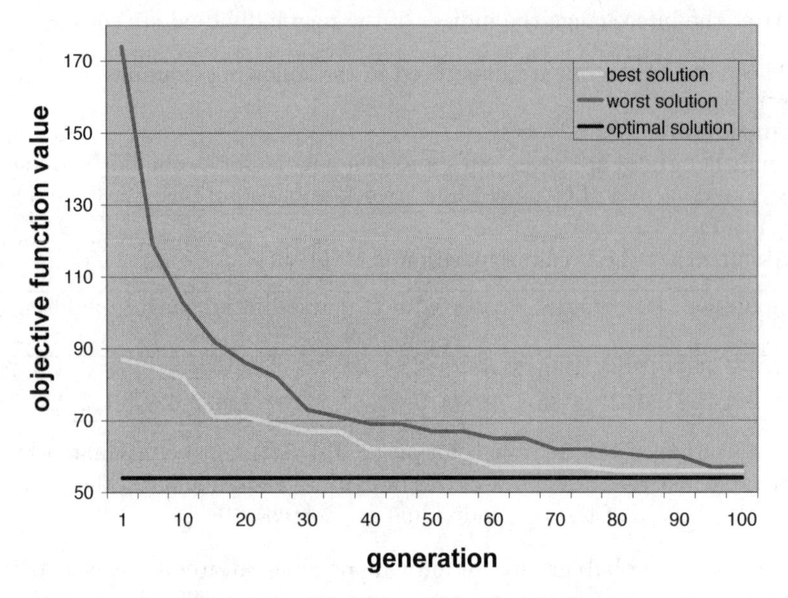

Fig. 15.1. Evolution of the population

The genetic algorithm depends very much on the tuning of a number of parameters. This issue is discussed in the next section.

Parameter Values

One of the most difficult tasks when attempting to obtain an efficient genetic algorithm is the determination of good parameters. In our case, some of them were chosen a priori such as the population size, $H = 25$, and the total number of iterations, 1000.

In the following we describe the experiments performed in order to set the probabilities of crossover and mutation operators for the individuals. We distinguish two cases depending on the selected initial population, i.e. considering or not a greedy solution of the OMP (see Section 15.2.1). To be able to compare the different values of probabilities, we always used the same seed for the random number generator. In this way, the solutions of the different problems depend only on the values of the parameters and not on the random character.

We considered two instances with $M = 30$ and $N = 10$ and solved them for eight different types of λ (in total 16 examples were tested).

- T1: $\lambda = (1, \ldots, 1)$, vector corresponding to the N-median problem.
- T2: $\lambda = (0, \ldots, 0, 1)$, vector corresponding to the N-center problem.
- T3: $\lambda = (0, \ldots, 0, \underbrace{1, \ldots, 1}_{k})$, vector corresponding to the k-centra problem, where $k = \lfloor \frac{M}{3} \rfloor = 10$.
- T4: $\lambda = (\underbrace{0, \ldots, 0}_{k_1}, 1, \ldots, 1, \underbrace{0, \ldots, 0}_{k_2})$, vector corresponding to the (k_1, k_2)-trimmed mean problem, where $k_1 = N + \lceil \frac{M}{10} \rceil = 13$ and $k_2 = \lceil \frac{M}{10} \rceil = 3$.
- T5: $\lambda = (0, 1, 0, 1, \ldots, 0, 1, 0, 1)$.
- T6: $\lambda = (1, 0, 1, 0, \ldots, 1, 0, 1, 0)$.
- T7: $\lambda = (0, 1, 1, 0, 1, 1, \ldots, 0, 1, 1, 0, 1, 1)$.
- T8: $\lambda = (0, 0, 1, 0, 0, 1, \ldots, 0, 0, 1, 0, 0, 1)$.

Based on preliminary tests, we decided to select different values for the probabilities of mutation (0.05, 0.075, 0.1, 0.125) and crossover (0.1, 0.3, 0.5). Then, we compared the gap between the optimal solution and the one given by the genetic algorithm, initialized either by *random* $P(0)$ or *greedy* $P(0)$:

$$\text{gap} = \frac{z_{\text{heu}} - z^*}{z^*} \times 100, \qquad (15.11)$$

being z^* the optimal objective function value and z_{heu} the objective function value of the best solution obtained by the evolution program. The optimal solutions were determined by the branch-and-bound (B&B) method presented in Section 15.1. Table 15.4 summarizes the computational results obtained for the different probabilities.

Based on the results presented in Table 15.4, we decided to fix the probability of mutation equal to 0.125 and the crossover probability equal to 0.1 for the EP initialized by both types of populations (*random* $P(0)$ and *greedy* $P(0)$). Note that even though the average performance of *random* $P(0)$, for the selected probability values, is better than that of *greedy* $P(0)$ (see corresponding row in Table 15.4), there are cases for which *greedy* $P(0)$ yields much better results (as Table 15.5 shows). That is, the behavior of both procedures can be seen as complementary.

Therefore, from the computational results given in Table 15.5, it seems advisable to solve the OMP by running the EP twice, once initialized by *random* $P(0)$ and once more initialized by *greedy* $P(0)$, and taking the best solution found. An extensive numerical study is presented in Section 15.2.3.

15.2.2 A Variable Neighborhood Search for the OMP

The second heuristic procedure we describe to solve the OMP is based on the Variable Neighborhood Search (VNS) proposed by Hansen & Mladenović [93] for the N-median problem. As mentioned above N-median is a particular case of OMP. However, computation of the objective function value is much harder for OMP than for N-median. Indeed, a major difficulty is to compute the

Table 15.4. Computational results obtained by using different values for the probabilities of mutation and crossover.

probabilities		EP *random* $P(0)$ gap(%)			EP *greedy* $P(0)$ gap(%)		
mut	*cross*	aver	min	max	aver	min	max
0.05	0.1	5.05	0.00	25.00	4.50	0.00	22.22
0.05	0.3	3.35	0.00	22.00	6.06	0.00	25.00
0.05	0.5	7.98	0.00	66.67	11.63	0.00	100.00
0.075	0.1	1.85	0.00	11.11	3.89	0.00	12.50
0.075	0.3	2.54	0.00	16.89	4.66	0.00	12.50
0.075	0.5	4.27	0.00	25.00	5.14	0.00	22.22
0.1	0.1	1.49	0.00	12.50	3.11	0.00	10.71
0.1	0.3	6.15	0.00	22.22	5.14	0.00	25.00
0.1	0.5	4.28	0.00	22.22	4.36	0.00	12.50
0.125	0.1	1.49	0.00	12.50	2.89	0.00	10.71
0.125	0.3	5.31	0.00	25.00	6.06	0.00	25.00
0.125	0.5	6.32	0.00	25.00	4.33	0.00	12.50

variation between the objective function values when an interchange between two facilities is performed. We are forced to update and sort the whole cost vector after this interchange takes place. For the N-median problem updating the value of the objective function can be done step by step. As a consequence, the complexity of the procedure applied to OMP is significantly higher.

In the following section we present a modified fast interchange algorithm, which is essential to describe the VNS developed to solve the OMP.

An Implementation of the Modified Fast Interchange Heuristic

In this section we present an implementation of the basic move of many heuristics, i.e. an interchange (or a change of location for one facility). This procedure is based on the fast interchange heuristic proposed by Whitaker [210] and implemented, among others, by Hansen & Mladenović [93] for the N-median problem. Two ingredients are incorporated in the interchange heuristic: *move evaluation*, where a best removal of a facility is found when the facility to be added is known, and *updating* the first and the second closest facility for each client.

Moreover, the variation of the ordered objective function value is computed after each interchange in *move*.

Thus, using this interchange only from a random initial solution gives a fairly good heuristic. Results are even better with an initial solution obtained with the Greedy Algorithm.

In the description of the heuristic we use the following notation:

Table 15.5. Computational results corresponding to the EP initialized either by *random* $P(0)$ or *greedy* $P(0)$.

			EP *random* $P(0)$		EP *greedy* $P(0)$	
	Problem Type	optimal value	best found	gap (%)	best found	gap (%)
example 1	T1	81	81	0.00	88	8.64
	T2	9	9	0.00	9	0.00
	T3	61	61	0.00	61	0.00
	T4	44	44	0.00	47	6.82
	T5	43	43	0.00	46	6.98
	T6	36	36	0.00	37	2.78
	T7	56	56	0.00	62	10.71
	T8	29	29	0.00	32	10.34
example 2	T1	78	78	0.00	78	0.00
	T2	8	9	12.50	8	0.00
	T3	54	54	0.00	54	0.00
	T4	46	46	0.00	46	0.00
	T5	41	42	2.44	41	0.00
	T6	36	36	0.00	36	0.00
	T7	55	58	5.45	55	0.00
	T8	28	29	3.57	28	0.00

- $d1(i)$: index of the closest facility with respect to client i, for each $i = 1, \ldots, M$;
- $d2(i)$: index of the second closest facility with respect to client i, for each $i = 1, \ldots, M$;
- $c(i, j)$: cost of satisfying the total demand of client i from facility j, for each $i, j = 1, \ldots, M$;
- $x_{cur}(i)$ for each $i = 1, \ldots, N$: current solution (new facilities);
- $cost_{cur}$: current cost vector;
- f_{cur}: current objective function value;
- $goin$: index of facility to be inserted in the current solution;
- $goout$: index of facility to be deleted from the current solution;
- g^*: change in the objective function value obtained by the best interchange;

In the following four subsections we describe components of our second heuristic for OMP.

Initial Solution

The heuristic is initialized with a solution constructed with the Greedy Algorithm, as done for the EP in Section 15.2.1.

The move evaluation is presented in the following section.

Move Evaluation

In the next procedure called *Modified Move*, the change in the objective function g^* is evaluated when the facility that is added (denoted by *goin*) to the current solution is known, while the best one to go out (denoted by *goout*) is to be found.

Algorithm 15.2: *Modified Move* ($d1$, $d2$, c, λ, x_{cur}, $cost_{cur}$, f_{cur}, $goin$, M, N, g^*, $goout^*$)

Initialization
Set $g^* \leftarrow \infty$
Best deletion
for $goout = x_{cur}(1)$ *to* $x_{cur}(N)$ **do**
  Set $cost_{new} \leftarrow cost_{cur}$
  for *each client* i $(i = 1, \ldots, M)$ **do**
    if $d1(i) = goout$ **then**
      $cost_{new}(i) \leftarrow \min\{c(i, goin), c(i, d2(i))\}$
    else
      if $c(i, goin) < c(i, d1(i))$ **then** $cost_{new}(i) \leftarrow c(i, goin)$
    end
  end
  Find the corresponding objective function value f_{new}
  $g \leftarrow f_{new} - f_{cur}$
  if $g < g^*$ **then** $g^* \leftarrow g$ and $goout^* \leftarrow goout$
end

Using algorithm *Modified Move*, each potential facility belonging to the current solution can be removed, i.e. be the facility *goout*. Furthermore, for each site we have to compute the objective function value corresponding to the new current solution for which facility *goout* is deleted and *goin* is inserted. Therefore, a new cost vector has to be sorted, which leads to a complexity of $O(M \log M)$ for each of the N values of *goout*. Thus, the number of operations needed for this algorithm is $O(MN \log M)$.

Updating First and Second Closest Facilities

In the *Modified Move* procedure both the closest (denoted by $d1(i)$) and the second closest facility (denoted by $d2(i)$) for each client i must be known in advance. Among formal variables in the description of algorithm *Modified Update* that follows, arrays $d1$ and $d2$ are both input and output variables. In this way, for each site i, if either $d1(i)$ or $d2(i)$ is removed from the current solution, we update their values. Furthermore, the current cost vector is also updated, being an input and an output variable, too. This is what distinguishes this procedure from the update approach presented in [93].

Algorithm 15.3: *Modified Update (c, goin, goout, M, N, d1, d2, cost)*

```
for each site i (i = 1, ..., M) do
    (* For clients whose closest facility is deleted, find a new one *)
    if d1(i) = goout then
        if c(i, goin) ≤ c(i, d2(i)) then
            d1(i) ← goin
            cost(i) ← c(i, goin)
        else
            d1(i) ← d2(i)
            cost(i) ← c(i, d2(i))
            (* Find second closest facility for client i *)
            find l* where c(i, l) is minimum (for l = 1, ..., N, l ≠ d1(i))
            d2(i) ← l*
        end
    else
        if c(i, d1(i)) > c(i, goin) then
            d2(i) ← d1(i) and d1(i) ← goin
            cost(i) ← c(i, goin)
        end
        else if c(i, goin) < c(i, d2(i)) then
            d2(i) ← goin
        end
        else if d2(i) = goout then
            find l* where c(i, l) is minimum (for l = 1, ..., N, l ≠ d1(i))
            d2(i) ← l*
        end
    end
end
```

The worst case complexity of the procedure *Modified Update* is $O(MN)$ as the index $d2(i)$ of the second closest facility must be recomputed without any additional information if it changes.

Modified Fast Interchange Heuristic

The modified fast interchange algorithm, that uses the procedures *Modified Move* and *Modified Update* described before, is as follows:

Algorithm 15.4: *Fast Interchange*

Initialization
Let x_{opt} be an initial solution
find the corresponding cost vector $cost_{opt}$ and objective function value f_{opt}
find closest and second closest facilities for each client $i = 1, \ldots, M$, i.e. find arrays $d1$ and $d2$
1 *Iteration step*
Set $g^* \leftarrow \infty$
for $goin = x_{opt}(N+1)$ *to* $x_{opt}(M)$ **do**
  (* Add facility $goin$ in the solution and find the best deletion *)
  Modified Move $(d1, d2, c, \lambda, x_{opt}, cost_{opt}, f_{opt}, goin, M, N, g, goout)$
  (* Algorithm 15.2 *)
  (* Keep the best pair of facilities to be interchanged *)
  if $g < g^*$ **then** $g^* \leftarrow g$, $goin^* \leftarrow goin$, $goout^* \leftarrow goout$
end
Termination
if $g^* \geq 0$ **then** Stop (* If no improvement in the neighborhood, Stop *)
Updating step
(* Update objective function value *)
$f_{opt} \leftarrow f_{opt} + g^*$
Update x_{opt}: interchange position of $x_{opt}(goout^*)$ with $x_{opt}(goin^*)$
(* Update closest, second closest facilities and cost vector *)
Modified Update $(c, goin^*, goout^*, M, N, d1, d2, cost_{opt})$
(* Algorithm 15.3 *)
Return to *Iteration step* (1)

The complexity of one iteration of this algorithm is $O(M^2 N \log M)$. This follows from the fact that procedure *Modified Move* is used $M - N$ times, its complexity is $O(MN \log M)$ and the complexity of *Modified Update* is $O(MN)$.

In the following section we describe a heuristic based on VNS that solves the OMP using the modified fast interchange algorithm.

Variable Neighborhood Search

The basic idea of VNS is to implement a systematic change of neighborhood within a local search algorithm (see Hansen & Mladenović [94], [96] and [95]). Exploration of these neighborhoods can be done in two ways. The first one consists of systematically exploring the smallest neighborhoods, i.e. those closest to the current solution, until a better solution is found. The second one consists of partially exploring the largest neighborhoods, i.e. those far from the current solution, by drawing a solution at random from them and beginning a (variable neighborhood) local search from there. The algorithm remains in the same solution until a better solution is found and then jumps there. We rank the neighborhoods to be explored in such a way that they are increasingly far from the current solution. We may view VNS as a way of escaping local optima, i.e. a "shaking" process, where movement to a neighborhood further from the current solution corresponds to a harder shake. In contrast to random restart, VNS allows a controlled increase in the level of the shake.

As in [93] let us denote by $S = \{s : s =$ set of N potential locations of the new facilities$\}$ a solution space of the problem. We say that the distance between two potential solutions s_1 and s_2 ($s_1, s_2 \in S$) is equal to k, if and only if they differ in k locations. Since S is a family of sets of equal cardinality, a (symmetric) distance function ρ can be defined as

$$\rho(s_1, s_2) = |s_1 \setminus s_2| = |s_2 \setminus s_1|, \forall s_1, s_2 \in S. \tag{15.12}$$

It can easily be checked that ρ is a metric function in S, thus S is a metric space. As in [93], the neighborhoods' structures are induced by metric ρ, i.e. k locations of facilities ($k \leq N$) from the current solution are replaced by k others. We denote by $\mathcal{N}_k$, $k = 1, \ldots, k_{max}$ ($k_{max} \leq N$) the set of such neighborhood structures and by $\mathcal{N}_k(s)$ the set of solutions forming neighborhood $\mathcal{N}_k$ of a current solution s. More formally

$$s_1 \in \mathcal{N}_k(s_2) \Leftrightarrow \rho(s_1, s_2) = k. \tag{15.13}$$

Note that the cardinality of $\mathcal{N}_k(s)$ is

$$|\mathcal{N}_k(s)| = \binom{N}{k}\binom{M-N}{k} \tag{15.14}$$

since k out of N facilities are dropped and k out of $M - N$ added into the solution. This number first increases and then decreases with k.

Note also that sets $\mathcal{N}_k(s)$ are disjoint, and their union, together with s, is S.

We now present the VNS algorithm for the OMP as pseudo-code:

Algorithm 15.5: *A VNS for the discrete OMP*

Initialization
Find arrays x_{opt}, $d1$ and $d2$, $cost_{opt}$ and f_{opt} as initialization of *Modified Fast Interchange*
the set of neighborhood structures $\mathcal{N}_k$, $k = 1, \ldots, k_{max}$ is induced by distance function ρ (see (15.12) and (15.13))
copy initial solution into the current one, i.e. copy x_{opt}, $d1$, $d2$, $cost_{opt}$ and f_{opt} into x_{cur}, $d1_{cur}$, $d2_{cur}$, $cost_{cur}$ and f_{cur}, respectively.
Choose stopping condition
Main step
$k \leftarrow 1$
repeat
  Shaking operator
  (* Generate a solution at random from the kth neighborhood, $\mathcal{N}_k$ *)
  for $j = 1$ *to* k **do**
    Take facility to be inserted *goin* at random
    Find facility to be deleted *goout* by using procedure *Modified* $Move(d1, d2, c, \lambda, x_{cur}, cost_{cur}, f_{cur}, goin, M, N, g, goout)$
    Find $d1_{cur}$, $d2_{cur}$ and $cost_{cur}$ for such interchange, i.e.
    Run *Modified* $Update(c, goin^*, goout^*, M, N, d1_{cur}, d2_{cur}, cost_{cur})$
    Update x_{cur} and f_{cur} accordingly
  end
  Local Search
  Apply algorithm *Modified Fast Interchange* (without *Initialization* step), with x_{cur}, f_{cur}, $cost_{cur}$, $d1_{cur}$ and $d2_{cur}$ as input and output values
  Move or not
  if $f_{cur} < f_{opt}$ **then**
    (* Save current solution to be incumbent; return to $\mathcal{N}_1$ *)
    $f_{opt} \leftarrow f_{cur}$; $x_{opt} \leftarrow x_{cur}$
    $d1 \leftarrow d1_{cur}$; $d2 \leftarrow d2_{cur}$; $cost_{opt} \leftarrow cost_{cur}$; and set $k \leftarrow 1$
  else
    (* Current solution is the incumbent; change neighborhood *)
    $f_{cur} \leftarrow f_{opt}$; $x_{cur} \leftarrow x_{opt}$
    $d1_{cur} \leftarrow d1$; $d2_{cur} \leftarrow d2$; $cost_{cur} \leftarrow cost_{opt}$; and set $k \leftarrow k + 1$
  end
until $(k = k_{max})$ *or (stopping condition is met)*

In the *Shaking operator* step the incumbent solution x_{opt} is perturbed in such a way that $\rho(x_{cur}, x_{opt}) = k$. Nevertheless, this step does not guarantee that x_{cur} belongs to $\mathcal{N}_k(x_{opt})$ due to randomization of the choice of *goin* and possible reinsertion of the same facility after it has left. Then, x_{cur} is used as initial solution for *Modified Fast Interchange* in *Local Search* step. If a better solution than x_{opt} is obtained, we move there and start again with small perturbations of this new best solution, i.e. $k \leftarrow 1$. Otherwise, we increase the distance between x_{opt} and the new randomly generated point, i.e. we set $k \leftarrow k + 1$. If k reaches k_{max} (this parameter can be chosen equal to N), we return to *Main* step, i.e. the main step can be iterated until some other stopping condition is met (e.g. maximum number of iterations, maximum CPU time allowed, or maximum number of iterations between two improvements). Note that the point x_{cur} is generated at random in *Shaking operator* step in order to avoid cycling which might occur if any deterministic rule were used.

In the following section computational results are reported which show the efficiency of these two heuristic approaches.

15.2.3 Computational Results

In order to test the heuristic procedures, we considered two groups of experiments. The instances belonging to the first group have been randomly generated with different combinations of the number of existing facilities, the number of new facilities, and the λ-vectors. The second group of experiments consists of N-median problems (whose optimal solutions are provided by Beasley [16]) and (k_1, k_2)-trimmed mean problems using the data publicly available electronically from *http://mscmga.ms.ic.ac.uk/info.html* (see Beasley [17]). The first group of experiments allows investigating the behavior of these heuristic approaches with different types of λ vectors. The second one helps in determining their capability to solve large problems.

In the following sections we describe in an exhaustive way these two groups of experiments and the corresponding computational results. All test problems were solved using a Pentium III 800 Mhz with 1 GB RAM.

The Experimental Design and Numerical Test

The first group of experimental data was designed considering four different values for the number of sites, $M = 15, 18, 25, 30$, four values for the number of new facilities, $N = \lceil \frac{M}{4} \rceil, \lceil \frac{M}{3} \rceil, \lceil \frac{M}{2} \rceil, \lceil \frac{M}{2} \rceil + 1$, and eight different λ-vectors (see Section 15.2.1). In total, 1920 problems were solved by both heuristic approaches (4 different values of $M \times 4$ values of $N \times 8$ values of $\lambda \times 15$ instances randomly generated for each combination of M, N and λ).

As mentioned in Subsection 15.2.1, the evolution program was ran twice (with *random* $P(0)$ and *greedy* $P(0)$). The solution of the evolution program is the best obtained by both procedures. Obviously, computation time increases with size but nevertheless, it is worthwhile due to the difficulty of solving the problem with a B&B method (see Section 15.1 and the quality of the solution obtained).

In order to compare the solutions given by the EP and the VNS with the optimal ones, the instances known from the literature have been used. These problems are of small size ($M = 15, 18, 25, 30$). The gap between the optimal solution and that one obtained by each heuristic algorithm is computed according to (15.11) with z^* denoting the optimal objective function value and z_{heu} denoting the objective function value of the best solution provided by the heuristic procedure.

Tables 15.6 and 15.7 show computational results for instances with $M = 30$, given by the EP and the VNS, respectively. In each row we present a summary of the outcomes for 15 replications of each combination (λ, N). Each row reports information about the frequency that the optimal solution is reached, the gap between the optimal solution and that provided by the corresponding heuristic approach and computing time.

Table 15.6. Computational results with $M = 30$ using EP.

Example		#opt.	Evolution program						B&B		
			gap (%)			CPU(s)			CPU(s)		
λ	N	found	aver	min	max	aver	min	max	aver	min	max
T1	8	11	2.45	0.00	21.78	22.57	22.30	22.88	167.19	40.92	408.73
	10	10	1.34	0.00	7.04	25.42	25.09	26.41	303.11	21.75	762.08
	15	11	0.87	0.00	5.00	32.25	31.97	32.67	274.74	28.03	562.25
	16	14	0.25	0.00	3.70	33.71	33.48	34.25	198.65	41.19	417.09
T2	8	10	7.79	0.00	55.56	22.46	22.03	22.73	13.54	2.14	44.19
	10	10	7.08	0.00	42.86	25.19	24.91	25.50	6.39	1.06	16.59
	15	12	4.44	0.00	25.00	32.16	31.81	32.45	4.81	0.00	14.84
	16	12	6.11	0.00	33.33	33.74	33.38	34.08	5.03	0.09	14.23
T3	8	11	1.09	0.00	7.92	22.35	22.16	22.69	86.44	39.11	197.08
	10	12	0.33	0.00	1.85	25.06	24.92	25.16	117.02	25.20	370.39
	15	11	0.88	0.00	3.70	31.99	31.72	32.27	125.19	15.36	246.95
	16	15	0.00	0.00	0.00	33.43	33.13	33.77	99.51	21.83	217.92
T4	8	12	0.37	0.00	2.47	22.36	22.14	22.72	56.37	5.06	125.25
	10	11	0.94	0.00	8.11	25.13	24.91	25.33	154.46	2.31	476.23
	15	15	0.00	0.00	0.00	32.04	31.80	32.22	649.56	20.81	1457.81
	16	15	0.00	0.00	0.00	33.43	33.22	33.78	498.32	42.88	1460.81
T5	8	11	1.00	0.00	5.88	22.41	22.17	22.73	160.75	45.42	332.02
	10	9	1.52	0.00	5.26	25.22	24.84	25.75	242.00	25.72	601.64
	15	14	0.83	0.00	12.50	32.20	31.63	32.81	209.68	15.55	502.16
	16	12	1.66	0.00	10.00	33.60	33.05	34.33	106.49	18.88	239.69

Continued on next page

Table 15.6 cont.: Computational results with $M = 30$ using EP.

Example		#opt.	Evolution program						B&B		
			gap (%)			CPU(s)			CPU(s)		
λ	N	found	aver	min	max	aver	min	max	aver	min	max
T6	8	10	1.30	0.00	9.52	22.54	22.25	22.83	131.82	15.64	272.22
	10	9	1.36	0.00	5.41	25.40	25.09	25.97	253.11	12.58	561.84
	15	14	0.39	0.00	5.88	32.38	31.97	32.91	327.44	26.34	777.64
	16	14	0.56	0.00	8.33	33.75	33.36	34.28	311.89	19.88	792.27
T7	8	8	1.66	0.00	4.94	22.53	22.33	22.72	170.57	40.69	368.23
	10	13	0.26	0.00	1.96	25.38	25.06	25.70	269.65	15.44	712.63
	15	12	1.08	0.00	8.00	32.41	32.02	33.41	179.30	12.47	442.06
	16	15	0.00	0.00	0.00	33.89	33.59	34.25	142.95	29.02	313.39
T8	8	10	2.83	0.00	21.05	22.60	22.33	23.30	151.38	38.84	328.09
	10	9	1.98	0.00	11.54	25.38	25.03	25.80	188.97	11.33	503.84
	15	14	0.44	0.00	6.67	32.39	31.98	32.69	101.19	0.91	436.17
	16	14	0.61	0.00	9.09	33.87	33.44	34.19	76.59	18.30	173.09

From Table 15.6 we can compute that the average gap over all instances is 1.61%. Moreover, in many cases, the optimal objective function value is reached, even for problems of type T2 (for which at least 60% of the 15 instances for each parameter combination were solved optimally). On average, the optimal solution is reached in 79.17% of the instances with $M = 30$. In addition, the required computing time is, in general (except for problems of type T2), shorter than that needed by the exact procedure. Observe that the average computing time required by the EP is 28.41 seconds, much shorter than 180.75 seconds given by the specific B&B method for the same instances.

Table 15.7. Computational results with $M = 30$ using VNS.

Example		#opt.	Var. Neigh. Search						B&B		
			gap (%)			CPU(s)			CPU(s)		
λ	N	found	aver	min	max	aver	min	max	aver	min	max
T1	8	10	2.13	0.00	11.86	0.31	0.20	0.47	167.19	40.92	408.73
	10	14	0.08	0.00	1.20	0.44	0.31	0.69	303.11	21.75	762.08
	15	13	0.39	0.00	3.13	0.71	0.52	0.86	274.74	28.03	562.25
	16	15	0.00	0.00	0.00	0.79	0.64	1.36	198.65	41.19	417.09

Continued on next page

Table 15.7 cont.: Computational results with $M = 30$ using VNS.

Example		#opt.	Var. Neigh. Search						B&B		
			gap (%)			CPU(s)			CPU(s)		
λ	N	found	aver	min	max	aver	min	max	aver	min	max
T2	8	5	14.74	0.00	55.56	0.30	0.11	0.50	13.54	2.14	44.19
	10	10	8.89	0.00	71.43	0.43	0.20	1.00	6.39	1.06	16.59
	15	12	5.78	0.00	50.00	0.60	0.38	1.05	4.81	0.00	14.84
	16	11	9.44	0.00	50.00	0.56	0.39	0.91	5.03	0.09	14.23
T3	8	9	2.61	0.00	11.29	0.27	0.22	0.41	86.44	39.11	197.08
	10	10	1.62	0.00	9.80	0.44	0.31	0.67	117.02	25.20	370.39
	15	14	0.19	0.00	2.78	0.76	0.52	1.06	125.19	15.36	246.95
	16	15	0.00	0.00	0.00	0.76	0.63	1.11	99.51	21.83	217.92
T4	8	13	0.28	0.00	2.47	0.27	0.19	0.38	56.37	5.06	125.25
	10	13	1.07	0.00	8.11	0.40	0.25	0.61	154.46	2.31	476.23
	15	15	0.00	0.00	0.00	0.60	0.42	0.78	649.56	20.81	1457.81
	16	15	0.00	0.00	0.00	0.62	0.45	0.78	498.32	42.88	1460.81
T5	8	8	2.61	0.00	8.00	0.29	0.20	0.47	160.75	45.42	332.02
	10	13	0.50	0.00	5.26	0.46	0.33	0.80	242.00	25.72	601.64
	15	15	0.00	0.00	0.00	0.64	0.48	0.73	209.68	15.55	502.16
	16	15	0.00	0.00	0.00	0.75	0.64	0.89	106.49	18.88	239.69
T6	8	13	0.60	0.00	6.67	0.32	0.19	0.56	131.82	15.64	272.22
	10	13	0.56	0.00	5.26	0.41	0.31	0.69	253.11	12.58	561.84
	15	15	0.00	0.00	0.00	0.70	0.47	1.17	327.44	26.34	777.64
	16	14	0.56	0.00	8.33	0.70	0.48	1.20	311.89	19.88	792.27
T7	8	9	1.50	0.00	7.46	0.26	0.20	0.41	170.57	40.69	368.23
	10	14	0.11	0.00	1.67	0.41	0.31	0.58	269.65	15.44	712.63
	15	14	0.32	0.00	4.76	0.73	0.50	1.11	179.30	12.47	442.06
	16	15	0.00	0.00	0.00	0.75	0.53	1.14	142.95	29.02	313.39
T8	8	10	1.46	0.00	7.69	0.28	0.19	0.41	151.38	38.84	328.09
	10	11	1.27	0.00	7.69	0.44	0.31	0.72	188.97	11.33	503.84
	15	15	0.00	0.00	0.00	0.73	0.45	1.06	101.19	0.91	436.17
	16	13	1.96	0.00	22.22	0.65	0.44	0.89	76.59	18.30	173.09

From Table 15.7 we can observe that the average gap provided by the VNS, 1.83%, over all instances is slightly higher than that given by the EP, 1.61%. Nevertheless, the optimal objective function is reached in 83.54% of the instances. In addition, the average computing time, 0.52 seconds, is much shorter than that required by the EP (28.41 seconds), and therefore, shorter than that given by the B&B algorithm (180.75 seconds).

It should be mentioned that the performance of both procedures on problems of type T2 (i.e. N-center problems) is rather poor, since the gap obtained is relatively large. However, the quality of the solution given by the EP for problems of type T2 is superior to that provided by the VNS. We point out that the new formulation proposed by Elloumi et al. [71] specifically developed for the N-center problem yields results considerably better than EP.

To compare both heuristic procedures on examples similar to real-life problems, instances of larger size ($M = 100$) have been generated and solved. The structure of these examples is similar to that presented for instances with $M = 30$, i.e. with four different values of N, eight of λ, and 15 replications for each combination (N, λ). The optimal solutions of these instances are not available; therefore, we can only compare the results given by both heuristic approaches. To this aim, we compute the relation between the solution given by the EP and that provided by the VNS, as follows

$$\text{ratio} = \frac{z_{\text{EP}}}{z_{\text{VNS}}}, \tag{15.15}$$

with z_{EP} denoting the objective function value of the best solution obtained by the evolution program, and z_{VNS} denoting the objective function value of the solution obtained by the variable neighborhood search. Therefore,

$$\text{ratio} \begin{cases} > 1 \text{ if VNS provides a solution better than that given by EP} \\ = 1 \text{ if VNS and EP provide the same solution} \\ < 1 \text{ if VNS provides a solution worse than that given by EP} \end{cases} \tag{15.16}$$

Figure 15.2 shows a summary of the results obtained among the 480 test problems.

From Figure 15.2 we can conclude that the quality of the solution provided by the VNS is usually superior to that given by the EP. Furthermore, the computing time required by VNS (63.22 seconds, on average) is also shorter than that required by EP (105.45 seconds, on average).

In the following section we investigate the behavior of both heuristic approaches for solving large N-median and (k_1, k_2)-trimmed mean problems.

Additional Tests for Large Problems

The exact procedures presented in Chapter 14 and Section 15.1 are not appropriate for solving large instances of the OMP. Therefore, we call upon the existing data often used in the literature for N-median (see [17]). In addition, these data have been used to solve the (k_1, k_2)-trimmed mean problem. To this aim we have set $k_1 = N + \lceil \frac{M}{10} \rceil$ and $k_2 = \lceil \frac{M}{10} \rceil$.

The second group of experiments consists of solving N-median and (k_1, k_2)-trimmed mean problems for 40 large instances. We denote these instances by

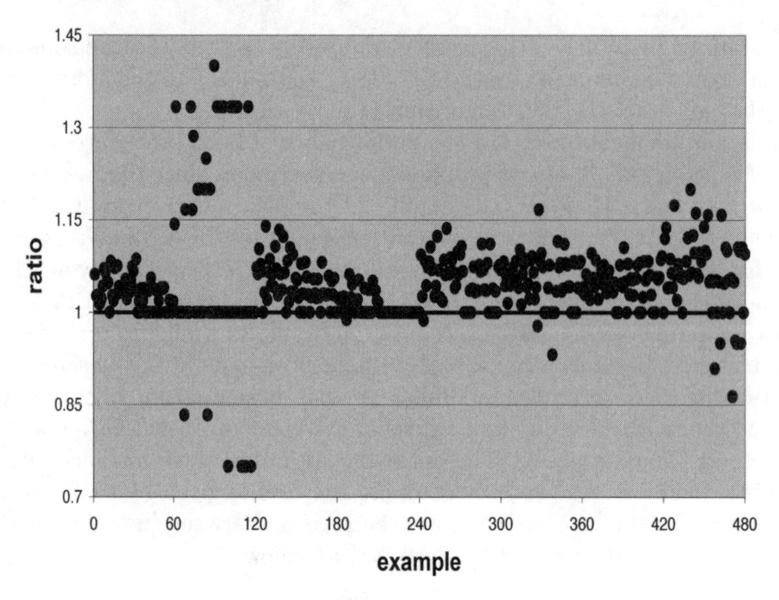

Fig. 15.2. Comparison between the solutions given by EP and VNS for problems with $M = 100$

pmed1,..., *pmed40*. The data is available at http://mscmga.ms.ic.ac.uk/info.html. Exhaustive information about how these problems were generated can be found in [16]. The number of existing facilities, M, in these instances varies from 100 to 900 in steps of 100, and the number of new facilities, N, takes values equal to 5, 10, $\frac{M}{10}$, $\frac{M}{5}$ and $\frac{M}{3}$ or, depending on the case, rounded to the nearest integer.

The optimal solutions of the N-median problems are given by [16], but those according to the (k_1, k_2)-trimmed mean problems are not available.

Therefore, using the first type of problem (N-median problems) we estimate the efficiency of our heuristic approaches comparing their results with the optimal solutions. The second type of problem allows pointing out the capability of these approaches to provide solutions for large instances of new facility location problems (such as the (k_1, k_2)-trimmed mean problem) in a reasonable computing time.

Large N-Median Problems

For solving large N-median problems, as before, the EP is run twice (once initialized with *random* $P(0)$ and once more with *greedy* $P(0)$), and the best solution obtained by both procedures is taken.

Due to the large increase in computing time required by VNS, a stopping condition was necessary. After some preliminary tests, the maximum number of iterations allowed was fixed at 50.

Figure 15.3 shows the behavior of the VNS against the required computing time (in seconds) on instance *pmed9* when the N-median problem is solved.

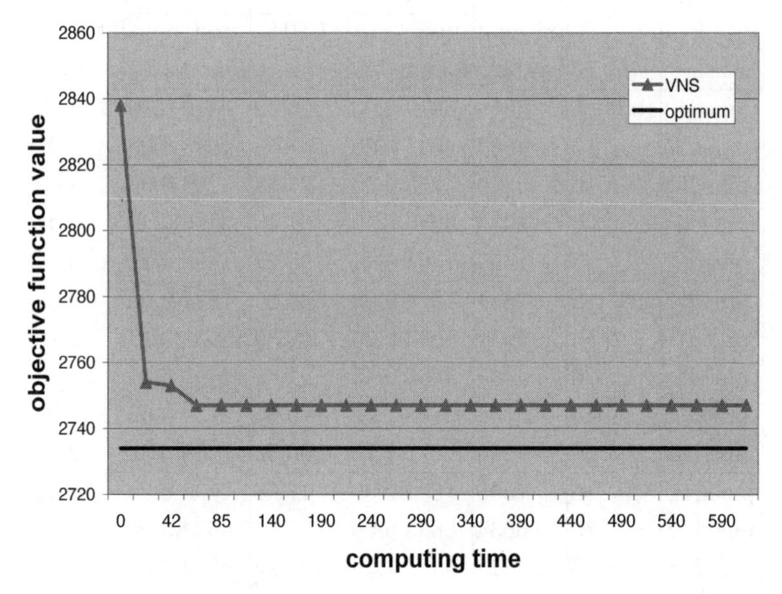

Fig. 15.3. Behavior of the VNS heuristic when solving the *pmed9* instance of the N-median problem

From Figure 15.3 we conclude that convergence of the VNS is very fast. After 18 seconds (in the first iteration) the solution is much improved (from 2838 to 2754, i.e. reducing the gap from 3.80% to 0.73%). Nevertheless, the VNS stops after 48 iterations, with a solution equal to 2747 (and a gap equal to 0.48%) but requiring almost 615 seconds.

We observed similar behavior for other instances, which require much more time (some of them have been solved after fifty iterations requiring more than eight hours). For this reason, the maximum number of iterations for solving large N-median problems was fixed at 5 instead of 50.

Computational results for large N-median problems are shown in Table 15.8.

Table 15.8. Computational results for large N-median problems ([16]).

Problem Name	M	N	optimal value	Evolution Program best found	gap (%)	CPU (s)	VNS best found	gap (%)	CPU (s)
pmed1	100	5	5819	5819	0.00	25.42	5819	0.00	1.19
pmed2		10	4093	4093	0.00	37.55	4093	0.00	2.97
pmed3		10	4250	4250	0.00	37.88	4250	0.00	3.00
pmed4		20	3034	3046	0.40	61.48	3046	0.40	5.98
pmed5		33	1355	1361	0.44	93.22	1358	0.22	6.81
pmed6	200	5	7824	7824	0.00	36.25	7824	0.00	7.95
pmed7		10	5631	5645	0.25	55.39	5639	0.14	12.72
pmed8		20	4445	4465	0.45	91.81	4457	0.27	21.05
pmed9		40	2734	2762	1.02	170.25	2753	0.69	41.98
pmed10		67	1255	1277	1.75	290.53	1259	0.32	72.22
pmed11	300	5	7696	7696	0.00	47.98	7696	0.00	12.52
pmed12		10	6634	6634	0.00	75.63	6634	0.00	26.02
pmed13		30	4374	4432	1.33	193.22	4374	0.00	87.92
pmed14		60	2968	2997	0.98	359.58	2969	0.03	241.95
pmed15		100	1729	1749	1.16	580.98	1739	0.58	363.39
pmed16	400	5	8162	8183	0.26	56.89	8162	0.00	24.36
pmed17		10	6999	6999	0.00	95.08	6999	0.00	47.30
pmed18		40	4809	4880	1.48	320.38	4811	0.04	275.69
pmed19		80	2845	2891	1.62	604.36	2864	0.67	469.30
pmed20		133	1789	1832	2.40	963.44	1790	0.06	915.17
pmed21	500	5	9138	9138	0.00	70.14	9138	0.00	27.39
pmed22		10	8579	8669	1.05	116.59	8669	1.05	64.25
pmed23		50	4619	4651	0.69	486.08	4619	0.00	443.23
pmed24		100	2961	3009	1.62	924.66	2967	0.20	1382.84
pmed25		167	1828	1890	3.39	1484.13	1841	0.71	2297.25
pmed26	600	5	9917	9919	0.02	84.34	9917	0.00	48.45
pmed27		10	8307	8330	0.28	136.53	8310	0.04	127.63
pmed28		60	4498	4573	1.67	673.30	4508	0.22	965.48
pmed29		120	3033	3099	2.18	1268.89	3036	0.10	2758.56
pmed30		200	1989	2036	2.36	2043.33	2009	1.01	3002.34

Continued on next page

Table 15.8 cont.: Computational results for large N-median problems.

Problem Name	M	N	optimal value	Evolution Program best found	gap (%)	CPU (s)	VNS best found	gap (%)	CPU (s)
pmed31	700	5	10086	10086	0.00	92.67	10086	0.00	56.02
pmed32		10	9297	9319	0.24	156.50	9301	0.04	165.27
pmed33		70	4700	4781	1.72	894.19	4705	0.11	2311.03
pmed34		140	3013	3100	2.89	1762.69	3024	0.37	5384.19
pmed35	800	5	10400	10400	0.00	109.86	10400	0.00	88.50
pmed36		10	9934	9947	0.13	182.06	9934	0.00	200.97
pmed37		80	5057	5126	1.36	1190.25	5066	0.18	2830.30
pmed38	900	5	11060	11060	0.00	120.14	11060	0.00	150.53
pmed39		10	9423	9423	0.00	207.75	9423	0.00	200.73
pmed40		90	5128	5188	1.17	1492.59	5141	0.25	4774.38

By analyzing this table we conclude that the average gap does not reach 1% for both heuristic procedures, being 0.86% for the EP and only 0.19% for the VNS. Moreover, the optimal objective function value is obtained 12 and 17 times, among the 40 test problems, for the EP and the VNS, respectively. Therefore, both approaches perform well on large N-median problems, the quality of the VNS being better. However, for both methods there is a tradeoff between quality of the solution and computing time required to obtain this solution: the average time is equal to 442.35 seconds for the EP and 747.97 seconds for the VNS. Notice that the maximal computing time required by the EP does not exceed 35 minutes, and that required by the VNS reaches almost 90 minutes.

Observe that the quality of the solutions provided by the VNS (0.19%, in average) is comparable with the one (0.18%, in average) given by the method specifically developed for N-median by Hansen et al. [97]. Nevertheless, as expected, computing time required by the VNS developed for the OMP is larger than that provided in [97].

Large (k_1, k_2)-Trimmed Mean Problems

As above, to solve large (k_1, k_2)-trimmed mean problems, the best solution obtained after running twice the EP (once initialized with *random* $P(0)$ and once more with *greedy* $P(0)$) is taken.

Furthermore, as before, a stopping condition based on fixing a maximum number of iterations was considered. Again, we observed that the VNS converged very fast when solving (k_1, k_2)-trimmed mean problems and again, computing time required for 50 iterations was too long. Therefore, the maximum number of iterations for solving large (k_1, k_2)-trimmed mean problems was fixed at 5 instead of 50.

Computational results obtained for such problems are shown in Table 15.9. Since the optimal solutions are not available, this table reports information about the relation between both heuristic approaches (see (15.15)).

Table 15.9. Comp. results for large (k_1, k_2)-trimmed mean problems ([16]).

Problem			Evolution Program		VNS		Ratio
			best	CPU	best	CPU	EP/
Name	M	N	found	(s)	found	(s)	VNS
pmed1	100	5	4523	25.20	4523	1.27	1.000
pmed2		10	2993	36.98	2987	3.80	1.002
pmed3		10	3067	36.91	3074	2.80	0.998
pmed4		20	2153	60.80	2142	6.98	1.005
pmed5		33	829	92.08	818	8.22	1.013
pmed6	200	5	6064	35.52	6079	7.88	0.998
pmed7		10	4225	54.17	4206	13.41	1.005
pmed8		20	3248	91.95	3182	28.30	1.021
pmed9		40	1831	167.61	1816	66.39	1.008
pmed10		67	849	274.09	829	75.91	1.024
pmed11	300	5	5979	47.75	5979	13.30	1.000
pmed12		10	5021	73.83	5021	25.86	1.000
pmed13		30	3175	183.25	3133	97.80	1.013
pmed14		60	2027	346.42	1957	303.64	1.036
pmed15		100	1181	549.67	1133	415.80	1.042
pmed16	400	5	6341	56.06	6341	24.13	1.000
pmed17		10	5440	89.30	5413	43.83	1.005
pmed18		40	3463	309.50	3443	261.86	1.006
pmed19		80	1973	618.88	1933	779.77	1.021
pmed20		133	1191	1000.41	1152	1108.48	1.034
pmed21	500	5	7245	71.69	7245	24.22	1.000
pmed22		10	6749	117.88	6722	58.58	1.004
pmed23		50	3379	461.50	3306	639.95	1.022
pmed24		100	2068	888.27	2005	1455.81	1.031
pmed25		167	1198	1524.86	1151	2552.02	1.041

Continued on next page

Table 15.9 cont.: Comp. results for large (k_1, k_2)-trimmed mean problems.

Problem Name	M	N	Evolution Program best found	Evolution Program CPU (s)	VNS best found	VNS CPU (s)	Ratio EP/ VNS
pmed26	600	5	7789	87.30	7787	48.11	1.000
pmed27		10	6481	141.97	6444	141.70	1.006
pmed28		60	3304	687.42	3210	1113.89	1.029
pmed29		120	2087	1249.78	2006	3178.69	1.040
pmed30		200	1359	1976.77	1308	4942.75	1.039
pmed31	700	5	8047	90.81	8046	66.16	1.000
pmed32		10	7318	148.77	7280	162.97	1.005
pmed33		70	3463	857.47	3413	2377.72	1.015
pmed34		140	2083	1624.61	2023	5657.56	1.030
pmed35	800	5	8191	102.58	8191	72.58	1.000
pmed36		10	7840	170.38	7820	201.64	1.003
pmed37		80	3684	1086.13	3604	3170.70	1.022
pmed38	900	5	8768	111.61	8720	140.84	1.006
pmed39		10	7398	189.19	7360	313.03	1.005
pmed40		90	3768	1372.98	3718	5422.73	1.013

From Table 15.9 we can observe that after a reasonable computing time the EP and the VNS solve large (k_1, k_2)-trimmed mean problems. The computing time required by the EP (427.81 seconds, on average) is much shorter than that needed by the VNS (875.78 seconds, on average). Nevertheless, in all but two cases the VNS provides a better solution than EP.

16

Related Problems and Outlook

16.1 The Discrete OMP with $\lambda \in \Lambda_M^{\leq}$

In this section we present a specific formulation for the discrete OMP with λ-weights in non-decreasing order, denoted by $\mathsf{OMP}_{\leq}$. This formulation is based in a linear programming formulation for the problem of minimizing the sum of the k-largest linear functions out of M, first proposed by Ogryczak & Tamir [154]. This formulation allows us to give a compact mixed-integer linear programming formulation for the discrete $\mathsf{OMP}_{\leq}$. In this chapter we study this reformulation, some of their properties and report some computational results that show its efficiency.

16.1.1 Problem Formulation

Let us first introduce the notation required to derive the formulation. Let $\lambda = (\lambda_1, \ldots, \lambda_M) \in \Lambda_M^{\leq}$ be the λ vector with components arranged in non-decreasing order; and for convenience define $\lambda_0 = 0$.

As shown in Section 13.1, for a given set of facilities $X \subseteq A$, we denote by $c(A, X) = (c_1(X), \ldots, c_M(X))$ the cost vector corresponding to satisfy the total demand of each client $k \in A$. In addition, we denote by $c_{\geq}(X) = (c_{(1)}(X), \ldots, c_{(M)}(X))$ the cost vector whose entries are in non-increasing order, i.e. $c_{(1)}(X) \geq \cdots \geq c_{(M)}(X)$.

Finally, let variables x_j and y_{kj} be defined as in (13.9) and (13.10), respectively, see Section 13.2.2. Thus, x_j is a binary variable which takes the value 1 if a new facility is built at site j and 0, otherwise, for each $j = 1, \ldots, M$. The variable y_{kj} is also binary and takes the value 1 if the demand of client k is satisfied by a facility located at site j and 0, otherwise, for each $j, k = 1, \ldots, M$.

Observe that the solution of the $\mathsf{OMP}_{\leq}$ coincides with that given by

$$\min_{X \subseteq A\,,\,|X|=N} \sum_{i=1}^{M} \lambda_{M-i+1}\, c_{(i)}(X). \tag{16.1}$$

Denote $r_k(X) = \sum_{i=1}^{k} c_{(i)}(X)$, the sum of the k largest allocation costs induced by set of facilities X. It is clear that Problem (16.1) can be equivalently written:

$$\min_{X \subseteq A\,,\,|X|=N} \sum_{i=1}^{M} (\lambda_{M-i+1} - \lambda_{M-i}) r_i(X). \tag{16.2}$$

Minimizing the function $r_i(X)$ can be done, according with [154], solving the following linear program:

$$\begin{aligned}
&\min\ it_i + \sum_{k=1}^{M} d_{ki} \\
&\text{s.t. } d_{ki} \geq \sum_{j=1}^{M} c_{kj} y_{kj} - t_i, \quad \forall\, i,k = 1,\dots,M, \\
&\quad d_{ki}, y_{kj} \geq 0, \qquad\qquad \forall\, i,j,k = 1,\dots,M, \\
&\quad t_i \text{ unrestricted.}
\end{aligned}$$

Combining the latter formulation with (16.2), we get the appropriate reformulation of Problem (16.1):

$$\begin{aligned}
(\mathsf{OMP}_{\leq}) \quad & \min \sum_{i=1}^{M} (\lambda_{M-i+1} - \lambda_{M-i})(i\,t_i + \sum_{k=1}^{M} d_{ki}) & \\
\text{s.t.} \quad & d_{ki} \geq \sum_{j=1}^{M} c_{kj}\, y_{kj} - t_i & \forall\, i,k = 1,\dots,M \\
& \sum_{j=1}^{M} x_j = N & \\
& \sum_{j=1}^{M} y_{kj} = 1 & \forall\, k = 1,\dots,M \\
& x_j \geq y_{kj} & \forall\, j,k = 1,\dots,M \\
& x_j \in \{0,1\}, \quad d_{ki}, y_{kj} \geq 0 & \forall\, i,j,k = 1,\dots,M \\
& t_i \text{ unrestricted} & \forall\, i = 1,\dots,M.
\end{aligned}$$

Notice that this is a valid formulation as a mixed-integer linear program for N-median, N-center or N-facility k-centra problems.

Computational results corresponding to ($\mathsf{OMP}_{\leq}$) (using standard software) obtained for N-median and N-center instances are reported in [55]. In the following section, we compare the computational results provided by ($\mathsf{OMP}_{\leq}$) against those given by the B&B method proposed in Section 15.1 for the general discrete OMP.

16.1.2 Computational Results

In this section, we compare the computational performance of $(\mathsf{OMP}_{\leq})$ against that of the B&B method with the max-regret branching rule, developed in Section 15.1.

As mentioned above, among the problems proposed in Section 14.3 only N-median, N-center and k-centra problems (problems type T1, T2, T3) can be solved with the formulation $(\mathsf{OMP}_{\leq})$. Table 16.1 shows information about the computational results obtained by solving these problems. In order to illustrate the performance of the new model, we report results for problems with $M = 30$ (with four values of N and 15 instances for each combination) in Table 16.1.

Table 16.1. Numerical results for problems with $M = 30$. All results are averages taken over 15 problem instances.

Problem		B&B Method			$(\mathsf{OMP}_{\leq})$		
Type	N	gap(%)	# nodes	CPU(s)	gap(%)	# nodes	CPU(s)
T1	8	50.8	376661.7	167.19	1.13	1.67	0.37
	10	43.1	698401.1	303.11	0.56	0.80	0.31
	15	28.2	710577.9	274.74	0.29	0.20	0.25
	16	25.0	529510.7	198.65	0.00	0.00	0.19
T2	8	55.9	30463.8	13.54	56.17	988.33	16.63
	10	47.0	15012.3	6.39	56.65	593.20	7.80
	15	37.2	12473.1	4.81	62.37	94.07	1.64
	16	35.9	13416.8	5.03	63.28	96.33	1.51

All examples were solved with the commercial package ILOG CPLEX 6.6 and with a Pentium III 800 Mhz with 1 GB RAM using the C++ modeling language ILOG Planner 3.3 (see [109]).

In general, computational results for the model $(\mathsf{OMP}_{\leq})$ are usually better than those obtained with the B&B method. We should stress that the B&B method developed in Section 15.1 is valid for each choice of $\lambda \in \mathbb{R}^M_+$, and $(\mathsf{OMP}_{\leq})$ has been specially developed for those cases with $\lambda \in \Lambda_M^{\leq}$.

From Table 16.1 we observe that for problems of type T1 (N-median problems), the model $(\mathsf{OMP}_{\leq})$ requires almost three orders of magnitude less computing time than the B&B method. Nevertheless, the differences in the required computing time are not so significant for problems of type T2 (N-center problems). Differences regarding the number of nodes are even more noteworthy, being for the B&B much larger than for $(\mathsf{OMP}_{\leq})$. We report the average root node gap, but note that whilst this may be indicative of the quality of an integer programming formulation, it is less meaningful for the B&B method, where the bounds are very weak high in the tree but improve rapidly deeper in the tree. However, the gap provided by ($\mathsf{OMP}_{\leq}$) is much better than that given by the B&B method for those instances of type T1. In contrast, for problems of type T2, the gap provided by the B&B method is slightly smaller than that given by ($\mathsf{OMP}_{\leq}$).

16.2 Conclusions and Further Research

In this chapter we have presented an interesting special case of the discrete OMP. The model was proposed by Ogryczak & Tamir [154] and can be applied to all cases for which $\lambda \in \Lambda_M^{\leq}$, i.e. for solving the $\mathsf{OMP}_{\leq}$. Computational results show its efficiency for solving N-median and N-center problems in contrast to the B&B method. The performance of ($\mathsf{OMP}_{\leq}$) for solving N-median problems is better than that observed for solving N-center problems. Moreover, additional tests on large N-median and N-center problems had been conducted, using the data publicly available electronically from *http://mscmga.ms.ic.ac.uk/info.html* (see Beasley [17]). However, due to the poor results obtained, they were not included in Section 16.1.2. On the one hand, for large N-median problems, only 15 among the 40 instances could be solved. On the other hand, for large N-center problems, none of the 40 instances could be solved. Thus, we think that further improvements on the formulation ($\mathsf{OMP}_{\leq}$) are needed in order to solve large instances. Further details on special cases and extensions, as well as extensive computational results on these models are reported in [55].

To conclude this chapter we would like to outlook some other extensions of the discrete OMP that are worth to be investigated: 1) new formulations giving rise to alternative algorithms and solution methods; 2) a polyhedral description of this family of problems; 3) capacitated versions of the discrete OMP; and 4) the multicriteria analysis of the discrete OMP. In our opinion these approaches will provide new avenues of research within the field of discrete location models.

References

1. P.K. Agarwal, B. Aronov, T.M. Chan, and M. Sharir. On levels in arrangements of lines, segments, planes and triangles. *Discrete and Computational Geometry*, 19:315–331, 1998.
2. M. Ajtai, J. Komlos, and E. Szemeredi. Sorting in $c \log n$ parallel steps. *Combinatorica*, 3:1–19, 1983.
3. S. Alstrup, P.W. Lauridsen, P. Sommerlund, and M. Thorup. Finding cores of limited length. In F. Dehne, A. Rau-Chaplin, J.R. Sack, and R. Tamassia, editors, *Algorithms and Data Structures*, number 1271 in Lecture Notes in Computer Science, pages 45–54. Springer, 1997.
4. G. Andreatta and F.M. Mason. Properties of the k-centra in a tree network. *Networks*, 15:21–25, 1985.
5. Y. P. Aneja and M. Parlar. Algorithms for weber facility location in the presence of forbidden regions and/or barriers to travel. *Transactions Science*, 28:70–76, 1994.
6. H. Attouch. *Variational convergence for functions and operators.* Pitman Advanced Publishing Program, 1984.
7. G. Aumann and K. Spitzmüller. *Computerorientierte Geometrie.* BI-Wissenschaftsverlag, 1993.
8. F. Aurenhammer and H. Edelsbrunner. An optimal algorithm for constructing the weighted voronoi diagram in the plane. *Pattern Recognition*, 17:251–257, 1984.
9. F. Aurenhammer and R. Klein. Voronoi diagrams. In J.-R. Sack and J. Urrutia, editors, *Handbook of Computational Geometry.* Elsevier Science Publishers B.V. North-Holland, 1998.
10. T. Bäck, F. Hoffmeister, and H.P. Schwefel. A survey of evolution strategies. In R.K. Belew and L.B. Booker, editors, *Genetic Algorithms, Proceedings of the Fourth International Conference*, pages 2–9. Morgan Kaufmann, San Mateo, CA, 1991.
11. M. Baronti, E. Casini, and P.L. Papini. Equilateral sets and their central points. *Rend. Mat. (VII)*, 13:133–148, 1993.
12. M. Baronti and G. Lewicki. On some constants in banach spaces. *Arch. Math.*, 62:438–444, 1994.

13. M. Baronti and P.L. Papini. Diameters, centers and diametrically maximal sets. *Supplemento ai Remdiconti del Circolo Matematico di Palermo, Serie II*, 38:11–24, 1995.
14. Y. Bartal. On approximating arbitrary metrics by tree metrics. In *Proceedings of the 30-th Ann. Symp. on Foundation of the Computer Sciences*, pages 161–168, 1998.
15. FL. Bauer, J. Stoer, and C. Witzgall. Absolute and monotonic norms. *Numer. Math.*, 3:257–264, 1961.
16. J.E. Beasley. Solving large p-median problems. *European Journal of Operational Research*, 21:270–273, 1985.
17. J.E. Beasley. OR-Library: Distributing test problems by electronic mail. *Journal of the Operational Research Society*, 41(11):1069–1072, 1990.
18. R.I. Becker, I. Lari, and A. Scozzari. Efficient algorithms for finding the (k, l)-core of tree networks. *Networks*, 40:208–215, 2003.
19. R.I. Becker and Y. Perl. Finding the two-core of a tree. *Discrete Applied Mathematics*, 11:103–113, 1985.
20. C. Benítez, M. Fernández, and M.L. Soriano. Location of the Fermat-Torricelli medians of three points. *Trans. Amer. Math. Soc.*, 354(12):5027–5038, (2002).
21. J.L. Bentley and T. Ottmann. Algorithms for reporting and counting geometric intersections. *IEEE Trans. on Computers*, 28:643–647, 1979.
22. M. Blum, R.W. Floyd, V. Pratt, R.L. Rivest, and R.E. Tarjan. Time bounds for selection. *Journal of Computer and Systems Sciences*, 7:448–461, 1973.
23. B. Boffey and J.A. Mesa. A review of extensive facility location in networks. *European Journal of Operational Research*, 95:592–600, 1996.
24. E. Bohne and W.-D. Klix. *Geometrie: Grundlagen für Anwendungen*. Fachbuchverlag, 1995.
25. N. Boland, P. Domínguez-Marín, S. Nickel, and J. Puerto. Exact procedures for solving the Discrete Ordered Median Problem. ITWM Bericht 47, Fraunhofer Institut für Techno– und Wirtschaftsmathematik (ITWM), Kaiserslautern, Germany, 2003. To appear in Computers and Operations Research.
26. J.A. Bondy and U.S.R. Murty. *Graph Theory with Applications*. North-Holland, 1976.
27. S.D. Brady and R.E. Rosenthal. Interactive computer graphical minimax location of multiple facilities with general constraints. *AIIE Transactions*, 15:242–252, 1983.
28. J. Bramel and D. Simchi-Levi. *The Logic of Logistics: Theory, Algorithms and Applications for Logistics Management*. Springer, 1997.
29. H. Buhl. Axiomatic considerations in multiobjective location theory. *European Journal of Operational Research*, 37:363–367, 1988.
30. R.E. Burkard. Locations with spatial interactions: The Quadratic Assignment Problem. In P.B. Mirchandani and R.L. Francis, editors, *Discrete Location Theory*, pages 387–437. Wiley, 1990.
31. R.E. Burkard, H.W. Hamacher, and R. Günter. Sandwich approximation of univariate convex functions with an application to separable convex programming. *Naval Research Logistics*, 38:911–924, 1991.
32. E. Carrizosa, E. Conde, F.R. Fernández, and J. Puerto. An axiomatic approach to the centdian criterion. *Location Science*, 3:165–171, 1994.
33. E. Carrizosa, F.R. Fernández, and J. Puerto. An axiomatic approach to location criteria. In Fr. Orban and J.P. Rasson, editors, *Proceedings of the 5th*

Meeting of the Euro Working Group in Locational, FUND, Namur, Belgium, 1990.
34. M. Charikar, C. Chekuri, A. Goel, and S. Guha. Rounding via trees: Deterministic approximation algorithms for group Steiner trees and k-median. In *Proceedings of the 30-th Ann. Symp. on Foundation of the Computer Sciences*, pages 114–123, 1998.
35. M. Charikar and S. Guha. Improved combinatorial algorithms for the facility location and k-median problems. In *Proceedings of the 31-st Ann. Symp. on Foundation of the Computer Sciences*, pages 378–388, 1999.
36. M. Charikar, S. Guha, E. Tardos, and D.B. Shmoys. A constant-factor approximation algorithm for the k-median problem. In *Proceedings of the 31-st Ann. ACM Symp. on Theory of Computing*, pages 1–10, 1999.
37. P.C. Chen, P. Hansen, and B. Jaumard. On-line and off-line vertex enumeration by adjacency lists. *Operations Research Letters*, 10:403–409, 1992.
38. P.C. Chen, P. Hansen, B. Jaumard, and H. Tuy. Weber's problem with attraction and repulsion. *Journal of regional science*, 32(4):467–486, 1992.
39. L. P. Chew and R. L Drysdale III. Voronoi diagrams based on convex distance functions. In *1st Annual ACM Symposium on Computational Geometry*, pages 235–244, 1985.
40. G. Cho and D.X. Shaw. A depth-first dynamic programming algorithm for the tree knapsack problem. *INFORMS Journal on Computing*, 9:431–438, 1997.
41. F.A. Claessens, F.A. Lotsma, and F.J. Vogt. An elementary proof of Paelinck's theorem on the convex hull of ranked criterion weights. *European Journal of Operational Research*, 52:255–258, 1991.
42. E. Cohen and N. Megiddo. Maximizing concave functions in fixed dimension. In P.M. Pardalos, editor, *Complexity in Numerical Optimization*, pages 74–87. World Scientific, 1993.
43. R. Cole. Slowing down sorting networks to obtain faster algorithms. *J. ACM*, 34:168–177, 1987.
44. M. Colebrook Santamaría. *Desirable and undesirable single facility location on networks with multiple criteria.* PhD thesis, University of La Laguna, Spain, 2003.
45. E. Conde. Optimización global en localización. In J. Puerto, editor, *Lecturas en Teoría de Localización*, pages 47–68. Secretariado de Publicaciones de la Universidad de Sevilla, 1996.
46. A. G. Corbalan, M. Mazon, and T. Recio. Geometry of bisectors for strictly convex distances. *International journal of computational geometry and applications*, 6:45–58, 1996.
47. G. Cornuejols, G.L. Nemhauser, and L.A. Wolsey. The Uncapacitated Facility Location Problem. In P.B. Mirchandani and R.L. Francis, editors, *Discrete Location Theory*, pages 119–171. Wiley, 1990.
48. J.P. Crouseix and R. Kebbour. On the convexity of some simple functions of ordered samples. *JORBEL*, 36:11–25, 1996.
49. M. S. Daskin. *Network and Discrete Location: Models, Algorithms, and Applications.* Wiley, 1995.
50. L. Davis, editor. *Handbook of Genetic Algorithms.* Van Nostrand Reinhold, New York, 1991.
51. R. Dawkins. *The selfish gene.* Oxford University Press, second edition, 1989.
52. P.M. Dearing, R.L. Francis, and T.J. Lowe. Convex location problems on tree networks. *Operations Research*, 24:628–642, 1976.

53. T.K. Dey. Improved bounds for planar k-sets and related problems. *Discrete and Computational Geometry*, 19:373–382, 1998.
54. P. Domínguez-Marín. A geometrical method to solve the planar 1-facility Ordered Weber Problem with polyhedral gauges. Diploma thesis, University of Kaiserslautern, Germany, 2000.
55. P. Domínguez-Marín. *The Discrete Ordered Median Problem: Models and Solution Methods.* Kluwer, 2003. PhD thesis.
56. P. Domínguez-Marín, S. Nickel, N. Mladenović, and P. Hansen. Heuristic procedures for solving the discrete ordered median problem. Technical report, Fraunhofer Institut für Techno– und Wirtschaftsmathematik (ITWM), Kaiserslautern, Germany, 2003, to appear in Computers & Operations Research.
57. Z. Drezner. Constrained location problems in the plane and on a sphere. *AIIE Transportation*, 12:300–304, 1983.
58. Z. Drezner. The Weber problem on the plane with some negative weights. *INFOR*, 29(2), 1991.
59. Z. Drezner, editor. *Facility Location. A survey of applications and methods.* Springer, 1995.
60. Z. Drezner and A.J. Goldman. On the set of optimal points to Weber problems. *Transportation Science*, 25:3–8, 1991.
61. Z. Drezner and H.W. Hamacher, editors. *Facility Location: Applications and Theory.* Springer, 2002.
62. R. Durier. A general framework for the one center location problem. In *Advances in Optimization, Lecture Notes in Economics and Mathematical Systems*, pages 441–457. Springer, 1992.
63. R. Durier. The general one center location problem. *Mathematics of Operations Research*, 20:400–414, 1995.
64. R. Durier. Constrained and unconstrained problems in location theory and inner product spaces. *Numer. Funct. Anal. Opt.*, 18:297–310, 1997.
65. R. Durier and C. Michelot. Geometrical properties of the Fermat–Weber problem. *European Journal of Operational Research*, 20:332–343, 1985.
66. R. Durier and C. Michelot. On the set of optimal points to Weber problems: Further results. *Transportation Science*, 28:141–149, 1994.
67. M.E. Dyer. Linear time algorithms for two-and-three-variable linear programs. *SIAM Journal on Computing*, 13(1):31–45, 1984.
68. H. Edelsbrunner. *Algorithms in combinatorial geometry.* Springer, 1987.
69. H. Edelsbrunner and R. Seidel. Voronoi diagrams and arrangements. *Discrete Comput. Geom.*, 1:25–44, 1986.
70. H. Edelsbrunner and E. Welzl. Constructing belts in two-dimensional arrangements with applications. *SIAM J. Computing*, 15:271–284, 1986.
71. S. Elloumi, M. Labbé, and Yves Pochet. A new formulation and resolution method for the p-center problem. *INFORMS Journal on Computing*, 16(1):84–94, 2004.
72. D. Eppstein. Geometric lower bounds for parametric matroid optimization. *Discrete and Computational Geometry*, 20:463–476, 1998.
73. U. Faigle and W. Kern. Computational complexity of some maximum average weight problems with precedence constraints. *Operations Research*, 42:688–693, 1994.
74. F.R. Fernández, S. Nickel, J. Puerto, and A.M. Rodríguez-Chía. Robustness in the Pareto-solutions for the multi-criteria minisum location problem. *Journal of Multi-Criteria Decision Analysis*, 4:191–203, 2001.

75. G. Fischer. *Lineare Algebra*. Vieweg, 11th edition, 1997.
76. M. Fischetti, H.W. Hamacher, K. Jornsten, and F. Maffioli. Weighted k-cardinality trees: complexity and polyhedral structure. *Networks*, 24:11–21, 1994.
77. R.L. Francis, H.W. Hamacher, C.Y. Lee, and S. Yeralan. On automating robotic assembly workplace planning. *IIE Transactions*, pages 47–59, 1994.
78. R.L. Francis, T.J. Lowe, and A. Tamir. Aggregation error bounds for a class of location models. *Operations Research*, 48:294–307, 2000.
79. R.L. Francis, T.J. Lowe, and A. Tamir. Worst-case incremental analysis for a class of p-facility location problems. *Networks*, 39:139–143, 2002.
80. M. Fredman and R.E. Tarjan. Fibonacci heaps and their uses in network optimization algorithms. *ACM*, 34:596–615, 1987.
81. A.L. Garkavi. On the Chebyshev center and the convex hull of a set. *Uspekhi Mat. Nauk. USSR*, 19:139–145, 1964.
82. Fred Glover and Gary A. Kochenberger, editors. *Handbook of metaheuristics*. International Series in Operations Research & Management Science. 57. Boston, MA: Kluwer Academic Publishers., 2003.
83. M. Groetschel, L. Lovasz, and A. Schrijver. *Geometric Algorithms and Combinatorial Optimization*. Springer, 1993.
84. S.L. Hakimi. Optimum location of switching centers and the absolute centers and medians of a graph. *Operations Research*, 12:450–459, 1964.
85. S.L. Hakimi, M. Labbé, and E. Schmeichel. The Voronoi partition of a network and its applications in location theory. *ORSA Journal of Computing*, 4(4):412–417, 1992.
86. S.L. Hakimi, E.F. Schmeichel, and M. Labbe. On locating path-or tree shaped facilities on networks. *Networks*, 23:543–555, 1993.
87. J. Halpern. Finding the minimal center-median convex combination (cent-dian) of a graph. *Management Science*, 24(5):535–547, 1978.
88. H.W. Hamacher, M. Labbé, S. Nickel, and A. Skriver. Multicriteria semi-obnoxious network location problems (msnlp) with sum and center objectives. *Annals of Operations Research*, 5(4):207–226, 1997.
89. H.W. Hamacher, M. Labbé, and S. Nickel. Multicriteria network location problems with sum objectives. *Networks*, 33:79–92, 1999.
90. H.W. Hamacher and S. Nickel. Combinatorial algorithms for some 1-facility median problems in the plane. *European Journal of Operational Research*, 79:340–351, 1994.
91. H.W. Hamacher and S. Nickel. Multicriteria planar location problems. *European Journal of Operational Research*, 94:66–86, 1996.
92. G.W. Handler and P.B. Mirchandani. *Location on Networks Theory and Algorithms*. MIT Press, 1979.
93. P. Hansen and N. Mladenović. Variable neighborhood search for the p-median. *Location Science*, 5(4):207–226, 1997.
94. P. Hansen and N. Mladenović. Developments of Variable Neighborhood Search. In C.C. Ribeiro and P. Hansen, editors, *Essays and Surveys in Metaheuristics*, pages 415–439. Kluwer Academic Publishers, 2001.
95. P. Hansen and N. Mladenović. Variable neighborhood search: Principles and applications. *European Journal of Operational Research*, 130(3):449–467, 2001.
96. P. Hansen and N. Mladenović. Variable Neighborhood Search. In F. Glover and G. Kochenberger, editors, *Handbook of Metaheuristics*. International Series in Operations Research & Management Science. 57. Boston, MA: Kluwer Academic Publishers., 2003.

97. P. Hansen, N. Mladenović, and D. Pérez-Brito. Variable neighborhood decomposition search. *Journal of Heuristics*, 7:335–350, 2001.
98. G.H. Hardy, J.E. Littlewood, and G. Polya. *Inequalities*. Cambridge University Press, 1952.
99. R. Hassin and A. Tamir. Improved complexity bounds for location problems on the realline. *Operations Research letters*, 10:395–402, 1991.
100. S.M. Hedetniemi, E.J. Cockaine, and S.T. Hedetniemi. Linear algorithms for finding the Jordan center and path center of a tree. *Transportation Science*, 15:98–114, 1981.
101. Y. Hinojosa and J. Puerto. Single facility location problems with unbounded unit balls. *Mathematical Methods of Operations Research*, 58:87–104, 2003.
102. J.B. Hiriart-Urruty and C. Lemarechal. *Convex Analysis and Minimization Algorithms*. Springer, 1993.
103. J.N. Hooker, R.S. Garfinkel, and C.K. Chen. Finite dominating sets for network location problems. *Operations Research*, 39(1):100–118, 1991.
104. R. Horst and H. Tuy. *Global optimization: Deterministic Approaches*. Springer, 1990.
105. C.M. Hosage and M.F. Goodchild. Discrete space location-allocation solutions from genetic algorithms. *Annals of Operations Research*, 6:35–46, 1986.
106. C. Icking, R. Klein, N.-M. Lê, and L. Ma. Convex distance functions in 3-space are different. *Fundamenta Informaticae*, 22:331–352, 1995.
107. C. Icking, R. Klein, L. Ma, S. Nickel, and A. Weißler. On bisectors for different distance functions. *Fifteenth Annual ACM Symposium on Computational Geometry*, 1999.
108. C. Icking, R. Klein, L. Ma, S. Nickel, and A. Weissler. On bisectors for different distance functions. *Discrete Applied Mathematics*, 109:139–161, 2001.
109. ILOG Optimization Suite. ILOG, Inc., Incline Village, Nevada, 2000. http://www.ilog.com/products/optimization.
110. K. Jain and V.V. Vazirani. Primal-dual approximation algorithms for the metric facility facility location problem and k-median problems. In *Proceedings of the 31-st Ann. Symp. on Foundation of the Computer Sciences*, 1999.
111. J.H. Jaramillo, J. Bhadury, and R. Batta. On the use of genetic algorithms to solve location problems. *Computers and Operations Research*, 29:761–779, 2002.
112. C.R. Johnson and P. Nylen. Monotonicity of norms. *Linear Algebra Appl.*, 148:43–58, 1991.
113. D.S. Johnson and K.A. Niemi. On knapsack, partitions and a new dynamic technique for trees. *Mathematics of Operations Research*, 8:1–14, 1983.
114. B. Käfer and S. Nickel. Error bounds for the approximative solution of restricted planar location problems. *European Journal of Operational Research*, 135:67–85, 2001.
115. J. Kalcsics. Geordnete Weber Probleme auf Netzwerken. Diplomarbeit, Universität Kaiserslautern, Fachbereich Mathematik, Universität Kaiserslautern, 1999.
116. J. Kalcsics, S. Nickel, and J. Puerto. Multi-facility ordered median problems: A further analysis. *Networks*, 41(1):1–12, 2003.
117. J. Kalcsics, S. Nickel, J. Puerto, and A. Tamir. Algorithmic results for ordered median problems defined on networks and the plane. *Operations Research Letters*, 30:149–158, 2002.

118. O. Kariv and S.L. Hakimi. An algorithmic approach to network location problems I: The p-centers. *SIAM Journal of applied mathematics*, 37:513–538, 1979.
119. O. Kariv and S.L. Hakimi. An algorithmic approach to network location problems II: The p-medians. *SIAM Journal of applied mathematics*, 37:539–560, 1979.
120. J. Karkazis. The general unweighted problem of locating obnoxious facilities on the plane. *Belgian Journal of Operations Research, Statistics and Computer Science*, 28:3–49, 1988.
121. L. Kaufman and F. Broeckx. An algorithm for the quadratic assignment problem using Benders' decomposition. *European Journal of Operational Research*, 2:204–211, 1978.
122. T.U. Kim, L.J. Lowe, A. Tamir, and J.E. Ward. On the location of a tree-shaped facility. *Networks*, 28:167–175, 1996.
123. V. Klee. Circumspheres and inner products. *Math. Scand.*, 8:363–370, 1960.
124. R. Klein. *Algorithmische Geometrie.* Addison-Wesley, 1997.
125. H. Knörrer. *Geometrie.* Vieweg, 1996.
126. H.W. Kuhn. On a pair of dual nonlinear programs. In J. Abadie, editor, *Nonlinear Programming.* North Holland, 1967.
127. M. Labbé, D. Peeters, and J.-F. Thisse. Location on networks. In M.O. Ball, T.L. Magnanti, C.L. Monma, and G.L. Nemhauser, editors, *Network Routing*, volume 8 of *Handbooks in OR & MS.* Elsevier, 1995.
128. E.L. Lawler. The Quadratic Assignment Problem. *Management Science*, 9:586–599, 1963.
129. D. G. Luemberger. *Optimization by vector space methods.* Wiley, 1969.
130. S.T. McCormick. Submodular function minimization. In A. Aardel, G. Nemhauser, and R. Weismantel, editors, *Handbooks in Operations Research and Management Science*, volume 12, chapter 7. Elsevier, 2005.
131. N. Megiddo. Linear-time algorithms for linear programming in $\mathbb{R}^3$ and related problems. *SIAM Journal on Computing*, 12:759–776, 1982.
132. N. Megiddo. Applying parallel computation algorithms in the design of serial algorithms. *Journal of the Association for Computing Machinery*, 30:852–865, 1983.
133. N. Megiddo. Linear programming in linear time when the dimension is fixed. *Journal of the Association for Computing Machinery*, 31:114–127, 1984.
134. N. Megiddo and A. Tamir. New results on the complexity of p-center problems. *SIAM J. Computing*, 12:751–758, 1983.
135. N. Megiddo and A. Tamir. Linear time algorithms for some separable quadratic programming problems. *Operations Research Letters*, 13:203–211, 1993.
136. N. Megiddo, A. Tamir, E. Zemel, and R. Chandrasekaran. An $O(n \log^2 n)$ algorithm for the $k-th$ longest path in a tree with applications to location problems. *SIAM J. on Computing*, 10:328–337, 1981.
137. Z. Michalewicz. *Genetic Algorithms + Data Structures = Evolution Programs.* Springer, 3rd edition, 1996.
138. J.M. Milnor. Games against nature in decision processes. In Thrall, Coombs, and Davis, editors, *Decision Processes.* Wiley, 1954.
139. E. Minieka. The optimal location of a path or tree in a tree network. *Networks*, 15:309–321, 1985.
140. E. Minieka and N.H. Patel. On finding the core of a tree with a specified length. *J. of Algorithms*, 4:345–352, 1983.

141. P.B. Mirchandani. The p-median problem and generalizations. In P.B. Mirchandani and R.L. Francis, editors, *Discrete Location Theory*, pages 55–117. Wiley, 1990.
142. P.B. Mirchandani and R.L. Francis, editors. *Discrete Location Theory*. Wiley, 1990.
143. J.A. Moreno Pérez, J.L. Roda García, and J.M. Moreno Vega. A parallel genetic algorithm for the discrete p-median problem. In *Studies in Locational Analysis*, volume 7, pages 131–141, 1994.
144. J.M. Moreno Vega. *Metaheurísticas en Localización: Análisis Teórico y Experimental*. PhD thesis, University of La Laguna, Spain, 1996. (In Spanish).
145. C.A. Morgan and J.P. Slater. A linear algorithm for a core of a tree. *J. of Algorithms*, 1:247–258, 1980.
146. J. Muñoz Pérez and J.J. Saameño Rodríguez. Location of an undesirable facility in a polygonal region with forbidden zones. *European Journal of Operational Research*, 114(2):372–379, 1999.
147. G.L. Nemhauser and L.A. Wolsey (Eds.). *Integer and Combinatorial Optimization*. Wiley, 1988.
148. S. Nickel. *Discretization of planar location problems*. Shaker, 1995.
149. S. Nickel. Discrete Ordered Weber problems. In B. Fleischmann, R. Lasch, U. Derigs, W. Domschke, and U. Rieder, editors, *Operations Research Proceedings 2000*, pages 71–76. Springer, 2001.
150. S. Nickel and J. Puerto. A unified approach to network location problems. *Networks*, 34:283–290, 1999.
151. S. Nickel, J. Puerto, and A.M. Rodríguez-Chía. An approach to location models involving sets as existing facilities. *Mathematics of Operations Research*, 28(4):693–715, 2003.
152. S. Nickel, J. Puerto, and A.M. Rodríguez-Chía. MCDM location problems. In J. Figueira, S. Greco, and M. Ehrgott, editors, *Multiple Criteria Decision Analysis: State of the Art Surveys*, International Series in Operations Research and Managemant Science, chapter 19. Springer, 2005.
153. S. Nickel, J. Puerto, A.M. Rodríguez-Chía, and A. Weißler. Multicriteria planar ordered median problems. *JOTA*, 2005. to appear.
154. W. Ogryczak and A. Tamir. Minimizing the sum of the k largest functions in linear time. *Information Processing Letters*, 85(3):117–122, 2003.
155. A. Okabe, B. Barry, and K. Sugihara. *Spatial Tessellations Concepts and Applications of Voronoi Diagrams*. Wiley, 1992.
156. A. Oudjit. *Median Locations on deterministic and probabilistic multidimensional networks*. PhD thesis, Rensselaer Polytechnic Institute, Troy, New York, 1981.
157. P.L. Papini. Existence of centers and medians. Preprint, 2001.
158. P.L. Papini and J. Puerto. Averaging the k largest distances among n: k-centra in Banach spaces. *Journal of Mathematical Analysis and Applications*, 291:477–487, 2004.
159. P.H. Peeters. Some new algorithms for location problems on networks. *European J. of Operational Research*, 104:299–309, 1998.
160. S. Peng and W. Lo. Efficient algorithms for finding a core of a tree with specified length. *J. of Algorithms*, 20:445–458, 1996.
161. S. Peng, A.B. Stephens, and Y. Yesha. Algorithms for a core and k-tree core of a tree. *J. of Algorithms*, 15:143–159, 1993.

162. D. Pérez-Brito, J. Moreno-Pérez, and I. Rodríguez-Marín. The finite dominating set for the p-facility cent-dian network location problem. *Studies in Locational Analysis*, 11:27–40, 1997.
163. F. Plastria. Continuous location problems: research, results and questions. In Z. Drezner, editor, *Facility Location: A Survey of Applications and Methods*, pages 85–127. Springer, 1995.
164. F.P. Preparata and M.I. Shamos. *Computational Geometry - An Introduction.* Springer, 1997.
165. J. Puerto. Lecturas en teoría de localización. Technical report, Universidad de Sevilla. Secretariado de Publicaciones. (Ed. Justo Puerto), 1996.
166. J. Puerto and F.R. Fernández. A convergent approximation scheme for efficient sets of the multicriteria weber location problem. *TOP, The Spanish Journal of Operations Research*, 6(2):195–204, 1988.
167. J. Puerto and F.R. Fernández. The symmetrical single facility location problem. Technical Report 34, Prepublicación de la Facultad de Matemáticas, Universidad de Sevilla, 1995.
168. J. Puerto and F.R. Fernández. Multicriteria minisum facility location problem. *Journal of Multi-Criteria Decision Analysis*, 8:268–280, 1999.
169. J. Puerto and F.R. Fernández. Geometrical properties of the symmetric single facility location problem. *Journal of Nonlinear and Convex Analysis*, 1(3):321–342, 2000.
170. J. Puerto and A.M. Rodríguez-Chía. On the exponential cardinality of finite dominating sets for the multifacility ordered median problem. to appear in Operations Research Letters, 2005.
171. J. Puerto, A.M. Rodríguez-Chía, and F. Fernández-Palacín. Ordered Weber problems with attraction and repulsion. *Studies in Locational Analysis*, 11:127–141, 1997.
172. J. Puerto, A.M. Rodríguez-Chía, D. Pérez-Brito, and J.A. Moreno. The p-facility ordered median problem on networks. to appear in TOP, 2005.
173. J. Puerto and A. Tamir. Locating tree-shaped facilities using the ordered median objective. *Mathematical Programming*, 2005. to appear.
174. N.M. Queyranne. Minimizing symmetric submodular functions. *Mathematical Programming*, 82:3–12, 1998.
175. F. Rendl. The Quadratic Assignment Problem. In Z. Drezner and H.W. Hamacher, editors, *Facility Location: Applications and Theory*, pages 439–457. Springer, 2002.
176. C.S. Revelle and R.W. Swain. Central facilities location. *Geographical Analysis*, 2:30–42, 1970.
177. A.M. Rodríguez-Chía, S. Nickel, J. Puerto, and F.R. Fernández. A flexible approach to location problems. *Mathematical Methods of Operations Research*, 51(1):69–89, 2000.
178. A.M. Rodríguez-Chía and J. Puerto. Geometrical description of the weakly efficient solution set for multicriteria location problems. *Annals of Operations Research*, 111:179–194, 2002.
179. R.A. Rosenbaum. Subadditive functions. *Duke Math. J.*, 17:227–247, 1950.
180. H.P. Schwefel. *Numerical Optimization of Computer Models.* Wiley, Chichester, 1981.
181. M. Sharir and P.K. Agarwal. *Davenport-Schinzel Sequences and Their Geometric Applications.* Cambridge University Press, 1995.

182. A. Shioura and M. Shigeno. The tree center problems and the relationship with the bottleneck knapsack problems. *Networks*, 29:107–110, 1997.
183. P.J. Slater. Locating central paths in a graph. *Transportation Science*, 16:1–18, 1982.
184. A. Tamir. A unifying location model on tree graphs based on submodularity properties. *Discrete Applied Mathematics*, 47:275–283, 1993.
185. A. Tamir. An $O(pn^2)$ algorithm for the p-median and related problems on tree graphs. *Operations Research Letters*, 19:59–64, 1996.
186. A. Tamir. Fully polynomial approximation schemes for locating a tree-shaped facility: a generalization of the knapsack problem. *Discrete Applied Mathematics*, 87:229–243, 1998.
187. A. Tamir. The k-centrum multi-facility location problem. *Discrete Applied Mathematics*, 109:292–307, 2000.
188. A. Tamir. Sorting weighted distances with applications to objective function evaluations in single facility location problems. *Operations Research Letters*, 32:249–257, 2004.
189. A. Tamir and T.J. Lowe. The generalized p-forest problem on a tree network. *Networks*, 22:217–230, 1992.
190. A. Tamir, D. Pérez-Brito, and J.A. Moreno-Pérez. A polynomial algorithm for the p-centdian problem on a tree. *Networks*, 32:255–262, 1998.
191. A. Tamir, J. Puerto, J.A. Mesa, and A.M. Rodríguez-Chía. Conditional location of path and tree shaped facilities on trees. Technical report, School of Mathematical Sciences, Tel-Aviv University, Tel-Aviv University, To appear in Journal of Algorithms, 2005.
192. A. Tamir, J. Puerto, and D. Perez-Brito. The centdian subtree on tree networks. *Discrete Applied Mathematics*, 118:263–278, 2002.
193. E. Tardos. A strongly polynomial algorithm to solve combinatorial linear programs. *Operations Research*, 34:250–256, 1986.
194. L.N. Tellier. *Economie spatiale: rationalité économique de l'space habité*. Gaétan Morin, Chicoutimi, Quebec, 1985.
195. L.N. Tellier and B. Polanski. The Weber problem: frecuency of different solution types and extension to repulsive forces and dynamic process. *Journal of Regional Science*, 29:387–405, 1989.
196. T. Tokuyama. Minimax parametric optimization problems in multidimensional parametric searching. In *Proceedings 33-rd Annual ACM Symposium on Theory of Computing*, pages 75–84, 2001.
197. P.M. Vaidya. An algorithm for linear programming which requires $O((m+n)n^2+(m+n)^{1.5}nl)$ arithmetic operations. *Mathematical Programming*, 47:175–201, 1990.
198. L. Veselý. Equality of two Banach space constants related to different notions of radius of a set. *Boll. Un. Mat. Ital.*, VII, 9-A:515–520, 1995.
199. L. Veselý. Generalized centers of finite sets in Banach spaces. *Acta Math. Univ. Comenian. N.S.*, 66:83–95, 1997.
200. L. Veselý. A Banach space in which all compact sets, but not bounded sets, admit Chebyshev centers. *Arch. Math.*, 79(6):499–506, 2002.
201. B.-F. Wang. Efficient parallel algorithms for optimally locating a path and a tree of a specified length in a weighted tree network. *J. of Algorithms*, 34:90–108, 2000.
202. B.-F. Wang. Finding a two-core of a tree in linear time. *SIAM J. on Discrete Mathematics*, to appear.

203. A.R. Warburton. Quasiconcave vector maximization: Connectedness of the sets of Pareto-optimal and weak Pareto-optimal alternatives. *Journal of Optimization Theory and Applications*, 40(4):537–557, 1983.
204. J.E. Ward and R.E. Wendell. Using block norms for location modeling. *Operations Research*, 33:1074–1090, 1985.
205. J.R. Weaver and R.L. Church. A median location model with non-closets facility service. *Transportation Science*, 7:18–23, 1985.
206. A. Weber. *Über den Standort der Industrien, Tübingen, 1909. (English translation by Friedrich C.J. (1929). Theory of the Location of Industries.* University of Chicago Press, 1929.
207. A. Weißler. *General bisectors and their application in planar location theory.* Shaker, 1999.
208. R.E. Wendell and A.P. Jr. Hurter. Location theory, dominance and convexity. *Oper. Res.*, 21:314–320, 1973.
209. G.O. Wesolowsky. The Weber Problem: History and Perspectives. *Location Science*, 1:5–23, 1993.
210. R.A. Whitaker. A fast algorithm for the greedy interchange for large-scale clustering and median location problems. *INFOR*, 21:95–108, 1983.
211. D.J. White. *Optimality and efficiency.* Wiley, 1982.
212. P. Widmayer, Y.F. Wu, and C.K. Wong. On some distance problems in fixed orientations. *SIAM Journal on Compututing*, 16:728–746, 1987.
213. A. Wouk. *A course of applied functional analysis.* Wiley, 1979.
214. R.R. Yager. On ordered weighted averaging aggregation operators in multicriteria decision making. *IEEE Trans. Sys. Man Cybern.*, 18:183–190, 1988.
215. E. Zemel. An $O(n)$ algorithm for the linear multiple choice knapsack problem and related problems. *Information Processing Letters*, 18:123–128, 1984.

Index

Printing and Binding: Strauss GmbH, Mörlenbach